AF389000

Targeted Cancer Therapy and Mechanisms of Resistance

Targeted Cancer Therapy and Mechanisms of Resistance

Editor

Valentina De Falco

MDPI • Basel • Beijing • Wuhan • Barcelona • Belgrade • Manchester • Tokyo • Cluj • Tianjin

Editor
Valentina De Falco
Institute of Endocrinology and
Experimental Oncology (IEOS)
National Research Council (CNR)
Naples
Italy

Editorial Office
MDPI
St. Alban-Anlage 66
4052 Basel, Switzerland

This is a reprint of articles from the Special Issue published online in the open access journal *International Journal of Molecular Sciences* (ISSN 1422-0067) (available at: www.mdpi.com/journal/ijms/special_issues/cancer_resistance).

For citation purposes, cite each article independently as indicated on the article page online and as indicated below:

LastName, A.A.; LastName, B.B.; LastName, C.C. Article Title. *Journal Name* **Year**, *Volume Number*, Page Range.

ISBN 978-3-0365-2857-1 (Hbk)
ISBN 978-3-0365-2856-4 (PDF)

Contents

About the Editor

Valentina De Falco

Valentina De Falco is currently a Senior Researcher at the Institute of Endocrinology and Experimental Oncology of the Italian National Research Council (CNR), and Adjunct Professor of General Pathology at the Federico II University of Naples (FIIU). De Falco graduated with the best grade and laude in Chemistry from FIIU in 2002, and received her Specialization in Clinical Pathology in 2007. She started her research activity in 2001 at the Department of Cellular and Molecular Biology and Pathology of FIIU where, in 2003, she became a research fellow. Subsequently, she became a research fellow at the Higher Institute of Oncology. In 2007, she moved to Claude Bernard University in Lyon, France, to collaborate with a French team on a project on RET-targeting in MTC. From 2010, she has been a research fellow for the IEOS Institute of the CNR, and became a full-time researcher at CNR in 2012. In 2017, De Falco received an International Prize on the Pathophysiology of the Thyroid from the National Academy of the Lincei, Italy. Her research activity was mainly focused on the study of the identification of new mediators and therapeutic targets in human cancers. She contributed to the characterization of RET/PTC and BRAF-mediated molecular mechanisms in thyroid carcinomas and was the scientific leader of a project concerning the generation and characterization of medullary thyroid carcinoma cells resistant to treatment with small-molecule kinase inhibitors. She is currently studying the involvement of the serine-threonine kinase p90RSK in resistance to anticancer treatments and its role in the pathogenesis and progression of several human cancers. De Falco became a Topical Advisory Panel Member for *IJMS* in 2021. She is a member of various medical associations, including the European Thyroid Association, the Italian Cancer Society, and the European Association for Cancer Research. She has authored 30 research articles published in international peer-reviewed journals, with a personal H-index of 22.

Preface to "Targeted Cancer Therapy and Mechanisms of Resistance"

Tumor cells commonly exhibit dependence on a single (often the initiating) activated oncogenic pathway or protein to maintain their malignant proliferation and survival, a phenomenon called "oncogene addiction". According to this concept, protein kinases have been elected as promising molecular targets for cancer therapy. There are several possibilities to target these proteins in cancer, including monoclonal antibodies that can bind to the extracellular domain of the RTK, compounds able to favor the proteolytic degradation of the kinase, and finally, small-molecule protein kinase inhibitors (PKIs). In addition to targeting oncogenes, new anticancer treatments have increasingly been developed for tumor suppressor genes and RNA interference.

Despite promising results in cancer treatment with targeted cancer drugs, clinical experience has shown that only a fraction of patients respond to targeted therapies, even if their tumors express the altered target. This kind of resistance is known as primary resistance. Otherwise, secondary or acquired resistance to the treatment arises, almost invariably, when tumors are treated with cancer drugs. Acquired resistance mechanisms can be divided into two main categories: target-dependent and target-independent mechanisms.

Target-dependent resistance typically occurs through genetic modifications of the target. Such genetic modifications may include point mutations and copy number amplifications. Evidence suggests that mutation may pre-exist in a minority of cancer cells, and it is then selected upon treatment. This suggests that secondary PKIs which can also bind the mutated kinase can be used to overcome resistance. Gene amplification is another major mechanism of target-dependent resistance. Selective pressure of the drug can drive amplification of the target gene, thus leading to additional overexpression of the encoded protein.

Instead, target-independent mechanisms occur through the activation of alternative pathways which enable bypassing of the drug-mediated block. In other words, cancer cells escape treatment by switching to an alternative signaling pathway which is not inhibited by the drug.

Other mechanisms of resistance can exploit the enormous genome plasticity of cancer cells by modulating miRNA expression or remodeling chromatin. Finally, although not as commonly as with classical cytotoxic drugs, other resistance mechanisms can cause a decrease in the effective intracellular concentration of the targeted cancer drug.

We set out to select some studies containing emerging developments on the subject. In essence, this collection aims to highlight some recent findings regarding resistance mechanisms and reviews of molecular targeting and resistance with 14 contributions, including 10 original research papers and 4 reviews. Aspects relating to solid cancers, such as breast, ovary, colon, and blood cancers such as leukemia, and the identification of resistance mechanisms and new molecular targets, help form the basis for the preclinical and clinical development of more effective next-generation drugs.

The collection is aimed at scientists in the field of molecular oncology and, specifically, for those who study innovative therapies to overcome the development of resistance to already approved treatments. The authors involved in the selected articles are all recognized experts in the field. I thank my colleague Dr. Aniello Cerrato and the graphic designer Fabrizio Fiorbianco of the Communication Services Center (Naples, Italy) for supporting me in the creation of the cover.

Valentina De Falco
Editor

International Journal of *Molecular Sciences*

MDPI

Review

Wee1 Kinase: A Potential Target to Overcome Tumor Resistance to Therapy

Francesca Esposito , Raffaella Giuffrida , Gabriele Raciti, Caterina Puglisi and Stefano Forte *

IOM Ricerca srl, Viagrande, I-95029 Catania, Italy; francesca.esposito@grupposamed.com (F.E.); raffaella.giuffrida@grupposamed.com (R.G.); gabriele.raciti@grupposamed.com (G.R.); caterina.puglisi@grupposamed.com (C.P.)
* Correspondence: stefano.forte@grupposamed.com

Abstract: During the cell cycle, DNA suffers several lesions that need to be repaired prior to entry into mitosis to preserve genome integrity in daughter cells. Toward this aim, cells have developed complex enzymatic machinery, the so-called DNA damage response (DDR), which is able to repair DNA, temporarily stopping the cell cycle to provide more time to repair, or if the damage is too severe, inducing apoptosis. This DDR mechanism is considered the main source of resistance to DNA-damaging therapeutic treatments in oncology. Recently, cancer stem cells (CSCs), which are a small subset of tumor cells, were identified as tumor-initiating cells. CSCs possess self-renewal potential and persistent tumorigenic capacity, allowing for tumor re-growth and relapse. Compared with cancer cells, CSCs are more resistant to therapeutic treatments. Wee1 is the principal gatekeeper for both G2/M and S-phase checkpoints, where it plays a key role in cell cycle regulation and DNA damage repair. From this perspective, Wee1 inhibition might increase the effectiveness of DNA-damaging treatments, such as radiotherapy, forcing tumor cells and CSCs to enter into mitosis, even with damaged DNA, leading to mitotic catastrophe and subsequent cell death.

Keywords: Wee1 kinase; cell cycle; tumor resistance

Citation: Esposito, F.; Giuffrida, R.; Raciti, G.; Puglisi, C.; Forte, S. Wee1 Kinase: A Potential Target to Overcome Tumor Resistance to Therapy. *Int. J. Mol. Sci.* **2021**, *22*, 10689. https://doi.org/10.3390/ijms221910689

Academic Editor: Valentina De Falco

Received: 8 September 2021
Accepted: 27 September 2021
Published: 1 October 2021

Publisher's Note: MDPI stays neutral with regard to jurisdictional claims in published maps and institutional affiliations.

1. Introduction

The cell cycle is a finely regulated process, where a series of growth and development steps alternate and several molecules are involved as negative or positive regulators. Proteins belonging to the highly evolutionarily conserved family, known as cyclin-dependent kinases (Cdks), interact with specific partner proteins named cyclins. The variation in cyclin protein levels is the mechanism through which cells progress through the cell cycle, alternating the phases that follow one another in a specific temporal sequence [1]. Besides the kinases, even their molecular counterparts, the phosphatases, take part in the cell cycle. Therefore, kinases and phosphatases alternatively regulate the same targets to allow for the correct conclusion of the process and to dynamically respond to the events.

The major events of the cell cycle are the S-phase, where the DNA replication occurs, and the M-phase, which finally leads to the generation of the two daughter cells. These phases are connected through "gap" steps (G1 and G2 respectively), during which, the cell prepares itself through modifications of its transcriptional activity to synthetize the molecules that are required for the following event.

Among the various steps, there are mechanisms of control, known as checkpoints, which temporarily stop the process when mistakes occur to avoid the transmission of mistakes that would compromise the result [2].

In normal cells, DNA damage is usually repaired via G1 phase arrest. In tumor cells, G1 checkpoint deficiencies can occur, especially in p53-deregulated cells. In these cells, the G2 checkpoint has the crucial role of repairing endogenous and exogenous DNA damage [3,4]. Consequently, targeting the effectors involved in the G2 checkpoint is a promising strategy for cancer therapy [5,6].

Wee1 belongs to a family of protein kinases that activate the G2/M checkpoint of the cell cycle in response to double-stranded DNA breaks (DDB) [7,8]. It is involved in the terminal phosphorylation and inactivation of Cdk1/Cdc2-bound cyclin B on its Tyr15 residue, resulting in cell cycle arrest at G2 and, therefore, a delayed entry into mitosis after DNA damage [9,10].

2. Wee1 Family

Wee1 is one of the most important gatekeepers for both the G2/M checkpoint and S-phase. While its role in regulating the entry into mitosis is well known since its discovery, its involvement in the S-phase was discovered only recently. Since DNA synthesis and mitosis are tightly connected, when replication errors occur, the regulatory mechanism can slow down or temporarily stop the S-phase. This allows for DNA repair before the onset of mitosis, which will fix genetic mutations and/or chromosomal aberrations in the genome of daughter cells.

Wee1 belongs to a family of protein kinases, consisting of three members in humans, including PKMYT1 (membrane-associated tyrosine- and threonine-specific cdc2-inhibitory kinase) and two Wee1 kinases (Wee1 and Wee1B). They show sequence similarity on their kinase domain but differ regarding their localization, temporal expression and regulation.

Human Wee1, also known as Wee1A or Wee1Hu, is a kinase of 646 amino acids with a molecular weight of 94 kDa comprising three domains: an N-terminal regulatory domain, a central kinase domain and a C-terminal regulatory domain [9,11,12]. The N-terminal domain is the longest and contains two phosphorylation sites (S53 and S123), which are involved in protein degradation through the proteasome. In fact, this domain presents three PEST (Pro-Glu-Ser-Thr) regions, which are found in many eukaryotic proteins and are characterized by a rapid turnover. A nuclear localization signal, which is in the same domain, is responsible for its subcellular localization (Figure 1).

Even if Wee1 acts as a tyrosine kinase, its catalytic domain is similar to the ones found on serine/threonine kinases rather than those found on tyrosine kinases. In fact, Wee1 acts on the Y15 residue on Cdk1-cyclin B. Its structural features suggest that Wee1 may have evolved from a serine/threonine kinase through mutations that somehow might have directed it to acquire the ability to phosphorylate tyrosine residues [12]. Finally, the C-terminal regulatory domain is very short and presents an S642, which constitutes a binding site for the chaperone 14-3-3.

Wee1B was identified for the first time in Xenopus oocytes. It shares with Wee1 its predominant nuclear localization and the ability to inactivate Cdk1-cyclin B through phosphorylation. However, higher levels of its mRNA were observed in mature oocytes with a marked decrease after fertilization, while Wee1 is more expressed in the zygotes. This suggests a different action during the early phase of embryogenesis, with Wee1B being more involved in the early steps and Wee1 in the latter steps [8].

While Wee1 and Wee1B are prevalently nuclear, PKMYT1 is not. Its C-terminal domain is anchored at the endoplasmic reticulum and Golgi apparatus; in addition, it binds Cdk1-cyclin B. Furthermore, it is a dual-specificity kinase and phosphorylates Cdk1-cyclin B in both Y15 and the adjacent threonine 14 [13,14]. PKMYT1 prevents cell entry in the M-phase in two ways: it maintains Cdk1-cyclin B inactivity and, at the same time, prevents the translocation of the complexes into the nucleus [15]. Moreover, it exhibits more restricted substrate specificity, acting on Cdk1 but only partially on Cdk2 [16].

The Wee1 gene, which is localized in the distal short arm of the human chromosome 11 [17], was first discovered in the fission yeast Schizosaccharomyces pombe. It is involved in the controlled onset of mitosis once the cell reaches the right size [18]. Moreover, a gene dosage effect was observed: several experiments on yeast indicated that Wee1 could delay progression through G2 into mitosis and this effect was more evident when its expression increased [9,19]. Orthologues of the yeast Wee1 gene were identified not only in human cells [20,21] but in several other eukaryotes, such as mouse, chicken, lizard, zebrafish and the African clawed frog [22–28]. In addition to the above-mentioned members of the Wee1 family, other paralog genes have been identified. They comprise four eukaryotic translation initiation factor genes (EIF2AK1, EIF2AK2, EIF2AK3, EIF2AK4), the serine/threonine kinase 35 (STK35) and the PDLIM1 interacting kinase 1 like (PDLIM1) genes.

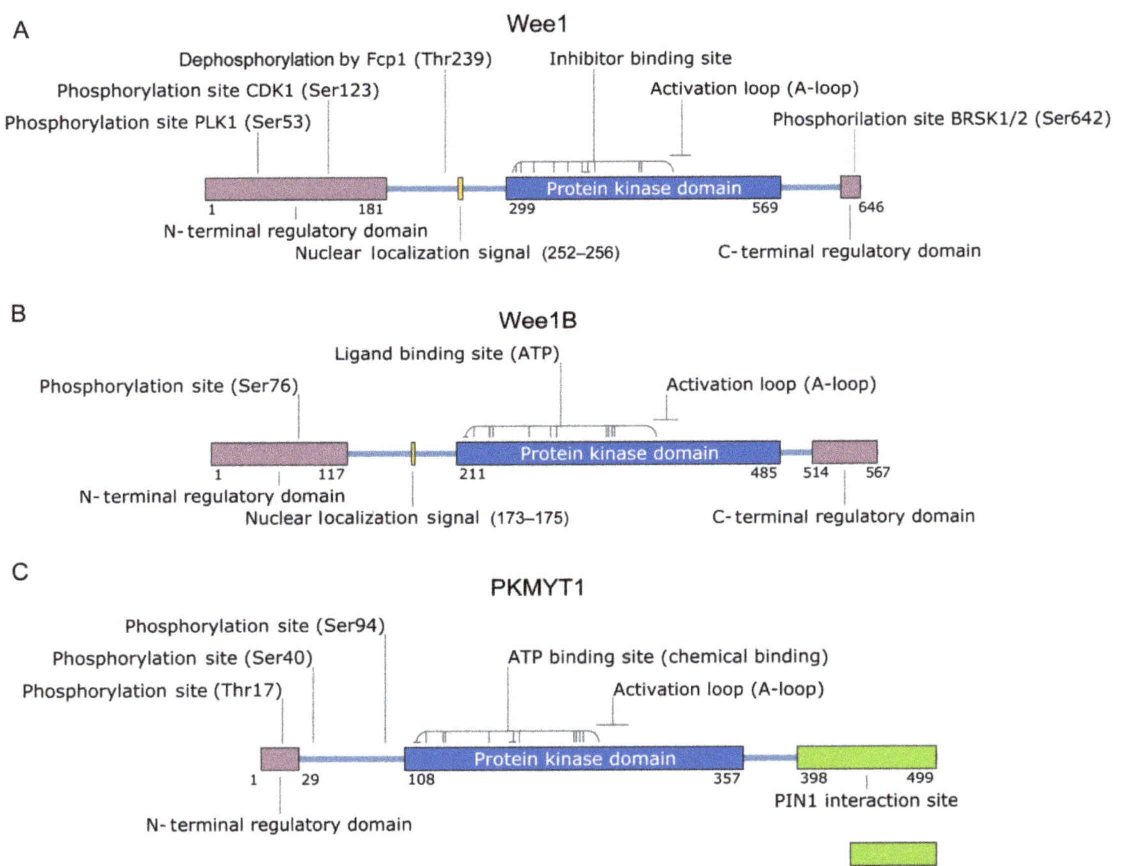

Figure 1. Schematic representation of Wee1 (**A**), Wee1B (**B**) and PKMYT1 (**C**) domain structures, interaction and post-translational modifications sites.

Its activity is sustained by various molecules to acquire more stability, a prolonged half-life and, consequently, increased biochemical activity. In particular 14-3-3 proteins, which are molecular chaperones that interact with many proteins that are involved in signal transduction, are likely the most important [29]. Their binding during the interphase induces a conformational change at the N-terminal domain of Wee1, hiding its site for degradation. Consequently, Wee1 becomes more stable and thus enhances kinase activity. During mitosis, 14-3-3 instead loses its binding with Wee1, allowing for an easier inactivation [30–32].

Moreover, Hsp90 and MIG6 support Wee1 stability. Heat shock protein 90 (Hsp90), which is one of the most abundant and evolutionarily conserved chaperones, interacts with Wee1 and maintains it in its native conformation [32–35]. MIG6 is a known EGFR inhibitor whose expression is upregulated with cell growth. Sasaki and his collaborators recently identified a new EGFR-independent role in cell cycle progression. They demonstrated that MIG6 stabilizes Wee1, hindering the recruitment of β-TrCP-SCF and the subsequent proteasomal degradation [36].

3. Wee1 in Cell Cycle Events

Wee1 is predominantly localized in the nucleus, where it coordinates and ensures the proper DNA replication and, at the same time, prevents premature mitosis [37,38].

Wee1 negatively regulates the G2/M transition that acts on the Cdk1-cyclin B complex (also known as mitosis promoting factor (MPF)), which, once activated, triggers all events leading to the onset of mitosis [39–41]. Wee1 phosphorylates Cdk1 (also known as cell division cycle 2 (Cdc2)) on the Y15 residue [21,42] within its ATP-binding site in the catalytic subunit, while PKMYT1 phosphorylates it on both the Y15 and T14 residues [13,14,16]. This phosphorylation inactivates Cdk1 during the interphase. Its activity is contrasted with Cdc25 phosphatases, whose levels change in a specular way compared to those of Wee1. Cdc25 dephosphorylates Y15 and T14 and, together with higher cyclin B levels, which promote the phosphorylation on its T161 residue by Cdk-activating kinase (CAK), activates Cdk1 [43,44].

The downregulation of Wee1 promotes entry into mitosis. This is usually achieved via both decreased synthesis and proteolytic degradation [38]. When Wee1 is phosphorylated by Plk1 (polo-like kinase 1) and Cdk1 in S53 and S123, respectively, it becomes a target of the β-TrCP F-box protein-containing SCF E3 ubiquitin ligase (SCFβ-TrCP) or Tome-1 F-box protein-containing SCF E3 ubiquitin ligase (SCFTome-1) complexes. Both β-TrCP and Tome-1 directly contribute to Wee1 inactivation [45–47]. Therefore, while phosphorylation does not directly inactivate Wee1, this event induces its proteasome-dependent degradation (Figure 2).

It is important to highlight that Wee1 inactivation is enhanced by Cdk1-cyclin B itself, which triggers a feedback loop: when Wee1 levels decrease, the balance between the tyrosine kinase and Cdk1-cyclin B complexes shifts toward the latter. These, in turn, phosphorylate Wee1, promoting its recognition for proteolytic degradation. Thus, the process ensures a rapid activation of Cdk1-cyclin B to allow for a quick transition to mitosis [11,46].

During the S-phase, Wee1 acts on Cdk2 to stabilize the replication machinery, preventing unscheduled replication origins and the resulting insurgence of abnormal structures that might cause genomic instability [48]. The exact molecular mechanism is unclear: it could downregulate both Cdk2 and the Mus81-Eme1 endonuclease or Cdk1 and, only indirectly, Mus81-Eme1 activity [49,50]. In the presence of stalled forks, DSBs are induced by Mus81-Eme1 complexes to initiate recombination-mediated replication fork recovery. When downregulation that is caused by Wee1 is turned off, Cdk2 becomes hyperphosphorylated and thereby activates the Mus81-Eme1 endonuclease. Moreover, the phosphorylated form of H2A histone, namely, γ-H2AX, increases the formation foci that recruit molecules that are responsible for the DNA repair pathway. As such, the endonuclease binds to DNA, even in the absence of damage, and generates DSBs, cleaving the DNA in an unscheduled manner, thus reducing replication fork speed [49,51]. Therefore, Wee1 activity stabilizes those replication forks, which are temporarily stalled without becoming a signal to activate the DNA damage response.

Furthermore, Wee1 prevents the multiple and unscheduled activation of replication forks, leading, in turn, to the rapid consumption of the deoxynucleotide triphosphates (dNTPs) pool. Therefore, more replication forks become stalled and S-phase arrest ensues [52,53].

Recently, Vassilopoulos et al. found a novel interaction between Wee1 and anaphase-promoting complex/cyclosome (APC/C). Activated APC/C is an E3 ubiquitin ligase that targets key proteins, including cyclin B and securin, causing them to degrade, thus promoting anaphase initiation and segregation of sister chromatids. In the presence of DNA damage in the S-phase, Wee1 phosphorylates APC/C, which, as a result, becomes inactive.

If Wee1 is inactivated, APC/C recruits polo-like kinase 1 (Plk-1), which degrades Wee1, and the cell cycle proceeds independently due to the integrity of the genetic information [54].

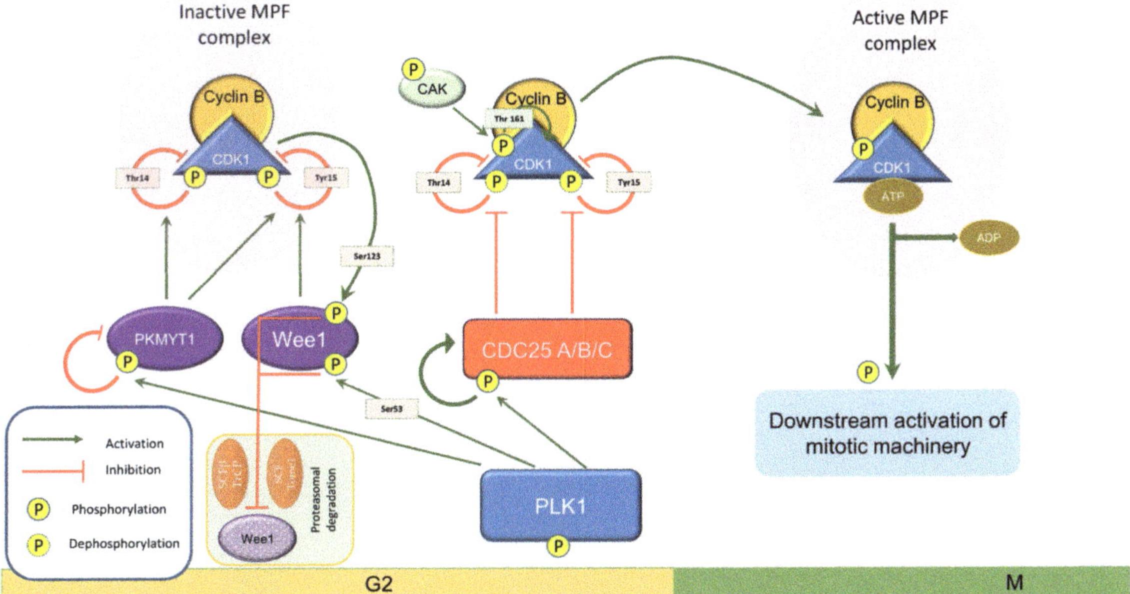

Figure 2. Schematic representation of the components involved in G2/M cell cycle transition.

Finally, Wee1 can act as an epigenetic modifier: in the late S-phase, when DNA synthesis is completed, it phosphorylates H2B histone at the Y37 residue, suppressing the transcription of Hist1, which is the major histone gene cluster. Thus, by modifying the chromatin structure, histone overproduction is avoided. This is an evolutionarily conserved process in the eukaryotic cells, as it is observed in yeast and mammals [55].

To better understand the role of Wee1 in mammalian tissue, in 2006, the group of Prof. Deng produced *Wee1* knockout mice via gene targeting, which resulted in embryonic lethality at the blastocyst stage [56], which explained the lack of in vivo models. In 2015, the same group established conditional and tissue-specific *Wee1* mutant mice and cells. Alongside the well-known role of Wee1 in preventing premature mitotic entry, *Wee1* mutant cells revealed impaired mitosis that was characterized by delayed progression and completion without normal cytokinesis. Moreover, they found a role of Wee1 in modulating the APC/C complex, i.e., it is responsible for the degradation of mitotic regulator protein by a proteasome. Thus, *Wee1* deficiency causes increased activity of APC/C, which allows for mutant cells to progress through mitosis at the expense of genomic integrity. Finally, in the animal model with a conditional expression of Wee1 in the mammary gland, they demonstrated that Wee1 is indispensable for maintaining genomic stability and it acts as a haploid tumor suppressor since a mutant mammary gland develops tumors. In conclusion, Wee1 coordinates distinct cell-division events to permit correct segregation of genetic information into daughter cells and preservation of genomic integrity [54].

4. Role of Wee1 Kinase in Cancer Progression and Therapy

The expression of Wee1 was investigated in few studies, including solid tumors and hematological malignancies. Wee1 downregulation in cancer tissues was observed in most of them, suggesting a tumor suppressor role for Wee1. In stromal breast cancer, Wee1 expression was investigated using immunohistochemistry. The lower Wee1 expression in malignant tumors compared to benign ones suggests that it acts as a tumor suppressor [57]. Similarly, a cDNA array performed on colon carcinoma cell lines and human tissues showed low Wee1 mRNA expression, leading to the speculation that genetic lesions targeting cell cycle regulation occur not only at the G1-S but also the G2-M transition checkpoint during tumor development [58]. Since the Wee1-mediated inhibition of G2-M transition is an essential step after DNA damage, permitting DNA repair prior to entry into mitosis [37], the decreased expression or loss of the Wee1 gene in colon carcinoma cells suggests its potential role as a tumor suppressor [58]. The association between Wee1 downmodulation and tumorigenesis was also found in non-small-cell lung cancer (NSCLC) [59]. In this work, the authors found a correlation between low Wee1 levels and a poor prognosis, as impaired Wee1 expression provided an advantage to neoplastic cells since it caused a faster progression through the cell cycle [59]. In their work, Yoshida et al. investigated the clinicopathological and prognostic significance of cyclin B and Wee1 expressions in 79 patients affected by NSCLC. It was found that in almost two-thirds of patients analyzed, Wee1 was not expressed and the patients had a poorer prognosis and a higher recurrence rate. Moreover, their tumors tended to have a higher Ki index and PCNA-LI values, which are typical parameters of cell proliferation and malignancy potential in NSCLC. Taken together, these data suggested that a loss of Wee1 expression may both promote tumor aggressiveness and be a useful prognostic indicator.

Few studies have evaluated the role of microRNAs (miRNAs), long non-coding RNAs (lncRNAs) and small-interference RNAs (siRNAs) that are able to modulate Wee1 expression in tumor cells and their involvement in tumor progression [60–64]. Aside from the evidence of an inverse correlation between Wee1 protein expression and the aggressiveness of melanoma cells, Bhattacharya et al. found a link with miR-195. They demonstrated that miR-195 has an inhibitory activity on Wee1 expression in melanoma metastases [61]. Data that was obtained by assessing cell cycle analyses during simultaneous Wee1 silencing (by siRNA or miR-195) and genotoxic agents' exposure suggested that when chemotherapy-mediated DNA damage occurs, miR-195 is able to significantly contrast the G2/M cell cycle arrest, downmodulating Wee1 in melanoma cells. [61]. In contrast, the upregulation of miR-101-3p via lncRNA NEAT1_2 in hepatocellular carcinoma cells decreased both the mRNA and protein levels of Wee1, inducing tumor radio-sensitization [60]. Comparable results were obtained in myeloid leukemia cell lines, whose Wee1 siRNA silencing resulted in a strong sensitization to cytarabine (Ara-C) [65]. While these data indicate that Wee1 acts as a tumor suppressor, several studies highlighted an increased expression of Wee1 in several types of cancer. This indicates that Wee1 may be important for cancer cells' viability under specific circumstances. A cell viability assessment with a kinase siRNA library in different cancer cell lines demonstrated that Wee1 gene expression correlated with the Wee1 gene copy number, potentially identifying a cause of increased expression [66]. Moreover, this study demonstrated that tumor cell lines that overexpress Wee1 are more sensitive to Wee1 inhibition by siRNA, leading to abrogation of the G2/M checkpoint and consequent tumor cell death via apoptosis. In particular, thanks to this approach, authors could identify a breast cancer patient subset (luminal breast cancer) that overexpressed Wee1, where its inhibition could be suggested as a potential therapeutic strategy [66]. Interestingly, by performing a stringent in silico analysis on data obtained comparing normal versus cancer tissue, Mir et al. found increased expression of Wee1 in most cancer types (27 samples in a 35-sample data set) [67]. In their dataset, the highest Wee1 mRNA expression was measured in glioblastoma, followed by non-small-cell lung carcinoma, (non-)seminoma and colon carcinoma, whereas the other cancer types mostly showed moderate overexpression as compared to the relevant non-neoplastic control tissue [67].

The activity of Wee1 was found to be increased in patients with advanced hepatocellular carcinoma when compared with noncancerous liver tissue [68]. Moreover, Wee1 was found to be overexpressed and functionally important in medulloblastoma [69], and a high expression of Wee1 was described in glioma [70,71]. Slipicevic et al. reported an expression of Wee1 in ovarian serous carcinoma effusions. Moreover, they observed a notable increase in Wee1 levels after exposure to chemotherapy, suggesting a role for this kinase in mediating the progression of the disease. Thus, increased Wee1 expression may represent an adaptive response to the chemotherapy that allows tumor cells to repair DNA damage and thereby survive [72].

Although Wee1 expression was found to increase in vulvar squamous cell carcinomas compared to normal tissue, Magnussen et al. did not observe any significant association between disease-specific survival and Wee1 expression in patients with vulvar carcinomas [73]. These findings did not directly support the tumor suppressor role of Wee1. Moreover, Wee1 silencing by siRNA did not translate to any major alteration in viability [73]. However, in their previous study, Magnussen and colleagues linked the expression of Wee1 with the activation of cellular pathways that are crucial for the specific disease [74]. In nasopharyngeal (NP) carcinoma cell lines, Wee1 was found to be overexpressed and, consequently, cells were found to be more sensitive to its inhibition compared to NP epithelial cells, although such inhibition was not very effective in sensitizing cells to radiotherapy [75]. In melanoma cells, Wee1 overexpression showed a strong, positive correlation with markers of proliferation: cyclin A, Ki67 and cyclin D3 [74]. Wee1 silencing caused an increase in phospho p38 protein levels, indicating a role in the regulation of p38/MAPK pathway activation during p53-independent DNA damage responses [74].

Aside from the reported use of siRNA for the inhibition of Wee1 expression in different cancer models, several pharmacological inhibitors were developed and validated, both as single agents or in combination with DNA damaging agents (chemotherapy/radiotherapy) [69,76–79]. Wee1 kinase inhibition causes a significant reduction in phospho- CDK1 (Tyr15), thus promoting the accumulation of the active CDK1-cyclin B1 complex and driving premature mitotic entry. Uncontrolled and deregulated mitosis is associated with a progressive DNA damage accumulation, culminating in cell death through a mechanism that is generally known as a mitotic catastrophe. The efficacy of Wee1 inhibitors as monotherapy was confirmed by a decrease in cell viability of ovarian cancer and sarcoma cell lines [80,81].

A large number of studies demonstrated that the cellular rate of response to treatment with Wee1 inhibitors or mimics was strictly dependent on a concomitant (i) presence of TP53 mutations [82–84] and/or (ii) administration of DNA-damaging agents (chemotherapy including doxorubicin, cytarabine, methotrexate, cisplatin, clofarabine, etoposide, 5-fluorouracil and radiotherapy) [85,86]. Moreover, some data suggest that cells with dysfunctional p53 are more sensitive to Wee1 inhibition combined with conventional chemotherapy than those with functional p53. A possible explanation for this is that the dysfunctional G1/S DNA damage checkpoint yields TP53-mutated cells that are more dependent on stopping in G2 to repair DNA damage before entering mitosis.

While some studies described ionizing radiation (IR) [87,88] and chemotherapy [89] sensitization mediated by Wee1 inhibition to be dependent on TP53 activity, the pharmacological inhibition of Wee1, in combination with cytarabine, was shown to be effective in AML cell lines with functional p53 [90]. Van Linden demonstrated that the functionality of p53 does not influence the sensitization to antimetabolite chemotherapeutics by Wee1 inhibitors in AML cells and lung cancer cells, suggesting that the use of p53 mutation as a predictive biomarker for response to Wee1 inhibition may be restricted to certain cancers and/or chemotherapeutics, as well as the preclinical data supporting the combination [91].

Table 1 summarizes the above reported evidence observed in different cancer types.

Table 1. Bibliographic evidence about Wee1 expression in different cancer types.

Wee1 Expression	Other Molecular Modulation	Cancer Type	Clinical Significance	Methods	Ref.
↓ Wee1		Stromal breast cancer (phyllodes tumor (PT)) human tissue samples	Wee1 reduction suggests a potential role of Wee1 as a tumor suppressor	Immunohistochemistry	[57]
↓ Wee1		Colon carcinoma cell lines and human tissue samples	Wee1 suppression suggests a potential role of Wee1 in tumorigenesis	cDNA array, Northern blotting and semi-quantitative reverse transcription-PCR (RT-PCR)	[58]
↓ Wee1	↑ Cyclin B1/cdc2 complex	NSCLC human tissue samples	Wee1 reduction is associated with a poorer prognosis and a higher recurrence rate	Immunohistochemistry	[59]
↓ Wee1	↑ miR-195	Metastatic melanoma cell lines and human tissue samples	Wee1 expression in malignant melanoma is directly regulated by miR-195	Immunoblotting, quantitative real-time PCR and immunohis-tochemistry	[61]
↑ Wee1	↓ miR-101-3p	Hepatocellular carcinoma (HCC) cell lines and human tissue samples	Downregulation of Wee1 enhances the radiosensitivity of HCC cells	Quantitative real-time PCR, Western blotting and flowcytometry	[60]
↑ Wee1		Leukemia cell lines	Wee1 kinase inhibition by siRNA silencing or by specific inhibitors potently sensitizes myeloid and lymphoid leukemia cells to Ara-C	High-throughput siRNA screen and Western blotting	[65]
↑ Wee1		Luminal breast cancer cell lines	Wee1 kinase inhibition by specific inhibitors has therapeutic potentials	High-throughput siRNA screen, Western blotting and immuno-histochemistry	[66]
↑ Wee1		Nasopharyngeal carcinoma (NPC) cell lines	Wee1 kinase inhibition by specific inhibitors has therapeutic potentials	Western blotting	[75]
↑ Wee1		Glioblastoma cell lines and human tissue samples	Wee1 kinase inhibition by siRNA or specific inhibitors causes cell death and sensitizes glioblastoma to ionizing radiation in vivo, suggesting it has a potential therapeutic target	In silico analysis of microarray data, immunofluorescence staining, immunohisto-chemistry and Western blotting	[67]
↑ Wee1 kinase activity	↑ Cyclin D1	Hepatocellular carcinoma (HCC) human tissue samples	Activation of cyclin D1, Cdk4, cyclin E, cyclin A and Wee1 may play important roles in the process of malignant transformation of cirrhosis to HCC	Evaluation of WEE1 kinase activity using autoradiography	[68]

Table 1. *Cont.*

Wee1 Expression	Other Molecular Modulation	Cancer Type	Clinical Significance	Methods	Ref.
↑ Wee1		Medulloblastoma cell lines and human tissue samples	Wee1 kinase inhibition by siRNA or specific inhibitors (MK-1775) potently inhibits tumor growth in vivo and sensitizes medulloblastoma cells to cisplatin in vitro, suggesting Wee1 as a potential therapeutic target	Gene expression analysis, high-throughput siRNA screen and Western blotting	[69]
↑ Wee1		Glioma human tissue samples	Wee1 expression is directly correlated with the malignancy grade in all types of gliomas, but it is inversely associated with prognosis in GBM	Immunohistochemistry	[71]
↑ Wee1		Pediatric high-grade glioma	Wee1 expression positively correlates with the glioma grade; the Wee1 inhibitor MK-1775 increases the radiation cytotoxic effect and prolongs survival for mice with engrafted, orthotopic glioma	Gene expression analysis and immuno-histochemistry	[70]
↑ Wee1		Melanoma cell lines and human tissue samples	High expression of WEE1 is associated with poor prognosis and Wee1 silencing increases tumor cell death; Wee1 represents a potential therapeutic target in melanoma	Immunohistochemistry and Western blotting	[74]
↑ Wee1		Vulvar squamous cell carcinomas human tissue samples	High Wee1 expression is associated with poor histological differentiation and lymph node metastases	Immunohistochemistry and Wee1 silencing using siRNA	[73]
↑ Wee1		Acute myeloid leukemia (AML) cell lines and AML primary cells	Wee1 inhibition sensitizes AML cells to cytarabine in vitro; Wee1 represents a potential therapeutic target in AML	Integrated genomic analyses	[90]
↑ Wee1		Osteosarcoma cell lines, human osteoblasts and tumor samples	Wee1 inhibition sensitizing OS cells to irradiation-induced cell death	Gene-expression data analysis, immunohisto-chemistry and Western blotting	[86]

Table 1. *Cont.*

Wee1 Expression	Other Molecular Modulation	Cancer Type	Clinical Significance	Methods	Ref.
↑ Wee1		Multiple myeloma (MM) tissue samples and cell lines	Wee1 inhibition in combination with Bortezomib induces cell death in all cell lines more efficiently compared to the single agents	Quantitative real-time PCR (qPCR)	[89]
↑ Wee1		Ovarian carcinoma (OC) peritoneal effusion samples and cell lines	Wee1 is overexpressed in post-chemotherapy disease, suggesting a role in mediating disease progression and as a prognostic marker of poor survival; Wee1 is a potential therapeutic target in OC	Immunohistochemistry and Western blotting	[72]

5. Radiotherapy and DNA Damage Response (DDR)

Radiation therapy is one of the crucial cancer treatment options along with surgery and chemotherapy [92]. The curative potential of radiotherapy depends on the amount of non-repairable DNA lesions that occurred in the exposed tumor tissue, thereby removing cancer cells from the clonogenic pool [93–96].

DNA integrity can be compromised by several types of exogenous and/or endogenous injuries. In particular, exposure to ionizing radiation is responsible for base and sugar damage, cross-links and both single- and double-strand breaks (SSBs and DSBs, respectively). Among these, DNA double-strand breaks (DSBs) principally contribute to radiation-induced cell death. However, cells can avoid this fate through DNA repair mechanisms. When DNA damage occurs, cells can arrest cycle progression to allow for DNA repair before cell division, which makes DNA alteration permanent. In fact, the accumulation of DNA lesions may lead to cell death and/or senescence if the damage is too severe to be repaired [97].

DNA repair mechanisms are the key determinant of tumor cell sensitivity (or resistance) to radiation and, for this reason, have gained a lot of interest in the oncology field. If the DNA repair capacity of tumor cells is not able to mitigate the severity of radiation-induced DSBs, this might result in the perpetuation of DNA damage, leading to irreparable genetic lesions that culminate in cell death. In contrast, if cells can repair radiation-induced damage, cancer cells continue to proliferate and tumors may recur [98].

In the presence of genotoxic stress, three different DNA damage repair (DDR) pathways are activated. Three phosphatidyl inositol 3-kinase-related protein kinases are the upstream molecules involved, though activated by different injury types. Ataxia telangiectasia mutated (ATM) responds to DNA-damage agents, such as IR, causing DSBs. Ataxia telangiectasia and Rad3-related (ATR) kinase detect alterations of replications (stalled replication forks and branched structures formation) that occur naturally or after ultraviolet light (UV) exposure. Like ATM, the DNA-dependent protein kinase (DNA-PK) signaling pathway is activated by DSBs under different cellular conditions, including IR exposure, environmental carcinogens and chemotherapeutic agents, or in cells with shortened telomeres.

ATM is recruited to DSBs by the Mre11–Rad50–Nbs1 (MRN) complex, whereas DNA-PK is recruited by the Ku70/Ku80 heterodimer [99]. ATR is recruited by the ATR-interacting protein (ATRIP) to replication protein A (RPA)-coated single-stranded DNA (ssDNA), which forms at stalled DNA replication forks or is generated by processing of the initial DNA damage [100–102]. Both ATM and ATR are able to induce chromatin modification in the presence of DNA damage through phosphorylation of H2AX, forming foci at the break sites. H2AX, once phosphorylated (creating γ-H2AX), allows for the recruitment of other proteins that take part in the repair mechanism [103]. ATM and ATR regulate the Werner syndrome protein (WRN), which is implicated in the recovery of stalled replication forks, to limit fork collapse [104] and act on BRCA-1, which serves as a scaffold to facilitate ATM and ATR to activate downstream substrates [105].

When recruited, ATM and ATR phosphorylate several substrates. Their principal downstream effectors are two kinases, namely, Chk2 and Chk1, which spread the signal to other molecules [106,107].

Chk2 acts on protein p53, which is the principal gatekeeper of the G1-phase, determining the arrest of the cell cycle. ATM can also directly regulate p53 stability, weakening its interaction with its negative regulator, namely, the MDM2 oncoprotein, whose gene is, in turn, activated by p53 itself [108].

Among Chk1 substrates, in addition to Wee1, whose function as the gatekeeper in G2/M checkpoint and S-phase was discussed above, there are the phosphatases Cdc25 (A, B and C). Their activities are directly inhibited, promoting their degradation and causing cell cycle delay not only in the G2/M checkpoint but even in other steps, providing time for DNA repair [109,110]. However, Chk1 regulates Cdc25 levels both in the presence of DNA damage and in physiological conditions, supporting its rapid turnover and avoiding an unscheduled massive DNA synthesis [111,112] (Figure 3).

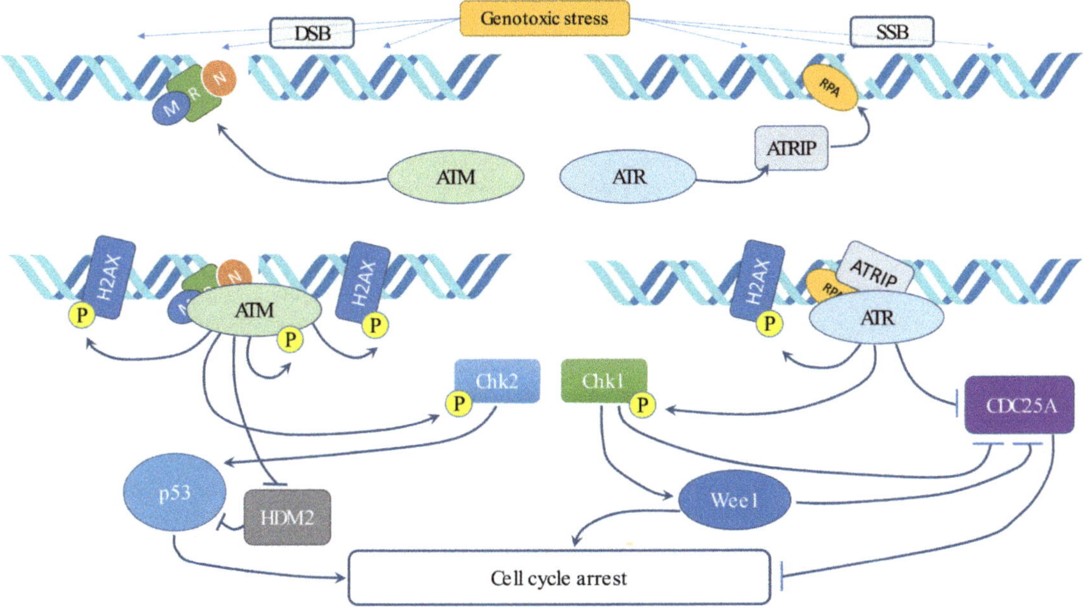

Figure 3. Schematic representation of the components involved in the DNA damage response.

Furthermore, Chk1 is required for homologous recombination repair (HRR), which is a mechanism that is essential in mammals for restoring DNA integrity after DSBs, as well as other lesions; in fact, it promotes the association of chromatin and phosphorylation of RAD51, which is a key protein in HRR [113].

Moreover, Chk1 is required for mitotic spindle checkpoint functioning. During the metaphase, chromosomes are hooked in both their sides by microtubules, forming a mitotic spindle coming from opposite poles of the cell; then, they are aligned in the central portion of the mitotic spindle and equally divided to the daughter cells, which receive a chromosomal copy each. If something at the level of alignment does not occur regularly and defects emerge, Chk1 is phosphorylated and, in turn, phosphorylates Aurora B protein kinase. This regulates BubR1 protein, which, together with Mad1 and Mad2, forms an inhibitory ternary complex with the E3 ligase APC and its activator Cdc20 in the mitotic spindle checkpoint. Thus, BubR1 is recruited at the kinetochore level and delays the passage from the metaphase to the anaphase, protecting against chromosomal segregation defects [114–116]. Following the DSB, DNA-PK-mediated DNA repair occurs via the non-homologous end joining (NHEJ) process. In such a process, each broken DNA end is first bound by one Ku70/80 heterodimer and two heterodimers interact together to bridge matching ends. Ku plays a crucial role in NHEJ, recruiting multiple downstream proteins, including the DNA-dependent protein kinase catalytic subunit (DNA-PKcs). In turn, DNA-PKcs interacts with Artemis and DNA ligase IV, enabling NHEJ [99,117].

6. Cancer Stem Cells and WEE1 Involvement in Radiation Oncology

Cancer cells are heterogeneous regarding their tumor-initiating properties. In particular, cancer stem cells (CSCs) are a small subset of a tumor population, normally representing 0.1–10% of all tumor cells, which were identified as tumor-initiating cells. CSCs possess self-renewal potential and persistent tumorigenic capacity that make them different from other tumor cells.

Experimental evidence on tumor cell transplantation, as well as in both isogenic murine models and human xenograft tumor models, demonstrated that the number of CSCs in these experimental tumors defines the therapeutic potential of radiotherapy, thus a higher proportion of CSCs correlates with a higher radio-resistance in the same histopathological tumor type [93,94,118–120]. Additionally, the intrinsic radiosensitivity of CSCs varies between tumors, thereby affecting their radio-curability [121,122]. A recent CSC model, which was developed by our research group, allowed for predicting the radiotherapy treatment efficacy based on the evident concordance between in vitro and in vivo CSC sensitivity to radiotherapy [121]. Notably, the specific CSC in vitro and in vivo sensitivity values correspond to patients' responses to radiotherapy. This approach may be useful for driving clinical decisions for correct therapeutic option management [121].

As a key source of resistance to DNA-damaging treatment in oncology, CSCs were reported to improve DNA repair capacity by enhancing the activation of the DNA damage response compared with differentiated cells [123,124]. CSCs contribute to radio- and chemo-resistance through a circular mechanism: repeated cycles of DNA-damaging treatments progressively kill the non-stem cells, causing an increase in the CSC fraction within the tumor cell population. The tumor microenvironment can also influence the tumor cell sensitivity or resistance to IR. In particular, it was demonstrated that CSCs are protected from the effects of IR by hypoxia that is mediated by specific hypoxic niches [93,125]. Hypoxia might affect CSC proliferation and viability via hypoxia-inducible factors (HIFs) that can induce the expression of OCT4, MYC and NOTCH1, which are crucial for stem cell maintenance in different tissues [126–128]. In addition, the cell–stroma interactions, by means of the integrin-mediated adhesion of cells to the extracellular matrix, increase the tumor cell resistance to IR [129]. As a result, classical therapies become progressively ineffective toward these CSC-enriched tumors [130].

Few studies reported Wee1 inhibition in the particular case of CSCs inducing radio-sensitization in glioblastoma [67,131]. However, these findings support the concept that the overexpression of checkpoint inhibitors, such as Wee1, could be an adopted protection mechanism in CSCs with sublethal damage induced by conventional radio- and chemotherapy. Therefore, targeting the DNA damage checkpoint response in CSC may sensitize these cells to DNA-damaging techniques and overcome tumor resistances [132].

7. Wee1 Inhibitors

Wee1 represents an optimal target for the inhibition of the G2-M checkpoint in order to potentiate both radio- and chemotherapy.

In fact, although the purpose of the majority of anti-cancer therapeutic strategies involves cell cycle arrest, Wee1 kinase inhibition triggers mitosis and induces genomic instability, driving cells to follow a replication cycle, with consequential apoptosis for mitotic catastrophe.

This strategy may allow for the dose reduction of conventional therapies, leading to a reduction in toxic side effects while maintaining clinical efficacy; moreover, it may sensitize tumors with poor prognosis to conventional therapies.

Recent studies demonstrated that Wee1 inhibition can be reached with low cytotoxicity through rational drug design [133,134]. The only Wee1 inhibitor that is currently used in clinical trials (query on ClinicalTrials.gov, July 2021) (Accessed on 22 September 2021) is AZD1775, combined with DNA damage agents or radiotherapy, which is tolerable and demonstrates promising anticancer activity.

In addition to AZD1775, hundreds of compounds were reported to have inhibitory activity against Wee1 kinase. A recent study classified the Wee1 inhibitors that are reported in scientific literature into five groups on the basis of their chemical core structure (pyridopyrimidine derivatives, pyrazolopyrimidinone derivatives, pyrrolocarbazole derivatives, pyrimidine-based tricyclic molecules and vanillates) [135].

The first small molecule to be reported bearing the pyridopyrimidine core was PD0166285 [87] Although it showed a potent Wee1 inhibition activity (IC50 = 24 nM) in various cancer cell lines and xenografts [67,87,136–138], its clinical application is limited because of its non-selective Wee1 inhibition. In fact, PD0166285 presents a broad spectrum of inhibitory activity on several tyrosine kinases, including CHK1, MYT1, c-Src, EGFR, FGFR1 and PDGFR [87,135]. Due to its poor selectivity, PD0166285's clinical application is limited. Furthermore, starting from the PD0166285 structure, several pyridopyrimidine derivates were synthesized while trying to increase the selectivity for Wee1, but most of them were found to preserve a potent inhibitory activity against c-Src [87,139–141].

Among the Wee1 inhibitors that have the pyrazolopyrimidinone scaffold, there is the known AZD1775. A series of pyrazolopyrimidinone analogs were synthetized [142] and, in 2016, Matheson and collaborators reported a compound (CJM-061) that showed the same Wee1 inhibitory efficacy of AZD1775, but had reduced single-agent cytotoxicity in medulloblastoma cells [133]. On this basis, Matheson et al. developed a series of potent pyrazolopyrimidinone-based Wee1 inhibitors; in particular, they found a compound that showed a stronger inhibition activity and reduced cytotoxicity compared to AZD1775 [134]. However, AZD1775 is the only Wee1 inhibitor that is currently used in clinical trials (query on ClinicalTrials.gov, July 2021).

The third class of Wee1 inhibitors is composed of the pyrrolocarbazole derivatives, all of which are analogs of the lead compound 4-phenylpyrrolecarbazole PD0407824, which is a powerful Wee1/Chk1 inhibitor. PD0407824 was a less potent Wee1 inhibitor compared to PD0166285 but was more selective for Wee1 and CHK1 than for c-Src [143]. Furthermore, it has only been tested on ovarian cancer, where it was shown to positively modulate the cis-platinum response [144]. On the basis of several structure–activity relationship studies [143,145,146], the analogs were designed in order to be more selective for Wee1 than for CHK1 [143]. Furthermore, the study of this class of Wee1 inhibitors has not been advanced into in vivo study, probably because of the broad spectrum of kinase inhibitor activity.

In 2014, Tong et al., aided by molecular modeling and structure–activity relationship studies, designed pyrimidine-based tricyclic molecules that showed potent Wee1 kinase inhibitor activity in both functional and mechanism-based cellular studies. The lead molecule, namely, 31, showed oral efficacy in the NCI-H1299 mouse xenograft model, potentiating the antiproliferative activity of irinotecan [147].

It is known that natural polyphenols, including polyphenols present in green tea (epigallocatechin gallate (EGCG)) [148], genistein [149] and curcumin [150], possess potential anticancer activity through inhibiting the proliferation, invasion and metastasis of tumoral cells and the induction of apoptosis, acting on different signaling pathways [148–155].

Starting from the core structure of polyphenols, Lamoral-Theys' group designed and synthesized several di- and trivanillate compounds, with some of them presenting WEE1 inhibitory and anti-tumor activities, probably due to their inhibitory activity against the Aurora A/B/C responsible [156].

Table 2 summarizes the above reported evidence.

Table 2. Wee1 inhibitors. Standard color coding for atoms is used (black—carbon, red—oxygen, blue—nitrogen, green—chlorine, brown—fluorine).

	Core Chemical Structure	Lead Compounds	CHK1 IC$_{50}$ (nM)	WEE1 IC$_{50}$ (nM)	References
Pyridopyrimidine derivatives		PD0166285	72	24	[86]
Pyrazolopyrimidinone derivatives		CJM-061	-	2.8	[133]
		AZD1775	-	1.9	[133]
Pyrrolocarbazole derivatives		PD0407824	47	97	[143]
Pyrimidine-based tricyclic molecules		31	-	H1299 EC = 230	[147]

Table 2. Cont.

	Core Chemical Structure	Lead Compounds	CHK1 IC$_{50}$ (nM)	WEE1 IC$_{50}$ (nM)	References
Vanillates	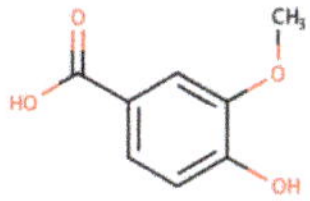	32 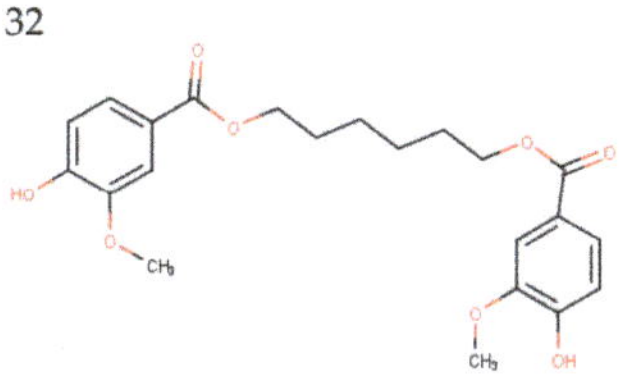	-	20,000	[156]
		33 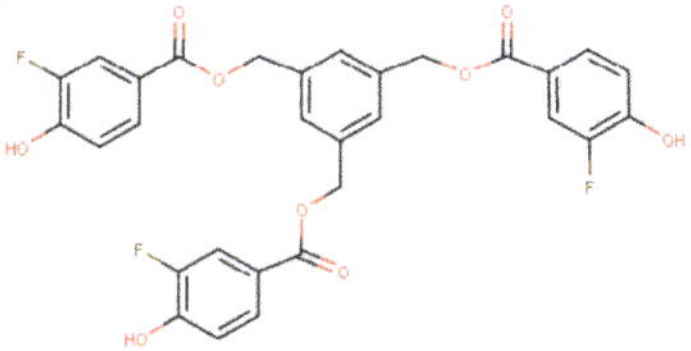	-	7120	[156]

8. Wee1 Inhibition Sensitizes Response to Chemo/Radiotherapy

Among the Wee1 inhibitors, only AZD1775 is currently used in clinical trials. Several studies show its ability to potentiate the activity of chemotherapeutic drugs and radiotherapy. Several studies indicate that AZD1775's sensitizing effect is selective only in p53-deficient tumors [142,157,158], although other evidence demonstrates that Wee1 inhibition can independently sensitize cancer cells to chemotherapeutics via p53 functionality [91].

In particular, in advanced squamous cell carcinoma of the head and neck (HNSCC), it was shown that a combination of AZD1775 and cisplatin can overcome cisplatin resistance, which is particularly high for patients whose tumor presents a mutation in the TP53 gene [159]. Other studies suggest a similar sensitizing effect in the radiation response in pontine gliomas [160], glioblastoma [131] and pancreatic cancer [161]. So far, there are 11 clinical trials testing AZD1775 in combination with various chemotherapeutic agents, 5 in combination with both chemo and radiotherapy and 2 in combination with only radiotherapy. Table 3 lists only the completed clinical trials with reported results.

Table 3. Registered clinical trial on Wee1 inhibitors.

	Title	Conditions	Interventions	ClinicalTrial.gov Identifier Number
1	WEE1 Inhibitor with Cisplatin and Radiotherapy: ATrial in Head and Neck Cancer	-Hypopharynx squamous cell carcinoma -Oral cavity squamous cell carcinoma -Larynx cancer	-AZD1775 -Cisplatin -Radiotherapy	NCT03028766
2	WEE1 Inhibitor AZD1775 with or without Cytarabine in Treating Patients with Advanced Acute Myeloid Leukemia or Myelodysplastic Syndrome	-Chronic myelomonocytic leukemia -Myelodysplastic syndrome with isolated del(5q) -Myelodysplastic/myeloproliferative neoplasm -Previously treated myelodysplastic syndrome -Recurrent adult acute myeloid leukemia -Untreated adult acute myeloid leukemia	-AZD1775 -Cytarabine	NCT02666950

Table 3. *Cont.*

	Title	Conditions	Interventions	ClinicalTrial.gov Identifier Number
3	Phase Ib Study AZD1775 in Combination with Carboplatin and Paclitaxel in Adult Asian Patients with Solid Tumours	Advanced solid tumors	-AZD1775 -Paclitaxel -Carboplatin	NCT02341456
4	Cisplatin with or without WEE1 Inhibitor MK-1775 in Treating Patients with Recurrent or Metastatic Head and Neck Cancer	-Recurrent hypopharyngeal squamous cell carcinoma -Recurrent laryngeal squamous cell carcinoma -Recurrent laryngeal verrucous carcinoma -Recurrent lip and oral cavity squamous cell carcinoma -Recurrent metastatic squamous cell carcinoma in the neck with an occult primary -another 24 tumors	-MK-1775 -Cisplatin	NCT02196168
5	Ph II Trial of Carboplatin and Pemetrexed with or without AZD1775 for Untreated Lung Cancer	Previously untreated stage IV non-squamous non-small cell lung cancer	-AZD1775 -Pemetrexed -Carboplatin	NCT02087241
6	Dose Escalation Trial of AZD1775 and Gemcitabine (+Radiation) for Unresectable Adenocarcinoma of the Pancreas	Adenocarcinoma of the pancreas	-AZD-1775 -Gemcitabine -Radiotherapy	NCT02037230
7	A Study of MK-1775 in Combination with Topotecan/Cisplatin in Participants with Cervical Cancer (MK-1775-008)	Cervical cancer	-MK-1775 -Topotecan -Cisplatin	NCT01076400
8	A Dose Escalation Study of MK-1775 in Combination with Either Gemcitabine, Cisplatin, or Carboplatin in Adults with Advanced Solid Tumors (MK-1775-001)	Solid tumors	-MK-1775 -Gemcitabine -Cisplatin -Carboplatin	NCT00648648

9. Conclusions

Developing novel strategies to contrast local tumor cell survival and to, consequently, avoid the insurgence of metastasis is the principal aim of the researchers operating in the oncology field. As functionally impaired p53 is one of the principal hallmarks of tumor cells, they rely on remaining checkpoints to survive and proliferate. From this perspective, Wee1 is surely a prominent target to use in clinical practice for its dual role as gatekeeper in both the S-phase and G2/M transition. While most kinase inhibition strategies aim to arrest the cell cycle to block proliferation, WEE1 inhibition allows for mitosis to occur in cells with heavy DNA damage. This phenomenon increases the amplification of genomic instability through cellular replication cycles, which makes replication errors permanent. Cell cycle progression, when DNA damage occurs, leads to cells quickly accumulating a great number of mutations that make them unable to survive. Thus, through its inhibition, it might be possible to increase the effectiveness of DNA-damaging treatments, such as radiotherapy, forcing tumor cells to enter into mitosis, even with damaged DNA. This, finally, leads them to mitotic catastrophe and subsequent cell death.

Wee1 inhibition may allow for the reduction of the dose of cytotoxic chemotherapy, thus improving the safety profiles. It may also be employed to sensitize resistant tumors to conventional therapies when they are not effective as monotherapy.

Author Contributions: F.E., R.G. wrote the manuscript and designed the tables, G.R. and C.P. contributed to the writing of the manuscript, S.F. designed the figures and made the critical revision of the manuscript. All authors have read and agreed to the published version of the manuscript.

Funding: This research received no external funding.

Data Availability Statement: Data supporting reported evidences can be obtained from the original sources (please see reference), data from clinical trials can be retrieved from www.ClinicalTrial.gov.

Acknowledgments: The authors thank Aurelio Lorico for help with English language editing.

Conflicts of Interest: The authors declare no conflict of interest.

References

1. Lim, S.; Kaldis, P. Cdks, cyclins and CKIs: Roles beyond cell cycle regulation. *Development* **2013**, *140*, 3079–3093. [CrossRef]
2. Poon, R.Y.C. Cell Cycle Control: A System of Interlinking Oscillators. *Methods Mol. Biol.* **2016**, *1342*, 3–19. [CrossRef]
3. Beetham, K.L.; Tolmach, L.J. The Action of Caffeine on X-Irradiated HeLa Cells. V. Identity of the Sector of Cells That Expresses Potentially Lethal Damage in G1 and G2. *Radiat. Res.* **1982**, *91*, 199–211. [CrossRef]
4. Dixon, H.; Norbury, C.J. Therapeutic Exploitation of Checkpoint Defects in Cancer Cells Lacking p53 Function. *Cell Cycle* **2002**, *1*, 362–368. [CrossRef] [PubMed]
5. Benada, J.; Macurek, L. Targeting the Checkpoint to Kill Cancer Cells. *Biomolecules* **2015**, *5*, 1912–1937. [CrossRef] [PubMed]
6. Santo, L.; Siu, K.T.; Raje, N. Targeting Cyclin-Dependent Kinases and Cell Cycle Progression in Human Cancers. *Semin. Oncol.* **2015**, *42*, 788–800. [CrossRef] [PubMed]
7. Schmidt, M.; Rohe, A.; Platzer, C.; Najjar, A.; Erdmann, F.; Sippl, W. Regulation of G2/M Transition by Inhibition of WEE1 and PKMYT1 Kinases. *Molecules* **2017**, *22*, 2045. [CrossRef]
8. Nakanishi, M.; Ando, H.; Watanabe, N.; Kitamura, K.; Ito, K.; Okayama, H.; Miyamoto, T.; Agui, T.; Sasaki, M. Identification and characterization of human Wee1B, a new member of the Wee1 family of Cdk-inhibitory kinases. *Genes Cells* **2000**, *5*, 839–847. [CrossRef]
9. Russell, P.; Nurse, P. Negative regulation of mitosis by wee1+, a gene encoding a protein kinase homolog. *Cell* **1987**, *49*, 559–567. [CrossRef]
10. Gould, K.; Nurse, P. Tyrosine phosphorylation of the fission yeast cdc2+ protein kinase regulates entry into mitosis. *Nature* **1989**, *342*, 39–45. [CrossRef]
11. McGowan, C.; Russell, P. Cell cycle regulation of human WEE1. *EMBO J.* **1995**, *14*, 2166–2175. [CrossRef]
12. Squire, C.; Dickson, J.M.; Ivanovic, I.; Baker, E. Structure and Inhibition of the Human Cell Cycle Checkpoint Kinase, Wee1A Kinase: An Atypical Tyrosine Kinase with a Key Role in CDK1 Regulation. *Structure* **2005**, *13*, 541–550. [CrossRef] [PubMed]
13. Mueller, P.R.; Coleman, T.R.; Kumagai, A.; Dunphy, W.G. Myt1: A Membrane-Associated Inhibitory Kinase That Phosphorylates Cdc2 on Both Threonine-14 and Tyrosine-15. *Science* **1995**, *270*, 86–90. [CrossRef] [PubMed]
14. Liu, F.; Stanton, J.J.; Wu, Z.; Piwnica-Worms, H. The human Myt1 kinase preferentially phosphorylates Cdc2 on threonine 14 and localizes to the endoplasmic reticulum and Golgi complex. *Mol. Cell. Biol.* **1997**, *17*, 571–583. [CrossRef] [PubMed]
15. Wells, N.; Watanabe, N.; Tokusumi, T.; Jiang, W.; Verdecia, M.; Hunter, T. The C-terminal domain of the Cdc2 inhibitory kinase Myt1 interacts with Cdc2 complexes and is required for inhibition of G(2)/M progression. *J. Cell Sci.* **1999**, *112*, 3361–3371. [CrossRef]
16. Booher, R.N.; Holman, P.S.; Fattaey, A. Human Myt1 Is a Cell Cycle-regulated Kinase That Inhibits Cdc2 but Not Cdk2 Activity. *J. Biol. Chem.* **1997**, *272*, 22300–22306. [CrossRef]
17. Richard, C.W.; Boehnke, M.; Berg, D.J.; Lichy, J.H.; Meeker, T.C.; Hauser, E.; Myers, R.M.; Cox, D.R. A radiation hybrid map of the distal short arm of human chromosome 11, containing the Beckwith-Weidemann and associated embryonal tumor disease loci. *Am. J. Hum. Genet.* **1993**, *52*, 915–921.
18. Thuriaux, P.; Nurse, P.; Carter, B. Mutants altered in the control co-ordinating cell division with cell growth in the fission yeast *Schizosaccharomyces pombe*. *MGG Mol. Gen. Genet.* **1978**, *161*, 215–220. [CrossRef] [PubMed]
19. Nurse, P.; Thuriaux, P. Regulatory genes controlling mitosis in the fission yeast *Schizosaccharomyces pombe*. *Genetics* **1980**, *96*, 627–637. [CrossRef]
20. Igarashi, M.; Nagata, A.; Jinno, S.; Suto, K.; Okayama, H. Wee1+-like gene in human cells. *Nature* **1991**, *353*, 80–83. [CrossRef]

21. McGowan, C.; Russell, P. Human Wee1 kinase inhibits cell division by phosphorylating p34cdc2 exclusively on Tyr15. *EMBO J.* **1993**, *12*, 75–85. [CrossRef]
22. Nurse, P. Universal control mechanism regulating onset of M-phase. *Nature* **1990**, *344*, 503–508. [CrossRef]
23. Mueller, P.R.; Coleman, T.R.; Dunphy, W.G. Cell cycle regulation of a Xenopus Wee1-like kinase. *Mol. Biol. Cell* **1995**, *6*, 119–134. [CrossRef]
24. Booher, R.; Deshaies, R.; Kirschner, M. Properties of Saccharomyces cerevisiae wee1 and its differential regulation of p34CDC28 in response to G1 and G2 cyclins. *EMBO J.* **1993**, *12*, 3417–3426. [CrossRef]
25. Campbell, S.D.; Sprenger, F.; Edgar, B.A.; O'Farrell, P.H. Drosophila Wee1 kinase rescues fission yeast from mitotic catastrophe and phosphorylates Drosophila Cdc2 in vitro. *Mol. Biol. Cell* **1995**, *6*, 1333–1347. [CrossRef] [PubMed]
26. Honda, R.; Tanaka, H.; Ohba, Y.; Yasuda, H. Mouse p87wee1 kinase is regulated by M-phase specific phosphorylation. *Chromosome Res.* **1995**, *3*, 300–308. [CrossRef] [PubMed]
27. Boardman, P.E.; Sanz-Ezquerro, J.; Overton, I.M.; Burt, D.W.; Bosch, E.; Fong, W.T.; Tickle, C.; Brown, W.R.; Wilson, S.A.; Hubbard, S.J. A comprehensive collection of chicken cDNAs. *Curr. Biol.* **2002**, *12*, 1965–1969. [CrossRef]
28. Hirayama, J.; Cardone, L.; Doi, M.; Sassone-Corsi, P. Common pathways in circadian and cell cycle clocks: Light-dependent activation of Fos/AP-1 in zebrafish controls CRY-1a and WEE-1. *Proc. Natl. Acad. Sci. USA* **2005**, *102*, 10194–10199. [CrossRef] [PubMed]
29. Obsilova, V.; Obsil, T. The 14-3-3 Proteins as Important Allosteric Regulators of Protein Kinases. *Int. J. Mol. Sci.* **2020**, *21*, 8824. [CrossRef]
30. Wang, Y.; Jacobs, C.; Hook, K.E.; Duan, H.; Booher, R.N.; Sun, Y. Binding of 14-3-3β to the carboxyl terminus of Wee1 increases Wee1 stability, kinase activity, and G2-M cell population. *Cell Growth Differ.* **2000**, *11*, 211–219. [PubMed]
31. Rothblum-Oviatt, C.J.; E Ryan, C.; Piwnica-Worms, H. 14-3-3 binding regulates catalytic activity of human Wee1 kinase. *Cell Growth Differ.* **2001**, *12*, 581–589.
32. Aligue, R.; Akhavan-Niak, H.; Russell, P. A role for Hsp90 in cell cycle control: Wee1 tyrosine kinase activity requires interaction with Hsp90. *EMBO J.* **1994**, *13*, 6099–6106. [CrossRef] [PubMed]
33. Goes, F.S.; Martin, J. Hsp90 chaperone complexes are required for the activity and stability of yeast protein kinases Mik1, Wee1 and Swe1. *Eur. J. Biochem.* **2001**, *268*, 2281–2289. [CrossRef] [PubMed]
34. Mollapour, M.; Tsutsumi, S.; Donnelly, A.C.; Beebe, K.; Tokita, M.J.; Lee, M.-J.; Lee, S.; Morra, G.; Bourboulia, D.; Scroggins, B.T.; et al. Swe1Wee1-Dependent Tyrosine Phosphorylation of Hsp90 Regulates Distinct Facets of Chaperone Function. *Mol. Cell* **2010**, *37*, 333–343. [CrossRef]
35. Iwai, A.; Bourboulia, D.; Mollapour, M.; Jensen-Taubman, S.; Lee, S.; Donnelly, A.C.; Yoshida, S.; Miyajima, N.; Tsutsumi, S.; Smith, A.K.; et al. Combined inhibition of Wee1 and Hsp90 activates intrinsic apoptosis in cancer cells. *Cell Cycle* **2012**, *11*, 3649–3655. [CrossRef]
36. Sasaki, M.; Terabayashi, T.; Weiss, S.M.; Ferby, I. The Tumor Suppressor MIG6 Controls Mitotic Progression and the G2/M DNA Damage Checkpoint by Stabilizing the WEE1 Kinase. *Cell Rep.* **2018**, *24*, 1278–1289. [CrossRef]
37. Heald, R.; McLoughlin, M.; McKeon, F. Human wee1 maintains mitotic timing by protecting the nucleus from cytoplasmically activated cdc2 kinase. *Cell* **1993**, *74*, 463–474. [CrossRef]
38. Watanabe, N.; Broome, M.; Hunter, T. Regulation of the human WEE1Hu CDK tyrosine 15-kinase during the cell cycle. *EMBO J.* **1995**, *14*, 1878–1891. [CrossRef]
39. Dunphy, W.G.; Brizuela, L.; Beach, D.; Newport, J. The Xenopus cdc2 protein is a component of MPF, a cytoplasmic regulator of mitosis. *Cell* **1988**, *54*, 423–431. [CrossRef]
40. Gautier, J.; Norbury, C.; Lohka, M.; Nurse, P.; Maller, J. Purified maturation-promoting factor contains the product of a Xenopus homolog of the fission yeast cell cycle control gene cdc2$^+$. *Cell* **1988**, *54*, 433–439. [CrossRef]
41. Murray, A.W. Turning on mitosis. *Curr. Biol.* **1993**, *3*, 291–293. [CrossRef]
42. Parker, L.L.; Piwnica-Worms, H. Inactivation of the p34cdc2-cyclin B complex by the human WEE1 tyrosine kinase. *Science* **1992**, *257*, 1955–1957. [CrossRef] [PubMed]
43. Solomon, M.J.; Lee, T.; Kirschner, M.W. Role of phosphorylation in p34cdc2 activation: Identification of an activating kinase. *Mol. Biol. Cell* **1992**, *3*, 13–27. [CrossRef]
44. Larochelle, S.; Merrick, K.A.; Terret, M.-E.; Wohlbold, L.; Barboza, N.M.; Zhang, C.; Shokat, K.M.; Jallepalli, P.V.; Fisher, R.P. Requirements for Cdk7 in the Assembly of Cdk1/Cyclin B and Activation of Cdk2 Revealed by Chemical Genetics in Human Cells. *Mol. Cell* **2007**, *25*, 839–850. [CrossRef] [PubMed]
45. Ayad, N.G.; Rankin, S.; Murakami, M.; Jebanathirajah, J.; Gygi, S.; Kirschner, M.W. Tome-1, a Trigger of Mitotic Entry, Is Degraded during G1 via the APC. *Cell* **2003**, *113*, 101–113. [CrossRef]
46. Watanabe, N.; Arai, H.; Nishihara, Y.; Taniguchi, M.; Watanabe, N.; Hunter, T.; Osada, H. M-phase kinases induce phospho-dependent ubiquitination of somatic Wee1 by SCF$^{β\text{-}TrCP}$. *Proc. Natl. Acad. Sci. USA* **2004**, *101*, 4419–4424. [CrossRef]
47. Smith, A.; Simanski, S.; Fallahi, M.; Ayad, N.G. Redundant Ubiquitin Ligase Activities Regulate Wee1 Degradation and Mitotic Entry. *Cell Cycle* **2007**, *6*, 2795–2799. [CrossRef]
48. Beck, H.; Nähse, V.; Larsen, M.S.Y.; Groth, P.; Clancy, T.; Lees, M.; Jørgensen, M.; Helleday, T.; Syljuåsen, R.G.; Sorensen, C. Regulators of cyclin-dependent kinases are crucial for maintaining genome integrity in S phase. *J. Cell Biol.* **2010**, *188*, 629–638. [CrossRef]

49. Martín, Y.; Domínguez-Kelly, R.; Freire, R. Novel insights into maintaining genomic integrity: Wee1 regulating Mus81/Eme1. *Cell Div.* **2011**, *6*, 21. [CrossRef]
50. Duda, H.; Arter, M.; Gloggnitzer, J.; Teloni, F.; Wild, P.; Blanco, M.G.; Altmeyer, M.; Matos, J. A Mechanism for Controlled Breakage of Under-replicated Chromosomes during Mitosis. *Dev. Cell* **2016**, *39*, 740–755. [CrossRef]
51. Domínguez-Kelly, R.; Martín, Y.; Koundrioukoff, S.; Tanenbaum, M.E.; Smits, V.A.; Medema, R.; Debatisse, M.; Freire, R. Wee1 controls genomic stability during replication by regulating the Mus81-Eme1 endonuclease. *J. Cell Biol.* **2011**, *194*, 567–579. [CrossRef]
52. Beck, H.; Nähse-Kumpf, V.; Larsen, M.S.Y.; O'Hanlon, K.A.; Patzke, S.; Holmberg, C.; Mejlvang, J.; Groth, A.; Nielsen, O.; Syljuåsen, R.G.; et al. Cyclin-Dependent Kinase Suppression by WEE1 Kinase Protects the Genome through Control of Replication Initiation and Nucleotide Consumption. *Mol. Cell. Biol.* **2012**, *32*, 4226–4236. [CrossRef]
53. Técher, H.; Koundrioukoff, S.; Carignon, S.; Wilhelm, T.; Millot, G.A.; Lopez, B.; Brison, O.; Debatisse, M. Signaling from Mus81-Eme2-Dependent DNA Damage Elicited by Chk1 Deficiency Modulates Replication Fork Speed and Origin Usage. *Cell Rep.* **2016**, *14*, 1114–1127. [CrossRef] [PubMed]
54. Vassilopoulos, A.; Tominaga, Y.; Kim, H.-S.; Lahusen, T.; Li, B.; Yu, H.; Gius, D.; Deng, C.-X. WEE1 murine deficiency induces hyper-activation of APC/C and results in genomic instability and carcinogenesis. *Oncogene* **2015**, *34*, 3023–3035. [CrossRef] [PubMed]
55. Mahajan, K.; Fang, B.; Koomen, J.M.; Mahajan, N.P. H2B Tyr37 phosphorylation suppresses expression of replication-dependent core histone genes. *Nat. Struct. Mol. Biol.* **2012**, *19*, 930–937. [CrossRef]
56. Tominaga, Y.; Li, C.; Wang, R.-H.; Deng, C.-X. Murine Wee1 Plays a Critical Role in Cell Cycle Regulation and Pre-Implantation Stages of Embryonic Development. *Int. J. Biol. Sci.* **2006**, *2*, 161–170. [CrossRef] [PubMed]
57. Eom, M.; Han, A.; Lee, M.J.; Park, K.H. Expressional Difference of RHEB, HDAC1, and WEE1 Proteins in the Stromal Tumors of the Breast and Their Significance in Tumorigenesis. *Korean J. Pathol.* **2012**, *46*, 324–330. [CrossRef]
58. Backert, S.; Gelos, M.; Kobalz, U.; Hanski, M.-L.; Mann, B.; Gratchev, A.; Mansmann, U.; Moyer, M.P.; Riecken, E.-O.; Hanski, C. Differential gene expression in colon carcinoma cells and tissues detected with a cDNA array. *Int. J. Cancer* **1999**, *82*, 868–874. [CrossRef]
59. Yoshida, T.; Tanaka, S.; Mogi, A.; Shitara, Y.; Kuwano, H. The clinical significance of Cyclin B1 and Wee1 expression innon-small-cell lung cancer. *Ann. Oncol.* **2004**, *15*, 252–256. [CrossRef]
60. Chen, X.; Zhang, N. Downregulation of lncRNA NEAT1_2 radiosensitizes hepatocellular carcinoma cells through regulation of miR-101-3p/WEE1 axis. *Cell Biol. Int.* **2019**, *43*, 44–55. [CrossRef]
61. Bhattacharya, A.; Schmitz, U.; Wolkenhauer, O.; Schönherr, M.; Raatz, Y.; Kunz, M. Regulation of cell cycle checkpoint kinase WEE1 by miR-195 in malignant melanoma. *Oncogene* **2013**, *32*, 3175–3183. [CrossRef] [PubMed]
62. Leary, A.; Auguste, A.; Mesnage, S. DNA damage response as a therapeutic target in gynecological cancers. *Curr. Opin. Oncol.* **2016**, *28*, 404–411. [CrossRef] [PubMed]
63. Wu, J.; Li, Y.; Dang, Y.-Z.; Gao, H.-X.; Jiang, J.-L.; Chen, Z.-N. HAb18G/CD147 Promotes Radioresistance in Hepatocellular Carcinoma Cells: A Potential Role for Integrin β1 Signaling. *Mol. Cancer Ther.* **2015**, *14*, 553–563. [CrossRef]
64. Yu, X.; Li, Z.; Zheng, H.; Chan, M.T.V.; Wu, W.K.K. NEAT1: A novel cancer-related long non-coding RNA. *Cell Prolif.* **2017**, *50*, e12329. [CrossRef]
65. Tibes, R.; Bogenberger, J.M.; Chaudhuri, L.; Hagelstrom, R.T.; Chow, D.; Buechel, M.E.; Gonzales, I.M.; Demuth, T.; Slack, J.; Mesa, R.A.; et al. RNAi screening of the kinome with cytarabine in leukemias. *Blood* **2012**, *119*, 2863–2872. [CrossRef]
66. Iorns, E.; Lord, C.; Grigoriadis, A.; McDonald, S.; Fenwick, K.; Mackay, A.; Mein, C.; Natrajan, R.; Savage, K.; Tamber, N.; et al. Integrated Functional, Gene Expression and Genomic Analysis for the Identification of Cancer Targets. *PLoS ONE* **2009**, *4*, e5120. [CrossRef]
67. Mir, S.E.; Hamer, P.C.D.W.; Krawczyk, P.M.; Balaj, L.; Claes, A.; Niers, J.M.; Van Tilborg, A.A.; Zwinderman, A.H.; Geerts, D.; Kaspers, G.J.; et al. In silico analysis of kinase expression identifies WEE1 as a gatekeeper against mitotic catastrophe in glioblastoma. *Cancer Cell* **2010**, *18*, 244–257. [CrossRef] [PubMed]
68. Masaki, T.; Shiratori, Y.; Rengifo, W.; Igarashi, K.; Yamagata, M.; Kurokohchi, K.; Uchida, N.; Miyauchi, Y.; Yoshiji, H.; Watanabe, S.; et al. Cyclins and cyclin-dependent kinases: Comparative study of hepatocellular carcinoma versus cirrhosis. *Hepatology* **2003**, *37*, 534–543. [CrossRef]
69. Harris, P.S.; Venkataraman, S.; Alimova, I.; Birks, D.K.; Balakrishnan, I.; Cristiano, B.; Donson, A.M.; Dubuc, A.M.; Taylor, M.D.; Foreman, N.K.; et al. Integrated genomic analysis identifies the mitotic checkpoint kinase WEE1 as a novel therapeutic target in medulloblastoma. *Mol. Cancer* **2014**, *13*, 72. [CrossRef] [PubMed]
70. Mueller, S.; Hashizume, R.; Yang, X.; Kolkowitz, I.; Olow, A.K.; Phillips, J.; Smirnov, I.; Tom, M.W.; Prados, M.D.; James, C.D.; et al. Targeting Wee1 for the treatment of pediatric high-grade gliomas. *Neuro Oncol.* **2014**, *16*, 352–360. [CrossRef]
71. Music, D.; Dahlrot, R.; Hermansen, S.K.; Hjelmborg, J.; De Stricker, K.; Hansen, S.; Kristensen, B.W. Expression and prognostic value of the WEE1 kinase in gliomas. *J. Neuro Oncol.* **2016**, *127*, 381–389. [CrossRef]
72. Slipicevic, A.; Holth, A.; Hellesylt, E.; Tropé, C.G.; Davidson, B.; Flørenes, V.A. Wee1 is a novel independent prognostic marker of poor survival in post-chemotherapy ovarian carcinoma effusions. *Gynecol. Oncol.* **2014**, *135*, 118–124. [CrossRef]
73. Magnussen, G.I.; Hellesylt, E.; Nesland, J.M.; Tropé, C.G.; Flørenes, V.A.; Holm, R. High expression of wee1 is associated with malignancy in vulvar squamous cell carcinoma patients. *BMC Cancer* **2013**, *13*, 288. [CrossRef]

74. Magnussen, G.I.; Holm, R.; Emilsen, E.; Rosnes, A.K.; Slipicevic, A.; Flørenes, V.A. High expression of Wee1 is associated with poor disease-free survival in malignant melanoma: Potential for targeted therapy. *PLoS ONE* **2012**, *7*, e38254. [CrossRef] [PubMed]

75. Mak, J.P.Y.; Man, W.Y.; Chow, J.P.; Ma, H.T.; Poon, R.Y. Pharmacological inactivation of CHK1 and WEE1 induces mitotic catastrophe in nasopharyngeal carcinoma cells. *Oncotarget* **2015**, *6*, 21074–21084. [CrossRef] [PubMed]

76. Aarts, M.; Sharpe, R.; Garcia-Murillas, I.; Gevensleben, H.; Hurd, M.S.; Shumway, S.D.; Toniatti, C.; Ashworth, A.; Turner, N.C. Forced Mitotic Entry of S-Phase Cells as a Therapeutic Strategy Induced by Inhibition of WEE1. *Cancer Discov.* **2012**, *2*, 524–539. [CrossRef] [PubMed]

77. Kogiso, T.; Nagahara, H.; Hashimoto, E.; Ariizumi, S.; Yamamoto, M.; Shiratori, K. Efficient Induction of Apoptosis by Wee1 Kinase Inhibition in Hepatocellular Carcinoma Cells. *PLoS ONE* **2014**, *9*, e100495. [CrossRef] [PubMed]

78. Pappano, W.N.; Zhang, Q.; Tucker, L.A.; Tse, C.; Wang, J. Genetic inhibition of the atypical kinase Wee1 selectively drives apoptosis of p53 inactive tumor cells. *BMC Cancer* **2014**, *14*, 430. [CrossRef]

79. Hamer, P.C.D.W.; Mir, S.E.; Noske, D.; Van Noorden, C.J.F.; Würdinger, T. WEE1 Kinase Targeting Combined with DNA-Damaging Cancer Therapy Catalyzes Mitotic Catastrophe. *Clin. Cancer Res.* **2011**, *17*, 4200–4207. [CrossRef] [PubMed]

80. Zhang, M.; Dominguez, D.; Chen, S.; Fan, J.; Qin, L.; Long, A.; Li, X.; Zhang, Y.; Shi, H.; Zhang, B. WEE1 inhibition by MK1775 as a single-agent therapy inhibits ovarian cancer viability. *Oncol. Lett.* **2017**, *14*, 3580–3586. [CrossRef] [PubMed]

81. Kreahling, J.M.; Gemmer, J.Y.; Reed, D.; Letson, D.; Bui, M.; Altiok, S. MK1775, a Selective Wee1 Inhibitor, Shows Single-Agent Antitumor Activity against Sarcoma Cells. *Mol. Cancer Ther.* **2012**, *11*, 174–182. [CrossRef] [PubMed]

82. Zhou, B.-B.S.; Bartek, J. Targeting the checkpoint kinases: Chemosensitization versus chemoprotection. *Nat. Rev. Cancer* **2004**, *4*, 216–225. [CrossRef]

83. Ku, B.M.; Bae, Y.-H.; Koh, J.; Sun, J.-M.; Lee, S.-H.; Ahn, J.S.; Park, K.; Ahn, M.-J. Mutational status of TP53 defines the efficacy of Wee1 inhibitor AZD1775 in KRAS-mutant non-small cell lung cancer. *Oncotarget* **2017**, *8*, 67526–67537. [CrossRef]

84. Bauman, J.E.; Chung, C.H. CHK It Out! Blocking WEE Kinase Routs TP53 Mutant Cancer. *Clin. Cancer Res.* **2014**, *20*, 4173–4175. [CrossRef]

85. Hirai, H.; Arai, T.; Okada, M.; Nishibata, T.; Kobayashi, M.; Sakai, N.; Imagaki, K.; Ohtani, J.; Sakai, T.; Yoshizumi, T.; et al. MK-1775, a small molecule Wee1 inhibitor, enhances anti-tumor efficacy of various DNA-damaging agents, including 5-fluorouracil. *Cancer Biol. Ther.* **2010**, *9*, 514–522. [CrossRef]

86. PosthumaDeBoer, J.; Würdinger, T.; Graat, H.C.; van Beusechem, V.W.; Helder, M.N.; van Royen, B.J.; Kaspers, G.J. WEE1 inhibition sensitizes osteosarcoma to radiotherapy. *BMC Cancer* **2011**, *11*, 156. [CrossRef] [PubMed]

87. Wang, Y.; Li, J.; Booher, R.N.; Kraker, A.; Lawrence, T.; Leopold, W.R.; Sun, Y. Radiosensitization of p53 mutant cells by PD0166285, a novel G(2) checkpoint abrogator. *Cancer Res.* **2001**, *61*, 8211–8217. [PubMed]

88. Li, J.; Wang, Y.; Sun, Y.; Lawrence, T.S. Wild-type TP53 inhibits G(2)-phase checkpoint abrogation and radiosensitization induced by PD0166285, a WEE1 kinase inhibitor. *Radiat. Res.* **2002**, *157*, 322–330. [CrossRef]

89. Barbosa, R.S.; Dantonio, P.M.; Guimarães, T.; de Oliveira, M.B.; Alves, V.L.F.; Sandes, A.F.; Fernando, R.C.; Colleoni, G.W. Sequential combination of bortezomib and WEE1 inhibitor, MK-1775, induced apoptosis in multiple myeloma cell lines. *Biochem. Biophys. Res. Commun.* **2019**, *519*, 597–604. [CrossRef]

90. Porter, C.C.; Kim, J.; Fosmire, S.; Gearheart, C.M.; Van Linden, A.; Baturin, D.; Zaberezhnyy, V.; Patel, P.R.; Gao, D.; Tan, A.C.; et al. Integrated genomic analyses identify WEE1 as a critical mediator of cell fate and a novel therapeutic target in acute myeloid leukemia. *Leukemia* **2012**, *26*, 1266–1276. [CrossRef]

91. Van Linden, A.A.; Baturin, D.; Ford, J.B.; Fosmire, S.P.; Gardner, L.; Korch, C.; Reigan, P.; Porter, C.C. Inhibition of Wee1 Sensitizes Cancer Cells to Antimetabolite Chemotherapeutics In Vitro and In Vivo, Independent of p53 Functionality. *Mol. Cancer Ther.* **2013**, *12*, 2675–2684. [CrossRef]

92. Atun, R.; Jaffray, D.; Barton, M.; Bray, F.; Baumann, M.; Vikram, B.; Hanna, T.; Knaul, F.M.; Lievens, Y.; Lui, T.; et al. Expanding global access to radiotherapy. *Lancet Oncol.* **2015**, *16*, 1153–1186. [CrossRef]

93. Baumann, M.; Krause, M.; Hill, R. Exploring the role of cancer stem cells in radioresistance. *Nat. Rev. Cancer* **2008**, *8*, 545–554. [CrossRef]

94. Krause, M.; Yaromina, A.; Eicheler, W.; Koch, U.; Baumann, M. Cancer Stem Cells: Targets and Potential Biomarkers for Radiotherapy. *Clin. Cancer Res.* **2011**, *17*, 7224–7229. [CrossRef] [PubMed]

95. Krause, M.; Dubrovska, A.; Linge, A.; Baumann, M. Cancer stem cells: Radioresistance, prediction of radiotherapy outcome and specific targets for combined treatments. *Adv. Drug Deliv. Rev.* **2017**, *109*, 63–73. [CrossRef] [PubMed]

96. Bütof, R.; Dubrovska, A.; Baumann, M. Clinical perspectives of cancer stem cell research in radiation oncology. *Radiother. Oncol.* **2013**, *108*, 388–396. [CrossRef] [PubMed]

97. Morgan, M.A.; Lawrence, T.S. Molecular Pathways: Overcoming Radiation Resistance by Targeting DNA Damage Response Pathways. *Clin. Cancer Res.* **2015**, *21*, 2898–2904. [CrossRef] [PubMed]

98. Baumann, M.; Krause, M.; Overgaard, J.; Debus, J.; Bentzen, S.M.; Daartz, J.; Richter, C.; Zips, D.; Bortfeld, T. Radiation oncology in the era of precision medicine. *Nat. Rev. Cancer* **2016**, *16*, 234–249. [CrossRef] [PubMed]

99. Hill, R.; Lee, P.W. The DNA-dependent protein kinase (DNA-PK): More than just a case of making ends meet? *Cell Cycle* **2010**, *9*, 3460–3469. [CrossRef] [PubMed]

100. Falck, J.; Coates, J.; Jackson, S.P. Conserved modes of recruitment of ATM, ATR and DNA-PKcs to sites of DNA damage. *Nature* **2005**, *434*, 605–611. [CrossRef]

101. Cortez, D.; Guntuku, S.; Qin, J.; Elledge, S.J. ATR and ATRIP: Partners in Checkpoint Signaling. *Science* **2001**, *294*, 1713–1716. [CrossRef]
102. Jazayeri, A.; Falck, J.; Lukas, C.; Bartek, J.; Smith, G.C.M.; Lukas, J.; Jackson, S.P. ATM- and cell cycle-dependent regulation of ATR in response to DNA double-strand breaks. *Nat. Cell Biol.* **2006**, *8*, 37–45. [CrossRef] [PubMed]
103. Burma, S.; Chen, B.P.; Murphy, M.; Kurimasa, A.; Chen, D.J. ATM Phosphorylates Histone H2AX in Response to DNA Double-strand Breaks. *J. Biol. Chem.* **2001**, *276*, 42462–42467. [CrossRef]
104. Ammazzalorso, F.; Pirzio, L.M.; Bignami, M.; Franchitto, A.; Pichierri, P. ATR and ATM differently regulate WRN to prevent DSBs at stalled replication forks and promote replication fork recovery. *EMBO J.* **2010**, *29*, 3156–3169. [CrossRef]
105. Johnson, N.; Cai, D.; Kennedy, R.D.; Pathania, S.; Arora, M.; Li, Y.-C.; D'Andrea, A.D.; Parvin, J.D.; Shapiro, G.I. Cdk1 Participates in BRCA1-Dependent S Phase Checkpoint Control in Response to DNA Damage. *Mol. Cell* **2009**, *35*, 327–339. [CrossRef]
106. Liu, Q.; Guntuku, S.; Cui, X.-S.; Matsuoka, S.; Cortez, D.; Tamai, K.; Luo, G.; Carattini-Rivera, S.; DeMayo, F.; Bradley, A.; et al. Chk1 is an essential kinase that is regulated by Atr and required for the G2/M DNA damage checkpoint. *Genes Dev.* **2000**, *14*, 1448–1459.
107. Matsuoka, S.; Rotman, G.; Ogawa, A.; Shiloh, Y.; Tamai, K.; Elledge, S.J. Ataxia telangiectasia-mutated phosphorylates Chk2 in vivo and in vitro. *Proc. Natl. Acad. Sci. USA* **2000**, *97*, 10389–10394. [CrossRef]
108. Khosravi, R.; Maya, R.; Gottlieb, T.; Oren, M.; Shiloh, Y.; Shkedy, D. Rapid ATM-dependent phosphorylation of MDM2 precedes p53 accumulation in response to DNA damage. *Proc. Natl. Acad. Sci. USA* **1999**, *96*, 14973–14977. [CrossRef] [PubMed]
109. Sanchez, Y.; Wong, C.; Thoma, R.S.; Richman, R.; Wu, Z.; Piwnica-Worms, H.; Elledge, S.J. Conservation of the Chk1 Checkpoint Pathway in Mammals: Linkage of DNA Damage to Cdk Regulation through Cdc25. *Science* **1997**, *277*, 1497–1501. [CrossRef]
110. Xiao, Z.; Chen, Z.; Gunasekera, A.H.; Sowin, T.J.; Rosenberg, S.H.; Fesik, S.; Zhang, H. Chk1 Mediates S and G2 Arrests through Cdc25A Degradation in Response to DNA-damaging Agents. *J. Biol. Chem.* **2003**, *278*, 21767–21773. [CrossRef] [PubMed]
111. Sorensen, C.; Syljuåsen, R.G.; Falck, J.; Schroeder, T.; Rönnstrand, L.; Khanna, K.K.; Zhou, B.-B.; Bartek, J.; Lukas, J. Chk1 regulates the S phase checkpoint by coupling the physiological turnover and ionizing radiation-induced accelerated proteolysis of Cdc25A. *Cancer Cell* **2003**, *3*, 247–258. [CrossRef]
112. Syljuåsen, R.G.; Sorensen, C.; Hansen, L.T.; Fugger, K.; Lundin, C.; Johansson, F.; Helleday, T.; Sehested, M.; Lukas, J.; Bartek, J. Inhibition of Human Chk1 Causes Increased Initiation of DNA Replication, Phosphorylation of ATR Targets, and DNA Breakage. *Mol. Cell. Biol.* **2005**, *25*, 3553–3562. [CrossRef] [PubMed]
113. Sorensen, C.; Hansen, L.T.; Dziegielewski, J.; Syljuåsen, R.G.; Lundin, C.; Bartek, J.; Helleday, T. The cell-cycle checkpoint kinase Chk1 is required for mammalian homologous recombination repair. *Nat. Cell Biol.* **2005**, *7*, 195–201. [CrossRef] [PubMed]
114. Zachos, G.; Black, E.J.; Walker, M.; Scott, M.; Vagnarelli, P.; Earnshaw, W.; Gillespie, D.A. Chk1 Is Required for Spindle Checkpoint Function. *Dev. Cell* **2007**, *12*, 247–260. [CrossRef]
115. Carrassa, L.; Sanchez, Y.; Erba, E.; Damia, G. U2OS cells lacking Chk1 undergo aberrant mitosis and fail to activate the spindle checkpoint. *J. Cell. Mol. Med.* **2009**, *13*, 1565–1576. [CrossRef]
116. Chilà, R.; Celenza, C.; Lupi, M.; Damia, G.; Carrassa, L. Chk1-Mad2 interaction: A crosslink between the DNA damage checkpoint and the mitotic spindle checkpoint. *Cell Cycle* **2013**, *12*, 1083–1090. [CrossRef]
117. Jette, N.; Lees-Miller, S.P. The DNA-dependent protein kinase: A multifunctional protein kinase with roles in DNA double strand break repair and mitosis. *Prog. Biophys. Mol. Biol.* **2015**, *117*, 194–205. [CrossRef]
118. Hill, R.; Milas, L. The proportion of stem cells in murine tumors. *Int. J. Radiat. Oncol.* **1989**, *16*, 513–518. [CrossRef]
119. Baumann, M.; Dubois, W.; Suit, H.D. Response of Human Squamous Cell Carcinoma Xenografts of Different Sizes to Irradiation: Relationship of Clonogenic Cells, Cellular Radiation Sensitivity in Vivo, and Tumor Rescuing Units. *Radiat. Res.* **1990**, *123*, 325–330. [CrossRef]
120. Koch, U.; Krause, M.; Baumann, M. Cancer stem cells at the crossroads of current cancer therapy failures—Radiation oncology perspective. *Semin. Cancer Biol.* **2010**, *20*, 116–124. [CrossRef]
121. Puglisi, C.; Giuffrida, R.; Borzì, G.; Di Mattia, P.; Costa, A.; Colarossi, C.; Deiana, E.; Picardo, M.C.; Colarossi, L.; Mare, M.; et al. Radiosensitivity of Cancer Stem Cells Has Potential Predictive Value for Individual Responses to Radiotherapy in Locally Advanced Rectal Cancer. *Cancers* **2020**, *12*, 3672. [CrossRef]
122. Yaromina, A.; Krause, M.; Thames, H.; Rosner, A.; Krause, M.; Hessel, F.; Grenman, R.; Zips, D.; Baumann, M. Pre-treatment number of clonogenic cells and their radiosensitivity are major determinants of local tumour control after fractionated irradiation. *Radiother. Oncol.* **2007**, *83*, 304–310. [CrossRef] [PubMed]
123. Maugeri-Saccà, M.; Bartucci, M.; De Maria, R. DNA Damage Repair Pathways in Cancer Stem Cells. *Mol. Cancer Ther.* **2012**, *11*, 1627–1636. [CrossRef]
124. Chang, C.-H.; Zhang, M.; Rajapakshe, K.; Coarfa, C.; Edwards, D.; Huang, S.; Rosen, J.M. Mammary Stem Cells and Tumor-Initiating Cells Are More Resistant to Apoptosis and Exhibit Increased DNA Repair Activity in Response to DNA Damage. *Stem Cell Rep.* **2015**, *5*, 378–391. [CrossRef]
125. Wilson, W.R.; Hay, M.P. Targeting hypoxia in cancer therapy. *Nat. Rev. Cancer* **2011**, *11*, 393–410. [CrossRef]
126. Hu, C.-J.; Wang, L.-Y.; Chodosh, L.A.; Keith, B.; Simon, M.C. Differential Roles of Hypoxia-Inducible Factor 1α (HIF-1α) and HIF-2α in Hypoxic Gene Regulation. *Mol. Cell. Biol.* **2003**, *23*, 9361–9374. [CrossRef]
127. Covello, K.L.; Kehler, J.; Yu, H.; Gordan, J.D.; Arsham, A.M.; Hu, C.-J.; Labosky, P.; Simon, M.C.; Keith, B. HIF-2 regulates Oct-4: Effects of hypoxia on stem cell function, embryonic development, and tumor growth. *Genes Dev.* **2006**, *20*, 557–570. [CrossRef]

128. Gordan, J.D.; Bertout, J.A.; Hu, C.-J.; Diehl, J.A.; Simon, M.C. HIF-2α Promotes Hypoxic Cell Proliferation by Enhancing c-Myc Transcriptional Activity. *Cancer Cell* **2007**, *11*, 335–347. [CrossRef] [PubMed]

129. Cordes, N.; Seidler, J.; Durzok, R.; Geinitz, H.; Brakebusch, C. β1-integrin-mediated signaling essentially contributes to cell survival after radiation-induced genotoxic injury. *Oncogene* **2006**, *25*, 1378–1390. [CrossRef] [PubMed]

130. Bao, S.; Wu, Q.; McLendon, R.E.; Hao, Y.; Shi, Q.; Hjelmeland, A.B.; Dewhirst, M.W.; Bigner, D.D.; Rich, J.N. Glioma stem cells promote radioresistance by preferential activation of the DNA damage response. *Nature* **2006**, *444*, 756–760. [CrossRef]

131. Sarcar, B.; Kahali, S.; Prabhu, A.H.; Shumway, S.D.; Xu, Y.; DeMuth, T.; Chinnaiyan, P. Targeting Radiation-Induced G2 Checkpoint Activation with the Wee-1 Inhibitor MK-1775 in Glioblastoma Cell Lines. *Mol. Cancer Ther.* **2011**, *10*, 2405–2414. [CrossRef]

132. Ronco, C.; Martin, A.R.; Demange, L.; Benhida, R. ATM, ATR, CHK1, CHK2 and WEE1 inhibitors in cancer and cancer stem cells. *MedChemComm* **2017**, *8*, 295–319. [CrossRef] [PubMed]

133. Matheson, C.J.; Venkataraman, S.; Amani, V.; Harris, P.S.; Backos, D.; Donson, A.M.; Wempe, M.F.; Foreman, N.; Vibhakar, R.; Reigan, P. A WEE1 Inhibitor Analog of AZD1775 Maintains Synergy with Cisplatin and Demonstrates Reduced Single-Agent Cytotoxicity in Medulloblastoma Cells. *ACS Chem. Biol.* **2016**, *11*, 921–930. [CrossRef] [PubMed]

134. Matheson, C.J.; Casalvieri, K.A.; Backos, D.S.; Reigan, P. Development of Potent Pyrazolopyrimidinone-Based WEE1 Inhibitors with Limited Single-Agent Cytotoxicity for Cancer Therapy. *ChemMedChem* **2018**, *13*, 1681–1694. [CrossRef]

135. Du, X.; Li, J.; Luo, X.; Li, R.; Li, F.; Zhang, Y.; Shi, J.; He, J. Structure-activity relationships of Wee1 inhibitors: A review. *Eur. J. Med. Chem.* **2020**, *203*, 112524. [CrossRef]

136. Hashimoto, O.; Shinkawa, M.; Torimura, T.; Nakamura, T.; Selvendiran, K.; Sakamoto, M.; Koga, H.; Ueno, T.; Sata, M. Cell cycle regulation by the Wee1 Inhibitor PD0166285, Pyrido [2,3-*d*] pyimidine, in the B16 mouse melanoma cell line. *BMC Cancer* **2006**, *6*, 292. [CrossRef]

137. Hashimoto, O.; Ueno, T.; Kimura, R.; Ohtsubo, M.; Nakamura, T.; Koga, H.; Torimura, T.; Uchida, S.; Yamashita, K.; Sata, M. Inhibition of proteasome-dependent degradation of Wee1 in G2-arrested Hep3B cells by TGFβ1. *Mol. Carcinog.* **2003**, *36*, 171–182. [CrossRef]

138. Panek, R.L.; Lu, G.H.; Klutchko, S.R.; Batley, B.L.; Dahring, T.K.; Hamby, J.M.; Hallak, H.; Doherty, A.M.; Keiser, J.A. In vitro pharmacological characterization of PD 166285, a new nanomolar potent and broadly active protein tyrosine kinase inhibitor. *J. Pharmacol. Exp. Ther.* **1997**, *283*, 1433–1444.

139. Palmer, B.D.; Smaill, J.B.; Rewcastle, G.W.; Dobrusin, E.M.; Kraker, A.; Moore, C.W.; Steinkampf, R.W.; Denny, W.A. Structure-activity relationships for 2-anilino-6-phenylpyrido[2,3-*d*]pyrimidin-7(8*H*)-ones as inhibitors of the cellular checkpoint kinase Wee1. *Bioorgan. Med. Chem. Lett.* **2005**, *15*, 1931–1935. [CrossRef] [PubMed]

140. Hamby, J.M.; Connolly, C.J.C.; Schroeder, M.C.; Winters, R.T.; Showalter, H.D.H.; Panek, R.L.; Major, T.C.; Olsewski, B.; Ryan, M.J.; Dahring, T.; et al. Structure-Activity Relationships for a Novel Series of Pyrido[2,3-*d*]pyrimidine Tyrosine Kinase Inhibitors. *J. Med. Chem.* **1997**, *40*, 2296–2303. [CrossRef]

141. Thompson, A.M.; Connolly, C.J.C.; Hamby, J.M.; Boushelle, S.; Hartl, B.G.; Amar, A.M.; Kraker, A.J.; Driscoll, D.L.; Steinkampf, R.W.; Patmore, S.J.; et al. 3-(3,5-Dimethoxyphenyl)-1,6-naphthyridine-2,7-diamines and Related 2-Urea Derivatives Are Potent and Selective Inhibitors of the FGF Receptor-1 Tyrosine Kinase. *J. Med. Chem.* **2000**, *43*, 4200–4211. [CrossRef] [PubMed]

142. Hirai, H.; Iwasawa, Y.; Okada, M.; Arai, T.; Nishibata, T.; Kobayashi, M.; Kimura, T.; Kaneko, N.; Ohtani, J.; Yamanaka, K.; et al. Small-molecule inhibition of Wee1 kinase by MK-1775 selectively sensitizes p53-deficient tumor cells to DNA-damaging agents. *Mol. Cancer Ther.* **2009**, *8*, 2992–3000. [CrossRef] [PubMed]

143. Palmer, B.D.; Thompson, A.M.; Booth, R.J.; Dobrusin, E.M.; Kraker, A.J.; Lee, H.H.; Lunney, E.A.; Mitchell, L.H.; Ortwine, D.F.; Smaill, J.B.; et al. 4-Phenylpyrrolo[3,4-*c*]carbazole-1,3(2*H*,6*H*)-dione Inhibitors of the Checkpoint Kinase Wee1. Structure-Activity Relationships for Chromophore Modification and Phenyl Ring Substitution. *J. Med. Chem.* **2006**, *49*, 4896–4911. [CrossRef] [PubMed]

144. Arora, S.; Bisanz, K.M.; Peralta, L.A.; Basu, G.D.; Choudhary, A.; Tibes, R.; Azorsa, D.O. RNAi screening of the kinome identifies modulators of cisplatin response in ovarian cancer cells. *Gynecol. Oncol.* **2010**, *118*, 220–227. [CrossRef]

145. Smaill, J.B.; Baker, E.N.; Booth, R.J.; Bridges, A.J.; Dickson, J.M.; Dobrusin, E.M.; Ivanovic, I.; Kraker, A.J.; Lee, H.H.; Lunney, E.A.; et al. Synthesis and structure-activity relationships of N-6 substituted analogues of 9-hydroxy-4-phenylpyrrolo[3,4-*c*]carbazole-1,3(2*H*,6*H*)-diones as inhibitors of Wee1 and Chk1 checkpoint kinases. *Eur. J. Med. Chem.* **2008**, *43*, 1276–1296. [CrossRef]

146. Smaill, J.B.; Lee, H.H.; Palmer, B.D.; Thompson, A.M.; Squire, C.J.; Baker, E.N.; Booth, R.J.; Kraker, A.; Hook, K.; Denny, W.A. Synthesis and structure-activity relationships of soluble 8-substituted 4-(2-chlorophenyl)-9-hydroxypyrrolo[3,4-*c*]carbazole-1,3(2*H*,6*H*)-diones as inhibitors of the Wee1 and Chk1 checkpoint kinases. *Bioorgan. Med. Chem. Lett.* **2008**, *18*, 929–933. [CrossRef]

147. Tong, Y.; Torrent, M.; Florjancic, A.S.; Bromberg, K.D.; Buchanan, F.G.; Ferguson, D.C.; Johnson, E.F.; Lasko, L.M.; Maag, D.; Merta, P.J.; et al. Pyrimidine-Based Tricyclic Molecules as Potent and Orally Efficacious Inhibitors of Wee1 Kinase. *ACS Med. Chem. Lett.* **2014**, *6*, 58–62. [CrossRef] [PubMed]

148. Khan, N.; Afaq, F.; Saleem, M.; Ahmad, N.; Mukhtar, H. Targeting Multiple Signaling Pathways by Green Tea Polyphenol (−)-Epigallocatechin-3-Gallate. *Cancer Res.* **2006**, *66*, 2500–2505. [CrossRef]

149. Banerjee, S.; Li, Y.; Wang, Z.; Sarkar, F.H. Multi-targeted therapy of cancer by genistein. *Cancer Lett.* **2008**, *269*, 226–242. [CrossRef] [PubMed]

150. Ravindran, J.; Prasad, S.; Aggarwal, B.B. Curcumin and Cancer Cells: How Many Ways Can Curry Kill Tumor Cells Selectively? *AAPS J.* **2009**, *11*, 495–510. [CrossRef]
151. Cuccioloni, M.; Mozzicafreddo, M.; Bonfili, L.; Cecarini, V.; Eleuteri, A.M.; Angeletti, M. Natural Occurring Polyphenols as Template for Drug Design. Focus on Serine Proteases. *Chem. Biol. Drug Des.* **2009**, *74*, 1–15. [CrossRef] [PubMed]
152. Korkina, L.; De Luca, C.; Kostyuk, V.; Pastore, S. Plant Polyphenols and Tumors: From Mechanisms to Therapies, Prevention, and Protection Against Toxicity of Anti-Cancer Treatments. *Curr. Med. Chem.* **2009**, *16*, 3943–3965. [CrossRef] [PubMed]
153. Lamoral-Theys, D.; Pottier, L.; Dufrasne, F.; Neve, J.; Dubois, J.; Kornienko, A.; Kiss, R.; Ingrassia, L. Natural Polyphenols that Display Anticancer Properties through Inhibition of Kinase Activity. *Curr. Med. Chem.* **2010**, *17*, 812–825. [CrossRef] [PubMed]
154. Reuter, S.; Eifes, S.; Dicato, M.; Aggarwal, B.B.; Diederich, M. Modulation of anti-apoptotic and survival pathways by curcumin as a strategy to induce apoptosis in cancer cells. *Biochem. Pharmacol.* **2008**, *76*, 1340–1351. [CrossRef] [PubMed]
155. Yang, C.S.; Wang, X.; Lu, G.; Picinich, S.C. Cancer prevention by tea: Animal studies, molecular mechanisms and human relevance. *Nat. Rev. Cancer* **2009**, *9*, 429–439. [CrossRef]
156. Lamoral-Theys, D.; Pottier, L.; Kerff, F.; Dufrasne, F.; Proutiere, F.; Wauthoz, N.; Neven, P.; Ingrassia, L.; Van Antwerpen, P.; Lefranc, F.; et al. Simple di- and trivanillates exhibit cytostatic properties toward cancer cells resistant to pro-apoptotic stimuli. *Bioorgan. Med. Chem.* **2010**, *18*, 3823–3833. [CrossRef]
157. Bridges, K.A.; Hirai, H.; Buser, C.A.; Brooks, C.; Liu, H.; Buchholz, T.; Molkentine, J.; Mason, K.A.; Meyn, R.E. MK-1775, a Novel Wee1 Kinase Inhibitor, Radiosensitizes p53-Defective Human Tumor Cells. *Clin. Cancer Res.* **2011**, *17*, 5638–5648. [CrossRef]
158. Do, K.; Doroshow, J.H.; Kummar, S. Wee1 kinase as a target for cancer therapy. *Cell Cycle* **2013**, *12*, 3348–3353. [CrossRef]
159. Osman, A.A.; Monroe, M.M.; Alves, M.V.O.; Patel, A.A.; Katsonis, P.; Fitzgerald, A.L.; Neskey, D.M.; Frederick, M.J.; Woo, S.H.; Caulin, C.; et al. Wee-1 Kinase Inhibition Overcomes Cisplatin Resistance Associated with High-Risk TP53 Mutations in Head and Neck Cancer through Mitotic Arrest Followed by Senescence. *Mol. Cancer Ther.* **2015**, *14*, 608–619. [CrossRef]
160. Caretti, V.; Hiddingh, L.; Lagerweij, T.; Schellen, P.; Koken, P.W.; Hulleman, E.; van Vuurden, D.G.; Vandertop, W.P.; Kaspers, G.J.; Noske, D.P.; et al. WEE1 Kinase Inhibition Enhances the Radiation Response of Diffuse Intrinsic Pontine Gliomas. *Mol. Cancer Ther.* **2013**, *12*, 141–150. [CrossRef]
161. Karnak, D.; Engelke, C.G.; Parsels, L.A.; Kausar, T.; Wei, D.; Robertson, J.R.; Marsh, K.B.; Davis, M.A.; Zhao, L.; Maybaum, J.; et al. Combined inhibition of Wee1 and PARP1/2 for radiosensitization in pancreatic cancer. *Clin. Cancer Res.* **2014**, *20*, 5085–5096. [CrossRef] [PubMed]

International Journal of
Molecular Sciences

Communication

Design and Synthesis of a Novel PLK1 Inhibitor Scaffold Using a Hybridized 3D-QSAR Model

Youri Oh [†], Hoyong Jung [†], Hyejin Kim, Jihyun Baek, Joonhong Jun, Hyunwook Cho, Daseul Im and Jung-Mi Hah *

College of Pharmacy and Institute of Pharmaceutical Science and Technology, Hanyang University, Ansan 426-791, Korea; apdlzld4477@hanyang.ac.kr (Y.O.); hyong@hanyang.ac.kr (H.J.); gpwls6121@hanyang.ac.kr (H.K.); jhb3534@hanyang.ac.kr (J.B.); jjh0328@hanyang.ac.kr (J.J.); lod0201@hanyang.ac.kr (H.C.); bestimda@hanyang.ac.kr (D.I.)
* Correspondence: jhah@hanyang.ac.kr; Tel.: +82-31-400-5803
† These authors contributed equally to this work.

Abstract: Polo-like kinase 1 (PLK1) plays an important role in cell cycle progression and proliferation in cancer cells. PLK1 also contributes to anticancer drug resistance and is a valuable target in anticancer therapeutics. To identify additional effective PLK1 inhibitors, we performed QSAR studies of two series of known PLK1 inhibitors and proposed a new structure based on a hybridized 3D-QSAR model. Given the hybridized 3D-QSAR models, we designed and synthesized 4-benzyloxy-1-(2-arylaminopyridin-4-yl)-1H-pyrazole-3-carboxamides, and we inspected its inhibitory activities to identify novel PLK1 inhibitors with decent potency and selectivity.

Keywords: polo-like kinase 1 (PLK1); pyrazole; quantitative structure-activity relationship; hybridization

Citation: Oh, Y.; Jung, H.; Kim, H.; Baek, J.; Jun, J.; Cho, H.; Im, D.; Hah, J.-M. Design and Synthesis of a Novel PLK1 Inhibitor Scaffold Using a Hybridized 3D-QSAR Model. *Int. J. Mol. Sci.* **2021**, *22*, 3865. https://doi.org/10.3390/ijms22083865

Academic Editor: Lorenzo Lo Muzio

Received: 23 December 2020
Accepted: 6 April 2021
Published: 8 April 2021

Publisher's Note: MDPI stays neutral with regard to jurisdictional claims in published maps and institutional affiliations.

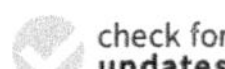

1. Introduction

Polo-like kinases (PLKs) are a family of five serine/threonine kinases [1,2] that have been identified in various eukaryotic organisms and play important roles in cell proliferation, especially cell cycle regulation [3]. PLK1 is involved in the mitotic cell cycle, peaking in the S phase, and gradually decreasing in the G phase. Thus, PLK1 appears to induce cell proliferation, and the overexpression of PLK1 is common in malignant tumors. Among PLKs, PLK1 is overexpressed in a wide range of human cancers, including leukemia [4], and is considered an attractive anticancer drug target [5]. In contrast, since PLK2 (also known as SNK) [6] and PLK3 (also known as FNK or PRK) [7,8] are tumor suppressors, development of a PLK1 inhibitor with isoform selectivity is very important. In addition, PLK1 is involved in resistance [9] to several anticancer drugs such as doxorubicin, paclitaxel, and gemcitabine. This is presumed to be the mechanism by which PLK1 promotes the cell cycle and, directly or indirectly, inactivates p53, suggesting that the inhibition of PLK1 can be very useful in single and combination anticancer therapy.

These properties have led many groups to focus on the development of PLK1 inhibitors, and developed PLK1 inhibitors have been able to uncover new cellular functions of PLK1 in many laboratories. Among them, BI 2536, developed by Boehringer Ingelheim, is a potent ATP competitive PLK1 inhibitor with IC_{50} of 0.83 nM [10]. It also inhibits the activities of PLK2 (IC_{50} = 3.5 nM) and PLK3 (IC_{50} = 9.0 nM). BI 2536 showed moderate selectivity for inhibiting kinase activity of PLK1 among PLK members. The level of PLK1 is high in dividing cells, whereas PLK2 and PLK3 are not specifically expressed in proliferating cells. However, its weak selectivity to the isoforms of PLK1 was pointed out as a possible hurdle. Treatment of BI 2536 induced mitotic arrest in prometaphase, forming aberrant mitotic spindles and, consequently, apoptosis. When BI 2536 was used in human cancer cells, it inhibited cell proliferation in several human cancer cell lines, showing an effect on diverse organ derivatives such as breast, colon, lung, pancreas, and prostate cancer preclinical data showed that BI 2536 could be a possible anticancer drug candidate, leading to an investigation of the clinical effects of BI 2536. Monotherapy with BI 2536, the first human study of PLK1 inhibitors, has been terminated now, but its combinational study is still available in several solid tumors. In subsequent pharmaceutical efforts, new PLK1 inhibitors such as GSK461364A [11], and PHA-680626 [12] were all developed to inhibit PLK1 with low nanomolar IC_{50}s and exert potent antiproliferative effects on various cancer cells. (Figure 1). Many of the PLK1 inhibitors discovered as a result of such studies have reached preclinical success and phase 1 clinical trials, but most have failed at this stage due to toxicity issues. Among them, PLK1 inhibitors such as BI 2536 [10] and GSK461364A [11] passed phase 1 clinical trials and progressed to phase 2 and phase 3 studies as treatment for specific malignancies. Nevertheless, the results of those studies were not as effective as expected when treatment was used as a second- or third-line therapy [13–16]. Only a small number of patients showed disease stabilization, indicating that further exploration of PLK1 inhibitors and related treatment plans is needed.

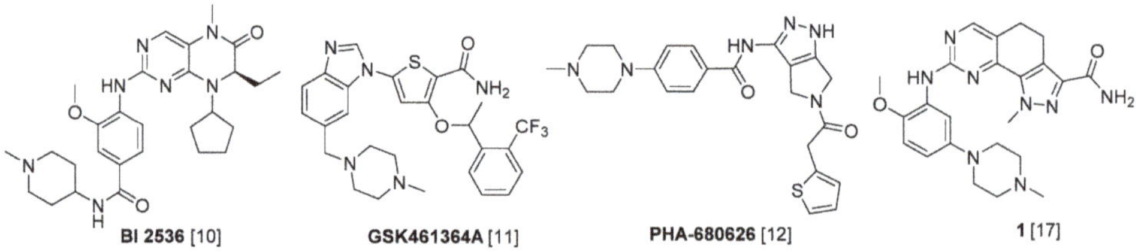

Figure 1. Representative PLK1 inhibitors.

Medicinal chemists rely on understanding structure-activity relationships (SAR) for drug discovery and development. However, the SAR are mostly deduced from syntheses of numerous compounds and the establishment of bioactivity assay data. These conventional processes are essential but time-consuming and labor intensive. In contrast, in silico approaches deduce SAR by statistical analysis of known active compounds. Calculations convert SAR into 3D quantitative structure-activity relationships (3D-QSAR), which are more efficient than conventional SAR in terms of time and cost. The QSAR results indicate the properties of a ligand with increased activity, such as steric, electrostatic interactions; hydrogen bonding affinity; hydrophobicity. In that context, we performed QSAR studies of two known PLK1 inhibitor series and proposed a new structure based on hybridization of the two resulting QSAR models for discovery of a novel effective PLK1 inhibitor. Of known 3D-QSAR methods, Comparative Molecular Field Analysis (CoMFA) and Comparative Molecular Similarity Index Analysis (CoMSIA) are the most common, and we used them for model validation of a hybridized QSAR model from two series of PLK1 inhibitors in the literature. The hybrid 3D-QSAR model suggested a guide for new scaffolds, based on which we designed and synthesized 4-benzyloxy-1-(2-arylaminopyridin-4-yl)-1*H*-pyrazole-3-carboxamide. Using these approaches, we could identify a novel PLK1 inhibitor with decent potency and improved selectivity.

2. Results and Discussion

Rational Design of a Hybridized 3D-QSAR Model

First, we selected two series of PLK1 inhibitors, series A of 44 8-amino-4, 5-dihydro-1*H*-pyrazolo [4, 3-*h*] quinazoline-3-carboxamide derivatives [17] and series B of 36 thiophene-2-carboxamide derivatives [18]. Then, we sorted 66 of 80 compounds, excluding 10 with low potency (IC_{50} > 3 μM) and four racemates to establish the hybridized 3D-QSAR model. We used pIC_{50} (= $-\log IC_{50}$) value as the dependent variable in the QSAR model and divided the 66 compounds into six groups (Supplementary Section S1). As a test set in each section, two compounds per group were selected. The remaining 54 compounds, excluding the 12 of the test set, were used as the training set. Using these, we generated CoMFA models with Gasteiger–Hückel charged descriptor through SYBYL-X 2.1.1 automatic PLS (Figure 2). All the compounds were prepared in 3D conformation with SYBYL-X Ligand preparation based on ligands bound to PDB 3KB7 [13–16]. The two series of compounds overlapped two functional groups of each series, *thiophene-2-carboxamide* and *pyrazol-3-carboxamide* (Figure 3a). The CoMSIA model was generated through a similar process. We select steric, electrostatic, hydrophobic, hydrogen bond donor, and hydrogen bond acceptor indices as descriptors. The statistical parameters of the CoMFA and CoMSIA models are in Table 1.

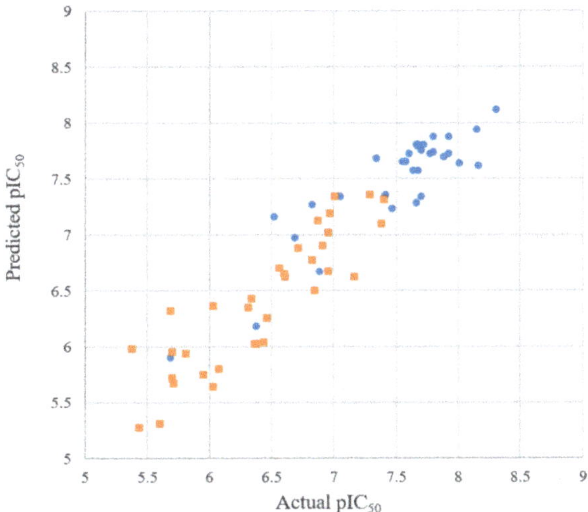

Figure 2. Observed vs. predicted values of the hybridized CoMFA model. 8-Amino-4,5-dihydro-1*H*-pyrazolo [4,3-*h*]quinazoline-3-carboxamides (series A; orange) and thiophene-2-carboxamide (series B; cyan).

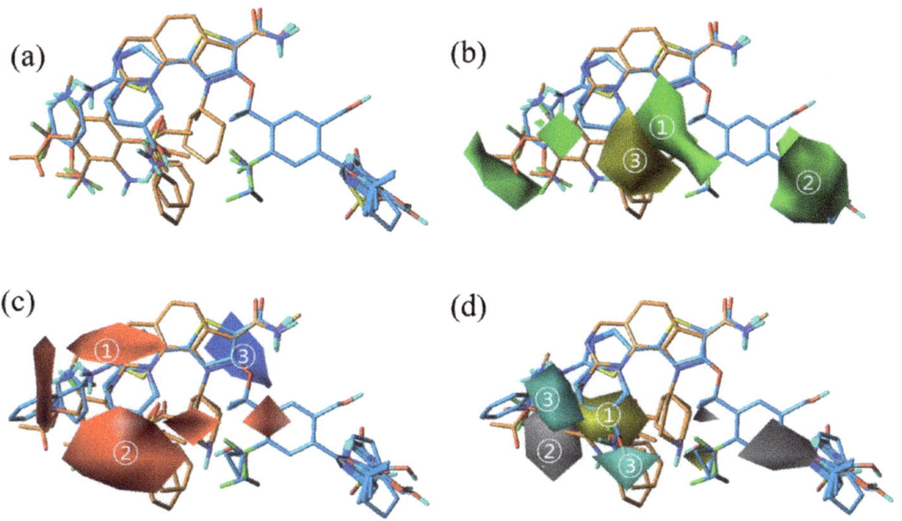

Figure 3. (**a**) Alignment of two series of compounds. Thiophene-2-carboxamide (cyan) and 8-amino-4, 5-dihydro-1*H*-pyrazolo[4,3-h]quinazoline-3-carbaldehyde derivatives (orange). (**b**) CoMFA steric contour map, (**c**) CoMFA electrostatic contour map, and (**d**) CoMSIA contour map.

Table 1. PLS results of final CoMFA and CoMSIA models.

Components		Receptor-Guided Aligned Model with MC Searching	
		CoMFA	CoMSIA
PLS statistics	q^2	0.657	0.641
	r^2	0.899	0.838
	N	3	3
	SEE	0.348	0.335
	F-value	125.847	125.847
	Predictive r^2	0.899	0.838
Field contribution	Steric	0.396	0.044
	Electrostatic	0.604	0.266
	Hydrophobic	-	0.262
	Donor	-	0.158
	Acceptor	-	0.270

q^2, LOO cross-validated correlation coefficient; r^2, non-cross-validated correlation coefficient; n, number of components used in the PLS analysis; SEE, standard error estimate; F value, F-statistic for the analysis (Supplementary Tables S1 and S2)

We obtained steric and electrostatic contour maps from the CoMFA model. In Figure 3b, the green field is the bulkiness-favored zone, and yellow is the bulkiness-disfavored zone. The cyclohexyl substituent in pyrazole (① series A; orange) and the substituents at the 4 position of the benzyloxy thiophene ring (② series B; cyan) were preferred in the green zone. The substituents on the imidazo[1,2-a]pyridine ring (the series B; cyan) and the amino pyrimidine ring (the series A; orange) were scattered in the mixed green and yellow zone (③). To better construct the models, we employed suitable rings with various substituents at this position. On electrostatic contour maps (c), the electrostatic field was visualized using blue for positive charge-favored zones and red for negative charge-favored zones. The aligned thiophene-2-carboxamides (the series B; cyan) and pyrazole-3-carboxamides (series A; orange) required positive charges at the aminopyrimidine (①), piperazine substitution (②) at the aniline in pyrimidine, and negative charges at the bottom of the thiophene and pyrazole rings (③). Additionally, the CoMSIA contour map containing hydrophobic, hydrogen bond donor fields were generated (d). The two major hydrophobic spots (yellow) on the substituents of the imidazo[1,2-a]pyridine ring (①) were found, which were near the hydrophilic gray field, aminopyrimidine (②). The two hydrogen bond donor fields (cyan) are shown on the map near the imidazo[1,2-a]pyridine ring and piperazine substitution (③). No hydrogen bond disfavored-field was observed.

From the hybridized CoMFA and CoMSIA models of the two series, several new chemical scaffolds were suggested by fusing unified alignments. First, we deduced a scaffold with a core, 1*H*-pyrrolo[2,3-c]pyridine ring with proper substituents satisfying the contour maps (Figure 3). Later, we modified the core, 1*H*-pyrrolo[2,3-c]pyridine ring into aminopyridine for flexibility, since some substituents were not properly positioned in hybridized CoMFA and CoMSIA models. The carboxamide group was unchanged, and the red fields in the electrostatic contour map (Figure 3c, ②) contributed to introducing aniline with electronegative substitutions to the pyridine ring and the 2-fluoro-4-trifluoromethyl benzene for the electronegative field. In the trimming step, we imposed bulky substituents (such as a dimethylamine) at perpendicular conformation of 2-carboxamide of the pyrazole, inducing the favored electrostatic field by rotation of the carboxamide group. Finally, we produced **4-((2-R^2-4-R^3-benzyl)oxy)-1-(2-(2-R^1-aminopyridin-4-yl)-1H-pyrazole-3-carboxamide** as a novel PLK1 inhibitor scaffold (Figure 4).

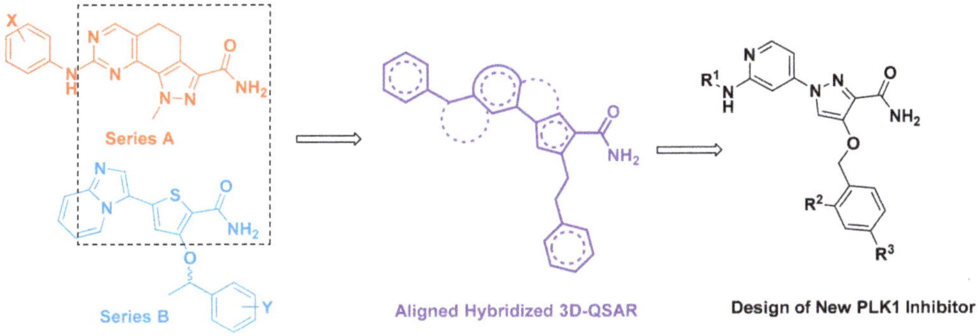

Figure 4. Design of a novel PLK1 inhibitor by the hybridization of two QSAR models.

The general synthesis of the designed 4-((2-R²-4-R³-benzyl)oxy)-1-(2-(2-R¹-aminopyridin-4-yl)-1*H*-pyrazole-3-carboxamide (**8a–10a**, **13a–b**, **14a–b**, **15–18**) is shown in Scheme 1. For synthesis of the 1, 3, 4-substituted pyrazole core, we azo coupled diazotized 2-fluoropyridin-4-amine (**1**) and ethyl 4-chloro-3-oxobutanoate and obtained ethyl (E)-4-chloro-2-(2-(2-fluoropyridin-4-yl)hydrazinylidene)-3-oxobutanoate (**2**) as an intermediate. Subsequent cyclization using potassium *t*-butoxide resulted in a proper pyrazole core (**3**) [19], of which the hydroxyl group was alkylated with three types of benzyl bromide to yield **4a–4c**. Next, the fluorine of pyridine was substituted via S_NAr to give **5a–7a**, **11a–c**, and **12a–c** according to R¹-NH₂. Then, the 3-ethyl ester group was converted to carboxamide using ammonia in methanol to give **8a–10a**, **13a–b**, and **14a–b**. When the 2-chloro-4-*tert*- butyldimethylsilyl oxybenzyloxy group (synthesized in two steps from methyl 3-chloro-4-methylbenzoate; Supplementary S2) was employed as R², further derivatizations were performed after conversion of ethyl ester (**11a–c**, **12a–c**) to desired product with benzylalcohol (**15**, **17**) and dimethylamino groups (**16**, **18**) by sequential deprotection and substitution.

All the 1-pyridyl-4-substituted-pyrazole-3-carboxamide derivatives **8a–10a**, **13a–b**, **14a–b**, and **15–18** were evaluated for inhibitory activity against PLK1 kinase, and the results are shown in Table 2. The synthesized compounds exhibited mild to potent inhibitory activity against PLK1, especially compounds that incorporated 2, 4-substituted benzyloxy moieties at the 4-position in the pyrazole core. Among the compounds evaluated, **15** showed the most potent activity against PLK1, with an IC_{50} value of 219 nM (Supplementary Section S3). We first noted that *ortho*-substituted phenyl is preferred as the R¹ group on the pyridine ring to the cycloalkylamino group (**8a**, **9a**, **10a**). We suspect the aminopyridine group acts as a hydrogen bond-hinge binder, and the X substituent impacts the angles of the two hydrogen bonds. However, it is not clear whether 2-methoxy aniline or simple aniline is preferred from the value of IC_{50}. Rather, the position of the bulky substituent on the pyrazole ring was found to play an important role in inhibitory activity. Specifically, pyrazole compounds incorporating a 2, 4-disubstituted benzyloxy group were 6–8 times more potent than compounds without substitution at the 2-position (**13a** vs. **13b**, **15**, **17**). Moreover, it seems that the hydroxyl substituent, rather than the dimethyl amino group, is preferred as R³ at the 4-position of the benzyloxy group, and the hydrogen bond, rather than the electrostatic interaction, could serve as an important interaction. For R², a large substituent is clearly preferred to hydrogen, but there was not much difference between -CF₃ and -Cl. We next performed a kinase panel screen for compound **15** against 30 protein kinases at 10 μM (Figure 5). Compound **15** achieved an excellent selectivity profile with an acquired inhibitory activity of 97% against PLK1 but with no significant activity against other protein kinases, especially PLK2 and PLK3. Additionally, compound **15** was not detected as PAINS [20]. Overall, this result indicates that the kinase activity profile and selectivity of the hybrid QSAR-driven inhibitor result in selective PLK1 inhibition, which could serve as a lead compound for the next step.

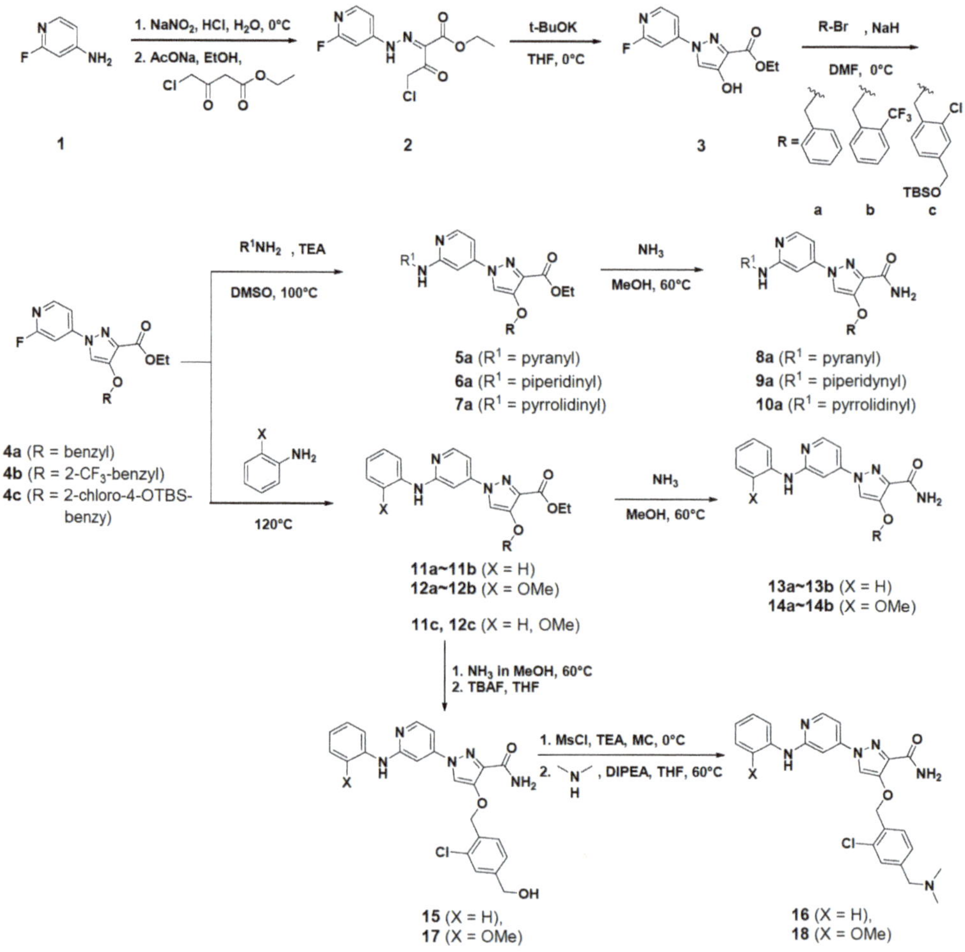

Scheme 1. Synthesis of 4-((2-R²-4-R³-benzyloxy)-1-(2-(2-R¹-amino)pyridin-4-yl)-1*H*-pyrazole-3-carboxamide.

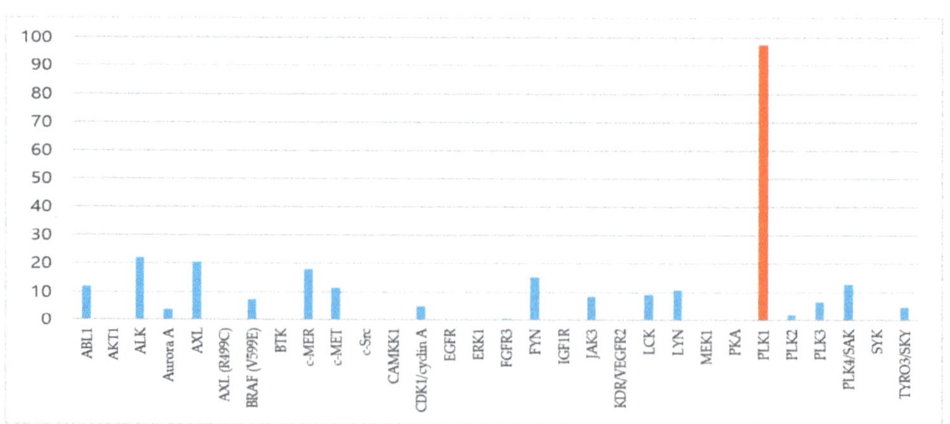

Figure 5. Percentages of enzymatic inhibition exerted by **15** (10 μM) on 30 selected protein kinases [21].

Table 2. Enzymatic activities of 4-benzyloxy-1-(2-(aryl/alkylamino)pyridin-4-yl)-1*H*-pyrazole-3-carboxamide.

	R^1	R^2	R^3	PLK1 IC$_{50}$ (µM)
8a		H	H	>10
9a				>10
10a				>10
13a		H	H	9.89
13b		CF$_3$	H	0.847
14a		H	H	1.02
14b		CF$_3$	H	1.31
15		Cl	OH	0.219
16		Cl	N(Me)$_2$	1.01
17		Cl	OH	0.312
18		Cl	N(Me)$_2$	0.998
Control	Staurosporine			0.143

To better understand the interactions between the newly synthesized compounds and PLK1, we performed molecular docking studies of compound **15** in the ATP binding pocket of PLK1 (PDB: 3KB7) [13–16] using Glide (SCHRODINGER software package Version 14.2), and the results are shown in Figure 6. In binding mode, compound **15** was tightly bound to the ATP-binding site of PLK1 via several hydrogen bonds and π-π interactions. The most important interactions seemed to be a pair of hydrogen bonds between carboxamide and Lys82 or Asp194. Further, the nitrogen of pyridine, as a hinge binder, formed a hydrogen bond with the amino hydrogen of Cys 133. The pyrazole core and benzyl ring in the 3-benzyloxy group of **15** formed two π- π interactions with Phe184. In addition, the 4-methylenehydroxy group on the benzyloxy tail exhibited a hydrogen bond with Glu140, explaining the better potency compared to a simple benzyl group (vs. **13a**, **14a**).

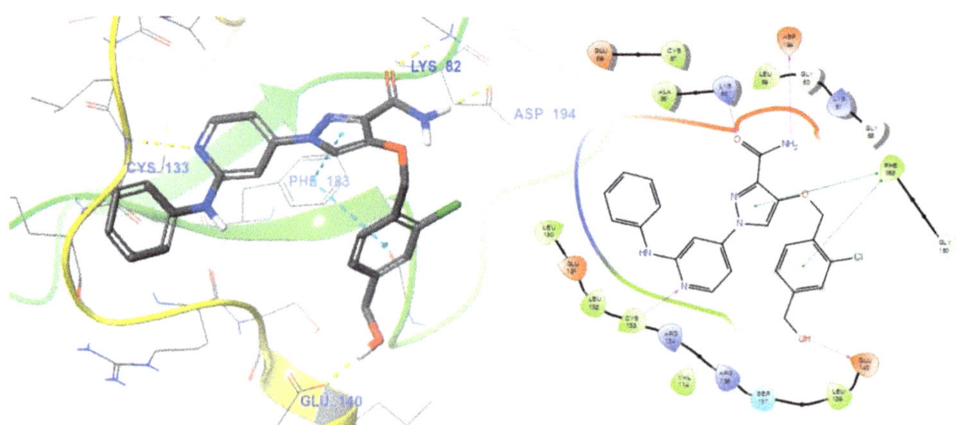

Figure 6. Docking structure of **15** at PLK1 (left, PDB: 3KB7) and the 2D-interaction map (right).

3. Materials and Methods

3.1. QSAR Studies

From thiophene-2-carboxamide derivatives and 8-amino-4,5-dihydro-1*H*-pyrazolo[4,3-h]quinazoline derivatives, two representative compounds, **s18** and **s49** (Supplementary Materials), were selected as standards for the appropriate series. Then, **s18** and **s49** were matched by Fit Atom in SYBYL-X 2.1.1 to integrate the QSAR models. In this step, we used the conformations of **s18** and **s49** from a previous study [13–16]. The structures and activities toward PLK1 of 66 compounds are presented in the Supplementary Materials.

3.2. Chemistry

3.2.1. General Chemical Methods

All chemicals were of reagent grade and were purchased from Aldrich (USA). Separation of the compounds by column chromatography was carried out with silica gel 60 (200–300 mesh ASTM, E. Merck, Germany). The quantity of silica gel used was 50–100 times the weight charged on the column. Thin layer chromatography (TLC) was run on silica gel-coated aluminum sheets (silica gel 60 GF254, E. Merck, Germany) and visualized under ultraviolet (UV) light (254 nm). Both ^{1}H NMR and ^{13}C NMR spectra were recorded on a Brucker model digital AVANCE III 400 MHz spectrometer at 25 °C using tetramethylsilane (TMS) as an internal standard. High-resolution MS (HR/MS) experiments were conducted with Q-TOF/Mass spectrometer 6530 (Agilent Technologies, Santa Clara, CA, USA) operated in positive-ion electrospray mode.

3.2.2. Synthesis of Ethyl 1-(2-fluoropyridin-4-yl)-4-hydroxy-1*H*-pyrazole-3-carboxylate (3)

To a solution of 2-fluoropyridin-4-amine (0.776 mmol) in 1.35 mL of H_2O and 0.206 mL (2.33 mmol) of 35% HCl at 0 °C, a solution of $NaNO_2$ (0.776 mmol) and 0.2 mL of H_2O was slowly added and reacted for 30 min. Then, a solution of ethyl 4-chloro-3-oxobutanoate (0.776 mmol) and 0.388 mL of EtOH was added to the reaction vessel, and NaOAc (4.66 mmol) was added, producing a solid. After stirring for 2 h, the organic layer was extracted with ethyl acetate and dried with anhydrous sodium sulphate (Na_2SO_4), and the solvent was evaporated to obtain compound **2**. Subsequently, compound **2** was dissolved in 0.776 mL of THF at 0 °C, and potassium t-butoxide (0.854 mmol) was added, followed by stirring for 1 h. After completion of the reaction, saturated NH_4Cl solution was added, the organic layer was extracted with ethyl acetate and dried with anhydrous sodium sulphate (Na_2SO_4), and the solvent was evaporated, followed by column chromatography and purification under EA:Hex (1:3) conditions to obtain compound **3** (74%). ^{1}**H NMR** (400 MHz, DMSO-d$_6$)) δ 9.58 (s, 1H), 8.32 (d, *J* = 5.7 Hz, 1H), 8.28 (d, *J* = 1.0 Hz, 1H), 7.84

(dd, *J* = 5.7, 1.3 Hz, 1H), 7.64 (d, *J* = 1.6 Hz, 1H), 4.32 (q, *J* = 7.1 Hz, 2H), 1.31 (t, *J* = 7.1 Hz, 3H).); **HRMS (ESI⁺)** m/z calculated for $C_{11}H_{11}FN_3O_3$ [M+H]+: 252.0779, found 252.0778.

3.2.3. General Procedure A (**4a–4c**)

Compound **3** (0.0199 mmol) was dissolved in 0.995 mL of DMF at 0 °C, and NaH (0.0239 mmol) and the appropriate benzyl bromide (0.0199 mmol) were added, followed by stirring for 1 h. After completion of the reaction, the reaction mixture was worked up 6 times with ethyl acetate and brine. The organic layer was dried with anhydrous sodium sulphate (Na_2SO_4), and the solvent was evaporated to give compound **4**.

4a as yellow solid (98%): General procedure A was followed, using benzyl bromide and **3**^{1}H NMR (400 MHz, DMSO-d₆): **¹H NMR** (400 MHz, DMSO) δ 8.81 (s, 1H), 8.37 (d, *J* = 5.7 Hz, 1H), 7.86 (d, *J* = 5.7 Hz, 1H), 7.65 (d, *J* = 1.7 Hz, 1H), 7.52–7.46 (m, 2H), 7.46–7.40 (m, 2H), 7.38 (dd, *J* = 5.0, 3.6 Hz, 1H), 5.12 (s, 2H), 4.32 (d, *J* = 7.1 Hz, 2H), 1.31 (t, *J* = 7.1 Hz, 3H).; **HRMS (ESI⁺)** m/z calculated for $C_{18}H_{17}FN_3O_3$ [M+H]+: 342.1248, found 342.1262.

4b (yellow solid, 92%): General procedure A was followed, using 1-(bromomethyl)-2-(trifluoromethyl)benzene and **3**.**¹H NMR** (400 MHz, DMSO-d₆) δ 8.89 (s, 1H), 8.38 (d, *J* = 5.7 Hz, 1H), 7.91 (dd, *J* = 12.4, 6.8 Hz, 2H), 7.80 (dd, *J* = 17.3, 7.9 Hz, 2H), 7.70 (d, *J* = 1.7 Hz, 1H), 7.62 (t, *J* = 7.8 Hz, 1H), 5.26 (s, 2H), 4.33 (q, *J* = 7.1 Hz, 2H), 1.30 (t, *J* = 7.1 Hz, 3H). **HRMS (ESI⁺)** m/z calculated for $C_{19}H_{16}F_4N_3O_3$ [M+H]⁺ 410.1122, found 410.1111.

4c (yellow solid, 56%): General procedure A was followed, using ((4-(bromomethyl)-3-chlorobenzyl)oxy)(tert-butyl)dimethylsilane and **3**. **¹H NMR** (400 MHz, DMSO-d₆) δ 8.88 (s, 1H), 8.37 (d, *J* = 5.7 Hz, 1H), 7.88 (d, *J* = 5.7 Hz, 1H), 7.67 (d, *J* = 8.1 Hz, 2H), 7.44 (s, 1H), 7.35 (d, *J* = 7.9 Hz, 1H), 5.16 (s, 2H), 4.75 (s, 2H), 4.32 (t, *J* = 7.1 Hz, 2H), 1.31 (t, *J* = 7.1 Hz, 3H), 0.91 (s, 9H), 0.09 (s, 6H). **HRMS (ESI⁺)** m/z calculated for $C_{25}H_{32}ClFN_3O_4Si$ [M+H]⁺: 520.1829, found 520.1789.

3.2.4. General Procedure B (**5a–7a**)

Compound **4** (0.0293 mmol) was dissolved in 0.293 mL of DMSO, and the appropriate amine (0.0586 mmol) and TEA (0.0586 mmol) were added, followed by stirring at 100 °C for 24 h. After completion of the reaction, the reaction mixture was cooled to room temperature, and work up was performed 6 times with ethyl acetate and washed with brine. The organic layer was dried with anhydrous sodium sulphate (Na_2SO_4), and the solvent was evaporated, followed by column chromatography and purification under EA:Hex (1:1) conditions to obtain a compound.

5a (38%): General procedure B was followed, using tetrahydro-2*H*-pyran-4-amine and **4**. **¹H NMR** (400 MHz, CDCl₃) δ 8.01 (d, *J* = 6.0 Hz, 1H), 7.55 (s, 1H), 7.45 (dd, *J* = 7.8, 1.1 Hz, 2H), 7.42–7.34 (m, 3H), 6.88 (d, *J* = 1.7 Hz, 1H), 6.76 (dd, *J* = 6.0, 2.0 Hz, 1H), 5.94 (s, 1H), 5.12 (s, 2H), 4.46 (q, *J* = 7.1 Hz, 2H), 4.01 (dt, *J* = 12.2, 3.9 Hz, 2H), 3.92 (s, 1H), 3.57 (td, *J* = 11.8, 2.3 Hz, 2H), 2.02 (s, 2H), 1.65–1.56 (m, 2H), 1.43 (t, *J* = 7.1 Hz, 3H).; **HRMS (ESI⁺)** m/z calculated for $C_{23}H_{27}N_4O_4$ [M+H]+: 423.2027, found 423.2129.

6a (37%): General procedure B was followed, using piperidin-4-amine and **4**. **¹H NMR** (400 MHz, DMSO-d₆) δ 8.60 (s, 1H), 8.10 (d, *J* = 6.1 Hz, 1H), 7.48 (d, *J* = 6.9 Hz, 2H), 7.44–7.34 (m, 3H), 7.11 (d, *J* = 30.0 Hz, 2H), 5.10 (d, *J* = 8.8 Hz, 2H), 4.31 (q, *J* = 7.1 Hz, 2H), 3.82 (s, 1H), 3.62 (s, 2H), 1.91 (s, 2H), 1.78 (d, *J* = 13.1 Hz, 2H), 1.65 (s, 2H), 1.38 (s, 9H), 1.29 (d, *J* = 5.3 Hz, 3H).; **HRMS (ESI⁺)** m/z calculated for $C_{28}H_{36}N_5O_5$ [M+H]⁺: 522.2711, found 522.2722.

7a (53%): General procedure B was followed, using pyrrolidin-3-amine and **4**. **¹H NMR** (400 MHz, CDCl₃) δ 8.06 (s, 1H), 7.55 (s, 1H), 7.47-7.43 (m, 2H), 7.42−7.33 (m, 3H), 6.82 (d, *J* = 12.8 Hz, 2H), 5.11 (s, 2H), 4.46 (q, *J* = 6.8 Hz, 2H), 4.42-4.34 (m, 1H), 3.73 (dd, *J* = 11.2, 6.0 Hz, 1H), 3.31 (s, 1H), 3.22 (s, 1H), 2.28–2.19 (m, 1H), 1.89 (d, *J* = 9.7 Hz, 2H), 1.46 (s, 9H), 1.42 (d, *J* = 7.1 Hz, 3H), 1.25-1.25 (m, 1H).; **HRMS (ESI⁺)** m/z calculated for $C_{27}H_{34}N_5O_3$ [M+H]⁺: 508.2554, found 508.2535.

3.2.5. General Procedure C (**8a–10a**, **13a–13b**, **14a–14b**)

After adding 0.47 mL of 7N NH$_3$ in MeOH to the appropriate ethyl 4-1*H*-pyrazole-3-carboxylate (0.0111 mmol), the mixture was stirred at 60 °C for 16 h. After completion of the reaction, the reaction mixture was cooled to room temperature and concentrated in vacuo. Column chromatography was performed and purified under MC:MeOH (20:1) conditions to obtain corresponding 4-1H-pyrazole-3-carboxamide.

8a as white solid (53%): General procedure C was followed, using **5a**. ^{1}**H NMR** (400 MHz, DMSO-d$_6$) δ 8.44 (s, 1H), 8.04 (d, *J* = 5.7 Hz, 1H), 7.53–7.45 (m, 3H), 7.45–7.39 (m, 2H), 7.38–7.33 (m, 1H), 7.25 (s, 1H), 6.95 (dd, *J* = 5.7, 2.0 Hz, 1H), 6.90 (d, *J* = 1.7 Hz, 1H), 6.75 (d, *J* = 7.7 Hz, 1H), 5.11 (s, 2H), 3.95 (d, *J* = 7.8 Hz, 1H), 3.86 (dd, *J* = 8.0, 3.3 Hz, 2H), 3.41 (td, *J* = 11.4, 2.1 Hz, 2H), 1.87 (d, *J* = 10.3 Hz, 2H), 1.49–1.40 (m, 2H); ^{13}**C NMR** (101 MHz, DMSO-d$_6$) δ 161.7, 159.2, 149.3, 146.2, 145.5, 136.2, 128.5, 128.2, 128.0, 114.3, 101.1, 95.3, 73.5, 66.0, 46.4, 32.7 (Supplementary S5); mp: 78.5–80.5 °C; **HRMS (ESI$^+$)** m/z calculated for C$_{21}$H$_{24}$N$_5$O$_3$ [M+H]+: 394.1874, found 394.1853.

9a (white solid, 30%): General procedure C was followed, using **6a**. ^{1}**H NMR** (400 MHz, DMSO-d$_6$) δ 8.82 (s, 2H), 8.50 (s, 1H), 8.11 (d, *J* = 5.7 Hz, 1H), 7.52-7.48 (m, 2H), 7.44–7.35 (m, 3H), 7.08 (dd, *J* = 5.8, 2.0 Hz, 1H), 6.97 (d, *J* = 1.7 Hz, 1H), 5.12 (s, 2H), 4.49–4.41 (m, 1H), 3.51–3.42 (m, 1H), 3.28-3.24 (m, 1H), 3.09 (dt, *J* = 11.0, 4.3 Hz, 1H), 2.23 (dt, *J* = 20.9, 7.3 Hz, 2H), 1.93 (dd, *J* = 15.9, 9.5 Hz, 2H); ^{13}**C NMR** (101 MHz, DMSO-d$_6$) δ 161.6, 159.1, 149.2, 141.1, 145.5, 137.3, 136.7, 133.5, 127.3, 124.4, 121.3, 99.3, 88.9, 70.5, 36.4, 33.6; mp: 58.5–60.5 °C; **HRMS (ESI$^+$)** calculated for C$_{21}$H$_{25}$N$_6$O$_2$ [M+H]+: 393.2034, found 393.2005.

10a (yellow solid, 57%): General procedure C was followed, using **7a**. ^{1}**H NMR** (400 MHz, DMSO-d$_6$) δ 8.48−8.43 (m, 1H), 8.06 (dd, *J* = 5.5, 3.6 Hz, 1H), 7.53−7.48 (m, 2H), 7.43–7.33 (m, 3H), 6.97 (d, *J* = 12.6 Hz, 1H), 6.85 (s, 1H), 5.11 (d, *J* = 5.8 Hz, 2H), 4.81−4.73 (m, 1H), 4.04 (s, 1H), 3.09 (s, 1H), 2.78 (s, 1H), 1.95 (d, *J* = 13.7 Hz, 2H), 1.47 (s, 2H), 1.30−1.25 (m, 2H); ^{13}**C NMR** (101 MHz, DMSO-d$_6$) δ 161.6, 159.1, 157.9, 149.4, 146.3, 145.5, 136.3; mp: 59.0–61.0 °C; **HRMS (ESI$^+$)** calculated for C$_{20}$H$_{23}$N$_6$O$_2$ [M+H]+: 379.1877, found 379.1848.

13a (brown solid, 73%): General procedure C was followed, using **11a**. ^{1}**H NMR** (400 MHz, DMSO-d$_6$) δ 9.27 (s, 1H), 8.53 (s, 1H), 8.24 (d, *J* = 5.7 Hz, 1H), 7.71–7.65 (m, 2H), 7.52 (dd, *J* = 6.9, 5.3 Hz, 3H), 7.45−7.40 (m, 2H), 7.39−7.34 (m, 1H), 7.31 (d, *J* = 1.6 Hz, 1H), 7.28 (dd, *J* = 8.5, 7.5 Hz, 3H), 7.22 (dd, *J* = 5.7, 2.0 Hz, 1H), 6.92 (t, *J* = 7.3 Hz, 1H), 5.14 (s, 2H) (Supplementary S4); ^{13}**C NMR** (101 MHz, DMSO-d$_6$) δ 161.6, 157.1, 149.2, 146.3, 145.6, 141.4, 136.7, 136.2, 128.6, 128.2, 128.0, 120.8, 118.3, 114.3, 103.5, 97.8, 73.5 (Supplementary S5); mp: 191 °C–193 °C; **HRMS (ESI$^+$)** m/z calculated for C$_{22}$H$_{20}$N$_5$O$_2$ [M+H]$^+$: 386.1612, found 386.1450.

13b (45%): General procedure C was followed, using **11b**. ^{1}**H NMR** (400 MHz, DMSO-d$_6$) δ 9.28 (s, 1H), 8.60 (s, 1H), 8.25 (d, *J* = 5.7 Hz, 1H), 7.89 (d, *J* = 7.6 Hz, 1H), 7.82 (d, *J* = 7.8 Hz, 1H), 7.77 (t, *J* = 7.6 Hz, 1H), 7.71–7.66 (m, 2H), 7.62 (t, *J* = 7.6 Hz, 1H), 7.53 (s, 1H), 7.32 (d, *J* = 1.6 Hz, 1H), 7.31−7.24 (m, 4H), 6.92 (t, *J* = 7.3 Hz, 1H), 5.28 (s, 2H); ^{13}**C NMR** (101 MHz, DMSO-d$_6$) δ 161.6 (s), 157.1 (d, *J* = 9.1 Hz), 149.1 (s), 146.3 (s), 145.4 (s), 141.4 (s), 136.7 (s), 134.2 (s), 132.9 (s), 130.5 (s), 129.0 (s), 128.6 (s), 126.9 (s), 126.1 (d, *J* = 5.5 Hz), 120.8 (s), 118.3 (d, *J* = 10.3 Hz), 114.7 (s), 103.6 (s), 97.9 (s), 70.2 (s); mp: 188.5–190.5 °C; **HRMS (ESI$^+$)** m/z calculated for C$_{23}$H$_{19}$F$_3$N$_5$O$_2$ [M+H]$^+$: 454.1485, found 454.1450.

14a (yellow solid 73%): General procedure C was followed, using **12a**. ^{1}**H NMR** (400 MHz, DMSO-d$_6$) δ 8.47 (s, 1H), 8.26 (s, 1H), 8.23-8.19 (m, 2H), 7.55–7.46 (m, 4H), 7.45–7.39 (m, 2H), 7.38 (dd, *J* = 5.0, 3.6 Hz, 1H), 7.28 (s, 1H), 7.22 (dd, *J* = 5.7, 1.9 Hz, 1H), 7.03 (dd, *J* = 8.0, 1.6 Hz, 1H), 6.97 (td, *J* = 7.7, 1.8 Hz, 1H), 6.93 (dd, *J* = 7.7, 1.7 Hz, 1H), 5.13 (s, 2H), 3.85 (s, 3H) (Supplementary S4); ^{13}**C NMR** (101 MHz, DMSO-d$_6$) δ 161.7, 157.3, 149.0, 146.3, 145.6, 136.3, 129.8, 128.5, 128.2, 128.0, 121.9, 120.4, 120.0, 114.2, 110.9, 103.5, 98.2, 73.5, 55.7 (Supplementary S5); mp: 197.0–199.5 °C; **HRMS (ESI$^+$)** m/z calculated for C$_{23}$H$_{22}$N$_5$O$_3$ [M+H]$^+$: 416.1717, found 416.1683.

14b (white solid 67%): General procedure C was followed, using **12b**. ^{1}**H NMR** (400 MHz, DMSO-d$_6$) δ 8.53 (s, 1H), 8.24 (dd, *J* = 8.6, 2.6 Hz, 2H), 8.21 (d, *J* = 5.7 Hz, 1H), 7.89 (d, *J* = 7.7 Hz, 1H), 7.82 (d, *J* = 7.8 Hz, 1H), 7.77 (t, *J* = 7.6 Hz, 1H), 7.62 (t, *J* = 7.7 Hz, 1H),

7.51 (d, *J* = 5.7 Hz, 1H), 7.49 (d, *J* = 1.8 Hz, 1H), 7.30 (s, 1H), 7.26 (dd, *J* = 5.7, 1.9 Hz, 1H), 7.03 (dd, *J* = 8.0, 1.6 Hz, 1H), 6.97 (td, *J* = 7.7, 1.8 Hz, 1H), 6.92 (td, *J* = 7.6, 1.7 Hz, 1H), 5.28 (s, 2H), 3.86 (s, 3H); ^{13}**C NMR** (101 MHz, DMSO-d$_6$) δ 161.7 (s), 157.2 (d, *J* = 10.4 Hz), 148.9 (d, *J* = 12.6 Hz), 146.3 (s), 145.4 (s), 136.5 (s), 134.2 (s), 132.9 (s), 130.5 (s), 129.8 (d, *J* = 7.2 Hz), 128.9 (s), 121.9 (s), 120.4 (s), 119.9 (d, *J* = 15.6 Hz), 114.6 (s), 110.9 (s), 103.7 (s), 98.4 (s), 70.2 (s), 55.7 (s). **mp**: 178.5–180.5 °C; **HRMS (ESI$^+$)** m/z calculated for C$_{24}$H$_{21}$F$_3$N$_5$O$_3$ [M+H]$^+$: 484.1591, found 484.1546.

3.2.6. General Procedure D (**11a–11c**, **12a–12c**)

Compound **4** (0.0293 mmol) was dissolved in the appropriate aniline (0.134 mL, 50 eq) and stirred at 120° C for 18 h. After completion of the reaction, the reaction mixture was cooled to room temperature, and work up was performed with ethyl acetate and water. The organic layer was dried with anhydrous sodium sulphate (Na$_2$SO$_4$) and the solvent was evaporated, followed by column chromatography and purification under EA:Hex (1:3.5) conditions to obtain product.

11a (62%): General procedure D was followed, using aniline and **4a** ^{1}H NMR (400 MHz, DMSO-d$_6$) δ 9.36 (s, 1H), 8.62 (s, 1H), 8.26 (d, *J* = 5.7 Hz, 1H), 7.70 (dd, *J* = 8.6, 1.0 Hz, 2H), 7.52–7.47 (m, 2H), 7.45–7.40 (m, 2H), 7.39–7.34 (m, 2H), 7.30–7.25 (m, 2H), 7.22 (dd, *J* = 5.8, 2.0 Hz, 1H), 6.92 (t, *J* = 7.3 Hz, 1H), 5.13 (s, 2H), 4.32 (q, *J* = 7.1 Hz, 2H), 1.31 (t, *J* = 7.1 Hz, 3H). **HRMS (ESI$^+$)** m/z calculated for C$_{24}$H$_{23}$N$_4$O$_3$ [M+H]$^+$: 415.1765, found 415.1821.

11b (42%): General procedure D was followed, using aniline and **4b**. ^{1}H NMR (400 MHz, DMSO-d$_6$) δ 9.37 (s, 1H), 8.71 (s, 1H), 8.27 (d, *J* = 5.7 Hz, 1H), 7.93 (d, *J* = 7.6 Hz, 1H), 7.79 (dd, *J* = 17.5, 7.9 Hz, 2H), 7.71 (dd, *J* = 8.6, 1.0 Hz, 2H), 7.61 (t, *J* = 7.6 Hz, 1H), 7.39 (d, *J* = 1.6 Hz, 1H), 7.31-7.25 (m, 3H), 6.92 (t, *J* = 7.3 Hz, 1H), 5.27 (s, 2H), 4.32 (q, *J* = 7.1 Hz, 2H), 1.30 (t, *J* = 7.1 Hz, 3H). **HRMS (ESI$^+$)** m/z calculated for C$_{25}$H$_{22}$F$_3$N$_4$O$_3$ [M+H]$^+$: 483.1639, found 483.1793

11c (63%): General procedure D was followed, using aniline and **4c**. ^{1}H NMR (400 MHz, DMSO-d$_6$) δ 9.36 (s, 1H), 8.71 (s, 1H), 8.27 (d, *J* = 5.7 Hz, 1H), 7.73−7.69 (m, 2H), 7.67 (d, *J* = 7.9 Hz, 1H), 7.44 (s, 1H), 7.39 (d, *J* = 1.7 Hz, 1H), 7.35 (d, *J* = 7.8 Hz, 1H), 7.31–7.24 (m, 3H), 6.92 (t, *J* = 7.3 Hz, 1H), 5.17 (s, 2H), 4.75 (s, 2H), 4.32 (q, *J* = 7.1 Hz, 2H), 1.31 (t, *J* = 7.1 Hz, 3H), 0.91 (s, 9H), 0.09 (s, 6H). **HRMS (ESI$^+$)** m/z calculated for C$_{31}$H$_{38}$ClN$_4$O$_4$Si [M+H]$^+$: 593.2345, found 593.2464.

12a (79%): General procedure D was followed, using 2-methoxyaniline and **4a**. ^{1}H NMR (400 MHz, DMSO-d$_6$) δ 8.56 (s, 1H), 8.45 (s, 1H), 8.21 (dd, *J* = 6.7, 5.2 Hz, 2H), 7.54–7.47 (m, 3H), 7.45–7.39 (m, 2H), 7.36 (dt, *J* = 9.6, 4.3 Hz, 1H), 7.19 (dd, *J* = 5.7, 2.0 Hz, 1H), 7.03 (dd, *J* = 8.0, 1.6 Hz, 1H), 6.97 (td, *J* = 7.7, 1.7 Hz, 1H), 6.91 (td, *J* = 7.6, 1.6 Hz, 1H), 5.12 (s, 2H), 4.32 (q, *J* = 7.1 Hz, 2H), 3.85 (s, 3H), 1.31 (t, *J* = 7.1 Hz, 3H). **HRMS (ESI$^+$)** m/z calculated for C$_{25}$H$_{25}$N$_4$O$_4$ [M+H]$^+$: 445.1870, found 445.1954.

12b (48%): General procedure D was followed, using 2-methoxyaniline and **4b**. ^{1}H NMR (400 MHz, DMSO-d$_6$) δ 8.64 (s, 1H), 8.46 (s, 1H), 8.22 (dd, *J* = 5.5, 3.9 Hz, 2H), 7.93 (d, *J* = 7.7 Hz, 1H), 7.79 (dd, *J* = 17.2, 8.0 Hz, 2H), 7.61 (t, *J* = 7.6 Hz, 1H), 7.55 (dd, *J* = 4.9, 1.7 Hz, 1H), 7.24 (dd, *J* = 5.7, 2.0 Hz, 1H), 7.04–7.01 (m, 1H), 6.97 (td, *J* = 7.7, 1.8 Hz, 1H), 6.92 (td, *J* = 7.6, 1.6 Hz, 1H), 5.27 (s, 2H), 4.32 (q, *J* = 7.1 Hz, 2H), 3.85 (s, 3H), 1.30 (t, *J* = 7.1 Hz, 3H). **HRMS (ESI$^+$)** m/z calculated for C$_{26}$H$_{24}$F$_3$N$_4$O$_4$ [M+H]$^+$: 513.1744, found 513.1765.

12c (81%): General procedure D was followed, using 2-methoxyaniline and **4c**. ^{1}H NMR (400 MHz, DMSO-d$_6$) δ 8.64 (s, 1H), 8.45 (s, 1H), 8.24−8.19 (m, 2H), 7.67 (d, *J* = 7.9 Hz, 1H), 7.54 (d, *J* = 1.8 Hz, 1H), 7.44 (s, 1H), 7.35 (d, *J* = 7.9 Hz, 1H), 7.23 (dd, *J* = 5.7, 2.0 Hz, 1H), 7.03 (dd, *J* = 8.0, 1.6 Hz, 1H), 6.97 (td, *J* = 7.7, 1.8 Hz, 1H), 6.94−6.89 (m, 1H), 5.17 (s, 2H), 4.75 (s, 2H), 4.32 (q, *J* = 7.1 Hz, 2H), 3.85 (s, 3H), 1.31 (t, *J* = 7.1 Hz, 3H), 0.91 (s, 9H), 0.09 (s, 6H). **HRMS (ESI$^+$)** m/z calculated for C$_{32}$H$_{40}$ClN$_4$O$_5$Si [M+H]$^+$: 623.2451, found 623.2416.

3.2.7. 4-((2-chloro-4-(hydroxymethyl)benzyl)oxy)-1-(2-(phenylamino)pyridin-4-yl)-1*H*-pyrazole-3-carboxamide (**15**)

After adding 1.45 mL of 7 N NH$_3$ in MeOH to **11c** (0.0244 mmol), the mixture was stirred at 60 °C for 24 h. After confirmation of completion of the reaction, the reaction mixture was cooled to room temperature and concentrated in vacuo to obtain compound ethyl 4-((4-(((tert-butyldimethylsilyl)oxy)methyl)-2-chlorobenzyl)oxy)-1-(2-(phenylamino)pyridin-4-yl)-1H-pyrazole-3-carboxylate. Next, the product (0.0244 mmol) was dissolved in 0.244 mL of THF, and teterabutylammonium fluoride 1M in THF (0.0244 mmol) was slowly added. After the reaction, work up was performed with ethyl acetate and saturated NH$_4$Cl solution. The organic layer was dried with anhydrous sodium sulphate (Na$_2$SO$_4$), and the solvent was evaporated, followed by column chromatography and purification under MC:MeOH (40:1) conditions to obtain compound **15** (60%): **^{1}H NMR** (400 MHz, DMSO-d$_6$) δ 9.27 (s, 1H), 8.61 (s, 1H), 8.24 (d, *J* = 5.7 Hz, 1H), 7.68 (dd, *J* = 8.6, 1.0 Hz, 2H), 7.63 (d, *J* = 7.9 Hz, 1H), 7.54 (s, 1H), 7.46 (s, 1H), 7.35−7.31 (m, 2H), 7.30−7.22 (m, 4H), 6.92 (t, *J* = 7.3 Hz, 1H), 5.35 (t, *J* = 5.8 Hz, 1H), 5.19 (s, 2H), 4.52 (d, *J* = 5.8 Hz, 2H) (Supplementary S4); ^{13}**C NMR** (101 MHz, DMSO-d$_6$) δ 161.6, 157.2, 149.2, 146.3, 145.3, 141.4, 136.6, 132.5, 131.7, 130.2, 128.7, 127.0, 125.1, 124.2, 120.8, 118.3, 114.5, 103.6, 97.9, 70.9, 61.9 (Supplementary S5); **mp**: 179.5–180.5°C; **HRMS (ESI$^+$)** m/z calculated for C$_{23}$H$_{21}$ClN$_5$O$_5$ [M+H]$^+$: 450.1327, found 450.1294.

3.2.8. 4-((2-chloro-4-(hydroxymethyl)benzyl)oxy)-1-(2-((2-methoxyphenyl)amino) pyridin -4-yl)-1*H*-pyrazole-3-carboxamide (**17**)

Compound **17** was prepared by an analogous procedure as described for the preparation of **15**. (67%): **^{1}H NMR** (400 MHz, DMSO-d$_6$) δ 8.54 (s, 1H), 8.27−8.19 (m, 3H), 7.63 (d, *J* = 7.9 Hz, 1H), 7.53 (s, 1H), 7.49 (d, *J* = 1.8 Hz, 1H), 7.46 (d, *J* = 1.3 Hz, 1H), 7.33 (d, *J* = 7.8 Hz, 1H), 7.30–7.23 (m, 2H), 7.03 (dd, *J* = 7.9, 1.6 Hz, 1H), 6.99-6.94 (m, 1H), 6.92 (dd, *J* = 7.5, 5.9 Hz, 1H), 5.35 (t, *J* = 5.8 Hz, 1H), 5.19 (s, 2H), 4.52 (d, *J* = 5.8 Hz, 2H), 3.86 (s, 3H) (Supplementary S4). ^{13}**C NMR** (101 MHz, DMSO-d$_6$) δ 161.6, 157.2, 149.2, 146.3, 145.4, 141.4, 137.2, 136.6, 132.4, 130.1, 129.4, 128.6, 120.8, 118.3, 118.3, 114.6, 103.6, 97.9, 70.9, 44.8, 29.0. **mp**: 198.5-200.5 °C; **HRMS (ESI$^+$)** m/z calculated for C$_{24}$H$_{23}$ClN$_5$O$_4$ [M+H]$^+$: 480.1433, found 480.1406.

3.2.9. General Procedure E (**16**, **18**)

After dissolving compound **15** (9.56 mmol) in 0.1 mL of MC, triethylamine (10.5 mmol) and methanesulfonyl chloride (14.3 mmol) were sequentially added dropwise at 0 °C and stirred for 3 h. After the reaction was complete, the organic layer was extracted using water and methylene chloride. The organic layer was dried with anhydrous sodium sulfate, the solvent was evaporated, and the next reaction proceeded. The obtained compound (0.00956 mmol) was dissolved in 0.1 mL of THF, and dimethylamine (19.1 mmol) and DIPEA (19.1 mmol) were added, followed by stirring at 60 °C for 18 h. After completion of the reaction, the mixture was cooled to room temperature, and work up was performed with ethyl acetate and water. The organic layer was dried with anhydrous sodium sulphate, and the solvent was evaporated, followed by column chromatography and purification under MC:MeOH (10:1) to obtain compound **16** as white solid (50%): **^{1}H NMR** (400 MHz, DMSO-d$_6$) δ 9.27 (s, 1H), 8.61 (s, 1H), 8.24 (d, *J* = 5.7 Hz, 1H), 7.71−7.66 (m, 2H), 7.63 (d, *J* = 7.8 Hz, 1H), 7.54 (s, 1H), 7.44 (s, 1H), 7.34-7.31 (m, 2H), 7.30–7.23 (m, 4H), 6.92 (t, *J* = 7.3 Hz, 1H), 5.19 (s, 2H), 3.42 (s, 2H), 2.16 (s, 6H) (Supplementary S4). ^{13}**C NMR** (101 MHz, DMSO-d$_6$) δ 161.6, 157.2, 149.2, 146.3, 145.4, 141.4, 137.2, 136.6, 132.4, 130.1, 129.4, 128.6, 120.8, 118.3, 118.3, 114.6, 103.6, 97.9, 70.9, 44.8, 29.0; **mp**: 138.5–140.5 °C; **HRMS (ESI$^+$)** m/z calculated for C$_{25}$H$_{26}$ClN$_6$O$_2$ [M+H]$^+$: 477.1800, found 477.1767.

18(. white solid, 39%): Compound **18** was prepared by the same procedure as described for the preparation of **16** using **17**. **^{1}H NMR** (400 MHz, DMSO-d$_6$) δ 8.54 (s, 1H), 8.27–8.19 (m, 3H), 7.63 (d, *J* = 7.8 Hz, 1H), 7.54–7.48 (m, 2H), 7.44 (d, *J* = 1.3 Hz, 1H), 7.32 (d, *J* = 7.8 Hz, 1H), 7.29 (s, 1H), 7.25 (dd, *J* = 5.7, 1.9 Hz, 1H), 7.03 (dd, *J* = 7.9, 1.6 Hz, 1H), 6.97

(td, *J* = 7.7, 1.8 Hz, 1H), 6.92 (td, *J* = 7.6, 1.7 Hz, 1H), 5.18 (s, 2H), 3.86 (s, 3H), 3.41 (s, 2H), 2.15 (s, 6H) (Supplementary S4); ^{13}**C NMR** (101 MHz, DMSO-d_6) δ 161.7, 157.2, 148.9, 145.9, 136.4, 132.4, 132.1, 130.1, 129.8, 129.3, 127.6, 121.9, 120.4, 119.9, 114.5, 110.9, 103.6, 98.3, 70.9, 62.2, 55.7, 44.9; **mp**: 160–162 °C; **HRMS (ESI$^+$)** m/z calculated for $C_{26}H_{28}ClN_6O_3$ [M+H]$^+$: 507.1906, found 507.1881.

3.3. Molecular Modelling

Compounds were docked into the PLK1 structure (PDB: 3KB7) [13–16]. Protein and ligand preparations were performed with Schrödinger's tools at standard settings, and Glide was used for docking and scoring. The 3D X-ray protein structures of PLK1 wildtype as a complex with a ligand were obtained from the PDB (code: 3KB7) and prepared using the Protein Preparation Wizard of the Schrödinger Maestro program. All water molecules were removed from the structure, and it was selected as a template. The structures of inhibitors were drawn using Chemdraw, and their 3D conformation was generated using the Schrödinger LigPrep program with the OPLS 2005 force field. Molecular docking of compound into the structure of PLK1 wildtype (PDB code: 3KB7) was carried out using Schrodinger Glide Version 12.2 (Schrödinger, LLC, New York, NY, USA).

3.4. Evaluation of IC$_{50}$ Values and Selected Kinase Profiling

We used Reaction Biology Corp. Kinase Hot Spot^{SM} service (www.reactionbiology. com, (accessed on 15 December 2020)) for screening of **15** (10 µM) and IC$_{50}$ Profiler Express for IC$_{50}$ measurement. Assay protocol: In a final reaction volume of 25 µL, substrate[Casein], 1 µM, ATP 10 µM, PLK1(h) (5–10 mU) is incubated with 25 mM Tris pH 7.5, 0.02 mM EGTA, 0.66 mg/mL myelin basic protein, 10 mM Mg acetate, and [^{33}P-ATP] (specific activity approx. 500 cpm/pmol, concentration as required). The reaction is initiated by addition of the Mg-ATP mix. After incubation for 40 min at room temperature, the reaction is stopped by addition of 5 µL of a 3% phosphoric acid solution. Next, 10 µL of the reaction is spotted onto a P30 filtremat and washed three times for 5 min in 75 mM phosphoric acid and once in methanol prior to drying and scintillation counting. The 30 selected protein kinases were ABL1, AKT1, ALK, Aurora A, AXL, AXL (R499C), BRAF (V599E), BTK, c-MER, c-MET, c-Src, CAMKK1, CDK1/cyclin A, EGFR, ERK1, FGFR3, FYN, IGF1R, JAK3, KDR/VEGFR2, LCK, LYN, MEK1, PKA, PLK1, PLK2, PLK3, PLK4/SAK, SYK, and TYRO3/SKY4.

4. Conclusions

In conclusion, two series of PLK1 inhibitors were aligned and combined using the CoMFA and CoMSIA 3D QSAR models, and several novel chemical scaffolds were suggested. We designed and synthesized a series of 4-benzyloxy-1-(2-(aryl/alkylamino)pyridin-4-yl)-1*H*-pyrazole-3-carboxamide derivatives based on the hybridized QSAR models. Of the suggested analogues, we synthesized 11 compounds and tested for inhibitory activity against PLK1. Compound **15**, i.e., 4-((2-chloro-4-(hydroxymethyl)benzyloxy)-1-(2-(phenylamino)pyridin-4-yl)-1*H*-pyrazole-3-carboxamide, displayed the most potent inhibitory activity against PLK1, with an IC$_{50}$ of 219 nM.

We successfully discovered a novel scaffold of PLK1 inhibitors by hybridizing two chemical series into a QSAR model. In addition, we performed a kinase panel screen using compound **15** for 30 kinases at a single dose of 10 µM in duplicate (Figure 5). The results of the screen showed that the newly synthesized PLK1 inhibitor had excellent selectivity profiles toward PLK1. Considering that PLK1 is significantly associated with various cancers, the unique chemical scaffold described in this study will be valuable for developing new molecules as potential therapeutic agents for this disease. Indeed, the above findings provide a theoretical basis for further structural modification of 4-benzyloxy-1-(2-(aryl/alkylamino)pyridin-4-yl)-1*H*-pyrazole-3-carboxamide derivatives as PLK1 inhibitors, and compound **15** is a promising lead for new therapeutics targeting cancer due to its strong kinase selectivity profile.

Supplementary Materials: The following are available online at https://www.mdpi.com/article/10.3390/ijms22083865/s1.

Author Contributions: Conceptualization, J.-M.H.; methodology, Y.O.; software, H.J.; validation, D.I.; formal analysis, H.K., J.B., J.J.; data curation, H.C.; writing—original draft preparation, writing; funding acquisition, J.-M.H.; All authors have read and agreed to the published version of the manuscript.

Funding: This work was financially supported by a National Research Foundation of Korea grant NRF-2019M3A9A8066500 (J.-M.H.), NRF-2020R1A6A1A03042854 (Center for Proteinopathy), and was supported by an Institute of Information & Communications Technology Planning & Evaluation (IITP) grant funded by the Korean government (MSIT) (No.2020-0-01343, Artificial Intelligence Convergence Research Centre (Hanyang University ERICA)).

Institutional Review Board Statement: Not applicable.

Informed Consent Statement: Not applicable.

Data Availability Statement: Data is contained within the article and Supplementary Materials.

Conflicts of Interest: The authors declare no conflict of interest.

References

1. Goroshchuk, O.; Kolosenko, I. Polo-like kinases and acute leukemia. *Oncogene* **2019**, *38*, 1–16. [CrossRef] [PubMed]
2. Fenton, B.; Glover, D.M. A conserved mitotic kinase active at late anaphase-telophase in syncytial Drosophila embryos. *Nature* **1993**, *363*, 637–640. [CrossRef] [PubMed]
3. Barr, F.A.; Herman, H.W.; Erich, A.N. Polo-like kinases and the orchestration of cell division. *Nat. Rev. Mol. Cell Biol.* **2004**, *6*, 429–441. [CrossRef] [PubMed]
4. Mito, K.; Kashima, K.; Kikuchi, H.; Daa, T.; Nakayama, I.; Yokoyama, S. Expression of Polo-Like Kinase (PLK1) in Non-Hodgkin's Lymphomas. *Leuk. Lymphoma* **2005**, *46*, 225–231. [CrossRef] [PubMed]
5. Yim, H. Current clinial trials with polo-like kinase 1 inhibitors in solid tumors. *Anticancer Drugs* **2013**, *24*, 999–1006. [CrossRef] [PubMed]
6. Shimizu-Yoshida, Y.; Sugiyama, K.; Rogounovitch, T.; Ohtsuru, A.; Namba, H.; Saenko, V.; Yamashita, S. Radiation-inducible hSNK gene is transcriptionally regulated by p53 binding homology element in human thyroid cells. *Biochem. Biophys. Res. Commun.* **2002**, *289*, 491–498. [CrossRef] [PubMed]
7. Ouyang, B.; Pan, H.; Lu, L.; Li, J.; Stambrook, P.; Li, B.; Dai, W. Human Prk is a conserved protein serine/threonine kinase involved in regulating M phase functions. *J. Biol. Chem.* **1997**, *272*, 28646–28651. [CrossRef] [PubMed]
8. Holtrich, U.; Wolf, G.; Yuan, J.; Bereiter-Hahn, J.; Karn, T.; Weiler, M.; Kauselmann, G.; Rehli, M.; Andreesen, R.; Kaufmann, M.; et al. Adhesion induced expression of the serine/threonine kinase Fnk in human macrophages. *Oncogene* **2000**, *19*, 4832–4839. [CrossRef] [PubMed]
9. Gutteridge, R.E.; Ndiaye, M.A. PLK1 Inhibitors in Cancer Therapy: From Laboratory to Clinics. *Mol. Cancer Ther.* **2016**, *15*, 1427–1435. [CrossRef] [PubMed]
10. Lénárt, P.; Petronczki, M.; Steegmaier, M.; Di Fiore, B.; Lipp, J.J.; Hoffmann, M.; Rettig, W.J.; Kraut, N.; Peters, J.M. The Small-Molecule Inhibitor BI 2536 Reveals Novel Insights into Mitotic Roles of Polo-like Kinase 1. *Curr. Biol.* **2007**, *17*, 304–315. [CrossRef] [PubMed]
11. Gilmartin, A.G.; Bleam, M.R.; Richter, M.C.; Erskine, S.G.; Kruger, R.G.; Madden, L.; Hassler, D.F.; Smith, G.K.; Gontarek, R.R.; Courtney, M.P.; et al. Distinct concentration-dependent effects of the polo-like kinase 1-specific inhibitor GSK461364A, including differential effect on apoptosis. *Cancer Res.* **2009**, *69*, 6969–6977. [CrossRef] [PubMed]
12. Gontarewicz, A.; Balabanov, S.; Keller, G.; Panse, J.; Schafhausen, P.; Bokemeyer, C.; Fiedler, W.; Moll, J. PHA-680626 exhibits anti-proliferative and pro-apoptotic activity on Imatinib-resistant chronic myeloid leukemia cell lines and primary CD34+ cells by inhibition of both Bcr-Abl tyrosine kinase and Aurora kinases. *Leuk. Res.* **2008**, *32*, 1857–1865. [CrossRef] [PubMed]
13. Awada, A.; Dumez, H.; Aftimos, P.G.; Costermans, J.; Bartholomeus, S.; Forceville, K.; Berghmans, T.; Meeus, M.-A.; Cescutti, J.; Munzert, G.; et al. Phase I trial of volasertib, a Polo-like kinase inhibitor, plus platinum agents in solid tumors: Safety, pharmacokinetics and activity. *Invest. New Drugs* **2015**, *33*, 611–620. [CrossRef] [PubMed]
14. Schöffski, P.; Awada, A.; Dumez, H.; Gil, T.; Bartholomeus, S.; Wolter, P.; Taton, M.; Fritsch, H.; Glomb, P.; Munzert, G. A phase I, dose-escalation study of the novel Polo-like kinase inhibitor volasertib (BI 6727) in patients with advanced solid tumours. *Eur. J. Cancer* **2012**, *48*, 179–186. [CrossRef] [PubMed]
15. Olmos, D.; Barker, D.; Sharma, R.; Brunetto, A.T.; Yap, T.A.; Taegtmeyer, A.B.; Barriuso, J.; Medani, H.; Degenhardt, Y.Y.; Allred, A.J.; et al. Phase I study of GSK461364, a specific and competitive polo-like kinase 1 inhibitor, in patients with advanced solid malignancies. *Clin. Cancer Res.* **2011**, *17*, 3420–3430. [CrossRef] [PubMed]

16. Döhner, H.; Lübbert, M.; Fiedler, W.; Fouillard, L.; Haaland, A.; Brandwein, J.M.; Lepretre, S.; Reman, O.; Turlure, P.; Ottmann, O.G.; et al. Randomized, phase 2 trial of low-dose cytarabine with or without volasertib in AML patients not suitable for induction therapy. *Blood* **2017**, *124*, 1426–1434. [CrossRef] [PubMed]
17. Beria, I.; Ballinari, D. Identification of 4, 5-Dihydro-1H-pyrazolo [4, 3-h]quinazoline Derivatives as a New Class of Orally and Selective Polo-Like Kinase 1 Inhibitors. *J. Med. Chem.* **2010**, *53*, 3532–3551. [CrossRef] [PubMed]
18. Sato, Y.; Onozaki, Y. Imidaozpyridine derivatives as potent and selective Polo-like kinase (PLK) inhibitors. *Bioorg. Med. Chem. Lett.* **2009**, *19*, 4673–4678. [CrossRef] [PubMed]
19. Swami, S.; Devi, N.; Agarwala, A.; Singh, V.; Shrivastava, R. ZnO nanoparticles as reusable heterogeneous catalyst for efficient one pot three component synthesis of imidazo-fused polyheterocycles. *Tetrahedron Lett.* **2016**, *57*, 1346–1350. [CrossRef]
20. PAINS Filtering was Done through. Available online: https://zinc20.docking.org/patterns/home (accessed on 25 March 2021).
21. We used Reaction Biology Corp. Kinase HotSpotSM Service (www.reactionbiology.com) for Screening of 15. Available online: www.reactionbiology.com (accessed on 25 March 2021).

International Journal of
Molecular Sciences

Review

The Cellular Prion Protein: A Promising Therapeutic Target for Cancer

Gyeongyun Go [1] and Sang Hun Lee [1,2,*] 

1   Department of Biochemistry, Soonchunhyang University College of Medicine, Cheonan 31151, Korea; ggy0227@naver.com
2   Medical Science Research Institute, Soonchunhyang University Seoul Hospital, Seoul 04401, Korea
*   Correspondence: ykckss1114@nate.com; Tel.: +82-2-709-9029

Received: 5 November 2020; Accepted: 30 November 2020; Published: 2 December 2020

Abstract: Studies on the cellular prion protein (PrP^C) have been actively conducted because misfolded PrP^C is known to cause transmissible spongiform encephalopathies or prion disease. PrP^C is a glycophosphatidylinositol-anchored cell surface glycoprotein that has been reported to affect several cellular functions such as stress protection, cellular differentiation, mitochondrial homeostasis, circadian rhythm, myelin homeostasis, and immune modulation. Recently, it has also been reported that PrP^C mediates tumor progression by enhancing the proliferation, metastasis, and drug resistance of cancer cells. In addition, PrP^C regulates cancer stem cell properties by interacting with cancer stem cell marker proteins. In this review, we summarize how PrP^C promotes tumor progression in terms of proliferation, metastasis, drug resistance, and cancer stem cell properties. In addition, we discuss strategies to treat tumors by modulating the function and expression of PrP^C via the regulation of HSPA1L/HIF-1α expression and using an anti-prion antibody.

Keywords: cellular prion protein; PrP^C; *PRNP*; cancer; cancer stem cell; targeted cancer therapy

1. Introduction

The cellular prion protein (PrP^C) is a cell surface glycophosphatidylinositol (GPI)-anchored protein consisting of 208 amino acids, and it is encoded by the *PRNP* gene located on chromosome 20. PrP^C has been intensively studied since it was proposed that misfolding of PrP^C plays a key role in the pathogenesis of neurodegenerative diseases called transmissible spongiform encephalopathies [1–3]. Studies have shown that PrP^C is not simply a cause of neurodegenerative diseases, but it is an important protein involved in many cellular functions such as stress protection, cellular differentiation, mitochondrial homeostasis, circadian rhythm, myelin homeostasis, and immune modulation [4–10]. Furthermore, several studies have shown that PrP^C expression is associated with tumor progression [11–15]. Before addressing the role of PrP^C in tumor progression, we briefly introduce herein some biochemical aspects of PrP^C.

PrP^C is first synthesized as a precursor protein (pre-pro-PrP) comprising 253 amino acids with a signal peptide at the N-terminus and a GPI anchor peptide signaling sequence (GPI-PSS) at the C-terminus. The signal peptide directs pre-pro-PrP into the endoplasmic reticulum (ER), wherein it is cleaved to generate pro-PrP. The pro-PrP is then translocated from the ER to the Golgi complex [16,17] to be further processed by the addition of N-linked glycans, removal of the GPI-PSS, and addition of the pre-assembled GPI anchor [18,19]. Finally, the mature PrP^C of 208 amino acids is translocated to the outer membrane leaflet of cells. However, not all PrP^Cs are present on the cell surface. They are constantly internalized through the recycling endosome and trafficked back repeatedly [20–22]. Through this recycling process, PrP^Cs are also found in the Golgi [22,23], in addition to the nucleus [24,25] and mitochondria [26,27].

The relationship between PrP^C and cancer progression was first discovered when *PRNP* was identified as one of the most-expressed genes in pancreatic cancer cells [28]. Around the same time, other researchers found that PrP^C was overexpressed in a drug-resistant cancer cell line compared to the parental cell line [29]. Based on several studies, it is now well established that PrP^C is involved in the main aspects of cancer biology: proliferation, metastasis, and drug resistance. Moreover, the relationship between PrP^C and cancer stem cell phenotypes has also been uncovered [30,31]. In this review, we summarize the role of PrP^C in tumor progression in terms of proliferation, metastasis, drug resistance, and cancer stem cell properties. Finally, we discuss strategies to control tumor growth by regulating the function and expression of PrP^C.

2. Overview of Physiological Functions of PrP^C

PrP^C is known to regulate several functions of cells, such as stress protection, cellular differentiation, mitochondrial homeostasis, circadian rhythm, myelin homeostasis, and immune modulation. In this review, we briefly summarize the effects of PrP^C on stress protection, cellular differentiation, and mitochondrial homeostasis.

Several studies have shown that PrP^C can directly inhibit apoptosis. PrP^C expression inhibited mitochondria-dependent apoptosis in Bax-overexpressing human primary neurons and MCF-7 breast cancer cells [32,33]. In addition, downregulation of PrP^C reduced the viability of MDA-MB-435 breast cancer cells after serum deprivation [34]. In primary hippocampal neurons, PrP^C protected the cells against staurosporine-induced cell death by interacting with stress-induced phosphoprotein 1 (STI1) [35–37]. PrP^C is also known to protect cells from oxidative stress. For example, the basal levels of ROS and lipid peroxidation were lower in PrP^C-transfected neuroblastoma and epithelial cell lines than in untransfected controls [38,39]. In addition, the expression of PrP^C by primary neurons and astrocytes has been associated with lower levels of damage caused by the addition of various oxidative toxins such as xanthine oxidase, kainic acid, and hydrogen peroxide [40,41]. PrP^C has also been found to be involved in the ER-stress response. When breast carcinoma cells were treated with the ER-stress inducing compounds such as brefeldin A, tunicamycin, and thapsigargin, the expression of PrP^C was induced. Downregulation of PrP^C in several cancer cell lines resulted in an increase in cell death in response to these toxins [13].

Neurite outgrowth is one of the characteristics of neuronal differentiation. Several studies have indicated that PrP^C promotes neurite outgrowth through interactions with other proteins such as neural cell adhesion molecule 1, epidermal growth factor receptor, integrins, laminin, and STI1 [35,42–45]. The downstream signaling of these interactions may include RhoA-Rho kinase-LIMK-cofilin pathway [44]. Activation of various signal pathways, including extracellular signal-regulated kinases 1 and 2 (ERK1/2), phosphatidylinositol-3-kinase (PI3K)/Akt, and mitogen-activated protein kinases (MAPKs), may also induce PrP^C-dependent neurite outgrowth [35,43,46]. It has been reported that PrP^C is also involved in the differentiation of embryonic stem cells. In human embryonic stem cells, downregulation of PrP^C delays spontaneous differentiation into the three germ layers [47]. Similarly, PrP^C expression promotes the differentiation of cultured human embryonic stem cells and multipotent neural precursors to mature neurons, astroglia, and oligodendroglia [47,48].

PrP^C expression also affects mitochondrial homeostasis. Transcriptomic and proteomic analyses of brain tissues and neurons of PrP^C-null and wild-type mice have identified differently expressed proteins. These proteins include cytochrome c oxidase subunits 1 and 2, which are involved in oxidative phosphorylation [49,50]. Furthermore, the absence of PrP^C reduces the number of total mitochondria and increases the number of mitochondria with unusual morphology [49].

3. PrP^C and Cancer Proliferation

PrP^C expression has been reported to promote cancer proliferation in several types of cancer cells, including gastric [51], pancreatic [52], and colon [53–55], as well as in glioblastoma (GBM) [56,57] and schwannanoma [58].

In gastric cancer, PrP^C promotes cell proliferation and metastasis of cancer cells and promotes tumor growth in xenograft mouse models [51]. PrP^C increases the expression of cyclin D1 and thereby promotes their transition from the G0/G1 phase to the S-phase. PrP^C expression also affects Akt signaling. Overexpression of PrP^C increases p-Akt levels, whereas PrP^C knockdown inhibits p-Akt expression [59]. Interestingly, it is known that certain regions of PrP^C influence cell proliferation. Specifically, deletion of amino acids 24–50 of PrP^C significantly reduced cell proliferation. Conversely, deletion of amino acids 51–91 did not affect apoptosis, metastasis cell proliferation, and multidrug resistance in gastric cancer [60].

In pancreatic ductal adenocarcinoma (PDAC), expression of PrP^C increases the proliferation and migration of the cells. In PDAC cell lines, PrP^C exists as a pro-PrP as it retains its GPI-PSS, which has a filamin A (FLNA) binding motif. It was found that the interaction between pro-PrP and FLNA, a cytoplasmic protein involved in actin organization, promotes cell migration [61]. In addition, other studies have shown that PrP^C promote pancreatic cell proliferation by activating the Notch signaling pathway [62].

PrP^C is known to interact with other membrane proteins or extracellular molecules to perform various cellular functions. In human GBM, PrP^C and heat shock 90/70 organizing protein (HOP) are upregulated, and their expression levels correlate with higher proliferation rates and poorer clinical outcomes [56]. Additionally, it has been demonstrated that the binding of HOP to PrP^C promotes proliferation of GBM cell lines and that disruption of PrP^C–HOP interaction inhibits tumor growth and improves the survival of mice [56].

In DLS-1 and SW480 colorectal cancer cells, knockdown of PrP^C significantly reduces the proliferation of cancer cells. It is known that the binding between HIF-2α and the GLUT1 promoter region decreases when PrP^C expression is suppressed, resulting in a decrease in the expression of GLUT1. This may reduce glucose uptake and glycolysis and inhibit cell proliferation [54].

4. PrP^C and Metastasis

It has been demonstrated that PrP^C promotes the invasion and migration of several types of cancer cells, such as gastric [63], pancreatic [62], colon [64], and melanoma [65] cells.

PrP^C expression is higher in metastatic gastric cancer than in non-metastatic gastric cancer. PrP^C increases the invasion and in vivo metastatic ability of gastric cancer cell lines SGC7901 and MKN45, and knockdown of PrP^C significantly reduces cancer cell invasion [63]. PrP^C seems to induce cancer cell invasion by activating the p-ERK1/2 signal and inducing the expression of MMP11. Interestingly, the N-terminal fragment (amino acids 24–90) of PrP^C has been proposed as the region for its invasion-promoting function.

PrP^C levels were found to increase in invasive melanoma, whereas in normal melanocytes, PrP^C was not detected [65]. In melanoma, PrP^C is known to exist as pro-PrP, retaining its GPI-PSS with an FLNA binding motif. As in PDAC earlier, PrP^C promoted migration by binding with FLNA and regulating cytoskeleton organization. PrP^C knockdown significantly reduced the migration of melanoma cells in a wound healing assay [65,66].

In colon cancer, PrP^C is known to promote migration by binding to HOP, also known as stress-induced phosphoprotein 1 (STI1) [64]. It was found that among the colon primary tumor cells, only PrP^C-positive cells were able to promote liver metastasis after injection into immunocompromised mice [67]. Metastasis is highly correlated with epithelial-mesenchymal transition (EMT), a process in which cells lose epithelial markers and interaction between the cell and the extracellular matrix change the cytoskeleton organization and differentiate into mesenchymal phenotype [68]. It is well known that transcription factors such as SNAI1, SNAI2, TWIST1, TWIST2, ZEB1, and ZEB2 induce EMT. *PRNP* expression is highly associated with the EMT signature in colon cancer patients, and PrP^C is known to control the expression of ZEB1 in colon cancer cells [53].

5. PrP^C and Drug Resistance

PrP^C levels were found to be higher in tumor necrosis factor α (TNF-α)-resistant breast cancer cells than in TNF-α-sensitive breast cancer cells [33]. After treatment of TNF-α, the resistant cells did not exhibit cytochrome c release or nuclear condensation. Moreover, PrP^C expression inhibited the Bax translocation to mitochondria and Bax-mediated cytochrome c release. In addition to TNF-α, PrP^C is also involved in the resistance to adriamycin (ADR) and TRAIL-mediated cell death in breast cancer cells as down-regulation of PrP^C increased the sensitivity to these molecules. [69]. Inhibition of PrP^C expression did not inhibit formation of death-inducing signaling complex (DISC); however, it inhibited Bcl-2 expression and promoted Bid cleavage, resulting in cell death [69]. In addition, it has been confirmed that PrP^C co-localization and coexpression with p-glycoprotein (P-gp) occur in ADR-resistant MCF-7 cells. When the expression of PrC^C was inhibited in these cells, the capability of paclitaxel, a P-gp substrate, to induce in vitro invasion of the cells decreased [52]. More importantly, tissue microarray analysis of 756 breast cancer tumors demonstrated that PrP^C was associated with ER-negative breast cancer subsets, and compared with ER-negative/PrP^C-positive cells, ER-negative/PrP^C-negative cells are more sensitive to adjuvant chemotherapy [70].

PrP^C is also involved in drug resistance in colon cancer. Hypoxia-induced PrP^C expression in colorectal cancer cells inhibits TRAIL-induced apoptosis [71]. PrP^C inhibited apoptosis of colon cancer cells by activating the PI3K-Akt pathway [72]. Conversely, deletion of PrP^C resulted in reduced Akt activation and enhanced caspase-3 activation [73]. Our group has demonstrated that PrP^C levels significantly increase in 5-FU-and oxaliplatin-resistant colorectal cancer [74–77]. In addition, knocking down PrP^C expression significantly reduces the drug resistance of colorectal cancer cells. Furthermore, we have shown that PrP^C suppresses the drug-induced activation of stress-associated proteins, such as p38, JNK, and p53. We have also demonstrated that PrP^C inhibits caspase-3 activation by drugs and PARP1 cleavage. These results suggest that the level of PrP^C plays an important role in the development of drug resistance in colorectal cancer cells.

6. PrP^C and Cancer Stem Cells

Cancer stem cells (CSCs) are a subpopulation of cells that are capable of self-renewal, differentiation into various cell types in a determined tumor, and tumor propagation when xenotransplanted into mice [78,79]. CSCs are resistant to conventional medical therapies and have been implicated in cancer recurrence, which has made these cells a key target for therapy [80–82].

Recently, studies on the correlation between PrP^C and CSCs have been conducted. PrP^C activation of Fyn-SP1 pathway in colon cancer cells promoted EMT and resulted in a more aggressive phenotype [83]. EMT is closely connected with CSC properties as EMT enhances metastasis and drug resistance of cancer and cancer microenvironment promotes activation of EMT program [84]. Du et al. demonstrated that PrP^C-positive primary colon cancer cells expressed high levels of the EMT-associated markers, TWIST and N-cadherin, and low levels of the epithelial marker, E-cadherin, as well as exhibiting CSC properties such as the expression of the CSC marker, CD44, and tumor-initiating capacity [67].

In line with this finding, PrP^C has been shown to interact with CD44 in multidrug-resistant breast cancer cells [85]. Furthermore, in primary GBM cells, PrP^C silencing reduces the expression of the CSC markers, Sox2 and Nanog, as well as the self-renewal and tumorigenic potential of CSCs [57]. Similar findings were demonstrated by Iglesia et al., who worked on GBM cell lines grown as neurospheres [86].

CSCs are one of the most-studied recent topics in cancer biology. They have emerged as pivotal components that can initiate and maintain tumors [79]. PrP^C is known to interact with CD44, and its expression correlates with resistance to chemotherapy in breast cancer cell lines [85]. Moreover, the CD44-positive and PrP^C-positive subpopulations of colorectal tumor cells have CSC properties, including tumorigenic and metastatic capacities [67], indicating that PrP^C contributes to tumor maintenance by modulating CSC behaviors.

Our group confirmed the correlation between PrP^C and CSC by demonstrating that the levels of PrP^C and CSC marker proteins such as Oct4, Nanog, Sox2, and ALDH1A1 significantly increased in human colorectal cancer tissues and colorectal cancer cells [31]. In addition, knockdown of PrP^C reduced the expression of CSC markers in the CSCs. More specifically, PrP^C inhibited the anticancer drug-induced degradation of Oct4, but did not inhibit the degradation of other stem cell markers such as Nanog, Sox2, and ALDH1A1. Oct4 is a master regulator involved in the self-renewal and pluripotency of CSCs. It has been reported that tumor sphere formation ability is activated in breast cancer cells overexpressing Oct4 [87,88]. Cancer cells overexpressing Oct4 also overexpress other CSC markers such as CD133, CD34, and ALDH1. Oct4 is also involved in the survival, self-renewal, metastasis, and drug resistance of CSCs [89–92]. These results indicate that PrP^C directly regulates Oct4 expression, whereas it indirectly regulates Nanog, Sox2, and ALDH1A to promote the self-renewal and survival of CSCs [31].

In summary, Figure 1 shows the proteins and signaling pathways that seem to be affected by PrP^C expression. The information on these interactions was retrieved from several studies that have been already mentioned in this review. Although PrP^C appear to interact and activate several interaction partners and signaling pathways to promote tumor progression, it is difficult to say that this applies to all cancer cells. The role of PrP^C in cancer needs to be interpreted differently depending on the cell type and interaction partner.

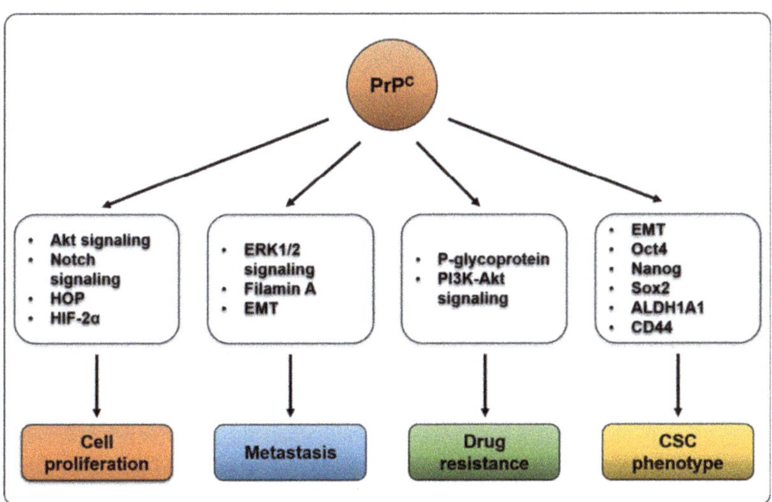

Figure 1. Proteins and signaling pathways that seem to be affected by PrP^C expression. Various proteins and signaling pathways reportedly interact with or are modulated by PrP^C. Some of the proteins shown are known to be regulated indirectly through other proteins rather than direct interaction with PrP^C. PrP^C: prion protein, ECM: extracellular matrix, CSC: cancer stem cell.

7. Cancer Treatment by Targeting PrP^C

Cancer growth can be inhibited by inhibiting the interaction between PrP^C and other proteins. Lopes et al. used a peptide named HOP/STI1230–245 corresponding to the prion binding site of HOP to inhibit the interaction between PrP^C and HOP [56]. Treatment with only the peptide did not inhibit cell proliferation, but co-treatment with HOP inhibited the interaction between HOP and PrP^C and HOP-induced cell proliferation. In addition, HOP/STI1230–245 treatment of orthotopic xenografts inhibited tumor growth and improved animal survival while maintaining cognitive performance [56]. It should be noted that the blockage of PrP^C and HOP may be deleterious, because long-term [93] but not short-term [94] intracranial infusion of antibodies against PrP^C, particularly those against its globular domain, can be neurotoxic [93].

Recently, our group identified that HSP family A (Hsp 70) member 1-like, HSPA1L, regulates the expression of PrP^C [77]. It was confirmed that the expression of HSPA1L increased in colon cancer cells and cancer tissues. We also demonstrated that HSPA1L increases the stability of HIF-1α by binding with HIF-1α and promotes the accumulation of PrP^C. When HSPA1L expression was knocked down, HIF-1α stability and PrP^C expression decreased [77]. In addition, we also showed that HSPA1L binds to GP78 and inhibits its ubiquitination activity, thereby reducing the ubiquitination of PrP^C. Several studies have indicated that deregulation of several E3 ligases affects the growth and metastasis of cancer and the growth of CSCs [95–98]. GP78 is an ER membrane-anchored E3 ligase that regulates the progression of cancer cells through ubiquitin ligase activity. For example, downregulation of GP78-mediated ubiquitination is known to inhibit metastasis in breast cancer cells [99]. We hypothesized that HIF-1α and HSPA1L are major therapeutic targets for colorectal cancer. Indeed, knockdown of HIF-1α and HSPA1L using siRNAs inhibited cancer sphere formation in HT-29 and S707 cells. In an in vivo xenograft model, knockdown of HIF-1α or HSPA1L inhibited tumor growth and liver metastasis (Figure 2). In addition, when both genes were knocked down simultaneously, cancer growth and liver metastasis were further suppressed [77]. These results indicated that PrP^C is important in tumor progression, and the suppression of PrP^C expression by targeting HIF-1α and HSPA1L could be a promising therapeutic strategy to treat cancer.

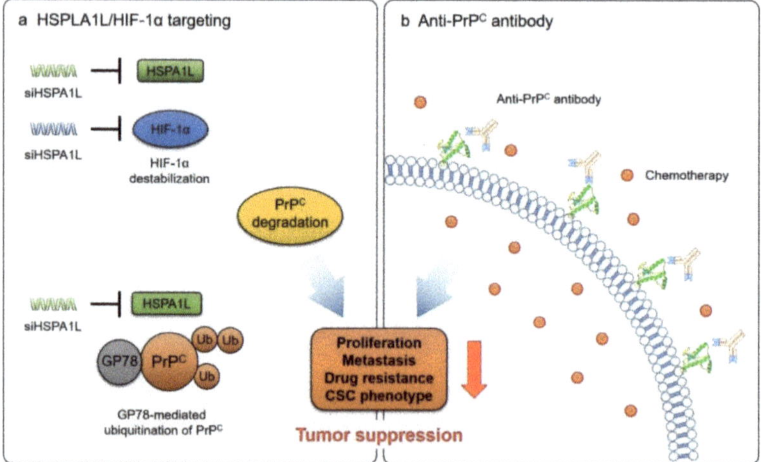

Figure 2. Strategies to suppress the tumor progression by regulating the expression and function of PrP^C. (**a**) In tumor niche, the expression of HSPA1L is increased. HSPA1L binds and stabilizes HIF-1α. HSPA1L directly binds to GP78 and inhibits its ubiquitination activity. Overall, PrP^C expression is increased in tumor cells. Therefore, knocking down HSPA1L and HIF-1α expression induces degradation of PrP^C and tumor suppression. (**b**) Direct targeting of PrP^C using anti-PrP^C antibodies has been demonstrated as a potent cancer therapy. Anti-PrP^C antibody also can be used as a combination therapy with conventional anticancer drugs.

Anti-prion antibodies can be utilized for the treatment of cancer. Antibody therapeutics have been used in the treatment of cancer, unlike existing anticancer drugs, as they have fewer side effects and exhibit high efficacy. For the development of effective antibody therapeutics, the discovery of specific molecular biomarkers in a wide range of solid malignancies is a key process [100]. The functions of PrP^C in the growth, metastasis, drug resistance, and CSC properties of various types of cancer suggest that it is a promising therapeutic target for cancer treatment. We confirmed that an anti-prion antibody showed anticancer effects in a xenograft model and that superior therapeutic effects appeared when the conventional anticancer drugs and anti-prion antibody were applied in combination (unpublished data) (Figure 2). Furthermore, compared to cetuximab, an EGFR-targeting antibody, the anti-prion antibody showed similar anticancer effects with 10 times lower dose in the xenograft mouse model (unpublished data).

Similar to our data, a previous study has revealed the effective epitope of PrP^C for antibody-mediated colon cancer therapy [101]. In colon cancer cell line HCT116 cells, epitope 139–142 and epitope 141–151 targeting anti-PrP antibodies highly inhibited the proliferative capacity of cells, compared with those of epitope 93–109 and epitope 101–112 [101]. Furthermore, epitope 141–151 targeting anti-PrP was approximately 10-fold more active than that of epitope 93–109 targeting anti-PrP [101]. These data indicate that effectiveness of anti-PrP antibodies might be related to the epitope-binding region. Further studies on anti-PrP structure and its targeting epitope site for cancer therapy are needed. One study has shown that N-terminal domain of PrP^C is a direct binding and sequestering site on anti-tumor drug, doxorubicin, in breast cancer [67]. This study suggests that N-terminal-domain-targeting anti-PrP antibodies might be effective antibody therapy when combined with anti-tumor drugs, for cancer treatment.

Although antibody therapy is a promising cancer treatment, resistance may arise due to the characteristics of cancer cells, such as intrinsic phenotypic variation and adaptive phenotypic modifications [100,102–105]. Antibody-drug conjugates (ADCs), which are novel antibody-based therapeutics, are another option for treating tumors. ADC is a technology that focuses on targeting only cancer cells by exploiting the advantages of antibodies: specificity, non-toxicity in circulation, and pharmacokinetics. ADC is known to enter cells through clathrin-mediated endocytosis [106]. The endosome that harbors ADC binds to other vesicles in the cell and forms an endo-lysosome. A protease cleaves the linker of the ADC and activates free drugs to move into the cytoplasm. The drugs bind to the molecular target, causing apoptosis of the tumor cells. A representative example of a successful ADC is Trastuzumab Emtansine (T-DM1) [107–109]. T-DM1 significantly prolonged progression-free and overall survival with less toxicity than lapatinib plus capecitabine in patients with HER2-positive advanced breast cancer who were previously treated with trastuzumab and a taxane. Anti-prion antibodies, such as T-DM1, are expected to be developed as anticancer agents in the form of ADCs.

8. Conclusions

Several studies have suggested that PrP^C promotes tumor progression. It has been demonstrated that PrP^C is overexpressed in various types of cancer cells and tumor tissues, including gastric, pancreatic, breast, and colon cancers, as well as melanoma, GBM, and schwannoma. In addition, it has been shown that PrP^C regulates cell proliferation, metastasis, drug resistance, and cancer stem cell properties through signaling pathways, such as PI3K-Akt and Notch, and interaction with ECM, cell surface molecules, and cancer stem cell markers. It should be noted that the function of PrP^C in cancer should be interpreted depending on the cell type and the molecule that interacts with it. Nevertheless, further research is needed to elucidate the function of PrP^C in tumor progression. However, there seems to be no disagreement that PrP^C is a promising target for cancer treatment.

In addition to PrP^C, the misfolded prion protein (PrP^{Sc}) may be highly expressed in cancer patients compared to the healthy people. Recently, somatic mutations in *PRNP* were analyzed in 10,967 cancer patients using the Cancer Genome atlas (TCGA) database [110]. A total of 48 mutations in *PRNP* gene were identified in cancer patient. Among them, eight somatic mutations— G131V, D167N, V180I, D202N, V203I, R208C, R208H, and E211Q—are known as pathogenic mutations of prion diseases. Interestingly, it has been reported that PrP^{Sc} was also detected in healthy people, who had not been diagnosed with prion diseases [111]. These results may indicate that cancer patients carrying pathogenic somatic mutations of *PRNP* may produce PrP^{Sc} and may not be diagnosed with prion disease.

In this review, we suggest that prion targeting is a promising strategy to treat cancer. Regulating the expression of HSPL1A and HIF-1α, which are involved in the stability and degradation of PrP^C, effectively inhibits cancer growth and metastasis. In addition, an anti-prion antibody has been used to inhibit the growth of cancer. Nevertheless, further studies are needed to verify the anticancer effect and safety of prion targeting in various cancer models. To date, no clinical trials using prion targeting have been conducted; therefore, the effectiveness and safety of cancer treatment strategies using prion targeting should also be verified.

Author Contributions: G.G.: designed and drafted the manuscript; S.H.L.: drafted the manuscript, procured funding, and supervised the study. All authors have read and agreed to the published version of the manuscript.

Funding: This work was funded by a National Research Foundation grant funded by the Korean government (2016R1D1A3B01007727).

Acknowledgments: This work was supported by a grant from the National Research Foundation funded by the Korean government (2016R1D1A3B01007727). The funders had no role in the manuscript design, preparation of the manuscript, or decision to publish.

Conflicts of Interest: The authors declare no conflict of interest.

References

1. Prusiner, S.B. Novel proteinaceous infectious particles cause scrapie. *Science* **1982**, *216*, 136–144. [CrossRef] [PubMed]
2. Caughey, B.; Chesebro, B. Prion protein and the transmissible spongiform encephalopathies. *Trends Cell Biol.* **1997**, *7*, 56–62. [CrossRef]
3. Aguzzi, A.; Heppner, F.L. Pathogenesis of prion diseases: A progress report. *Cell Death Differ.* **2000**, *7*, 889–902. [CrossRef] [PubMed]
4. Castle, A.R.; Gill, A.C. Physiological Functions of the Cellular Prion Protein. *Front. Mol. Biosci.* **2017**, *4*, 19. [CrossRef] [PubMed]
5. Dupiereux, I.; Falisse-Poirrier, N.; Zorzi, W.; Watt, N.T.; Thellin, O.; Zorzi, D.; Pierard, O.; Hooper, N.M.; Heinen, E.; Elmoualij, B. Protective effect of prion protein via the N-terminal region in mediating a protective effect on paraquat-induced oxidative injury in neuronal cells. *J. Neurosci. Res.* **2008**, *86*, 653–659. [CrossRef] [PubMed]
6. Graner, E.; Mercadante, A.F.; Zanata, S.M.; Martins, V.R.; Jay, D.G.; Brentani, R.R. Laminin-induced PC-12 cell differentiation is inhibited following laser inactivation of cellular prion protein. *FEBS Lett.* **2000**, *482*, 257–260. [CrossRef]
7. Ramljak, S.; Asif, A.R.; Armstrong, V.W.; Wrede, A.; Groschup, M.H.; Buschmann, A.; Schulz-Schaeffer, W.; Bodemer, W.; Zerr, I. Physiological role of the cellular prion protein (PrPc): Protein profiling study in two cell culture systems. *J. Proteome Res.* **2008**, *7*, 2681–2695. [CrossRef]
8. Tobler, I.; Gaus, S.E.; Deboer, T.; Achermann, P.; Fischer, M.; Rulicke, T.; Moser, M.; Oesch, B.; McBride, P.A.; Manson, J.C. Altered circadian activity rhythms and sleep in mice devoid of prion protein. *Nature* **1996**, *380*, 639–642. [CrossRef]
9. Bremer, J.; Baumann, F.; Tiberi, C.; Wessig, C.; Fischer, H.; Schwarz, P.; Steele, A.D.; Toyka, K.V.; Nave, K.A.; Weis, J.; et al. Axonal prion protein is required for peripheral myelin maintenance. *Nat. Neurosci.* **2010**, *13*, 310–318. [CrossRef]
10. Haddon, D.J.; Hughes, M.R.; Antignano, F.; Westaway, D.; Cashman, N.R.; McNagny, K.M. Prion protein expression and release by mast cells after activation. *J. Infect. Dis* **2009**, *200*, 827–831. [CrossRef]
11. Santos, T.G.; Lopes, M.H.; Martins, V.R. Targeting prion protein interactions in cancer. *Prion* **2015**, *9*, 165–173. [CrossRef] [PubMed]
12. Gao, Z.; Peng, M.; Chen, L.; Yang, X.; Li, H.; Shi, R.; Wu, G.; Cai, L.; Song, Q.; Li, C. Prion Protein Protects Cancer Cells against Endoplasmic Reticulum Stress Induced Apoptosis. *Virol. Sin.* **2019**, *34*, 222–234. [CrossRef] [PubMed]
13. Dery, M.A.; Jodoin, J.; Ursini-Siegel, J.; Aleynikova, O.; Ferrario, C.; Hassan, S.; Basik, M.; LeBlanc, A.C. Endoplasmic reticulum stress induces PRNP prion protein gene expression in breast cancer. *Breast Cancer Res.* **2013**, *15*, R22. [CrossRef]

14. Mehrpour, M.; Codogno, P. Prion protein: From physiology to cancer biology. *Cancer Lett.* **2010**, *290*, 1–23. [CrossRef] [PubMed]

15. Tang, Z.; Ma, J.; Zhang, W.; Gong, C.; He, J.; Wang, Y.; Yu, G.; Yuan, C.; Wang, X.; Sun, Y.; et al. The Role of Prion Protein Expression in Predicting Gastric Cancer Prognosis. *J. Cancer* **2016**, *7*, 984–990. [CrossRef] [PubMed]

16. Tanaka, S.; Maeda, Y.; Tashima, Y.; Kinoshita, T. Inositol deacylation of glycosylphosphatidylinositol-anchored proteins is mediated by mammalian PGAP1 and yeast Bst1p. *J. Biol. Chem.* **2004**, *279*, 14256–14263. [CrossRef] [PubMed]

17. Bonnon, C.; Wendeler, M.W.; Paccaud, J.P.; Hauri, H.P. Selective export of human GPI-anchored proteins from the endoplasmic reticulum. *J. Cell Sci.* **2010**, *123*, 1705–1715. [CrossRef]

18. Sarnataro, D.; Campana, V.; Paladino, S.; Stornaiuolo, M.; Nitsch, L.; Zurzolo, C. PrP(C) association with lipid rafts in the early secretory pathway stabilizes its cellular conformation. *Mol. Biol. Cell* **2004**, *15*, 4031–4042. [CrossRef]

19. Campana, V.; Sarnataro, D.; Zurzolo, C. The highways and byways of prion protein trafficking. *Trends Cell Biol.* **2005**, *15*, 102–111. [CrossRef]

20. Shyng, S.L.; Huber, M.T.; Harris, D.A. A prion protein cycles between the cell surface and an endocytic compartment in cultured neuroblastoma cells. *J. Biol. Chem.* **1993**, *268*, 15922–15928.

21. Sunyach, C.; Jen, A.; Deng, J.; Fitzgerald, K.T.; Frobert, Y.; Grassi, J.; McCaffrey, M.W.; Morris, R. The mechanism of internalization of glycosylphosphatidylinositol-anchored prion protein. *EMBO J.* **2003**, *22*, 3591–3601. [CrossRef] [PubMed]

22. Magalhaes, A.C.; Silva, J.A.; Lee, K.S.; Martins, V.R.; Prado, V.F.; Ferguson, S.S.; Gomez, M.V.; Brentani, R.R.; Prado, M.A. Endocytic intermediates involved with the intracellular trafficking of a fluorescent cellular prion protein. *J. Biol. Chem.* **2002**, *277*, 33311–33318. [CrossRef] [PubMed]

23. Lee, K.S.; Magalhaes, A.C.; Zanata, S.M.; Brentani, R.R.; Martins, V.R.; Prado, M.A. Internalization of mammalian fluorescent cellular prion protein and N-terminal deletion mutants in living cells. *J. Neurochem.* **2001**, *79*, 79–87. [CrossRef] [PubMed]

24. Gu, Y.; Hinnerwisch, J.; Fredricks, R.; Kalepu, S.; Mishra, R.S.; Singh, N. Identification of cryptic nuclear localization signals in the prion protein. *NeuroBiol. Dis.* **2003**, *12*, 133–149. [CrossRef]

25. Morel, E.; Fouquet, S.; Strup-Perrot, C.; Pichol Thievend, C.; Petit, C.; Loew, D.; Faussat, A.M.; Yvernault, L.; Pincon-Raymond, M.; Chambaz, J.; et al. The cellular prion protein PrP(c) is involved in the proliferation of epithelial cells and in the distribution of junction-associated proteins. *PLoS ONE* **2008**, *3*, e3000. [CrossRef]

26. Hachiya, N.S.; Yamada, M.; Watanabe, K.; Jozuka, A.; Ohkubo, T.; Sano, K.; Takeuchi, Y.; Kozuka, Y.; Sakasegawa, Y.; Kaneko, K. Mitochondrial localization of cellular prion protein (PrPC) invokes neuronal apoptosis in aged transgenic mice overexpressing PrPC. *Neurosci. Lett.* **2005**, *374*, 98–103. [CrossRef]

27. Satoh, J.; Onoue, H.; Arima, K.; Yamamura, T. The 14-3-3 protein forms a molecular complex with heat shock protein Hsp60 and cellular prion protein. *J. Neuropathol. Exp. Neurol* **2005**, *64*, 858–868. [CrossRef]

28. Han, H.; Bearss, D.J.; Browne, L.W.; Calaluce, R.; Nagle, R.B.; Von Hoff, D.D. Identification of differentially expressed genes in pancreatic cancer cells using cDNA microarray. *Cancer Res.* **2002**, *62*, 2890–2896.

29. Zhao, Y.; You, H.; Liu, F.; An, H.; Shi, Y.; Yu, Q.; Fan, D. Differentially expressed gene profiles between multidrug resistant gastric adenocarcinoma cells and their parental cells. *Cancer Lett.* **2002**, *185*, 211–218. [CrossRef]

30. Domingues, P.H.; Nanduri, L.S.Y.; Seget, K.; Venkateswaran, S.V.; Agorku, D.; Vigano, C.; von Schubert, C.; Nigg, E.A.; Swanton, C.; Sotillo, R.; et al. Cellular Prion Protein PrP(C) and Ecto-5′-Nucleotidase Are Markers of the Cellular Stress Response to Aneuploidy. *Cancer Res.* **2017**, *77*, 2914–2926. [CrossRef]

31. Lee, J.H.; Yun, C.W.; Han, Y.S.; Kim, S.; Jeong, D.; Kwon, H.Y.; Kim, H.; Baek, M.J.; Lee, S.H. Melatonin and 5-fluorouracil co-suppress colon cancer stem cells by regulating cellular prion protein-Oct4 axis. *J. Pineal. Res.* **2018**, *65*, e12519. [CrossRef] [PubMed]

32. Bounhar, Y.; Zhang, Y.; Goodyer, C.G.; LeBlanc, A. Prion protein protects human neurons against Bax-mediated apoptosis. *J. Biol. Chem.* **2001**, *276*, 39145–39149. [CrossRef] [PubMed]

33. Roucou, X.; Giannopoulos, P.N.; Zhang, Y.; Jodoin, J.; Goodyer, C.G.; LeBlanc, A. Cellular prion protein inhibits proapoptotic Bax conformational change in human neurons and in breast carcinoma MCF-7 cells. *Cell Death Differ.* **2005**, *12*, 783–795. [CrossRef] [PubMed]

34. Yu, G.; Jiang, L.; Xu, Y.; Guo, H.; Liu, H.; Zhang, Y.; Yang, H.; Yuan, C.; Ma, J. Silencing prion protein in MDA-MB-435 breast cancer cells leads to pleiotropic cellular responses to cytotoxic stimuli. *PLoS ONE* **2012**, *7*, e48146. [CrossRef] [PubMed]

35. Lopes, M.H.; Hajj, G.N.; Muras, A.G.; Mancini, G.L.; Castro, R.M.; Ribeiro, K.C.; Brentani, R.R.; Linden, R.; Martins, V.R. Interaction of cellular prion and stress-inducible protein 1 promotes neuritogenesis and neuroprotection by distinct signaling pathways. *J. Neurosci.* **2005**, *25*, 11330–11339. [CrossRef] [PubMed]

36. Beraldo, F.H.; Arantes, C.P.; Santos, T.G.; Queiroz, N.G.; Young, K.; Rylett, R.J.; Markus, R.P.; Prado, M.A.; Martins, V.R. Role of alpha7 nicotinic acetylcholine receptor in calcium signaling induced by prion protein interaction with stress-inducible protein 1. *J. Biol. Chem.* **2010**, *285*, 36542–36550. [CrossRef]

37. Ostapchenko, V.G.; Beraldo, F.H.; Mohammad, A.H.; Xie, Y.F.; Hirata, P.H.; Magalhaes, A.C.; Lamour, G.; Li, H.; Maciejewski, A.; Belrose, J.C.; et al. The prion protein ligand, stress-inducible phosphoprotein 1, regulates amyloid-beta oligomer toxicity. *J. Neurosci.* **2013**, *33*, 16552–16564. [CrossRef]

38. Rachidi, W.; Vilette, D.; Guiraud, P.; Arlotto, M.; Riondel, J.; Laude, H.; Lehmann, S.; Favier, A. Expression of prion protein increases cellular copper binding and antioxidant enzyme activities but not copper delivery. *J. Biol. Chem.* **2003**, *278*, 9064–9072. [CrossRef]

39. Zeng, F.; Watt, N.T.; Walmsley, A.R.; Hooper, N.M. Tethering the N-terminus of the prion protein compromises the cellular response to oxidative stress. *J. Neurochem.* **2003**, *84*, 480–490. [CrossRef]

40. Brown, D.R.; Nicholas, R.S.; Canevari, L. Lack of prion protein expression results in a neuronal phenotype sensitive to stress. *J. Neurosci. Res.* **2002**, *67*, 211–224. [CrossRef]

41. Anantharam, V.; Kanthasamy, A.; Choi, C.J.; Martin, D.P.; Latchoumycandane, C.; Richt, J.A.; Kanthasamy, A.G. Opposing roles of prion protein in oxidative stress- and ER stress-induced apoptotic signaling. *Free Radic. Biol. Med.* **2008**, *45*, 1530–1541. [CrossRef] [PubMed]

42. Santuccione, A.; Sytnyk, V.; Leshchyns'ka, I.; Schachner, M. Prion protein recruits its neuronal receptor NCAM to lipid rafts to activate p59fyn and to enhance neurite outgrowth. *J. Cell Biol.* **2005**, *169*, 341–354. [CrossRef] [PubMed]

43. Llorens, F.; Carulla, P.; Villa, A.; Torres, J.M.; Fortes, P.; Ferrer, I.; del Rio, J.A. PrP(C) regulates epidermal growth factor receptor function and cell shape dynamics in Neuro2a cells. *J. Neurochem.* **2013**, *127*, 124–138. [CrossRef] [PubMed]

44. Loubet, D.; Dakowski, C.; Pietri, M.; Pradines, E.; Bernard, S.; Callebert, J.; Ardila-Osorio, H.; Mouillet-Richard, S.; Launay, J.M.; Kellermann, O.; et al. Neuritogenesis: The prion protein controls beta1 integrin signaling activity. *FASEB J.* **2012**, *26*, 678–690. [CrossRef]

45. Graner, E.; Mercadante, A.F.; Zanata, S.M.; Forlenza, O.V.; Cabral, A.L.; Veiga, S.S.; Juliano, M.A.; Roesler, R.; Walz, R.; Minetti, A.; et al. Cellular prion protein binds laminin and mediates neuritogenesis. *Brain Res. Mol. Brain Res.* **2000**, *76*, 85–92. [CrossRef]

46. Caetano, F.A.; Lopes, M.H.; Hajj, G.N.; Machado, C.F.; Pinto Arantes, C.; Magalhaes, A.C.; Vieira Mde, P.; Americo, T.A.; Massensini, A.R.; Priola, S.A.; et al. Endocytosis of prion protein is required for ERK1/2 signaling induced by stress-inducible protein 1. *J. Neurosci.* **2008**, *28*, 6691–6702. [CrossRef]

47. Lee, Y.J.; Baskakov, I.V. The cellular form of the prion protein guides the differentiation of human embryonic stem cells into neuron-, oligodendrocyte-, and astrocyte-committed lineages. *Prion* **2014**, *8*, 266–275. [CrossRef]

48. Steele, A.D.; Emsley, J.G.; Ozdinler, P.H.; Lindquist, S.; Macklis, J.D. Prion protein (PrPc) positively regulates neural precursor proliferation during developmental and adult mammalian neurogenesis. *Proc. Natl. Acad. Sci. USA* **2006**, *103*, 3416–3421. [CrossRef]

49. Miele, G.; Jeffrey, M.; Turnbull, D.; Manson, J.; Clinton, M. Ablation of cellular prion protein expression affects mitochondrial numbers and morphology. *Biochem. Biophys. Res. Commun.* **2002**, *291*, 372–377. [CrossRef]

50. Stella, R.; Cifani, P.; Peggion, C.; Hansson, K.; Lazzari, C.; Bendz, M.; Levander, F.; Sorgato, M.C.; Bertoli, A.; James, P. Relative quantification of membrane proteins in wild-type and prion protein (PrP)-knockout cerebellar granule neurons. *J. Proteome Res.* **2012**, *11*, 523–536. [CrossRef]

51. Liang, J.; Luo, G.; Ning, X.; Shi, Y.; Zhai, H.; Sun, S.; Jin, H.; Liu, Z.; Zhang, F.; Lu, Y.; et al. Differential expression of calcium-related genes in gastric cancer cells transfected with cellular prion protein. *Biochem. Cell Biol.* **2007**, *85*, 375–383. [CrossRef] [PubMed]

52. Li, Q.Q.; Cao, X.X.; Xu, J.D.; Chen, Q.; Wang, W.J.; Tang, F.; Chen, Z.Q.; Liu, X.P.; Xu, Z.D. The role of P-glycoprotein/cellular prion protein interaction in multidrug-resistant breast cancer cells treated with paclitaxel. *Cell Mol. Life Sci.* **2009**, *66*, 504–515. [CrossRef] [PubMed]

53. Le Corre, D.; Ghazi, A.; Balogoun, R.; Pilati, C.; Aparicio, T.; Martin-Lanneree, S.; Marisa, L.; Djouadi, F.; Poindessous, V.; Crozet, C.; et al. The cellular prion protein controls the mesenchymal-like molecular subtype and predicts disease outcome in colorectal cancer. *EBioMedicine* **2019**, *46*, 94–104. [CrossRef] [PubMed]

54. Li, Q.Q.; Sun, Y.P.; Ruan, C.P.; Xu, X.Y.; Ge, J.H.; He, J.; Xu, Z.D.; Wang, Q.; Gao, W.C. Cellular prion protein promotes glucose uptake through the Fyn-HIF-2alpha-Glut1 pathway to support colorectal cancer cell survival. *Cancer Sci.* **2011**, *102*, 400–406. [CrossRef]

55. Chieng, C.K.; Say, Y.H. Cellular prion protein contributes to LS 174T colon cancer cell carcinogenesis by increasing invasiveness and resistance against doxorubicin-induced apoptosis. *Tumour. Biol.* **2015**, *36*, 8107–8120. [CrossRef]

56. Lopes, M.H.; Santos, T.G.; Rodrigues, B.R.; Queiroz-Hazarbassanov, N.; Cunha, I.W.; Wasilewska-Sampaio, A.P.; Costa-Silva, B.; Marchi, F.A.; Bleggi-Torres, L.F.; Sanematsu, P.I.; et al. Disruption of prion protein-HOP engagement impairs glioblastoma growth and cognitive decline and improves overall survival. *Oncogene* **2015**, *34*, 3305–3314. [CrossRef]

57. Corsaro, A.; Bajetto, A.; Thellung, S.; Begani, G.; Villa, V.; Nizzari, M.; Pattarozzi, A.; Solari, A.; Gatti, M.; Pagano, A.; et al. Cellular prion protein controls stem cell-like properties of human glioblastoma tumor-initiating cells. *Oncotarget* **2016**, *7*, 38638–38657. [CrossRef]

58. Provenzano, L.; Ryan, Y.; Hilton, D.A.; Lyons-Rimmer, J.; Dave, F.; Maze, E.A.; Adams, C.L.; Rigby-Jones, R.; Ammoun, S.; Hanemann, C.O. Cellular prion protein (PrP(C)) in the development of Merlin-deficient tumours. *Oncogene* **2017**, *36*, 6132–6142. [CrossRef]

59. Liang, J.; Ge, F.; Guo, C.; Luo, G.; Wang, X.; Han, G.; Zhang, D.; Wang, J.; Li, K.; Pan, Y.; et al. Inhibition of PI3K/Akt partially leads to the inhibition of PrP(C)-induced drug resistance in gastric cancer cells. *FEBS J.* **2009**, *276*, 685–694. [CrossRef]

60. Liang, J.; Wang, J.; Luo, G.; Pan, Y.; Wang, X.; Guo, C.; Zhang, D.; Yin, F.; Zhang, X.; Liu, J.; et al. Function of PrPC (1-OPRD) in biological activities of gastric cancer cell lines. *J. Cell Mol. Med.* **2009**, *13*, 4453–4464. [CrossRef]

61. Li, C.; Xin, W.; Sy, M.S. Binding of pro-prion to filamin A: By design or an unfortunate blunder. *Oncogene* **2010**, *29*, 5329–5345. [CrossRef] [PubMed]

62. Wang, Y.; Yu, S.; Huang, D.; Cui, M.; Hu, H.; Zhang, L.; Wang, W.; Parameswaran, N.; Jackson, M.; Osborne, B.; et al. Cellular Prion Protein Mediates Pancreatic Cancer Cell Survival and Invasion through Association with and Enhanced Signaling of Notch1. *Am. J. Pathol.* **2016**, *186*, 2945–2956. [CrossRef] [PubMed]

63. Pan, Y.; Zhao, L.; Liang, J.; Liu, J.; Shi, Y.; Liu, N.; Zhang, G.; Jin, H.; Gao, J.; Xie, H.; et al. Cellular prion protein promotes invasion and metastasis of gastric cancer. *FASEB J.* **2006**, *20*, 1886–1888. [CrossRef]

64. De Lacerda, T.C.; Costa-Silva, B.; Giudice, F.S.; Dias, M.V.; de Oliveira, G.P.; Teixeira, B.L.; Dos Santos, T.G.; Martins, V.R. Prion protein binding to HOP modulates the migration and invasion of colorectal cancer cells. *Clin. Exp. Metastasis.* **2016**, *33*, 441–451. [CrossRef] [PubMed]

65. Li, C.; Yu, S.; Nakamura, F.; Pentikainen, O.T.; Singh, N.; Yin, S.; Xin, W.; Sy, M.S. Pro-prion binds filamin A, facilitating its interaction with integrin beta1, and contributes to melanomagenesis. *J. Biol. Chem.* **2010**, *285*, 30328–30339. [CrossRef] [PubMed]

66. Yang, L.; Gao, Z.; Hu, L.; Wu, G.; Yang, X.; Zhang, L.; Zhu, Y.; Wong, B.S.; Xin, W.; Sy, M.S.; et al. Glycosylphosphatidylinositol anchor modification machinery deficiency is responsible for the formation of pro-prion protein (PrP) in BxPC-3 cells and increases cancer cell motility. *J. Biol. Chem.* **2016**, *291*, 6785. [CrossRef] [PubMed]

67. Du, L.; Rao, G.; Wang, H.; Li, B.; Tian, W.; Cui, J.; He, L.; Laffin, B.; Tian, X.; Hao, C.; et al. CD44-positive cancer stem cells expressing cellular prion protein contribute to metastatic capacity in colorectal cancer. *Cancer Res.* **2013**, *73*, 2682–2694. [CrossRef]

68. Lu, W.; Kang, Y. Epithelial-Mesenchymal Plasticity in Cancer Progression and Metastasis. *Dev. Cell* **2019**, *49*, 361–374. [CrossRef]

69. Meslin, F.; Hamai, A.; Gao, P.; Jalil, A.; Cahuzac, N.; Chouaib, S.; Mehrpour, M. Silencing of prion protein sensitizes breast adriamycin-resistant carcinoma cells to TRAIL-mediated cell death. *Cancer Res.* **2007**, *67*, 10910–10919. [CrossRef]

70. Meslin, F.; Conforti, R.; Mazouni, C.; Morel, N.; Tomasic, G.; Drusch, F.; Yacoub, M.; Sabourin, J.C.; Grassi, J.; Delaloge, S.; et al. Efficacy of adjuvant chemotherapy according to Prion protein expression in patients with estrogen receptor-negative breast cancer. *Ann. Oncol.* **2007**, *18*, 1793–1798. [CrossRef]

71. Park, J.Y.; Jeong, J.K.; Lee, J.H.; Moon, J.H.; Kim, S.W.; Lee, Y.J.; Park, S.Y. Induction of cellular prion protein (PrPc) under hypoxia inhibits apoptosis caused by TRAIL treatment. *Oncotarget* **2015**, *6*, 5342–5353. [CrossRef] [PubMed]

72. Vassallo, N.; Herms, J.; Behrens, C.; Krebs, B.; Saeki, K.; Onodera, T.; Windl, O.; Kretzschmar, H.A. Activation of phosphatidylinositol 3-kinase by cellular prion protein and its role in cell survival. *Biochem. Biophys. Res. Commun.* **2005**, *332*, 75–82. [CrossRef] [PubMed]

73. Weise, J.; Sandau, R.; Schwarting, S.; Crome, O.; Wrede, A.; Schulz-Schaeffer, W.; Zerr, I.; Bahr, M. Deletion of cellular prion protein results in reduced Akt activation, enhanced postischemic caspase-3 activation, and exacerbation of ischemic brain injury. *Stroke* **2006**, *37*, 1296–1300. [CrossRef] [PubMed]

74. Lee, J.H.; Yun, C.W.; Lee, S.H. Cellular Prion Protein Enhances Drug Resistance of Colorectal Cancer Cells via Regulation of a Survival Signal Pathway. *Biomol. Ther. (Seoul)* **2018**, *26*, 313–321. [CrossRef] [PubMed]

75. Lee, J.H.; Yoon, Y.M.; Han, Y.S.; Yun, C.W.; Lee, S.H. Melatonin Promotes Apoptosis of Oxaliplatin-resistant Colorectal Cancer Cells Through Inhibition of Cellular Prion Protein. *Anticancer Res.* **2018**, *38*, 1993–2000. [CrossRef] [PubMed]

76. Go, G.; Yun, C.W.; Yoon, Y.M.; Lim, J.H.; Lee, J.H.; Lee, S.H. Role of PrP(C) in Cancer Stem Cell Characteristics and Drug Resistance in Colon Cancer Cells. *Anticancer Res.* **2020**, *40*, 5611–5620. [CrossRef]

77. Lee, J.H.; Han, Y.S.; Yoon, Y.M.; Yun, C.W.; Yun, S.P.; Kim, S.M.; Kwon, H.Y.; Jeong, D.; Baek, M.J.; Lee, H.J.; et al. Role of HSPA1L as a cellular prion protein stabilizer in tumor progression via HIF-1alpha/GP78 axis. *Oncogene* **2017**, *36*, 6555–6567. [CrossRef]

78. Al-Hajj, M.; Clarke, M.F. Self-renewal and solid tumor stem cells. *Oncogene* **2004**, *23*, 7274–7282. [CrossRef]

79. Beck, B.; Blanpain, C. Unravelling cancer stem cell potential. *Nat. Rev. Cancer* **2013**, *13*, 727–738. [CrossRef]

80. Ho, M.M.; Ng, A.V.; Lam, S.; Hung, J.Y. Side population in human lung cancer cell lines and tumors is enriched with stem-like cancer cells. *Cancer Res.* **2007**, *67*, 4827–4833. [CrossRef]

81. Tehranchi, R.; Woll, P.S.; Anderson, K.; Buza-Vidas, N.; Mizukami, T.; Mead, A.J.; Astrand-Grundstrom, I.; Strombeck, B.; Horvat, A.; Ferry, H.; et al. Persistent malignant stem cells in del(5q) myelodysplasia in remission. *N. Engl. J. Med.* **2010**, *363*, 1025–1037. [CrossRef]

82. Dean, M.; Fojo, T.; Bates, S. Tumour stem cells and drug resistance. *Nat. Rev. Cancer* **2005**, *5*, 275–284. [CrossRef] [PubMed]

83. Wang, Q.; Qian, J.; Wang, F.; Ma, Z. Cellular prion protein accelerates colorectal cancer metastasis via the Fyn-SP1-SATB1 axis. *Oncol. Rep.* **2012**, *28*, 2029–2034. [CrossRef] [PubMed]

84. Shibue, T.; Weinberg, R.A. EMT, CSCs, and drug resistance: The mechanistic link and clinical implications. *Nat. Rev. Clin. Oncol.* **2017**, *14*, 611–629. [CrossRef] [PubMed]

85. Cheng, Y.; Tao, L.; Xu, J.; Li, Q.; Yu, J.; Jin, Y.; Chen, Q.; Xu, Z.; Zou, Q.; Liu, X. CD44/cellular prion protein interact in multidrug resistant breast cancer cells and correlate with responses to neoadjuvant chemotherapy in breast cancer patients. *Mol. Carcinog.* **2014**, *53*, 686–697. [CrossRef]

86. Iglesia, R.P.; Prado, M.B.; Cruz, L.; Martins, V.R.; Santos, T.G.; Lopes, M.H. Engagement of cellular prion protein with the co-chaperone Hsp70/90 organizing protein regulates the proliferation of glioblastoma stem-like cells. *Stem Cell Res. Ther.* **2017**, *8*, 76. [CrossRef]

87. Kreso, A.; Dick, J.E. Evolution of the cancer stem cell model. *Cell Stem Cell* **2014**, *14*, 275–291. [CrossRef]

88. Kim, R.J.; Nam, J.S. OCT4 Expression Enhances Features of Cancer Stem Cells in a Mouse Model of Breast Cancer. *Lab. Anim. Res.* **2011**, *27*, 147–152. [CrossRef]

89. Yin, X.; Zhang, B.H.; Zheng, S.S.; Gao, D.M.; Qiu, S.J.; Wu, W.Z.; Ren, Z.G. Coexpression of gene Oct4 and Nanog initiates stem cell characteristics in hepatocellular carcinoma and promotes epithelial-mesenchymal transition through activation of Stat3/Snail signaling. *J. Hematol. Oncol.* **2015**, *8*, 23. [CrossRef]

90. Wang, Y.J.; Herlyn, M. The emerging roles of Oct4 in tumor-initiating cells. *Am. J. Physiol. Cell Physiol.* **2015**, *309*, C709–C718. [CrossRef]

91. Lin, Y.; Yang, Y.; Li, W.; Chen, Q.; Li, J.; Pan, X.; Zhou, L.; Liu, C.; Chen, C.; He, J.; et al. Reciprocal regulation of Akt and Oct4 promotes the self-renewal and survival of embryonal carcinoma cells. *Mol. Cell* **2012**, *48*, 627–640. [CrossRef] [PubMed]

92. Wang, X.Q.; Ongkeko, W.M.; Chen, L.; Yang, Z.F.; Lu, P.; Chen, K.K.; Lopez, J.P.; Poon, R.T.; Fan, S.T. Octamer 4 (Oct4) mediates chemotherapeutic drug resistance in liver cancer cells through a potential Oct4-AKT-ATP-binding cassette G2 pathway. *Hepatology* **2010**, *52*, 528–539. [CrossRef] [PubMed]

93. Sonati, T.; Reimann, R.R.; Falsig, J.; Baral, P.K.; O'Connor, T.; Hornemann, S.; Yaganoglu, S.; Li, B.; Herrmann, U.S.; Wieland, B.; et al. The toxicity of antiprion antibodies is mediated by the flexible tail of the prion protein. *Nature* **2013**, *501*, 102–106. [CrossRef]

94. Coitinho, A.S.; Lopes, M.H.; Hajj, G.N.; Rossato, J.I.; Freitas, A.R.; Castro, C.C.; Cammarota, M.; Brentani, R.R.; Izquierdo, I.; Martins, V.R. Short-term memory formation and long-term memory consolidation are enhanced by cellular prion association to stress-inducible protein 1. *NeuroBiol. Dis.* **2007**, *26*, 282–290. [CrossRef] [PubMed]

95. Jones, M.F.; Hara, T.; Francis, P.; Li, X.L.; Bilke, S.; Zhu, Y.; Pineda, M.; Subramanian, M.; Bodmer, W.F.; Lal, A. The CDX1-microRNA-215 axis regulates colorectal cancer stem cell differentiation. *Proc. Natl. Acad. Sci. USA* **2015**, *112*, E1550–E1558. [CrossRef] [PubMed]

96. Li, N.; Lorenzi, F.; Kalakouti, E.; Normatova, M.; Babaei-Jadidi, R.; Tomlinson, I.; Nateri, A.S. FBXW7-mutated colorectal cancer cells exhibit aberrant expression of phosphorylated-p53 at Serine-15. *Oncotarget* **2015**, *6*, 9240–9256. [CrossRef]

97. Zhang, T.; Kho, D.H.; Wang, Y.; Harazono, Y.; Nakajima, K.; Xie, Y.; Raz, A. Gp78, an E3 ubiquitin ligase acts as a gatekeeper suppressing nonalcoholic steatohepatitis (NASH) and liver cancer. *PLoS ONE* **2015**, *10*, e0118448. [CrossRef]

98. Tsai, Y.C.; Mendoza, A.; Mariano, J.M.; Zhou, M.; Kostova, Z.; Chen, B.; Veenstra, T.; Hewitt, S.M.; Helman, L.J.; Khanna, C.; et al. The ubiquitin ligase gp78 promotes sarcoma metastasis by targeting KAI1 for degradation. *Nat. Med.* **2007**, *13*, 1504–1509. [CrossRef]

99. Chang, Y.W.; Tseng, C.F.; Wang, M.Y.; Chang, W.C.; Lee, C.C.; Chen, L.T.; Hung, M.C.; Su, J.L. Deacetylation of HSPA5 by HDAC6 leads to GP78-mediated HSPA5 ubiquitination at K447 and suppresses metastasis of breast cancer. *Oncogene* **2016**, *35*, 1517–1528. [CrossRef]

100. Cruz, E.; Kayser, V. Monoclonal antibody therapy of solid tumors: Clinical limitations and novel strategies to enhance treatment efficacy. *Biologics* **2019**, *13*, 33–51. [CrossRef]

101. McEwan, J.F.; Windsor, M.L.; Cullis-Hill, S.D. Antibodies to prion protein inhibit human colon cancer cell growth. *Tumour. Biol.* **2009**, *30*, 141–147. [CrossRef] [PubMed]

102. Slamon, D.J.; Leyland-Jones, B.; Shak, S.; Fuchs, H.; Paton, V.; Bajamonde, A.; Fleming, T.; Eiermann, W.; Wolter, J.; Pegram, M.; et al. Use of chemotherapy plus a monoclonal antibody against HER2 for metastatic breast cancer that overexpresses HER2. *N. Engl. J. Med.* **2001**, *344*, 783–792. [CrossRef] [PubMed]

103. Graham, J.; Muhsin, M.; Kirkpatrick, P. Cetuximab. *Nat. Rev. Drug Discov.* **2004**, *3*, 549–550. [CrossRef] [PubMed]

104. Reslan, L.; Dalle, S.; Dumontet, C. Understanding and circumventing resistance to anticancer monoclonal antibodies. *MAbs* **2009**, *1*, 222–229. [CrossRef]

105. Viloria-Petit, A.; Crombet, T.; Jothy, S.; Hicklin, D.; Bohlen, P.; Schlaeppi, J.M.; Rak, J.; Kerbel, R.S. Acquired resistance to the antitumor effect of epidermal growth factor receptor-blocking antibodies in vivo: A role for altered tumor angiogenesis. *Cancer Res.* **2001**, *61*, 5090–5101.

106. Sutherland, M.S.; Sanderson, R.J.; Gordon, K.A.; Andreyka, J.; Cerveny, C.G.; Yu, C.; Lewis, T.S.; Meyer, D.L.; Zabinski, R.F.; Doronina, S.O.; et al. Lysosomal trafficking and cysteine protease metabolism confer target-specific cytotoxicity by peptide-linked anti-CD30-auristatin conjugates. *J. Biol. Chem.* **2006**, *281*, 10540–10547. [CrossRef]

107. Verma, S.; Miles, D.; Gianni, L.; Krop, I.E.; Welslau, M.; Baselga, J.; Pegram, M.; Oh, D.Y.; Dieras, V.; Guardino, E.; et al. Trastuzumab emtansine for HER2-positive advanced breast cancer. *N. Engl. J. Med.* **2012**, *367*, 1783–1791. [CrossRef]

108. Hardy-Werbin, M.; Quiroga, V.; Cirauqui, B.; Romeo, M.; Felip, E.; Teruel, I.; Garcia, J.J.; Erasun, C.; Espana, S.; Cucurull, M.; et al. Real-world data on T-DM1 efficacy - results of a single-center retrospective study of HER2-positive breast cancer patients. *Sci. Rep.* **2019**, *9*, 12760. [CrossRef]

109. Von Minckwitz, G.; Huang, C.S.; Mano, M.S.; Loibl, S.; Mamounas, E.P.; Untch, M.; Wolmark, N.; Rastogi, P.; Schneeweiss, A.; Redondo, A.; et al. Trastuzumab Emtansine for Residual Invasive HER2-Positive Breast Cancer. *N. Engl. J. Med.* **2019**, *380*, 617–628. [CrossRef]

110. Kim, Y.C.; Won, S.Y.; Jeong, B.H. Identification of Prion Disease-Related Somatic Mutations in the Prion Protein Gene (PRNP) in Cancer Patients. *Cells* **2020**, *9*, 1480. [CrossRef]
111. Gill, O.N.; Spencer, Y.; Richard-Loendt, A.; Kelly, C.; Brown, D.; Sinka, K.; Andrews, N.; Dabaghian, R.; Simmons, M.; Edwards, P.; et al. Prevalence in Britain of abnormal prion protein in human appendices before and after exposure to the cattle BSE epizootic. *Acta Neuropathol.* **2020**, *139*, 965–976. [CrossRef] [PubMed]

Publisher's Note: MDPI stays neutral with regard to jurisdictional claims in published maps and institutional affiliations.

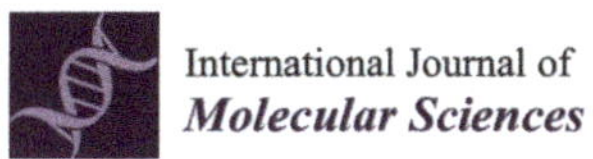

International Journal of
Molecular Sciences

Review

Moving Beyond PARP Inhibition: Current State and Future Perspectives in Breast Cancer

Michela Palleschi [1], Gianluca Tedaldi [2,*], Marianna Sirico [1], Alessandra Virga [2], Paola Ulivi [2,†]
and Ugo De Giorgi [1,†]

1   Department of Medical Oncology, IRCCS Istituto Romagnolo per lo Studio dei Tumori (IRST)
    "Dino Amadori", 47014 Meldola, Italy; michela.palleschi@irst.emr.it (M.P.); marianna.sirico@irst.emr.it (M.S.);
    ugo.degiorgi@irst.emr.it (U.D.G.)
2   Biosciences Laboratory, IRCCS Istituto Romagnolo per lo Studio dei Tumori (IRST) "Dino Amadori",
    47014 Meldola, Italy; alessandra.virga@irst.emr.it (A.V.); paola.ulivi@irst.emr.it (P.U.)
*   Correspondence: gianluca.tedaldi@irst.emr.it; Tel.: +39-0543-739232; Fax: +39-0543-739221
†   These authors contributed equally to this work.

Abstract: Breast cancer is the most frequent and lethal tumor in women and finding the best therapeutic strategy for each patient is an important challenge. PARP inhibitors (PARPis) are the first, clinically approved drugs designed to exploit synthetic lethality in tumors harboring *BRCA1/2* mutations. Recent evidence indicates that PARPis have the potential to be used both in monotherapy and combination strategies in breast cancer treatment. In this review, we show the mechanism of action of PARPis and discuss the latest clinical applications in different breast cancer treatment settings, including the use as neoadjuvant and adjuvant approaches. Furthermore, as a class, PARPis show many similarities but also certain critical differences which can have essential clinical implications. Finally, we report the current knowledge about the resistance mechanisms to PARPis. A systematic PubMed search, using the entry terms "PARP inhibitors" and "breast cancer", was performed to identify all published clinical trials (Phase I-II-III) and ongoing trials (ClinicalTrials.gov), that have been reported and discussed in this review.

Keywords: PARP inhibitors; breast cancer; PARP inhibitor resistance

Citation: Palleschi, M.; Tedaldi, G.; Sirico, M.; Virga, A.; Ulivi, P.; De Giorgi, U. Moving Beyond PARP Inhibition: Current State and Future Perspectives in Breast Cancer. *Int. J. Mol. Sci.* **2021**, *22*, 7884. https://doi.org/10.3390/ijms22157884

Academic Editor: Valentina De Falco

Received: 15 June 2021
Accepted: 20 July 2021
Published: 23 July 2021

Publisher's Note: MDPI stays neutral with regard to jurisdictional claims in published maps and institutional affiliations.

1. Introduction

Breast cancer (BC) is the most commonly diagnosed cancer and the leading cause of cancer death in women [1]. One of the several risk factors for BC development is genetic predisposition, mainly linked to mutations in *BRCA1* and *BRCA2* genes [2]. In the early years of the current millennium, the evidence that *BRCA1/2*-mutant cells could be sensitive to PARP inhibitors (PARPis) emerged and paved the way for new therapeutic opportunities in different tumors, including BC [3–6].

Olaparib (Lynparza®) was the first PARPi approved by the Food and Drug Administration (FDA) in 2014 for the treatment of patients with advanced ovarian cancer and germline or somatic mutations in *BRCA1/2* genes who have been previously treated with three or more lines of chemotherapy [7]. Specific approval for advanced epithelial ovarian, fallopian tube or primary peritoneal cancer maintenance treatments occurred after few years, based on SOLO-1 clinical trial [8]. Selected patients for this treatment are in complete or partial response to first-line platinum-based therapy and carriers of germline or somatic *BRCA1/2* mutations [9]. In 2020 the use of olaparib in combination with bevacizumab was approved as a maintenance treatment for patients with advanced ovarian cancer [10].

Olaparib efficacy was demonstrated also in other tumors, such as pancreatic, prostate and breast cancers [11]. In particular, olaparib was approved by the FDA in 2019 as a maintenance treatment for patients with germline *BRCA1/2* mutations and metastatic pancreatic adenocarcinoma [12], after the demonstration of its efficacy in the multi-center trial POLO [13]. In addition, in 2018, olaparib has been authorized also in the treatment of metastatic BC, germline *BRCA1/2*-mutated and HER2-negative, based on OlympiAD clinical trial [14]. Finally, in 2020, the FDA approved the use of olaparib also in metastatic castration-resistant prostate cancers (mCRPC) with germline or somatic *BRCA1/2* mutations previously treated with Enzalutamide and/or Abiraterone, as demonstrated in the PROfound clinical trial [15,16].

Since 2016, other PARPis were authorized for the treatment of ovarian, prostate and breast cancers. In 2016, FDA approved rucaparib (Rubraca®) for the treatment of *BRCA1/2*-mutated patients with advanced ovarian carcinoma [17]. TRITON2 and TRITON3 clinical trials evaluated the efficacy of rucaparib also in mCRPC patients with germline or somatic *BRCA1/2* mutations [18].

Moreover, niraparib (Zejula®) has recently been approved for maintenance treatment of ovarian cancer patients based on the PRIMA clinical trial [19,20]. Finally, talazoparib (Talzenna®) has been authorized in 2018 for the treatment of BCs with germline *BRCA1/2* mutations based on the EMBRACA clinical trial [21].

Finally, several phase-II and phase-III trials are in progress focusing on the efficacy of new PARPis in the treatment of advanced ovarian cancer, advanced prostate cancer and non-small cell lung cancer (NSCLC) [22]. Current clinical trials on olaparib for the treatment of triple-negative breast cancer (TNBC) are evaluating also the PARPi efficacy in patients who carry mutations in other homologous recombination genes such as *PALB2, ATM, RAD51* [23].

2. DNA Damage Repair and Mechanism of Action of PARPis

DNA instability is an important characteristic of carcinogenesis. Endogenous and exogenous factors are responsible for DNA damage, such as chemical and physical agents including ROS or ultraviolet radiations [24]. Tumor takes advantage of these damaging agents, bypassing cellular repair mechanisms and upsetting correct signaling networks. In addition, chemotherapy and radiation therapy, despite their benefits, can be considered among DNA mutational agents [25].

Once the DNA damage has occurred, cells activate different repair systems. Some of them act on single base mutation, such as base excision repair (BER), nucleotide excision repair (NER) and DNA mismatch repair (MMR) [26–28]. Regarding double-strand breaks (DSB), the damage repair is mediated by nonhomologous end-joining (NHEJ) and homologous recombination repair (HRR) systems [29]. The first DNA repair system directly binds DNA breaks but it can introduce alterations in the sequence. On the contrary, in the second case, there is the necessity to have a strand to guide the repair and replicate the sequence correctly. Schwart et al. demonstrated that downregulation of NHEJ and HRR key components generated fragile sites on DNA, proving the two systems are complementary and essential to maintain chromosome stability [30].

The HRR process consists of three main phases: damage recognition, strand preparation and junction resolution (Figure 1). The damage identification is the result of the activity of the MRN complex, composed of Mre11, Rad50, and Nbs1 proteins, that starts the degradation of the $5'$ strand, in collaboration with CtBP-interacting protein (CtIP) nuclease, recruits and activates the upstream ATM kinase [31–33]. Subsequently, ATM phosphorylates the BRCA1 protein, which is involved in the third process phase [34,35].

The second phase involves Replication Protein A (RPA), a heterotrimeric protein that binds to just generated $3'$ single-strand DNA (ssDNA) [36]. RPA tightly cooperates with RAD51 protein, a DNA-dependent ATPase, binding ssDNA within a gap and pairs linear dsDNA with a small circular ssDNA (called intermediate joint). Meanwhile, RPA promotes the RAD51-DNA link by removing secondary structures and stabilizing the intermediate joint through the non-complementary DNA strand sequestration [37–40].

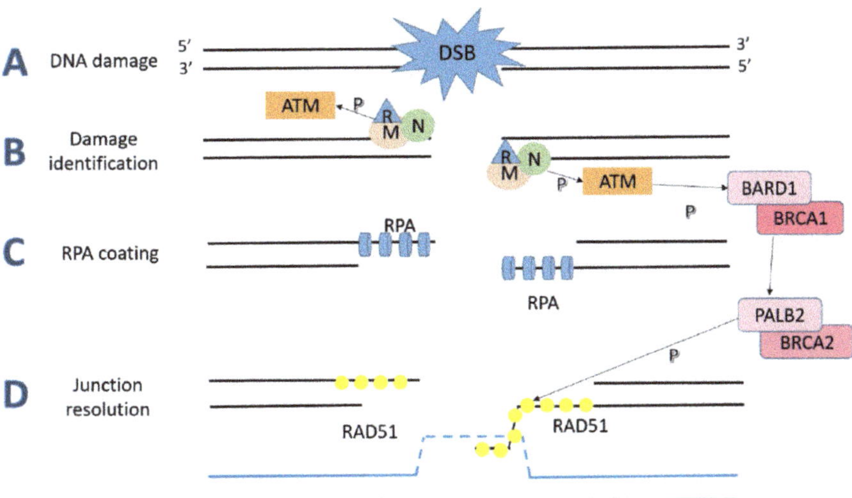

Figure 1. When a DSB occurs in the DNA (**A**), the MRN complex recognizes the damage (**B**) and starts 5′ strand degradation to get a 3′ ssDNA free to be coated by RPA and ATM is activated by phosphorylation (**C**). The interaction of RPA and RAD51 repairs the damage with the support of BRCA1-BARD1 and BRCA2-PALB2 complexes (**D**).

In the third phase, RAD51 recombinase activity is enhanced by the activated BRCA1-BARD1 complex, which interacts with both ssDNA and RAD51 [41,42]. The coiled-coil domain of BRCA1 binds the PALB2-BRCA2 complex, which stabilizes RAD51 nucleofilament and promotes the interaction between RAD51 and RPA [43,44]. CDK12, cyclinD1, CHK1 and many other players cooperate to complete the structural mechanism of the HRR pathway [45,46].

Mutations in genes coding the HRR system components can alter the process and generate a homologous recombination deficiency (HRD). The HRD is a frequent driver of tumorigenesis in many tumors, in particular ovarian, pancreatic, prostate and breast cancers [2,47–50].

However, the HRD signature can be exploited for personalized medicine since it is predictive of the sensitivity to targeted therapy with inhibitors of the poly ADP-ribose polymerase (PARP) enzyme, as well as DNA damaging reagents [5,47,51–53].

PARP is a family of 17 proteins essential in the BER system, involved in DNA single-strand break (SSB) repair [54]. In particular, PARP-1, PARP-2, and PARP-3 are primarily involved in DNA damage repair, otherwise PARP-5a and PARP-5b act in the regulation of mitosis and telomere maintenance for the conservation of chromosome stability [55]. As DNA damage sensors, PARP proteins bind DNA and start the synthesis of a poly (ADP-Ribose) chain (PARylation), which acts as a recruitment signal for other scaffold and regulatory proteins, such as DNA Ligase III (LigIII), DNA polymerase beta (polβ) and X-ray Cross Complementing Protein 1 (XRCC1) to repair the DNA damage (Figure 2A) [56].

The use of PARPis generates an impairment of the BER system and the inability to repair SSBs. In this situation, SSBs are converted into DSBs and the HRR system becomes essential to repair the damage [57,58].

This process can be exploited in cancer therapy through the mechanism of synthetic lethality. Indeed, if the cell is characterized by an HRD, caused by a mutation in *BRCA1/2* or other HRR genes, the treatment with a PARPi generates a genetic instability that leads to cell death (Figure 2B) [3]. Substantially, the PARPi treatment is a targeted therapy that selects HRD cancer cells and brings exclusively them to death [59,60].

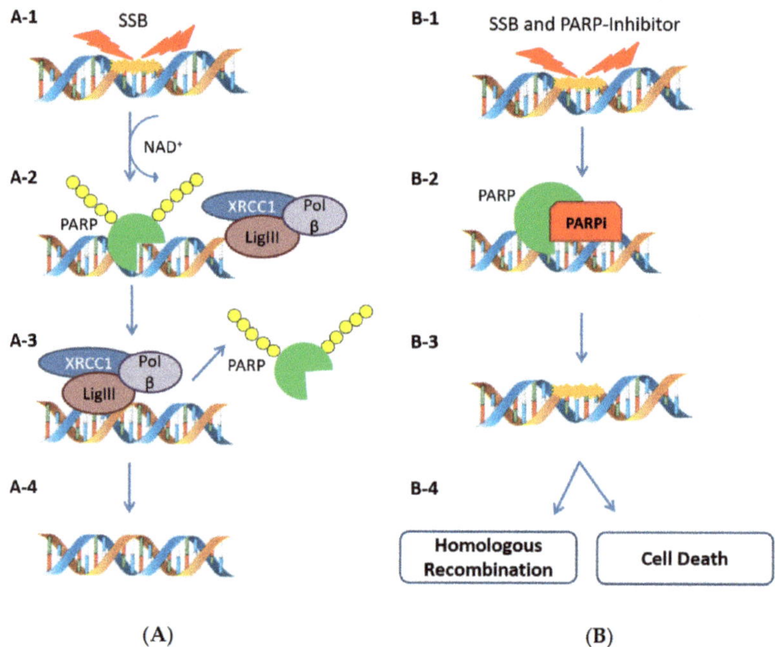

Figure 2. When an SSB occurs (**A-1**), PARP-1 recruits scaffold proteins and forms an ADP-ribose chain on itself (PARylation) using NAD+ (**A-2**). The PARylation also promotes PARP-1 dissociation and leads scaffold proteins to repair the SSB (**A-3** and **A-4**). When an SSB occurs in presence of a PARPi (**B-1**), the PARPi binds the NAD+ binding site in a competitive way (**B-2**). Since the SSB cannot be repaired by PARP enzyme, the HRR system can repair the damage but the coexistence of HRD prevents the repair and induces cell death (**B-3** and **B-4**).

3. Clinical Development of PARPis as a Single Agent in Breast Cancer Treatment

3.1. Advanced and/or Metastatic Breast Cancer

The first phase-I trial evaluating olaparib in metastatic breast cancer (mBC) was reported in 2009 by Fong et al. [53]. This trial enrolled 60 patients with refractory disease to standard therapies or those for whom there were no suitable or effective standard treatments. They were treated using a dose-escalation strategy ranging from 10 mg per day to 600 mg twice per day. Of these patients, nine had mBC and three of them carried a *BRCA1/2* mutation. One of the latter three patients showed a complete response according to the RECIST classification, lasting for 60 weeks. Mainly grade I and II adverse events were reported and included digestive (anorexia, nausea, vomiting) and hematological (anemia, thrombocytopenia) toxicities. These initial results led to the completion of three phase-II trials evaluating olaparib in BC. The first, published in 2010, by Tutt et al. [52], recruited 54 patients with locally advanced or mBC carrying a *BRCA1/2* mutation. They were divided into two cohorts of 27 patients; the first cohort received 400 mg olaparib twice daily (which was found to be the maximum tolerated dose in the phase-I trial) and the second was treated with 100 mg olaparib twice daily (minimum pharmacodynamically effective dose as identified in the phase-I trial). The main goal of this study was to determine the objective response rate (ORR). The 41% (11 out of 27) in the first cohort responded to olaparib therapy, with one patient in complete response and ten in partial response, while the 22% (6 out of 27) in the second cohort, did not exhibit complete response. Again, mainly grade I and II adverse effects, similar to those experienced in the phase-I trial, were reported. The second study by Gelmon et al. [61], treated patients with 400 mg olaparib twice daily but unfortunately could not determine a conclusive ORR after treatment. Finally, Kaufman et al. [62] reported an ORR of 12.9% in their cohort of 62 heavily pretreated patients, all bearing *BRCA1/2*-mutated mBC. In 2017, Robson et al. [14] reported the first randomized phase-III trial comparing olaparib with standard chemotherapy in patients with HER2-negative mBC carrying a *BRCA1/2* germline mutation with resistance to hormone treatment and having received no more than two lines of chemotherapy for the treatment of their mBC. A total of 302 patients were randomized (2:1), 205 received olaparib at a dose of 300 mg twice daily (new dosage form in film-coated tablets) and 97 received clinician's choice chemotherapy (vinorelbine, eribulin or capecitabine). The study met its primary endpoint of a statistically significant increase in progression-free survival (PFS); however, there was no significant difference in overall survival (OS) between the two groups. The most frequently observed hematological adverse reaction was anemia [63]. These results led to the approval of olaparib by the FDA and the European Medicines Agency (EMA), for locally advanced or HER2-negative mBCs, as a monotherapy.

In addition to the inhibition of catalytic activity common to all PARPis, talazoparib has exhibited in vitro "trapping" of the PARP complex at the most important breakage site. In 2017, De Bono et al. [64] published the results of the first phase-I trial evaluating talazoparib in the treatment of advanced solid tumors. The trial consisted of two phases: a dose-escalation phase in which patients received talazoparib at a dose between 0.025 mg/day and 1.1 mg/day and an expansion phase comprising 71 patients receiving talazoparib at the recommended dose of 1 mg/day. Significant PARP inhibition was observed for a dose of 0.60 mg/day, with hematological toxicity reported to be the most frequent adverse effect, although reversible with the suspension of treatment and reduction of doses. The reported ORR was 50% and included one patient with complete response. Further to this, Turner et al. [65] reported the results of the ABRAZO trial: a phase-II trial evaluating talazoparib at a dose of 1 mg/day in patients with locally advanced or mBC and with a *BRCA1/2* germline mutation. In this trial, the population under investigation consisted of 2 cohorts of individuals: the first cohort included patients who had previously been treated with platinum salts while the second cohort enrolled patients without prior exposure to platinum. The ORR for cohort 1 was 21% and 37% for cohort 2, and the most common adverse reaction was anemia. In 2018, Litton et al. [66] conducted the first randomized phase-III trial comparing talazoparib with chemotherapy (capecitabine, eribulin, gemcitabine or vinorelbine) in germline *BRCA1/2*-mutated HER2-negative advanced BC. All patients randomized to talazoparib started at 1 mg once daily. This trial met its primary endpoint of a statistically significant increase in PFS from 5.6 months to 8.6 months and the ORR was found to be twice as high in patients treated with talazoparib (62.6% vs. 27.2%). The most frequent adverse event was haematological toxicity for all talazoparib treatment arms and approximately 50% of patients in the chemotherapy arm. The other predominantly reported adverse reactions were asthenia and nausea. Finally, patients reported an improvement in the quality of life (mean change in QLQ-C30 score), greater with talazoparib (3.0) vs. chemotherapy (−5.4), and the time to deterioration in the quality of life was significantly lengthened in the talazoparib arm. Recently, talazoparib obtained the marketing authorization for the treatment of locally advanced *BRCA1/2*-mutated HER2-negative mBCs. It has been approved for use as a monotherapy in patients who have already received treatment with an anthracycline and/or a taxane, as (neo) adjuvant therapy, for locally advanced or metastatic cancers, unless they were not eligible for this type of treatment. Patients with hormone receptor-positive BC must have previously received hormone therapy or be considered ineligible for hormone therapy to receive this form of treatment.

Rucaparib has been evaluated by Drew et al. [67] in a phase-II trial including 78 patients with *BRCA1/2*-mutated advanced breast or ovarian cancers. This trial had two cohorts, an oral treatment cohort and an intravenous treatment cohort. In each cohort, the first stage of the study included a dose-escalation phase. No objective response was observed in BC patients, both in the oral and intravenous cohorts. In contrast, 44% of patients in the intravenous cohort and 20% in the oral cohort exhibited disease stabilization over a

12-week period. Rucaparib was generally well tolerated. The Hoosier Oncology BRE09-146 group [68] reported a randomized phase-II trial evaluating rucaparib in combination with cisplatin in adjuvant therapy in TNBC or *BRCA1/2*-mutated BC with residual disease after neoadjuvant therapy (anthracyclines and/or taxanes). Patients were randomized (1:1) to receive cisplatin alone or in combination with rucaparib, with the main goal being two-year disease-free survival. A total of 128 patients were randomized, 22% of them carrying a *BRCA1/2* germline mutation. The toxicity profile was similar in both arms and the addition of rucaparib did not show significant improvement in two-year disease-free survival, regardless of *BRCA1/2* mutational status.

Sandhu et al. reported a phase-I trial evaluating niraparib in patients with sporadic advanced solid tumors or with a *BRCA1/2* mutation. The maximum tolerated dose was 300 mg of niraparib orally per day. Sixty patients were included in phase-I dose-escalation including 12 patients with mBC [20]. Among these patients, four carried a germline *BRCA1/2* mutation and 2/4 had a partial response to treatment. The toxicity was predominantly hematological with few grade 3 adverse effects. A phase-III randomized trial (the BRAVO trial, NCT01905592) is currently underway, comparing niraparib at a dose of 300 mg per day with chemotherapy of the physician's choice in patients with locally advanced or metastatic HER2-negative BC, carrying a *BRCA1/2* germline mutation. Clinical trials on the use of PARPis in the treatment of mBC are summarized in Table 1.

Table 1. Clinical trials with PARP inhibitors in the treatment of mBC.

Trial	Study Design	No. of Pts	Phase	Patients' Population	Primary Endpoint	Results	Approval
OlympiAD [63]	Olaparib vs. PCT	302	III	Advanced/metastatic gBRCA ≤2 prior lines	PFS	7.0 vs. 4.2 months	FDA/EMA approved
BROCADE3 [69]	CP + veliparib/placebo vs. Temozolomide + veliparib	337	II	Metastatic gBRCA ≤2 prior lines	PFS	14.1 vs. 12.3 months	-
EMBRACA [21]	Talazoparib vs. PCT	431	III	Advanced/metastatic gBRCA ≤3 prior lines	PFS	8.6 vs. 5.6 months	FDA/EMA approved

Abbreviations: C (carboplatin); P (paclitaxel); gBRCA (germline *BRCA1/2* mutation); PFS (progression free-survival); PCT (physician's choice therapy).

3.2. Neoadjuvant Setting

The I-SPY 2 trial [70] was one of the first trials to evaluate the PARPi veliparib as a neoadjuvant treatment for localized BC. This trial included patients with stage II or III BC who had never received cytotoxic treatment for their BC. A total of 72 patients were randomized to receive veliparib-carboplatin in addition to standard neoadjuvant chemotherapy (weekly paclitaxel followed by doxorubicin plus cyclophosphamide), and 44 received standard therapy. Among the patients treated with a PARPi, 17% of them carried a germline *BRCA1/2* mutation. The addition of the veliparib-carboplatin combination to the standard neoadjuvant chemotherapy regimen allowed a doubling of the complete response rate in the TNBC subtype (51% vs. 26%). Another trial, confirming the effectiveness of PARPis in the neoadjuvant setting was the BrighTNess trial [71], which included women with stage II or III TNBC with or without a *BRCA1/2* germline mutation. Patients were randomized (2:1:1) to receive one of the following three treatment regimens: either paclitaxel (80 mg/m2, weekly for a total of 12 doses) plus carboplatin (area under the curve (AUC) 6 every three weeks for a total of four cycles) and veliparib (50 mg orally twice daily), either paclitaxel plus carboplatin and a placebo of veliparib or paclitaxel plus a placebo of carboplatin and a placebo of veliparib. After the weekly paclitaxel sequence, all patients received doxorubicin-cyclophosphamide chemotherapy every two or three weeks. A total of 634 patients were randomized to one of the three arms of the study, among them 15% carried a *BRCA1/2* germline mutation. Overall, the histologic complete response rate was significantly increased in the carboplatin-veliparib paclitaxel arm compared to the paclitaxel alone arm (53% vs. 31%).

In contrast, the addition of veliparib to the carboplatin-paclitaxel combination did not increase the proportion of patients achieving a complete tumor response. This trial corroborates the previous results on the benefit of carboplatin as a neoadjuvant treatment in TNBC. However, the presence of a *BRCA1/2* germline mutation did not appear to be associated with a greater benefit from the use of platinum and/or veliparib.

Furthermore, the GeparOLA study evaluated neoadjuvant olaparib in patients with non-metastatic HER2-negative T2-T4 or T1c BCs with lymph node involvement with either a *BRCA1/2* mutation (germline or somatic) or a high HRD score. Patients were randomized to receive chemotherapy combining weekly paclitaxel 80 mg/m2 and olaparib 100 mg twice daily or a weekly combination of paclitaxel 80 mg/m2 and carboplatin AUC 2, for 12 weeks. All patients received chemotherapy with epirubicin/cyclophosphamide thereafter. A total of 107 patients were randomized, 72% of which had TNBCs and 60% of patients had a *BRCA1/2* mutation. The histologic complete response rates were 55% with olaparib and 48% with carboplatin; however, the study could not rule out a 55% histological complete response rate in the olaparib arm, which was the primary outcome, but the results seemed more favorable in the subgroup of patients aged over 40 years or with hormone receptor-positive BC [72].

Regarding the use of talazoparib in this setting, Litton et al. [73] reported the results of a pilot study evaluating talazoparib as a neoadjuvant monotherapy in the treatment of localized (stage I to III) HER2-negative *BRCA1/2*-mutated BCs. Twenty patients (including 15 with a TNBC) were randomized to receive talazoparib at a dose of 1 mg/day for six months. The main objective was the histological complete response rate at six months, assessed on the surgical specimen. A total of 19 patients completed the six months of treatment and ten of them (53%) achieved a complete tumor response. The main toxicity was found to be hematological with grade I to III anemia leading to a need for dose reduction in nine patients and to erythrocyte transfusions in eight patients. The most commonly reported non-hematological adverse reaction was nausea, and all of these toxicities were easily managed with appropriate supportive care. Despite its small patients sample, this trial is the first to demonstrate the effectiveness of targeted therapy in the neoadjuvant setting in patients with *BRCA1/2* mutations without the addition of chemotherapy.

3.3. Adjuvant Setting

OlympiA is a phase-III, randomized, double-blind, placebo-controlled trial that evaluated olaparib as an adjuvant monotherapy in localized, HER2-negative *BRCA1/2*-mutated BCs [74]. The recruited patients had to have completed local treatment and adjuvant or neoadjuvant chemotherapy. Eligible patients should have had a TNBC (pT2 or pN1 for patients operated on straight away or residual disease for those receiving neoadjuvant chemotherapy) or expressing hormone receptors and should be without *HER2* gene amplification, pN2 for patients treated with adjuvant chemotherapy or high-risk residual disease after neoadjuvant chemotherapy, as defined by the CSP + EG score (clinical stage and post-treatment pathological stage incorporating estrogen receptor status and tumor grade). They were randomized in a 1:1 ratio to receive 300 mg olaparib twice daily or a placebo for 12 months. The main goal was invasive disease-free survival. Among the high-risk early BC patients with germline *BRCA1/2* mutations, 1 year of adjuvant olaparib was found to significantly reduce the risk of recurrence and prevent progression to metastasis (85.9% in the olaparib group and 77.1% in the placebo) [75]. Clinical trials on the use of PARPis in the neoadjuvant/adjuvant treatment of BC are summarized in Table 2.

Table 2. Clinical trials with PARP inhibitors in the neoadjuvant/adjuvant treatment of BC.

Trial	Study Design	No. of Pts	Phase	Patients' Population	Primary Endpoint	Results
OlympiA [75]	Olaparib vs. placebo	1836	III	Early-stage gBRCA adjuvant therapy	IDFS	85.9% vs. 77.1%
BrighTNess [71]	CP + veliparib or CP + placebo or P + placebo + placebo→AC	634	III	Stage II or III TNBC gBRCA neoadjuvant therapy	pCR	58% vs. 53% vs. 31%
I-SPY 2 [70]	CP + veliparib/placebo→AC	116	II	Stage II or III TNBC neoadjuvant therapy	pCR	51% vs. 26%

Abbreviations: AC (doxorubicin plus cyclophosphamide); C (carboplatin); P (paclitaxel); gBRCA (germline *BRCA1/2* mutation); IDFS (invasive disease-free survival); pCR (pathological complete response).

4. Combination Strategies with PARPis in BC Treatment

4.1. PARPis and Chemotherapy

The mechanism of action of the majority of chemotherapeutic drugs is the damage to the DNA of cancer cells. PARPis alter DNA repair mechanisms and may be used as chemo-sensitizers. This hypothesis was examined by the trials cited above, as the BrighTNess, which combined veliparib with carboplatin and paclitaxel, or the GeparOLA, which combined olaparib and paclitaxel. However, despite obtaining promising results at the preclinical data analysis stage, the concomitant combination of a PARPi with a cytotoxic agent, including temozolomide, platinum salts, gemcitabine, or topoisomerase inhibitors, has been shown to be tricky in terms of toxicities, with an increase of hematological toxicity, leading to control by PARPi or chemotherapy dose reduction [76]. Regarding the effectiveness, the benefit of these combinations still remains to be established in populations with *BRCA1/2* or HRD mutations as well as in patients without such anomalies.

Veliparib (ABT-888) has primarily been evaluated in combination with platinum salt chemotherapy in advanced mBC. Indeed, its low activity in terms of "trapping" of PARP-1 allows for its development in combination with chemotherapy. Several phase-I trials have been reported evaluating the maximum tolerated dose and the potential efficacy of veliparib in solid tumors, alone [77] or in combination with other systemic treatments [78]. In 2017, Han et al. [79] reported the results of the BROCADE trial, a randomized phase-II trial evaluating the combination of veliparib with carboplatin and paclitaxel in the treatment of locally advanced or *BRCA1/2*-mutated HER2-negative BC. A third arm studied the combination of veliparib with temozolomide, an alkylating agent, that has demonstrated potential synergistic efficacy in combination with PARPis [80]. In each arm, veliparib was administered at low doses and both intermittently and concomitantly with the administration of chemotherapy (carboplatin or temozolomide). No significant difference was observed in the mean PFS with the addition of veliparib to carboplatin/paclitaxel. In contrast, there was a significant increase in the ORR in the experimental arms (ORR 77.8% vs. 61.3%, $p = 0.027$). The ORR of 61.3% in the carboplatin and paclitaxel arm confirmed previous data on the efficacy of platinum salts in the treatment of *BRCA1/2* mutated cancers [81]. As already described for the other PARPis, hematological adverse events such as neutropenia and thrombocytopenia were the most reported. Moreover, disappointing results were reported in the veliparib plus temozolomide arm with an ORR of 28.6%. In 2020, Dieras et al. [69] reported the results of the BROCADE3 trial, a randomized (2:1) phase-III double-blind trial, evaluating veliparib vs. placebo in combination with carboplatin (administered every 3 weeks) and paclitaxel (administered weekly) in the treatment of locally advanced or metastatic HER2-negative BC with a *BRCA1/2* germline mutation. Veliparib was administered orally at a low dose (120 mg twice daily) from day 2 to day 5, carboplatin with AUC 6 on day 1 every three weeks and paclitaxel on a weekly basis. If chemotherapy was discontinued, veliparib was continued as a full-dose maintenance monotherapy (300–400 mg/day). A crossover was planned for progression to veliparib monotherapy in patients receiving a placebo. The addition of veliparib showed an increase in the median PFS from 13.5 to 19.3 months after Blinded Central Review.

4.2. PARPis and Immunotherapy

Checkpoint inhibitors, such as anti-CTLA4 and anti-PD-1/PD-L1, seem to be more effective against cancer with a high mutagenic burden, probably because they have an increased propensity to produce neo-antigens for immune activations [82]. Therefore, tumors with *BRCA1/2* mutations or BRCAness, given the potentially high mutagenic burden, might be particularly responsive to checkpoint inhibitors [83].

Moreover, preclinical studies demonstrated an increased expression of immunologic markers such as tumor-infiltrating lymphocytes (TILs), especially in *BRCA1/2*-mutated TNBC [84]. In addition, in this specific subtype, olaparib has been shown to stimulate PD-L1 expression in tumor cells, improving anti-PD-1 antibody efficacy. Considering these data, clinical trials are underway to assess the combination of PARPis and immunological checkpoint inhibitors [85]. MEDIOLA is a phase-II trial evaluating the combination of olaparib plus durvalumab (anti-PD-L1 monoclonal antibody) in mBC HER2-negative with a *BRCA1/2* germline mutation. Patients were randomized to receive olaparib as monotherapy at a dose of 300 mg, twice daily for four weeks, followed by durvalumab 1500 mg intravenously for four-week cycles, with disease assessment every eight weeks. Domchek et al. published the results concerning the first 25 patients who showed a disease control rate at 24 weeks of nearly 50% without differences, according to hormonal or mutational status, and without an increase in toxicities during dual therapy [86]. MEDIOLA is the first study to report encouraging results for hormone receptor (HR)-positive germline *BRCA1/2*-mutated mBC treated with PARPis in combination with immune checkpoint inhibitors. Similarly, TOPACIO [87] is a phase-I/II trial evaluating the combination of niraparib with pembrolizumab (anti-PD-1 monoclonal antibody), particularly in metastatic TNBCs and in ovarian cancers. A total of 55 women were included in the BC cohort to receive niraparib at the recommended dose in phase I, i.e., 200 mg per day, in combination with pembrolizumab (200 mg intravenous on day 1/21). Of the 47 evaluable patients, 21% presented an ORR with five patients in complete response and five in partial response. Among 15 patients with a *BRCA1/2* germline mutation, 32% showed an ORR.

4.3. Other Innovative Combinations

Recent preclinical studies are interested in combining PARPis with other molecular targeted therapies to interfere with oncogenic pathways such as VEGF, IGF, PI3K and EGFR. For instance, the PI3K/mTOR signaling pathway is essential, particularly in detecting DNA DSBs, and it might be involved in the expression of BRCA1 and BRCA2 [76]. Therefore, PI3K inhibition could potentially weaken the HRR mechanism, resulting in a "BRCAness" tumor phenotype, regardless of *BRCA1/2* mutational status, increasing the effect of PARPis. Accordingly, early-phase clinical trials have been initiated evaluating the combination of PI3K or mTOR inhibitors with PARPis. In addition, several studies are currently underway evaluating treatments targeting molecules involved in cell cycle regulation or DNA repair, in association with PARPis [88]. These trials included a phase-II study of olaparib and ATR inhibitor (NCT02264678), a phase-IB study of olaparib and WEE1 inhibitor (NCT02511795), and a phase-II study randomizing olaparib as monotherapy vs. olaparib and WEE1 or ATR inhibitor in TNBC (VIOLETTE test, NCT03330847). A deeper knowledge of the biology of the HR-positive mBC with germline *BRCA1/2* mutations is necessary to define further studies with PARPis and immune checkpoint inhibitors, based on known lower response to anti-PD-L1 drugs and the potential sensitivity to CDK4/6 inhibitors; further studeis are needed to investigate the combination or the sequence of PARP inhibitors and CDK4/6 inhibitors (e.g., olaparib, fulvestrant and palbociclib–NCT03685331). Ongoing clinical trials of PARPis in combination with molecular targeted therapies are summarized in Table 3.

Table 3. Selected ongoing trials with PARP inhibitors in combinations with molecular targeted therapies.

Clinical Trial Identifier	Study Design	Intervention/s	Setting	Primary Endpoint	Phase	Status
NCT02264678	330 Participants, Interventional, Parallel Assignment, open-label, Non-Randomized, Multi-center	Olaparib + Ceralasertib	Advanced solid malignancies not considered appropriate for further standard treatment	AE and SAE	I	Recruiting
NCT02511795	128 Participants, Interventional, Parallel Assignment, open-label, Non-randomized	Olaparib + Adavosertib	Refractory solid tumor	DLT, MTD, TEAEs	Ib	Completed
NCT03330847 (VIOLETTE)	273 Participants Interventional, Parallel Assignment, open-label, Randomized, Multi-center	Olaparib + Ceralasertib Olaparib + Adavosertib	Second or third line	PFS, ORR, DoR	II	Active, Not Recruiting
NCT03685331 (HOPE)	54 Participants, Interventional, Sequential Assignment, open-label, Non randomized	Olaparib + Palbociclib + Fulvestrant	First, second and third line	PFS	I/II	Recruiting
NCT01905592 (BRAVO)	215 Participants, Interventional, Parallel Assignment, open-label, Randomized, Multi-center	Niraparib	First, second line and third line	PFS	III	Active, Not recruiting

Abbreviations: Adverse Events, AE; Dose Limiting Toxicity, DLT; Duration of response, DoR; Maximum Tolerated Dose, MTD; Objective Response Rate, ORR; Progression Free Survival, PFS; Serious Adverse Events, SAEs; Treatment-emergent adverse events, TEAEs. The information was extracted from www.clinicaltrials.gov (accessed on 15 July 2021).

5. Acquired Resistance to PARPis

As PARPi therapy entered the clinical practice, resistance mechanisms to the treatment have emerged. Mutations in *BRCA1* and *BRCA2* genes are excellent targets for PARPis, but it has been shown that in 40–70% of patients the therapy is not effective [61].

The resistance to PARPis can be innate, when PARPis are ineffective from the beginning of the treatment for the presence of intrinsic resistance mechanisms, or acquired, when PARPis become ineffective after an initial benefit for the patient [89].

While innate resistance to PARPis is poorly known, several mechanisms of acquired resistance have been observed during the treatment with PARPis.

5.1. Restoration of BRCA1 and BRCA2 Functionality

Among the resistance mechanisms occurring during the treatment with PARPis, the restoration of BRCA1 and BRCA2 functionality by reversion mutations is the most common one [60,90–99].

Quigley et al. demonstrated the acquisition of a multi-nucleotide deletion that removes pathogenic mutation in *BRCA2* gene, restoring the correct open reading frame (ORF). Thus after some months of treatment, the recovered protein brings tumor cells to the loss of sensitivity to the PARPi Talazoparib [100]. Another group identified a deleterious germline mutation in *BRCA2*, whose carrier patient was treated with carboplatin and Rucaparib. After therapies, they identified in cfDNA twelve new somatic mutations that have occurred. Six new variants determined the restoration of the correct ORF [101]. Secondary restoration mutations have been observed also in other genes involved in the HRR system, for example in *RAD51B* and *RAD51C* [102].

5.2. Hypomorphic Forms of BRCA1

Another resistance mechanism observed during PARPi treatment is the partial restoration of HRR through increased activity of hypomorphic mutant BRCA1/2 proteins [103,104]. In particular, it has been observed that the knockout of 53BP1, a protein involved in DSB repair, can cause partial rescuing of the HRR in hypomorphic BRCA1 cells making them resistant to PARPi treatment [105].

5.3. Epigenetic Changes in HRR Genes

PARPi resistance can be the result of epigenetic changes in HRR genes. In particular, promoter hypermethylation of genes such as *BRCA1/2* determines a reduced expression of the corresponding mRNAs that results in HRD and PARPi sensitivity. On the opposite, the demethylation of these genes is associated with the restoration of protein expression and resistance to PARPi treatment [102,106,107].

5.4. Loss of End Resection Regulation

Another player involved in the PARPi resistance mechanism is the 53BP1 protein, whose loss in mice with a *BRCA1* mutation determines resistance to PARPi treatment restoring the HRR system. Furthermore, this mechanism seems to be dependent on ATM, another possible target in PARPi-resistant tumors [108]. PARPi resistance can also occur in presence of *BRCA1* mutations disrupting the N-term RING domain, taking advantage of the residual DNA repair activity of the protein [109,110].

Moreover, another player involved in the PARPi resistance mechanism is the PTEN protein, since the concomitant presence of PTEN loss and *BRCA1* mutation rewires the functionality of the HRR system [111].

5.5. Restoration of PARP-1 Activity

The PARP-1 enzyme is another player in PARPi resistance mechanisms, since mutations in the Zinc Finger Domain (ZFD) of PARP-1 that abolish DNA binding cause PARPi resistance [112]. Previous studies have demonstrated that a decrease in PARP-1 expression during PARPi treatment can promote the onset of a resistance mechanism [113]. In a different study, phosphorylation of PARP-1 by receptor tyrosine kinase c-Met has been shown to determine an increase in its enzymatic activity and reduce the binding to PARPi, thereby rendering cancer cells resistant to the treatment [114].

Another important player is the poly(ADP-ribose glycohydrolase (PARG), an enzyme the degrades the ADP-ribose chain synthetized by PARP-1. The depletion of PARG leads to the partial restoration of PARP-1 activity, inducing less sensitivity to PARPis [115].

5.6. NHEJ Suppression

Different repair pathways act during the cell cycle in the repair of DNA damage: NHEJ and HRR are complementary arms of the same system. When DNA strand breaks occur, a precise balance between these two regulatory systems is maintained by competition in binding broken strand ends, by Ku complex or MRN complex, respectively. The miR-662 can induce resistance to PARPis and platinum in *BRCA1*-mutant cells by targeting the Ku complex and restoring HRR. Indeed, in *BRCA1*-mutated ovarian tumors, the overexpression of miR-622 is associated with a reduced response to PARPi and platinum therapy [116,117].

5.7. Replication Fork Protection

Multiple studies showed that HRD tumor cells can develop PARPi resistance through the protection of the replication forks during DNA replication [118,119]. In particular, BRCA1/2-deficient cells have been shown to be able of reducing the recruitment of nucleases, such as MRE11 and MUS81, to the stalled forks, becoming resistant to the PARPi treatment without restoring the HRR [120,121].

5.8. Drug Concentration

The *ABCB1* gene encodes a transmembrane transporter P-glycoprotein that pumps out from cells a wide range of xenobiotic compounds, including drugs such as PARPis [122,123]. The overexpression of *ABCB1* has been associated with PARPi resistance [124]. P-glycoprotein inhibitors are able to restore the PARPi sensitivity in BRCA1-deficient cells [125].

6. Conclusions

Going forward, it has been well established that PARPis should be a component of the therapeutic strategy for BC arising in *BRCA1/2* mutation carriers. Furthermore, their application will likely move beyond metastatic setting to the adjuvant and neoadjuvant settings. In both OlympiAD and EMBRACA trials, PARPis demonstrated survival and quality of life benefit compared to chemotherapy. However, the side effects associated with platinum salts are known and are considered clinically significant. The PARPis talazoparib and olaparib as maintenance therapy, after initial cytotoxic chemotherapy (with or without platinum), have not been clearly evaluated, although patients with stable disease after chemotherapy may be included in EMBRACA. On the other hand, this strategy has been clearly provided in the BROCADE3 trial and may have played a major role in the PFS benefit observed in this trial.

Nowadays, both talazoparib and olaparib are registered in the treatment of metastatic or locally advanced *BRCA1/2*-mutated HER2-negative BC. However, patients must have received prior treatment with anthracyclines and taxanes in an adjuvant, neoadjuvant or metastatic setting. Moreover, PARPis might be used from the first line of cytotoxic treatment, but in tumors expressing hormone receptors, prior hormone therapy must have been administered or the patients must not be candidates for it. At this point, it is noteworthy to discuss different clinical strategies: in TNBCs with a *BRCA1/2* germline mutation and PD-L1 expression, which strategy should be adopted between taxanes-atezolizumab combination and PARPis? The OS benefit, in the IMpassion 130 study, favoring the chemo-immunotherapy combination does not seem to be impacted by the *BRCA1/2* mutational status [82,126]. On the other hand, in hormone receptor-positive BC patients, who have already been exposed to hormone therapy but not in combination with CDK4/6 inhibitors, should PARPis be preferred over a combination of hormone therapy plus cell cycle inhibitors? Given the increased OS associated with the latter combination [83,127,128], this is probably the most reasonable option. In the absence of visceral crisis, PARPis might be useful when hormone therapy is no longer effective. The is no direct comparison beteween PARPis and platinum salt-based chemotherapy: the efficacy of platinum salts in germline *BRCA1/2*-mutated patients is already underscored in the metastatic setting [129]. Finally, reinforcing the role of already approved PARPis, in a meta-analysis, Schettini et al. showed that PARPi regimens are correlated with an overall reduction in the instantaneous risk of progression of 41%, and about 14% reduction in the instantaneous risk of death. In addition, based on the results of subgroup analysis, they found an association between PARPis and PFS in ovarian cancer, prostate cancer, pancreatic cancer, melanoma and small-cell lung cancer, but also a statistically significant PFS improvement in BC, as it was already described in olaparib and talazoparib pivotal trial [130].

Author Contributions: M.P. and M.S. wrote the clinical parts; G.T. and A.V. wrote the biological parts; U.D.G. and P.U. supervised and reviewed the work. All authors have read and agreed to the published version of the manuscript.

Funding: This research received no external funding.

Institutional Review Board Statement: Not applicable.

Informed Consent Statement: Not applicable.

Conflicts of Interest: M.P. has received advisory board fees from Novartis; U.D.G. has received advisory board or consultant fees from Merck Sharp & Dohme, Bristol Myers Squibb, Janssen, Astellas, Sanofi, Bayer, Pfizer, Ipsen, Novartis, and Pharmamar and institutional research grants from Astrazeneca, Sanofi, and Roche.

References

1. Sung, H.; Ferlay, J.; Siegel, R.L.; Laversanne, M.; Soerjomataram, I.; Jemal, A.; Bray, F. Global cancer statistics 2020: GLOBOCAN estimates of incidence and mortality worldwide for 36 cancers in 185 countries. *CA Cancer J. Clin.* **2021**, *71*, 209–249. [CrossRef]

2. Angeli, D.; Salvi, S.; Tedaldi, G. Genetic Predisposition to Breast and Ovarian Cancers: How Many and Which Genes to Test? *Int. J. Mol. Sci.* **2020**, *21*, 1128. [CrossRef]

3. Farmer, H.; McCabe, N.; Lord, C.J.; Tutt, A.N.J.; Johnson, D.A.; Richardson, T.B.; Santarosa, M.; Dillon, K.J.; Hickson, I.; Knights, C.; et al. Targeting the DNA repair defect in BRCA mutant cells as a therapeutic strategy. *Nature* **2005**, *434*, 917–921. [CrossRef]

4. Haince, J.-F.; Rouleau, M.; Hendzel, M.J.; Masson, J.-Y.; Poirier, G.G. Targeting poly(ADP-ribosyl)ation: A promising approach in cancer therapy. *Trends Mol. Med.* **2005**, *11*, 456–463. [CrossRef] [PubMed]

5. De Soto, J.A.; Wang, X.; Tominaga, Y.; Wang, R.-H.; Cao, L.; Qiao, W.; Li, C.; Xu, X.; Skoumbourdis, A.P.; Prindiville, S.A.; et al. The inhibition and treatment of breast cancer with poly (ADP-ribose) polymerase (PARP-1) inhibitors. *Int. J. Biol. Sci.* **2006**, *2*, 179–185. [CrossRef] [PubMed]

6. Cortesi, L.; Rugo, H.S.; Jackisch, C. An Overview of PARP Inhibitors for the Treatment of Breast Cancer. *Target. Oncol.* **2021**, *16*, 255–282. [CrossRef] [PubMed]

7. Lynparza (Olaparib) Drug Approval Package. Available online: https://www.accessdata.fda.gov/drugsatfda_docs/nda/2014/206162Orig1s000TOC.cfm (accessed on 15 July 2021).

8. Moore, K.; Colombo, N.; Scambia, G.; Kim, B.-G.; Oaknin, A.; Friedlander, M.; Lisyanskaya, A.; Floquet, A.; Leary, A.; Sonke, G.S.; et al. Maintenance Olaparib in Patients with Newly Diagnosed Advanced Ovarian Cancer. *N. Engl. J. Med.* **2018**, *379*, 2495–2505. [CrossRef] [PubMed]

9. Fong, P.C.; Yap, T.A.; Boss, D.S.; Carden, C.P.; Mergui-Roelvink, M.; Gourley, C.; De Greve, J.; Lubinski, J.; Shanley, S.; Messiou, C.; et al. Poly(ADP)-ribose polymerase inhibition: Frequent durable responses in BRCA carrier ovarian cancer correlating with platinum-free interval. *J. Clin. Oncol.* **2010**, *28*, 2512–2519. [CrossRef]

10. Arora, S.; Balasubramaniam, S.; Zhang, H.; Berman, T.; Narayan, P.; Suzman, D.; Bloomquist, E.; Tang, S.; Gong, Y.; Sridhara, R.; et al. FDA Approval Summary: Olaparib Monotherapy or in Combination with Bevacizumab for the Maintenance Treatment of Patients with Advanced Ovarian Cancer. *Oncologist* **2021**, *26*, e164–e172. [CrossRef] [PubMed]

11. Haddad, F.G.; Karam, E.; Moujaess, E.; Kourie, H.R. Poly-(ADP-ribose) polymerase inhibitors: Paradigm shift in the first-line treatment of newly diagnosed advanced ovarian cancer. *Pharmacogenomics* **2020**, *21*, 721–727. [CrossRef]

12. FDA. Approves Olaparib for gBRCAm Metastatic Pancreatic Adenocarcinoma. Available online: https://www.fda.gov/drugs/resources-information-approved-drugs/fda-approves-olaparib-gbrcam-metastatic-pancreatic-adenocarcinoma (accessed on 15 July 2021).

13. Golan, T.; Hammel, P.; Reni, M.; Van Cutsem, E.; Macarulla, T.; Hall, M.J.; Park, J.-O.; Hochhauser, D.; Arnold, D.; Oh, D.-Y.; et al. Maintenance Olaparib for Germline BRCA-Mutated Metastatic Pancreatic Cancer. *N. Engl. J. Med.* **2019**, *381*, 317–327. [CrossRef]

14. Robson, M.; Im, S.-A.; Senkus, E.; Xu, B.; Domchek, S.M.; Masuda, N.; Delaloge, S.; Li, W.; Tung, N.; Armstrong, A.; et al. Olaparib for Metastatic Breast Cancer in Patients with a Germline BRCA Mutation. *N. Engl. J. Med.* **2017**, *377*, 523–533. [CrossRef]

15. de Bono, J.; Mateo, J.; Fizazi, K.; Saad, F.; Shore, N.; Sandhu, S.; Chi, K.N.; Sartor, O.; Agarwal, N.; Olmos, D.; et al. Olaparib for Metastatic Castration-Resistant Prostate Cancer. *N. Engl. J. Med.* **2020**, *382*, 2091–2102. [CrossRef]

16. Hussain, M.; Mateo, J.; Fizazi, K.; Saad, F.; Shore, N.; Sandhu, S.; Chi, K.N.; Sartor, O.; Agarwal, N.; Olmos, D.; et al. Survival with Olaparib in Metastatic Castration-Resistant Prostate Cancer. *N. Engl. J. Med.* **2020**, *383*, 2345–2357. [CrossRef]

17. FDA. Grants Accelerated Approval to New Treatment for Advanced Ovarian Cancer. Available online: https://www.fda.gov/news-events/press-announcements/fda-grants-accelerated-approval-new-treatment-advanced-ovarian-cancer (accessed on 15 July 2021).

18. FDA. Grants Accelerated Approval to Rucaparib for BRCA-Mutated Metastatic Castration-Resistant Prostate Cancer. Available online: https://www.fda.gov/drugs/fda-grants-accelerated-approval-rucaparib-brca-mutated-metastatic-castration-resistant-prostate (accessed on 15 July 2021).

19. González-Martín, A.; Pothuri, B.; Vergote, I.; DePont Christensen, R.; Graybill, W.; Mirza, M.R.; McCormick, C.; Lorusso, D.; Hoskins, P.; Freyer, G.; et al. Niraparib in Patients with Newly Diagnosed Advanced Ovarian Cancer. *N. Engl. J. Med.* **2019**, *381*, 2391–2402. [CrossRef]

20. Sandhu, S.K.; Schelman, W.R.; Wilding, G.; Moreno, V.; Baird, R.D.; Miranda, S.; Hylands, L.; Riisnaes, R.; Forster, M.; Omlin, A.; et al. The poly(ADP-ribose) polymerase inhibitor niraparib (MK4827) in BRCA mutation carriers and patients with sporadic cancer: A phase 1 dose-escalation trial. *Lancet. Oncol.* **2013**, *14*, 882–892. [CrossRef]

21. Litton, J.K.; Hurvitz, S.A.; Mina, L.A.; Rugo, H.S.; Lee, K.-H.; Gonçalves, A.; Diab, S.; Woodward, N.; Goodwin, A.; Yerushalmi, R.; et al. Talazoparib versus chemotherapy in patients with germline BRCA1/2-mutated HER2-negative advanced breast cancer: Final overall survival results from the EMBRACA trial. *Ann. Oncol. Off. J. Eur. Soc. Med. Oncol.* **2020**, *31*, 1526–1535. [CrossRef]

22. Paluch-Shimon, S.; Cardoso, F. PARP inhibitors coming of age. *Nat. Rev. Clin. Oncol.* **2021**, *18*, 69–70. [CrossRef]

23. ClinicalTrials.gov. Identifier: NCT03344965 Olaparib in Metastatic Breast Cancer. Available online: https://clinicaltrials.gov/ct2/show/NCT03344965 (accessed on 15 July 2021).

24. Giglia-Mari, G.; Zotter, A.; Vermeulen, W. DNA damage response. *Cold Spring Harb. Perspect. Biol.* **2011**, *3*, a000745. [CrossRef]

25. O'Connor, M.J. Targeting the DNA Damage Response in Cancer. *Mol. Cell* **2015**, *60*, 547–560. [CrossRef]

26. Weber, S. Light-driven enzymatic catalysis of DNA repair: A review of recent biophysical studies on photolyase. *Biochim. Biophys. Acta* **2005**, *1707*, 1–23. [CrossRef] [PubMed]

27. Hegde, M.L.; Dutta, A.; Yang, C.; Mantha, A.K.; Hegde, P.M.; Pandey, A.; Sengupta, S.; Yu, Y.; Calsou, P.; Chen, D.; et al. Scaffold attachment factor A (SAF-A) and Ku temporally regulate repair of radiation-induced clustered genome lesions. *Oncotarget* **2016**, *7*, 54430–54444. [CrossRef] [PubMed]

28. Hegde, M.L.; Hazra, T.K.; Mitra, S. Early steps in the DNA base excision/single-strand interruption repair pathway in mammalian cells. *Cell Res.* **2008**, *18*, 27–47. [CrossRef] [PubMed]

29. Lama-Sherpa, T.D.; Shevde, L.A. An Emerging Regulatory Role for the Tumor Microenvironment in the DNA Damage Response to Double-Strand Breaks. *Mol. Cancer Res.* **2020**, *18*, 185–193. [CrossRef]

30. Schwartz, M.; Zlotorynski, E.; Goldberg, M.; Ozeri, E.; Rahat, A.; le Sage, C.; Chen, B.P.C.; Chen, D.J.; Agami, R.; Kerem, B. Homologous recombination and nonhomologous end-joining repair pathways regulate fragile site stability. *Genes Dev.* **2005**, *19*, 2715–2726. [CrossRef] [PubMed]

31. Makharashvili, N.; Tubbs, A.T.; Yang, S.-H.; Wang, H.; Barton, O.; Zhou, Y.; Deshpande, R.A.; Lee, J.-H.; Lobrich, M.; Sleckman, B.P.; et al. Catalytic and noncatalytic roles of the CtIP endonuclease in double-strand break end resection. *Mol. Cell* **2014**, *54*, 1022–1033. [CrossRef]

32. Quennet, V.; Beucher, A.; Barton, O.; Takeda, S.; Löbrich, M. CtIP and MRN promote non-homologous end-joining of etoposide-induced DNA double-strand breaks in G1. *Nucleic Acids Res.* **2011**, *39*, 2144–2152. [CrossRef]

33. Lee, J.-H.; Paull, T.T. Activation and regulation of ATM kinase activity in response to DNA double-strand breaks. *Oncogene* **2007**, *26*, 7741–7748. [CrossRef]

34. Krajewska, M.; Fehrmann, R.S.N.; de Vries, E.G.E.; van Vugt, M.A.T.M. Regulators of homologous recombination repair as novel targets for cancer treatment. *Front. Genet.* **2015**, *6*, 96. [CrossRef]

35. Peng, G.; Lin, S.-Y. Exploiting the homologous recombination DNA repair network for targeted cancer therapy. *World J. Clin. Oncol.* **2011**, *2*, 73–79. [CrossRef]

36. Li, X.; Heyer, W.-D. Homologous recombination in DNA repair and DNA damage tolerance. *Cell Res.* **2008**, *18*, 99–113. [CrossRef] [PubMed]

37. Eggler, A.L.; Inman, R.B.; Cox, M.M. The Rad51-dependent pairing of long DNA substrates is stabilized by replication protein A. *J. Biol. Chem.* **2002**, *277*, 39280–39288. [CrossRef] [PubMed]

38. Sugiyama, T.; Zaitseva, E.M.; Kowalczykowski, S.C. A single-stranded DNA-binding protein is needed for efficient presynaptic complex formation by the Saccharomyces cerevisiae Rad51 protein. *J. Biol. Chem.* **1997**, *272*, 7940–7945. [CrossRef]

39. Sleeth, K.M.; Sørensen, C.S.; Issaeva, N.; Dziegielewski, J.; Bartek, J.; Helleday, T. RPA mediates recombination repair during replication stress and is displaced from DNA by checkpoint signalling in human cells. *J. Mol. Biol.* **2007**, *373*, 38–47. [CrossRef] [PubMed]

40. San Filippo, J.; Sung, P.; Klein, H. Mechanism of eukaryotic homologous recombination. *Annu. Rev. Biochem.* **2008**, *77*, 229–257. [CrossRef] [PubMed]

41. Zhao, W.; Steinfeld, J.B.; Liang, F.; Chen, X.; Maranon, D.G.; Jian Ma, C.; Kwon, Y.; Rao, T.; Wang, W.; Sheng, C.; et al. BRCA1-BARD1 promotes RAD51-mediated homologous DNA pairing. *Nature* **2017**, *550*, 360–365. [CrossRef]

42. Bakr, A.; Oing, C.; Köcher, S.; Borgmann, K.; Dornreiter, I.; Petersen, C.; Dikomey, E.; Mansour, W.Y. Involvement of ATM in homologous recombination after end resection and RAD51 nucleofilament formation. *Nucleic Acids Res.* **2015**, *43*, 3154–3166. [CrossRef]

43. Prakash, R.; Zhang, Y.; Feng, W.; Jasin, M. Homologous recombination and human health: The roles of BRCA1, BRCA2, and associated proteins. *Cold Spring Harb. Perspect. Biol.* **2015**, *7*, a016600. [CrossRef] [PubMed]

44. Hartford, S.A.; Chittela, R.; Ding, X.; Vyas, A.; Martin, B.; Burkett, S.; Haines, D.C.; Southon, E.; Tessarollo, L.; Sharan, S.K. Interaction with PALB2 Is Essential for Maintenance of Genomic Integrity by BRCA2. *PLoS Genet.* **2016**, *12*, e1006236. [CrossRef]

45. Jirawatnotai, S.; Hu, Y.; Michowski, W.; Elias, J.E.; Becks, L.; Bienvenu, F.; Zagozdzon, A.; Goswami, T.; Wang, Y.E.; Clark, A.B.; et al. A function for cyclin D1 in DNA repair uncovered by protein interactome analyses in human cancers. *Nature* **2011**, *474*, 230–234. [CrossRef]

46. Esashi, F.; Christ, N.; Gannon, J.; Liu, Y.; Hunt, T.; Jasin, M.; West, S.C. CDK-dependent phosphorylation of BRCA2 as a regulatory mechanism for recombinational repair. *Nature* **2005**, *434*, 598–604. [CrossRef]

47. Mateo, J.; Carreira, S.; Sandhu, S.; Miranda, S.; Mossop, H.; Perez-Lopez, R.; Nava Rodrigues, D.; Robinson, D.; Omlin, A.; Tunariu, N.; et al. DNA-Repair Defects and Olaparib in Metastatic Prostate Cancer. *N. Engl. J. Med.* **2015**, *373*, 1697–1708. [CrossRef]

48. Lord, C.J.; Ashworth, A. BRCAness revisited. *Nat. Rev. Cancer* **2016**, *16*, 110–120. [CrossRef]

49. Nguyen, L.; Martens, J.; Van Hoeck, A.; Cuppen, E.; Martens, J.; Van Hoeck, A.; Cuppen, E.; Martens, J.; Van Hoeck, A.; Cuppen, E. Pan-cancer landscape of homologous recombination deficiency. *Nat. Commun.* **2020**, *11*, 1–12. [CrossRef] [PubMed]

50. Pilarski, R. The Role of BRCA Testing in Hereditary Pancreatic and Prostate Cancer Families. *Am. Soc. Clin. Oncol. Educ. Book* **2019**, *39*, 79–86. [CrossRef]

51. Min, A.; Kim, K.; Jeong, K.; Choi, S.; Kim, S.; Suh, K.J.; Lee, K.-H.; Kim, S.; Im, S.-A. Homologous repair deficiency score for identifying breast cancers with defective DNA damage response. *Sci. Rep.* **2020**, *10*, 12506. [CrossRef] [PubMed]

52. Tutt, A.; Robson, M.; Garber, J.E.; Domchek, S.M.; Audeh, M.W.; Weitzel, J.N.; Friedlander, M.; Arun, B.; Loman, N.; Schmutzler, R.K.; et al. Oral poly(ADP-ribose) polymerase inhibitor olaparib in patients with BRCA1 or BRCA2 mutations and advanced breast cancer: A proof-of-concept trial. *Lancet* **2010**, *376*, 235–244. [CrossRef]

53. Fong, P.C.; Boss, D.S.; Yap, T.A.; Tutt, A.; Wu, P.; Mergui-Roelvink, M.; Mortimer, P.; Swaisland, H.; Lau, A.; O'Connor, M.J.; et al. Inhibition of poly(ADP-ribose) polymerase in tumors from BRCA mutation carriers. *N. Engl. J. Med.* **2009**, *361*, 123–134. [CrossRef]

54. Haince, J.-F.; McDonald, D.; Rodrigue, A.; Déry, U.; Masson, J.-Y.; Hendzel, M.J.; Poirier, G.G. PARP1-dependent kinetics of recruitment of MRE11 and NBS1 proteins to multiple DNA damage sites. *J. Biol. Chem.* **2008**, *283*, 1197–1208. [CrossRef] [PubMed]

55. Pascal, J.M. The comings and goings of PARP-1 in response to DNA damage. *DNA Repair* **2018**, *71*, 177–182. [CrossRef]

56. Do, K.; Chen, A.P. Molecular pathways: Targeting PARP in cancer treatment. *Clin. Cancer Res.* **2013**, *19*, 977–984. [CrossRef]

57. Comen, E.A.; Robson, M. Poly(ADP-ribose) polymerase inhibitors in triple-negative breast cancer. *Cancer J.* **2010**, *16*, 48–52. [CrossRef]

58. McGlynn, P.; Lloyd, R.G. Recombinational repair and restart of damaged replication forks. *Nat. Rev. Mol. Cell Biol.* **2002**, *3*, 859–870. [CrossRef]

59. Min, A.; Im, S.-A. PARP Inhibitors as Therapeutics: Beyond Modulation of PARylation. *Cancers* **2020**, *12*, 394. [CrossRef]

60. Jiang, X.; Li, X.; Li, W.; Bai, H.; Zhang, Z. PARP inhibitors in ovarian cancer: Sensitivity prediction and resistance mechanisms. *J. Cell. Mol. Med.* **2019**, *23*, 2303–2313. [CrossRef]

61. Gelmon, K.A.; Tischkowitz, M.; Mackay, H.; Swenerton, K.; Robidoux, A.; Tonkin, K.; Hirte, H.; Huntsman, D.; Clemons, M.; Gilks, B.; et al. Olaparib in patients with recurrent high-grade serous or poorly differentiated ovarian carcinoma or triple-negative breast cancer: A phase 2, multicentre, open-label, non-randomised study. *Lancet. Oncol.* **2011**, *12*, 852–861. [CrossRef]

62. Kaufman, B.; Shapira-Frommer, R.; Schmutzler, R.K.; Audeh, M.W.; Friedlander, M.; Balmaña, J.; Mitchell, G.; Fried, G.; Stemmer, S.M.; Hubert, A.; et al. Olaparib monotherapy in patients with advanced cancer and a germline BRCA1/2 mutation. *J. Clin. Oncol.* **2015**, *33*, 244–250. [CrossRef] [PubMed]

63. Robson, M.E.; Tung, N.; Conte, P.; Im, S.-A.; Senkus, E.; Xu, B.; Masuda, N.; Delaloge, S.; Li, W.; Armstrong, A.; et al. OlympiAD final overall survival and tolerability results: Olaparib versus chemotherapy treatment of physician's choice in patients with a germline BRCA mutation and HER2-negative metastatic breast cancer. *Ann. Oncol. Off. J. Eur. Soc. Med. Oncol.* **2019**, *30*, 558–566. [CrossRef] [PubMed]

64. de Bono, J.; Ramanathan, R.K.; Mina, L.; Chugh, R.; Glaspy, J.; Rafii, S.; Kaye, S.; Sachdev, J.; Heymach, J.; Smith, D.C.; et al. Phase I, Dose-Escalation, Two-Part Trial of the PARP Inhibitor Talazoparib in Patients with Advanced Germline BRCA1/2 Mutations and Selected Sporadic Cancers. *Cancer Discov.* **2017**, *7*, 620–629. [CrossRef] [PubMed]

65. Turner, N.C.; Telli, M.L.; Rugo, H.S.; Mailliez, A.; Ettl, J.; Grischke, E.-M.; Mina, L.A.; Balmana Gelpi, J.; Fasching, P.A.; Hurvitz, S.A.; et al. Final results of a phase 2 study of talazoparib (TALA) following platinum or multiple cytotoxic regimens in advanced breast cancer patients (pts) with germline BRCA1/2 mutations (ABRAZO). *J. Clin. Oncol.* **2017**, *35*, 1007. [CrossRef]

66. Litton, J.K.; Rugo, H.S.; Ettl, J.; Hurvitz, S.A.; Gonçalves, A.; Lee, K.-H.; Fehrenbacher, L.; Yerushalmi, R.; Mina, L.A.; Martin, M.; et al. Talazoparib in Patients with Advanced Breast Cancer and a Germline BRCA Mutation. *N. Engl. J. Med.* **2018**, *379*, 753–763. [CrossRef]

67. Drew, Y.; Ledermann, J.; Hall, G.; Rea, D.; Glasspool, R.; Highley, M.; Jayson, G.; Sludden, J.; Murray, J.; Jamieson, D.; et al. Phase 2 multicentre trial investigating intermittent and continuous dosing schedules of the poly(ADP-ribose) polymerase inhibitor rucaparib in germline BRCA mutation carriers with advanced ovarian and breast cancer. *Br. J. Cancer* **2016**, *114*, 723–730. [CrossRef]

68. Miller, K.; Tong, Y.; Jones, D.R.; Walsh, T.; Danso, M.A.; Ma, C.X.; Silverman, P.; King, M.-C.; Badve, S.S.; Perkins, S.M. Cisplatin with or without rucaparib after preoperative chemotherapy in patients with triple negative breast cancer: Final efficacy results of Hoosier Oncology Group BRE09-146. *J. Clin. Oncol.* **2015**, *33*, 1082. [CrossRef]

69. Rugo, H.S.; Olopade, O.I.; DeMichele, A.; Yau, C.; van 't Veer, L.J.; Buxton, M.B.; Hogarth, M.; Hylton, N.M.; Paoloni, M.; Perlmutter, J.; et al. Adaptive Randomization of Veliparib-Carboplatin Treatment in Breast Cancer. *N. Engl. J. Med.* **2016**, *375*, 23–34. [CrossRef]

70. Loibl, S.; O'Shaughnessy, J.; Untch, M.; Sikov, W.M.; Rugo, H.S.; McKee, M.D.; Huober, J.; Golshan, M.; von Minckwitz, G.; Maag, D.; et al. Addition of the PARP inhibitor veliparib plus carboplatin or carboplatin alone to standard neoadjuvant chemotherapy in triple-negative breast cancer (BrighTNess): A randomised, phase 3 trial. *Lancet. Oncol.* **2018**, *19*, 497–509. [CrossRef]

71. Fasching, P.A.; Jackisch, C.; Rhiem, K.; Schneeweiss, A.; Klare, P.; Hanusch, C.; Huober, J.B.; Link, T.; Untch, M.; Schmatloch, S.; et al. GeparOLA: A randomized phase II trial to assess the efficacy of paclitaxel and olaparib in comparison to paclitaxel/carboplatin followed by epirubicin/cyclophosphamide as neoadjuvant chemotherapy in patients (pts) with HER2-negative early breast cancer (BC) and homologous recombination deficiency (HRD). *J. Clin. Oncol.* **2019**, *37*, 506.

72. Litton, J.K.; Scoggins, M.E.; Hess, K.R.; Adrada, B.E.; Murthy, R.K.; Damodaran, S.; DeSnyder, S.M.; Brewster, A.M.; Barcenas, C.H.; Valero, V.; et al. Neoadjuvant Talazoparib for Patients With Operable Breast Cancer With a Germline BRCA Pathogenic Variant. *J. Clin. Oncol.* **2020**, *38*, 388–394. [CrossRef] [PubMed]

73. Tutt, A.N.J.; Kaufman, B.; Gelber, R.D.; Mc Fadden, E.; Goessl, C.D.; Viale, G.; Arahmani, A.; Fumagalli, D.; Azim, H.A.; Wu, W.; et al. OlympiA: A randomized phase III trial of olaparib as adjuvant therapy in patients with high-risk HER2-negative breast cancer (BC) and a germline BRCA 1/2 mutation (g BRCA m). *J. Clin. Oncol.* **2015**, *33*, TPS1109. [CrossRef]

74. Diéras, V.; Han, H.S.; Kaufman, B.; Wildiers, H.; Friedlander, M.; Ayoub, J.-P.; Puhalla, S.L.; Bondarenko, I.; Campone, M.; Jakobsen, E.H.; et al. Veliparib with carboplatin and paclitaxel in BRCA-mutated advanced breast cancer (BROCADE3): A randomised, double-blind, placebo-controlled, phase 3 trial. *Lancet. Oncol.* **2020**, *21*, 1269–1282. [CrossRef]

75. Tutt, A.N.J.; Garber, J.E.; Kaufman, B.; Viale, G.; Fumagalli, D.; Rastogi, P.; Gelber, R.D.; de Azambuja, E.; Fielding, A.; Balmaña, J.; et al. Adjuvant Olaparib for Patients with BRCA1- or BRCA2-Mutated Breast Cancer. *N. Engl. J. Med.* **2021**, *384*, 2394–2405. [CrossRef]

76. Dréan, A.; Lord, C.J.; Ashworth, A. PARP inhibitor combination therapy. *Crit. Rev. Oncol. Hematol.* **2016**, *108*, 73–85. [CrossRef]

77. Huggins-Puhalla, S.L.; Beumer, J.H.; Appleman, L.J.; Tawbi, H.A.-H.; Stoller, R.G.; Lin, Y.; Kiesel, B.; Tan, A.R.; Gibbon, D.; Jiang, Y.; et al. A phase I study of chronically dosed, single-agent veliparib (ABT-888) in patients (pts) with either BRCA 1/2-mutated cancer (BRCA+), platinum-refractory ovarian cancer, or basal-like breast cancer (BRCA-wt). *J. Clin. Oncol.* **2012**, *30*, 3054. [CrossRef]

78. Appleman, L.J.; Beumer, J.H.; Jiang, Y.; Lin, Y.; Ding, F.; Puhalla, S.; Swartz, L.; Owonikoko, T.K.; Donald Harvey, R.; Stoller, R.; et al. Phase 1 study of veliparib (ABT-888), a poly (ADP-ribose) polymerase inhibitor, with carboplatin and paclitaxel in advanced solid malignancies. *Cancer Chemother. Pharmacol.* **2019**, *84*, 1289–1301. [CrossRef] [PubMed]

79. Han, H.S.; Diéras, V.; Robson, M.; Palácová, M.; Marcom, P.K.; Jager, A.; Bondarenko, I.; Citrin, D.; Campone, M.; Telli, M.L.; et al. Veliparib with temozolomide or carboplatin/paclitaxel versus placebo with carboplatin/paclitaxel in patients with BRCA1/2 locally recurrent/metastatic breast cancer: Randomized phase II study. *Ann. Oncol. Off. J. Eur. Soc. Med. Oncol.* **2018**, *29*, 154–161. [CrossRef]

80. Isakoff, S.J.; Overmoyer, B.; Tung, N.M.; Gelman, R.S.; Giranda, V.L.; Bernhard, K.M.; Habin, K.R.; Ellisen, L.W.; Winer, E.P.; Goss, P.E. A phase II trial of the PARP inhibitor veliparib (ABT888) and temozolomide for metastatic breast cancer. *J. Clin. Oncol.* **2010**, *28*, 1019. [CrossRef]

81. Nicolas, E.; Bertucci, F.; Sabatier, R.; Gonçalves, A. Targeting BRCA Deficiency in Breast Cancer: What are the Clinical Evidences and the Next Perspectives? *Cancers* **2018**, *10*, 506. [CrossRef] [PubMed]

82. Emens, L.; Loi, S.; Rugo, H.; Schneeweiss, A.; Diéras, V.; Iwata, H.; Barrios, C.; Nechaeva, M.; Molinero, L.; Nguyen Duc, A.; et al. Abstract GS1-04: IMpassion130: Efficacy in immune biomarker subgroups from the global, randomized, double-blind, placebo-controlled, phase III study of atezolizumab + nab-paclitaxel in patients with treatment-naïve, locally advanced or metastatic triple-negative breast cancer. In Proceedings of the 2018 San Antonio Breast Cancer Symposium, San Antonio, TX, USA, 4–8 December 2018; Volume 79. [CrossRef]

83. Im, S.-A.; Lu, Y.-S.; Bardia, A.; Harbeck, N.; Colleoni, M.; Franke, F.; Chow, L.; Sohn, J.; Lee, K.-S.; Campos-Gomez, S.; et al. Overall Survival with Ribociclib plus Endocrine Therapy in Breast Cancer. *N. Engl. J. Med.* **2019**, *381*, 307–316. [CrossRef]

84. Nolan, E.; Savas, P.; Policheni, A.N.; Darcy, P.K.; Vaillant, F.; Mintoff, C.P.; Dushyanthen, S.; Mansour, M.; Pang, J.-M.B.; Fox, S.B.; et al. Combined immune checkpoint blockade as a therapeutic strategy for BRCA1-mutated breast cancer. *Sci. Transl. Med.* **2017**, *9*, eaal4922. [CrossRef]

85. Jiao, S.; Xia, W.; Yamaguchi, H.; Wei, Y.; Chen, M.-K.; Hsu, J.-M.; Hsu, J.L.; Yu, W.-H.; Du, Y.; Lee, H.-H.; et al. PARP Inhibitor Upregulates PD-L1 Expression and Enhances Cancer-Associated Immunosuppression. *Clin. Cancer Res.* **2017**, *23*, 3711–3720. [CrossRef] [PubMed]

86. Domchek, S.M.; Postel-Vinay, S.; Im, S.-A.; Park, Y.H.; Delord, J.-P.; Italiano, A.; Alexandre, J.; You, B.; Bastian, S.; Krebs, M.G.; et al. Olaparib and durvalumab in patients with germline BRCA-mutated metastatic breast cancer (MEDIOLA): An open-label, multicentre, phase 1/2, basket study. *Lancet. Oncol.* **2020**, *21*, 1155–1164. [CrossRef]

87. Vinayak, S.; Tolaney, S.M.; Schwartzberg, L.; Mita, M.; McCann, G.; Tan, A.R.; Wahner-Hendrickson, A.E.; Forero, A.; Anders, C.; Wulf, G.M.; et al. Open-label Clinical Trial of Niraparib Combined With Pembrolizumab for Treatment of Advanced or Metastatic Triple-Negative Breast Cancer. *JAMA Oncol.* **2019**, *5*, 1132–1140. [CrossRef] [PubMed]

88. Gourley, C.; Balmaña, J.; Ledermann, J.A.; Serra, V.; Dent, R.; Loibl, S.; Pujade-Lauraine, E.; Boulton, S.J. Moving From Poly (ADP-Ribose) Polymerase Inhibition to Targeting DNA Repair and DNA Damage Response in Cancer Therapy. *J. Clin. Oncol.* **2019**, *37*, 2257–2269. [CrossRef] [PubMed]

89. De Soto, J.A.; Deng, C.-X. PARP-1 inhibitors: Are they the long-sought genetically specific drugs for BRCA1/2-associated breast cancers? *Int. J. Med. Sci.* **2006**, *3*, 117–123. [CrossRef] [PubMed]

90. Vidula, N.; Rich, T.A.; Sartor, O.; Yen, J.; Hardin, A.; Nance, T.; Lilly, M.B.; Nezami, M.A.; Patel, S.P.; Carneiro, B.A.; et al. Routine Plasma-Based Genotyping to Comprehensively Detect Germline, Somatic, and Reversion BRCA Mutations among Patients with Advanced Solid Tumors. *Clin. Cancer Res.* **2020**, *26*, 2546–2555. [CrossRef]

91. Sakai, W.; Swisher, E.M.; Karlan, B.Y.; Agarwal, M.K.; Higgins, J.; Friedman, C.; Villegas, E.; Jacquemont, C.; Farrugia, D.J.; Couch, F.J.; et al. Secondary mutations as a mechanism of cisplatin resistance in BRCA2-mutated cancers. *Nature* **2008**, *451*, 1116–1120. [CrossRef]

92. Edwards, S.L.; Brough, R.; Lord, C.J.; Natrajan, R.; Vatcheva, R.; Levine, D.A.; Boyd, J.; Reis-Filho, J.S.; Ashworth, A. Resistance to therapy caused by intragenic deletion in BRCA2. *Nature* **2008**, *451*, 1111–1115. [CrossRef] [PubMed]

93. Lheureux, S.; Lai, Z.; Dougherty, B.A.; Runswick, S.; Hodgson, D.R.; Timms, K.M.; Lanchbury, J.S.; Kaye, S.; Gourley, C.; Bowtell, D.; et al. Long-Term Responders on Olaparib Maintenance in High-Grade Serous Ovarian Cancer: Clinical and Molecular Characterization. *Clin. Cancer Res.* **2017**, *23*, 4086–4094. [CrossRef]

94. Barber, L.J.; Sandhu, S.; Chen, L.; Campbell, J.; Kozarewa, I.; Fenwick, K.; Assiotis, I.; Rodrigues, D.N.; Reis Filho, J.S.; Moreno, V.; et al. Secondary mutations in BRCA2 associated with clinical resistance to a PARP inhibitor. *J. Pathol.* **2013**, *229*, 422–429. [CrossRef] [PubMed]

95. Norquist, B.; Wurz, K.A.; Pennil, C.C.; Garcia, R.; Gross, J.; Sakai, W.; Karlan, B.Y.; Taniguchi, T.; Swisher, E.M. Secondary somatic mutations restoring BRCA1/2 predict chemotherapy resistance in hereditary ovarian carcinomas. *J. Clin. Oncol.* **2011**, *29*, 3008–3015. [CrossRef] [PubMed]

96. Goodall, J.; Mateo, J.; Yuan, W.; Mossop, H.; Porta, N.; Miranda, S.; Perez-Lopez, R.; Dolling, D.; Robinson, D.R.; Sandhu, S.; et al. Circulating Cell-Free DNA to Guide Prostate Cancer Treatment with PARP Inhibition. *Cancer Discov.* **2017**, *7*, 1006–1017. [CrossRef]

97. Weigelt, B.; Comino-Méndez, I.; de Bruijn, I.; Tian, L.; Meisel, J.L.; García-Murillas, I.; Fribbens, C.; Cutts, R.; Martelotto, L.G.; Ng, C.K.Y.; et al. Diverse BRCA1 and BRCA2 Reversion Mutations in Circulating Cell-Free DNA of Therapy-Resistant Breast or Ovarian Cancer. *Clin. Cancer Res.* **2017**, *23*, 6708–6720. [CrossRef]

98. Christie, E.L.; Fereday, S.; Doig, K.; Pattnaik, S.; Dawson, S.-J.; Bowtell, D.D.L. Reversion of BRCA1/2 Germline Mutations Detected in Circulating Tumor DNA From Patients With High-Grade Serous Ovarian Cancer. *J. Clin. Oncol.* **2017**, *35*, 1274–1280. [CrossRef]

99. Waks, A.G.; Cohen, O.; Kochupurakkal, B.; Kim, D.; Dunn, C.E.; Buendia Buendia, J.; Wander, S.; Helvie, K.; Lloyd, M.R.; Marini, L.; et al. Reversion and non-reversion mechanisms of resistance to PARP inhibitor or platinum chemotherapy in BRCA1/2-mutant metastatic breast cancer. *Ann. Oncol. Off. J. Eur. Soc. Med. Oncol.* **2020**, *31*, 590–598. [CrossRef]

100. Quigley, D.; Alumkal, J.J.; Wyatt, A.W.; Kothari, V.; Foye, A.; Lloyd, P.; Aggarwal, R.; Kim, W.; Lu, E.; Schwartzman, J.; et al. Analysis of Circulating Cell-Free DNA Identifies Multiclonal Heterogeneity of BRCA2 Reversion Mutations Associated with Resistance to PARP Inhibitors. *Cancer Discov.* **2017**, *7*, 999–1005. [CrossRef] [PubMed]

101. Simmons, A.D.; Nguyen, M.; Pintus, E. Polyclonal BRCA2 mutations following carboplatin treatment confer resistance to the PARP inhibitor rucaparib in a patient with mCRPC: A case report. *BMC Cancer* **2020**, *20*, 215. [CrossRef] [PubMed]

102. Kondrashova, O.; Nguyen, M.; Shield-Artin, K.; Tinker, A.V.; Teng, N.N.H.; Harrell, M.I.; Kuiper, M.J.; Ho, G.-Y.; Barker, H.; Jasin, M.; et al. Secondary Somatic Mutations Restoring RAD51C and RAD51D Associated with Acquired Resistance to the PARP Inhibitor Rucaparib in High-Grade Ovarian Carcinoma. *Cancer Discov.* **2017**, *7*, 984–998. [CrossRef] [PubMed]

103. Cruz, C.; Castroviejo-Bermejo, M.; Gutiérrez-Enríquez, S.; Llop-Guevara, A.; Ibrahim, Y.H.; Gris-Oliver, A.; Bonache, S.; Morancho, B.; Bruna, A.; Rueda, O.M.; et al. RAD51 foci as a functional biomarker of homologous recombination repair and PARP inhibitor resistance in germline BRCA-mutated breast cancer. *Ann. Oncol. Off. J. Eur. Soc. Med. Oncol.* **2018**, *29*, 1203–1210. [CrossRef] [PubMed]

104. Bouwman, P.; Jonkers, J. Molecular pathways: How can BRCA-mutated tumors become resistant to PARP inhibitors? *Clin. Cancer Res.* **2014**, *20*, 540–547. [CrossRef]

105. Nacson, J.; Krais, J.J.; Bernhardy, A.J.; Clausen, E.; Feng, W.; Wang, Y.; Nicolas, E.; Cai, K.Q.; Tricarico, R.; Hua, X.; et al. BRCA1 Mutation-Specific Responses to 53BP1 Loss-Induced Homologous Recombination and PARP Inhibitor Resistance. *Cell Rep.* **2018**, *24*, 3513–3527.e7. [CrossRef]

106. Castroviejo-Bermejo, M.; Cruz, C.; Llop-Guevara, A.; Gutiérrez-Enríquez, S.; Ducy, M.; Ibrahim, Y.H.; Gris-Oliver, A.; Pellegrino, B.; Bruna, A.; Guzmán, M.; et al. A RAD51 assay feasible in routine tumor samples calls PARP inhibitor response beyond BRCA mutation. *EMBO Mol. Med.* **2018**, *10*, e9172. [CrossRef]

107. Ter Brugge, P.; Kristel, P.; van der Burg, E.; Boon, U.; de Maaker, M.; Lips, E.; Mulder, L.; de Ruiter, J.; Moutinho, C.; Gevensleben, H.; et al. Mechanisms of Therapy Resistance in Patient-Derived Xenograft Models of BRCA1-Deficient Breast Cancer. *J. Natl. Cancer Inst.* **2016**, *108*, djw148. [CrossRef] [PubMed]

108. Bunting, S.F.; Callén, E.; Wong, N.; Chen, H.-T.; Polato, F.; Gunn, A.; Bothmer, A.; Feldhahn, N.; Fernandez-Capetillo, O.; Cao, L.; et al. 53BP1 inhibits homologous recombination in Brca1-deficient cells by blocking resection of DNA breaks. *Cell* **2010**, *141*, 243–254. [CrossRef]

109. Kim, D.S.; Camacho, C.V.; Kraus, W.L. Alternate therapeutic pathways for PARP inhibitors and potential mechanisms of resistance. *Exp. Mol. Med.* **2021**, *53*, 42–51. [CrossRef]

110. Drost, R.; Bouwman, P.; Rottenberg, S.; Boon, U.; Schut, E.; Klarenbeek, S.; Klijn, C.; van der Heijden, I.; van der Gulden, H.; Wientjens, E.; et al. BRCA1 RING function is essential for tumor suppression but dispensable for therapy resistance. *Cancer Cell* **2011**, *20*, 797–809. [CrossRef] [PubMed]

111. Peng, G.; Chun-Jen Lin, C.; Mo, W.; Dai, H.; Park, Y.-Y.; Kim, S.M.; Peng, Y.; Mo, Q.; Siwko, S.; Hu, R.; et al. Genome-wide transcriptome profiling of homologous recombination DNA repair. *Nat. Commun.* **2014**, *5*, 3361. [CrossRef] [PubMed]

112. Pettitt, S.J.; Krastev, D.B.; Brandsma, I.; Dréan, A.; Song, F.; Aleksandrov, R.; Harrell, M.I.; Menon, M.; Brough, R.; Campbell, J.; et al. Genome-wide and high-density CRISPR-Cas9 screens identify point mutations in PARP1 causing PARP inhibitor resistance. *Nat. Commun.* **2018**, *9*, 1849. [CrossRef] [PubMed]

113. Makvandi, M.; Pantel, A.; Schwartz, L.; Schubert, E.; Xu, K.; Hsieh, C.-J.; Hou, C.; Kim, H.; Weng, C.-C.; Winters, H.; et al. A PET imaging agent for evaluating PARP-1 expression in ovarian cancer. *J. Clin. Investig.* **2018**, *128*, 2116–2126. [CrossRef] [PubMed]

114. Du, Y.; Yamaguchi, H.; Wei, Y.; Hsu, J.L.; Wang, H.-L.; Hsu, Y.-H.; Lin, W.-C.; Yu, W.-H.; Leonard, P.G.; Lee, G.R.; et al. Blocking c-Met-mediated PARP1 phosphorylation enhances anti-tumor effects of PARP inhibitors. *Nat. Med.* **2016**, *22*, 194–201. [CrossRef]

115. Gogola, E.; Duarte, A.A.; de Ruiter, J.R.; Wiegant, W.W.; Schmid, J.A.; de Bruijn, R.; James, D.I.; Guerrero Llobet, S.; Vis, D.J.; Annunziato, S.; et al. Selective Loss of PARG Restores PARylation and Counteracts PARP Inhibitor-Mediated Synthetic Lethality. *Cancer Cell* **2018**, *33*, 1078–1093.e12. [CrossRef]

116. Choi, Y.E.; Meghani, K.; Brault, M.-E.; Leclerc, L.; He, Y.J.; Day, T.A.; Elias, K.M.; Drapkin, R.; Weinstock, D.M.; Dao, F.; et al. Platinum and PARP Inhibitor Resistance Due to Overexpression of MicroRNA-622 in BRCA1-Mutant Ovarian Cancer. *Cell Rep.* **2016**, *14*, 429–439. [CrossRef]

117. Liu, G.; Yang, D.; Rupaimoole, R.; Pecot, C.V.; Sun, Y.; Mangala, L.S.; Li, X.; Ji, P.; Cogdell, D.; Hu, L.; et al. Augmentation of response to chemotherapy by microRNA-506 through regulation of RAD51 in serous ovarian cancers. *J. Natl. Cancer Inst.* **2015**, *107*, djv108. [CrossRef]

118. Schlacher, K.; Wu, H.; Jasin, M. A distinct replication fork protection pathway connects Fanconi anemia tumor suppressors to RAD51-BRCA1/2. *Cancer Cell* **2012**, *22*, 106–116. [CrossRef] [PubMed]

119. Lomonosov, M.; Anand, S.; Sangrithi, M.; Davies, R.; Venkitaraman, A.R. Stabilization of stalled DNA replication forks by the BRCA2 breast cancer susceptibility protein. *Genes Dev.* **2003**, *17*, 3017–3022. [CrossRef]

120. Ray Chaudhuri, A.; Callen, E.; Ding, X.; Gogola, E.; Duarte, A.A.; Lee, J.-E.; Wong, N.; Lafarga, V.; Calvo, J.A.; Panzarino, N.J.; et al. Replication fork stability confers chemoresistance in BRCA-deficient cells. *Nature* **2016**, *535*, 382–387. [CrossRef] [PubMed]

121. Rondinelli, B.; Gogola, E.; Yücel, H.; Duarte, A.A.; van de Ven, M.; van der Sluijs, R.; Konstantinopoulos, P.A.; Jonkers, J.; Ceccaldi, R.; Rottenberg, S.; et al. EZH2 promotes degradation of stalled replication forks by recruiting MUS81 through histone H3 trimethylation. *Nat. Cell Biol.* **2017**, *19*, 1371–1378. [CrossRef] [PubMed]

122. Fojo, A.; Lebo, R.; Shimizu, N.; Chin, J.E.; Roninson, I.B.; Merlino, G.T.; Gottesman, M.M.; Pastan, I. Localization of multidrug resistance-associated DNA sequences to human chromosome 7. *Somat. Cell Mol. Genet.* **1986**, *12*, 415–420. [CrossRef]

123. Riordan, J.R.; Deuchars, K.; Kartner, N.; Alon, N.; Trent, J.; Ling, V. Amplification of P-glycoprotein genes in multidrug-resistant mammalian cell lines. *Nature* **1985**, *316*, 817–819. [CrossRef]

124. Durmus, S.; Sparidans, R.W.; van Esch, A.; Wagenaar, E.; Beijnen, J.H.; Schinkel, A.H. Breast cancer resistance protein (BCRP/ABCG2) and P-glycoprotein (P-GP/ABCB1) restrict oral availability and brain accumulation of the PARP inhibitor rucaparib (AG-014699). *Pharm. Res.* **2015**, *32*, 37–46. [CrossRef]

125. Rottenberg, S.; Jaspers, J.E.; Kersbergen, A.; van der Burg, E.; Nygren, A.O.H.; Zander, S.A.L.; Derksen, P.W.B.; de Bruin, M.; Zevenhoven, J.; Lau, A.; et al. High sensitivity of BRCA1-deficient mammary tumors to the PARP inhibitor AZD2281 alone and in combination with platinum drugs. *Proc. Natl. Acad. Sci. USA* **2008**, *105*, 17079–17084. [CrossRef]

126. Schmid, P.; Adams, S.; Rugo, H.S.; Schneeweiss, A.; Barrios, C.H.; Iwata, H.; Diéras, V.; Hegg, R.; Im, S.-A.; Shaw Wright, G.; et al. Atezolizumab and Nab-Paclitaxel in Advanced Triple-Negative Breast Cancer. *N. Engl. J. Med.* **2018**, *379*, 2108–2121. [CrossRef]

127. Slamon, D.J.; Neven, P.; Chia, S.; Fasching, P.A.; De Laurentiis, M.; Im, S.-A.; Petrakova, K.; Bianchi, G.V.; Esteva, F.J.; Martín, M.; et al. Overall Survival with Ribociclib plus Fulvestrant in Advanced Breast Cancer. *N. Engl. J. Med.* **2020**, *382*, 514–524. [CrossRef] [PubMed]

128. Sledge, G.W.; Toi, M.; Neven, P.; Sohn, J.; Inoue, K.; Pivot, X.; Burdaeva, O.; Okera, M.; Masuda, N.; Kaufman, P.A.; et al. The Effect of Abemaciclib Plus Fulvestrant on Overall Survival in Hormone Receptor-Positive, ERBB2-Negative Breast Cancer That Progressed on Endocrine Therapy-MONARCH 2: A Randomized Clinical Trial. *JAMA Oncol.* **2020**, *6*, 116–124. [CrossRef] [PubMed]

129. Tutt, A.; Tovey, H.; Cheang, M.C.U.; Kernaghan, S.; Kilburn, L.; Gazinska, P.; Owen, J.; Abraham, J.; Barrett, S.; Barrett-Lee, P.; et al. Carboplatin in BRCA1/2-mutated and triple-negative breast cancer BRCAness subgroups: The TNT Trial. *Nat. Med.* **2018**, *24*, 628–637. [CrossRef] [PubMed]

130. Schettini, F.; Giudici, F.; Bernocchi, O.; Sirico, M.; Corona, S.P.; Giuliano, M.; Locci, M.; Paris, I.; Scambia, G.; De Placido, S.; et al. Poly (ADP-ribose) polymerase inhibitors in solid tumours: Systematic review and meta-analysis. *Eur. J. Cancer* **2021**, *149*, 134–152. [CrossRef] [PubMed]

International Journal of
Molecular Sciences

Article

Activated ERK Signaling Is One of the Major Hub Signals Related to the Acquisition of Radiotherapy-Resistant MDA-MB-231 Breast Cancer Cells

Anjugam Paramanantham [1,2], Eun Joo Jung [1], Se-IL Go [1], Bae Kwon Jeong [3], Jin-Myung Jung [4], Soon Chan Hong [5], Gon Sup Kim [2,*] and Won Sup Lee [1,*]

1   Departments of Internal Medicine, Institute of Health Sciences and Gyeongsang National University Hospital, Gyeongsang National University College of Medicine, 90 Chilam-dong, Jinju 660-702, Korea; anju.udhay@gmail.com (A.P.); eunjoojung@gnu.ac.kr (E.J.J.); gose1@hanmail.net (S.-I.G.)
2   School of Veterinary and Institute of Life Science, Gyeongsang National University, 900 Gajwadong, Jinju 660-701, Korea
3   Departments of Radiation Oncology, Institute of Health Sciences and Gyeongsang National University Hospital, Gyeongsang National University College of Medicine, 90 Chilam-dong, Jinju 660-702, Korea; blue129j@hanmail.net
4   Departments of Neurosurgery, Institute of Health Sciences and Gyeongsang National University Hospital, Gyeongsang National University College of Medicine, 90 Chilam-dong, Jinju 660-702, Korea; gnuhjjm@gnu.ac.kr
5   Departments of Surgery, Institute of Health Sciences and Gyeongsang National University Hospital, Gyeongsang National University College of Medicine, 90 Chilam-dong, Jinju 660-702, Korea; hongsc@gnu.ac.kr
*   Correspondence: gonskim@gnu.ac.kr (G.S.K.); lwshmo@gnu.ac.kr (W.S.L.); Tel.: +82-55-772-2356 (G.S.K.); +82-55-750-8733 (W.S.L.); Fax: +82-55-758-9122 (W.S.L.)

Citation: Paramanantham, A.; Jung, E.J.; Go, S.-I.; Jeong, B.K.; Jung, J.-M.; Hong, S.C.; Kim, G.S.; Lee, W.S. Activated ERK Signaling Is One of the Major Hub Signals Related to the Acquisition of Radiotherapy-Resistant MDA-MB-231 Breast Cancer Cells. *Int. J. Mol. Sci.* **2021**, *22*, 4940. https://doi.org/10.3390/ijms22094940

Academic Editor: Valentina De Falco

Received: 8 March 2021
Accepted: 30 April 2021
Published: 6 May 2021

Publisher's Note: MDPI stays neutral with regard to jurisdictional claims in published maps and institutional affiliations.

Abstract: Breast cancer is one of the major causes of deaths due to cancer, especially in women. The crucial barrier for breast cancer treatment is resistance to radiation therapy, one of the important local regional therapies. We previously established and characterized radio-resistant MDA-MB-231 breast cancer cells (RT-R-MDA-MB-231 cells) that harbor a high expression of cancer stem cells (CSCs) and the EMT phenotype. In this study, we performed antibody array analysis to identify the hub signaling mechanism for the radiation resistance of RT-R-MDA-MB-231 cells by comparing parental MDA-MB-231 (p-MDA-MB-231) and RT-R-MDA-MB-231 cells. Antibody array analysis unveiled that the MAPK1 protein was the most upregulated protein in RT-R-MDA-MB-231 cells compared to in p-MDA-MB-231 cells. The pathway enrichment analysis also revealed the presence of MAPK1 in almost all enriched pathways. Thus, we used an MEK/ERK inhibitor, PD98059, to block the MEK/ERK pathway and to identify the role of MAPK1 in the radio-resistance of RT-R-MDA-MB-231 cells. MEK/ERK inhibition induced cell death in both p-MDA-MB-231 and RT-R-MDA-MB-231 cells, but the death mechanism for each cell was different; p-MDA-MB-231 cells underwent apoptosis, showing cell shrinkage and PARP-1 cleavage, while RT-R-MDA-MB-231 cells underwent necroptosis, showing mitochondrial dissipation, nuclear swelling, and an increase in the expressions of CypA and AIF. In addition, MEK/ERK inhibition reversed the radio-resistance of RT-R-MDA-MB-231 cells and suppressed the increased expression of CSC markers (CD44 and OCT3/4) and the EMT phenotype (β-catenin and N-cadherin/E-cadherin). Taken together, this study suggests that activated ERK signaling is one of the major hub signals related to the radio-resistance of MDA-MB-231 breast cancer cells.

Keywords: radiation-resistant; breast cancer; cell death; ERK; EMT; cancer stem cells (CSCs); PD98059

1. Introduction

Breast cancer is one of the major causes of death due to cancer worldwide, especially in women [1]. For breast cancer, many therapies are available such as surgical resection, with or without lymph node dissection, radiation, and chemotherapy [2]. Radiation therapy is one of the important local regional therapies for breast cancer treatment [3]. Radiotherapy is applied for most breast cancer patients after surgical resection, but not all patients obtain the same benefits, because some of them suffer from a loco-regional relapse. Radio-resistance is the primary reason for this relapse [4].

Radio-resistance is a process in which the tumor cells or tissues adapt to radio therapy-induced damage [5] and survive irradiation (IR) [6,7]. Radiation can induce a DNA damage response (DDR), which causes cell cycle arrest and the induction of DNA repair, even though the cells with more severe damage from the radiation are induced to undergo apoptosis. The DDR may help the cells survive the IR-induced DNA damage, eventually developing radio-resistance by increasing the DDR rate. In addition, repopulation, hypoxic tumor areas, and cancer stemness are involved in radio-resistance. In DDR to IR and cancer stemness, several signaling pathways are reportedly involved: phosphatidylinositol 3-kinase (PI3K), mitogen-activated protein kinase (MAPK), SIRT pathways, Wnt/β-catenin signaling, IL22RA1/STAT3 signaling, and sonic Hedgehog signaling [8]. In addition to the suggested signaling pathways, IR resistance comprises the involvement of a large number of other proteins and their pathways [9,10]. Particularly, the radiation-induced ERK activation allows cancer cells to overcome the G2/M phase, which is considered the most vulnerable phase during IR, thereby causing radio-resistance [11–13].

MAPK pathways are key signaling pathways involved in the regulation of normal cell proliferation, survival, and differentiation. Aberrant regulation of the MAPK pathways contributes to the development of cancer; particularly, the extracellular signal-regulated kinase (ERK) is crucially involved in cancer cell proliferation, survival, and metastasis [14]. ERK consists of the p44 ERK1 and p42 ERK2. The ERK is the only known substrate of MEK. MEK1/2 activates ERK through dual tyrosine and threonine phosphorylation [15]. Thus, blocking the ERK pathway has proved to be an efficient mechanism to force cells into a cell death pathway. To potentiate the anti-tumoral effects of various cytotoxic agents, many trials of MEK1/2 pharmacological inhibitors (PD98059 [16], UO0126 [17], and PD184352 [18]) have been used. In addition, recent studies have shown the synergetic effect of the MEK/ERK inhibitor and radiation therapy [19].

Radio-resistant MDA-MB-231 breast cancer cells (RT-R-MDA-MB-231 cells) are reported to have a high proliferation rate, metastatic activity, and adhesion to endothelial cells compared with the parental MDA-MB-231 (p-MDA-MB-231) breast cancer cell line. RT-R-MDA-MB-231 cells harbor increased expressions of cancer stem cell (CSC) markers and the epithelial–mesenchymal transition (EMT) phenotype [8]. We hypothesized that there is a key altered signaling (driving oncogenic signaling) involved in developing RT-R-MDA-MB-231 cells. Here, we performed antibody microarray analysis to identify the hub proteins involved in the radio-resistance of RT-R-MDA-MB 231 cells by comparing p-MDA-MB-231 and RT-R-MDA-MB-231 cells because antibody microarray analysis is one of the technologies used for high-throughput protein characterization and discovery [10]. Antibody array analysis is used to measure the expression level of proteins between two different samples [9].

This study was designed to decipher the proteomic differences between p-MDA-MB-231 and RT-R-MDA-MB-231 cells, enabling us to corroborate our findings at the molecular level. In addition, we also aimed to investigate the importance of the key altered signaling in the reversal of radio-resistance and the regulation of the CSC and EMT phenotypes that are strongly associated with radio-resistance.

2. Results

2.1. Proteomic Profiling of RT-R-MDA-MB-231 Cell Lines

To determine the key altered expressions of proteins involved in the radio-resistance of RT-R-MDA-MB 231 cells, we performed and analyzed antibody microarrays to assess the difference in protein expressions between p-MDA-MB-231 and RT-R-MDA-MB-231 cells. The internal normalization ratio (INR) was kept as INR > 1.0 and INR < 1.0. With respect to this value, we selected around 10 upregulated proteins and 16 downregulated proteins, which are specified in Figure 1A. The highly expressed proteins included mitogen-activated protein kinase 1 (MAPK1), which exhibited about a 2.81-fold increase compared to the p-MDA-MB-231 cells. Next to MAPK1, the highly expressed protein was dual-specificity protein kinase CDC-like kinases (CLK1), which exhibited a 1.32-fold increase.

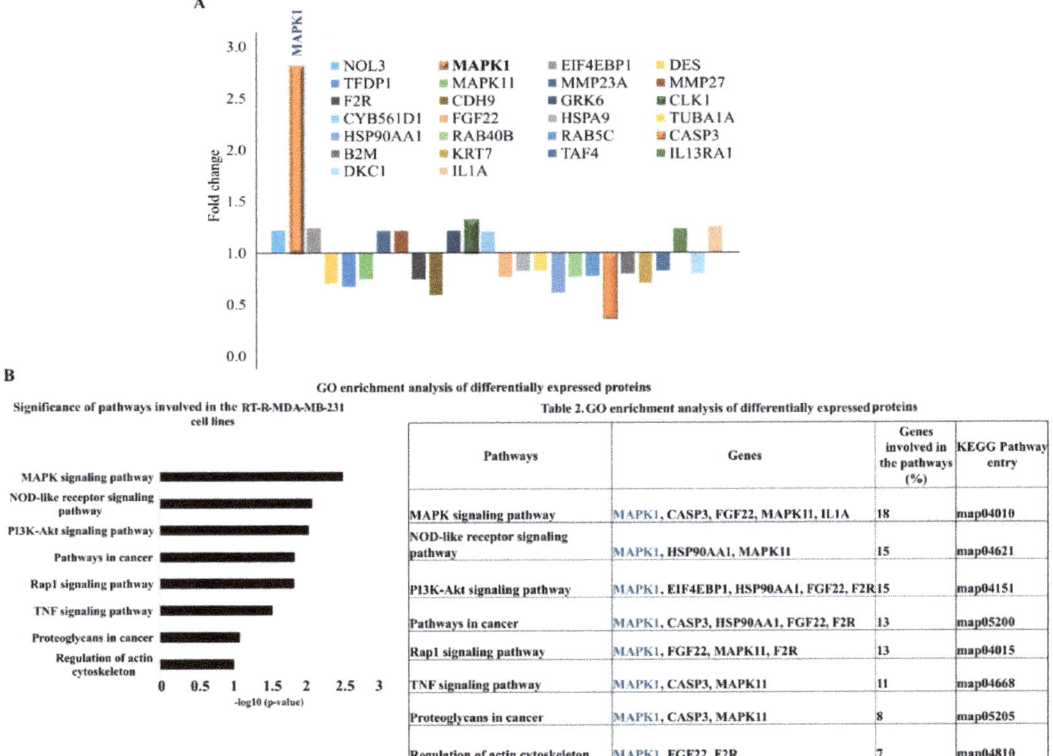

Table 2. GO enrichment analysis of differentially expressed proteins

Pathways	Genes	Genes involved in the pathways (%)	KEGG Pathway entry
MAPK signaling pathway	MAPK1, CASP3, FGF22, MAPK11, IL1A	18	map04010
NOD-like receptor signaling pathway	MAPK1, HSP90AA1, MAPK11	15	map04621
PI3K-Akt signaling pathway	MAPK1, EIF4EBP1, HSP90AA1, FGF22, F2R	15	map04151
Pathways in cancer	MAPK1, CASP3, HSP90AA1, FGF22, F2R	13	map05200
Rap1 signaling pathway	MAPK1, FGF22, MAPK11, F2R	13	map04015
TNF signaling pathway	MAPK1, CASP3, MAPK11	11	map04668
Proteoglycans in cancer	MAPK1, CASP3, MAPK11	8	map05205
Regulation of actin cytoskeleton	MAPK1, FGF22, F2R	7	map04810

Figure 1. Antibody array analysis of radiation-resistant MDA-MB 231 breast cancer cells (RT-R-MDA-MB-231 cells). (**A**) Graphical representation of differentially expressed proteins with respect to the fold change in RT-R-MDA-MB-231 cells compared to parental MDA-MB-231 (p-MDA-MB-231) cells. (**B**) Gene ontology enrichment analysis of differentially expressed proteins in RT-R-MDA-MB-231 cells by KEGG analysis. GO enrichment analysis showed that MAPK1 is related to all the suggested signaling pathways involved in the radio-resistance of RT-R-MDA-MB-231 cells (Table 1).

Among the 16 downregulated proteins, caspase 3 was the most downregulated in the RT-R-MDA-MB-231 cells, which is suggested as one of the mechanisms for RT resistance [20]. Figure 1A shows a graphical representation of the proteins concerning the fold change. These findings suggested that the upregulation of MAPK1, CLK1, and FGF22 and the downregulation of caspase 3 might be involved in the acquisition of radio-resistant MDA-MB-231 cells.

Table 1. GO enrichment analysis of differentially expressed proteins. GO enrichment analysis showed that MAPK1 is related to all the suggested signaling pathways involved in the radio-resistance of RT-R-MDA-MB-231 cells.

Pathways	Genes	Genes Involved in the Pathways (%)	KEGG Pathway Entry
MAPK signaling pathway	MAPK1, CASP3, FGF22, MAPK11, IL1A	18	map04010
NOD-like receptor signaling pathway	MAPK1, HSP90AA1, MAPK11	15	map04621
PI3K-Akt signaling pathway	MAPK1, EIF4EBP1, HSP90AA1, FGF22, F2R	15	map04151
Pathways in cancer	MAPK1, CASP3, HSP90AA1, FGF22, F2R	13	map05200
Rap1 signaling pathway	MAPK1, FGF22, MAPK11, F2R	13	map04015
TNF signaling pathway	MAPK1, CASP3, MAPK11	11	map04668
Proteoglycans in cancer	MAPK1, CASP3, MAPK11	8	map05205
Regulation of actin cytoskeleton	MAPK1, FGF22, F2R	7	map04810

2.2. MAPK1 Is the Most Important Signaling Pathway in Acquiring Radio-Resistant RT-R-MDA-MB-231 Cells

Gene ontology (GO) enrichment analysis of differentially expressed proteins was carried out with the use of KEGG (Kyoto Encyclopedia of Genes and Genomes) pathway analysis. The KEGG pathway analysis showed that the most significant pathway involved in the RT-R-MDA-MB-231 cells was the MAPK1 signaling pathway (Figure 1B). GO enrichment analysis suggested that MAPK1 is the most important signaling pathway in acquiring radio-resistant RT-R-MDA-MB-231 cells. The differentially expressed proteins were interrogated using the STRING database for the protein–protein interaction network analysis. String analysis of the protein–protein interaction (PPI) network generated an interconnected protein network with a medium confidence level of 0.04, which created a single module. The PPI network analysis of differentially expressed proteins showed a single module with 15 proteins such as MAPK1, CASP3, FGF22, MAPK11, HSP90AA1, and F2R. They are involved in MAPK signaling, NOD-like receptor signaling, PI3K-Akt signaling, and Pathways in cancer. The highly increased MAPK1 is related to all the suggested pathways. In addition, this module revealed that MAPK1 harbored a direct protein–protein interaction with caspase 3, which is crucial in inducing programmed cell death type 1 (apoptosis) (Figure 2). These findings support MAPK1 as being one of the important proteins involved in the acquisition of radio-resistant MDA-MB-231 cells.

2.3. ERK Signaling Was Important in the Cell Survival of RT-R-MDA-MB-231 Cells, and the Inhibition of MEK/ERK Signaling Reversed the Radio-Resistance of MDA-MB-231 Cells

To investigate the inhibitory effect of a MEK/ERK inhibitor, PD98059, in RT-R-MDA-MB-231 cells, we performed cell viability assays in p-MDA-MB-231 cells and RT-R-MDA-MB-231 cells. Morphological analysis (Figure 3A) revealed that the RT-R-MDA-MB-231 cells were highly proliferative compared to the p-MDA-MB-231 cells even in the low dose of PD98059-treated cells (less than 20 μM). However, the proliferation rate of RT-R-MDA-MB-231 cells was not higher when they were treated with 20 μM of PD98059. The MTT assay showed that the anti-cancer effect of PD98059 was greater on RT-R-MDA-MB-231 cells than on p-MDA-MB-231 cells (Figure 3B). Morphological analysis (Figure 3A) also revealed that more cell deaths and cellular collapse were observed in RT-R-MDA-MB-231 cells than in p-MDA-MB-231 cells. In addition, there was a difference in morphology of the cell death between p-MDA-MB-231 cells and RT-R-MDA-MB-231 cells (Figure 3A). These findings suggested that ERK signaling should be important in the cell survival of RT-R-MDA-MB-231 cells, and that the inhibition of ERK signaling might reverse the radio-resistance of MDA-MB-231 cells.

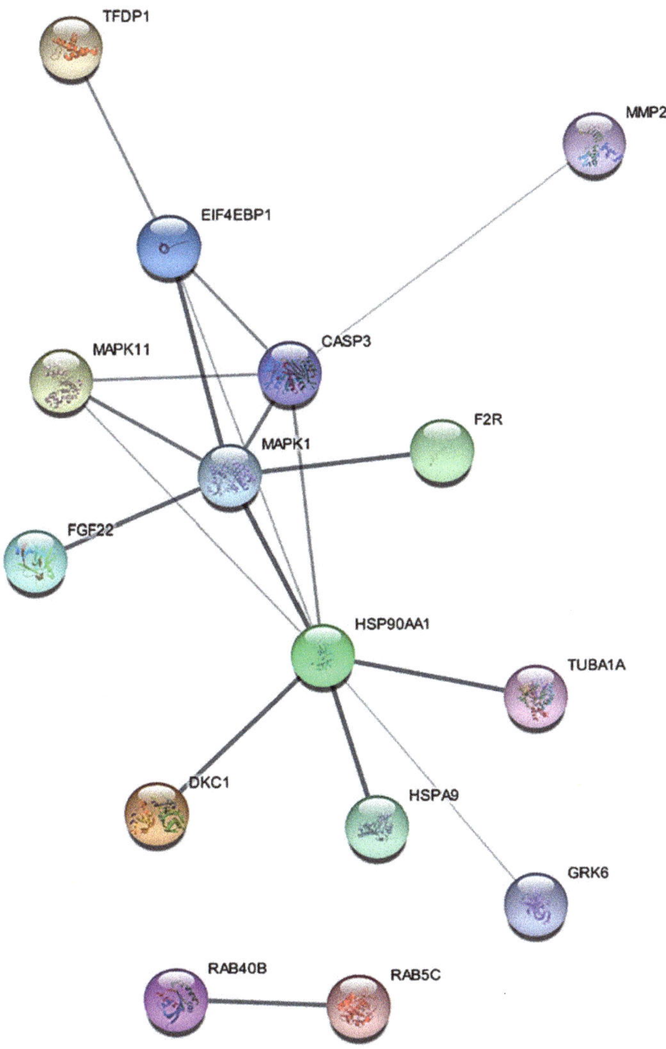

Figure 2. STRING (Cytoscape 3.6) analysis showed the protein–protein interaction (PPI) network of 13 proteins such as MAPK1, CASP3, FGF22, MAPK11, HSP90AA1, and F2R. They are involved in MAPK signaling, NOD-like receptor signaling, PI3K-Akt signaling, and Pathways in cancer in RT-R-MDA-MB-231 cells. The lines connecting the proteins depict "known" or "predicted" interactions. The thickness of the line corresponds to the strength of the interaction between the proteins. A total of 30 edges (protein–protein relationships) were discovered from 22 expected edges.

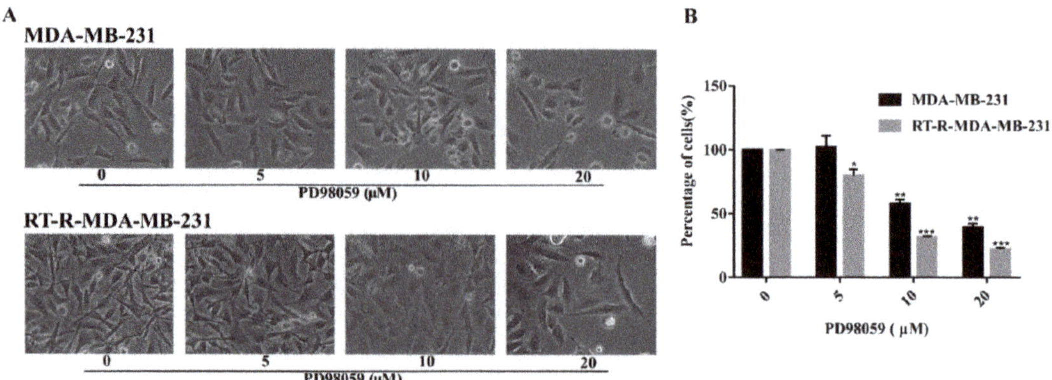

Figure 3. A MEK/ERK inhibitor (PD98059) inhibited the proliferation and promoted cell death in both p-MDA-MB-231 and RT-R-MDA-MB-231 cells. (**A**) p-MDA-MB-231 and RT-R-MDA-MB-231 cells were treated with PD98059 at the given concentrations. Morphological analyses were performed under a light microscope at 48 h. MEK/ERK inhibition resulted in significant morphological changes in both p-MDA-MB-231 and RT-R-MDA-MB-231 cells with a loss of cell integrity, as well as the reduction in cell population compared to the untreated intact cells (magnification, ×200; scale bar, 200 µm). (**B**) Cell viability was determined via MTT assay. The graph represents the % of viable cells after MEK/ERK inhibition. The values are expressed as mean ± standard deviation (SD) (n = 5) (* $p < 0.05$ vs. each control; ** $p < 0.01$ vs. each control; *** $p < 0.005$ vs. each control).

2.4. Inhibition of ERK Signaling Reversed the Radio-Resistance of RT-R-MDA-MB-231 Cells

To explore the radio-sensitivity of both p-MDA-MB-231 and RT-R-MDA-MB-231 cells, we performed a colony formation assay. This revealed that RT-R-MDA-MB-231 cells were resistant to radiation (RT) until 4 Gy, whereas p-MDA-MB-231 cells were sensitive to RT treatment (Figure 4A,B). The colony number of RT-R-MDA-MB-231 cells was higher than that of p-MDA-MB-231 cells, which suggested that the RT-R-MDA-MB-231 cells were highly proliferative compared to p-MDA-MB-231 cells (Figure 4A,B). To investigate the correlation between activated ERK signaling and radio-resistance in RT-R-MDA-MB-231 cells, we performed an ERK inhibition test with a colony formation assay. As shown in Figure 4C,D, the inhibition of MEK/ERK (at 20 µM of PD98059) reversed the radio-resistance of RT-R-MDA-MB-231 cells. These findings support the importance of activated ERK signaling for the radio-resistance of RT-R-MDA-MB-231 cells.

2.5. Inhibition of ERK Signaling-Induced Necroptosis of RT-R-MDA-MB-231 Cells While It Induced the Apoptosis of p-MDA-MB-231 Cells

In Figure 3A, we found differences in the morphology between p-MDA-MB-231 cells and RT-R-MDA-MB-231 cells after ERK inhibition. To elucidate the differences in cell morphology between the two types of cells, we performed mitochondria staining, Mayer's hematoxylin staining for the cell structure, and DAPI for the nucleus. MitoTracker® Red staining is used to show the live time status of mitochondria [21]. The staining revealed that, with the treatment of the MEK/ERK inhibitor, mitochondrial fragmentation was seen in RT-R-MDA-MB-231 cells at the 24 h-inhibition of ERK signaling (Figure 5A). With the inhibition of ERK signaling, RT-R-MDA-MB-231 cells showed more fragmentation and swollen mitochondria than p-MDA-MB-231 cells did, suggesting that ERK inhibition contributes to the mitochondrial fission in RT-R-MDA-MB-231 cells. Mayer's hematoxylin staining revealed that 24 h-MEK/ERK inhibition induced the cell swelling of nuclei and cytoplasm in RT-R-MDA-MB-231, while it induced the shrinkage of nuclei in the p-MDA-MB-231 cell (Figure 5B). These results were also confirmed with DAPI staining. The DAPI staining revealed a high level of nuclear swelling in RT-R-MDA-MB-231 cells treated with the MEK/ERK inhibitor, and it revealed nuclear fragmentation in p-MDA-MB-231 cells

(Figure 5C). These results suggest that the ERK inhibition promotes cell death in both p-MDA-MB-231 and RT-R-MDA-MB-231 cells, but that the mechanisms for the cell death of the two cells were different.

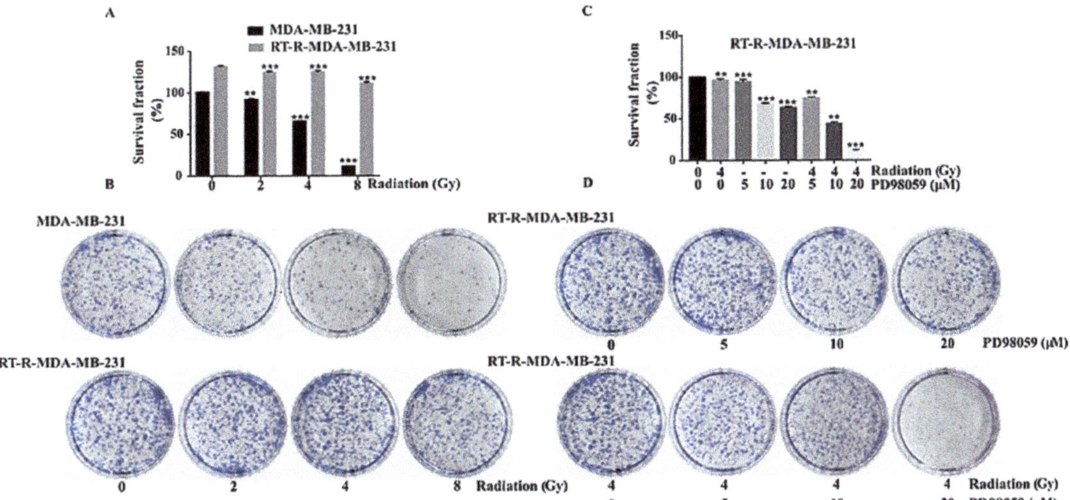

Figure 4. Clonogenic assay for effects of ERK inhibition on radio-resistance of RT-R-MDA-MB-231 cells. (**A**) Graphical representation of survival fraction of p-MDA-MB-231 cells and RT-R-MDA-MB-231 cells in % with the number of colonies after RT treatment. (**B**) Colony formation assay of RT-R-MDA-MB-231 cells and MDA-MB-231 cells. The cells were irradiated with different doses of RT (as indicated), they were grown for 2 weeks, and they were then stained with 0.1% Giemsa stain. Images were captured by a CCD (charge-coupled device) camera and the figures are representative of three independent experiments. (**C**) Graphical representation of RT-R-MDA-MB-231 cells survival fraction in % with the number of colonies after MEK/ERK inhibition with and without IR. (**D**) Colony formation assay of RT-R-MDA-MB-231 cells after MEK/ERK inhibition with and without IR and recorded as specified in (**B**). The values are represented as mean $\pm$ standard deviation (SD) ($n = 5$). ** $p < 0.01$; *** $p < 0.005$.

2.6. ERK Inhibition Induced Caspase Activation and PARP-1 Cleavages in p-MDA-MB-231 Cells, While It Did Increase the Expression of Cyclophilin A (CypA) and AIF in RT-R-MDA-MB-231 Cells

To molecularly confirm the difference in cell death between p-MDA-MB-231 and RT-R-MDA-MB-231 cells, we performed Western blot analysis. Figure 6 demonstrates that, in MDA-MB-231 cells, ERK inhibition induced the cleavage PARP-1 and caspase-3, which is a hallmark for caspase-dependent apoptosis, but that, in the RT-R-MDA-MB-231 cells, ERK inhibition induced AIF (apoptosis-inducing factor), which positively regulates the CypA protein, which is considered a biomarker of necroptosis [22]. These findings support ERK inhibition inducing the apoptosis of p-MDA-MB-231 and the necroptosis of RT-R-MDA-MB-231 cells.

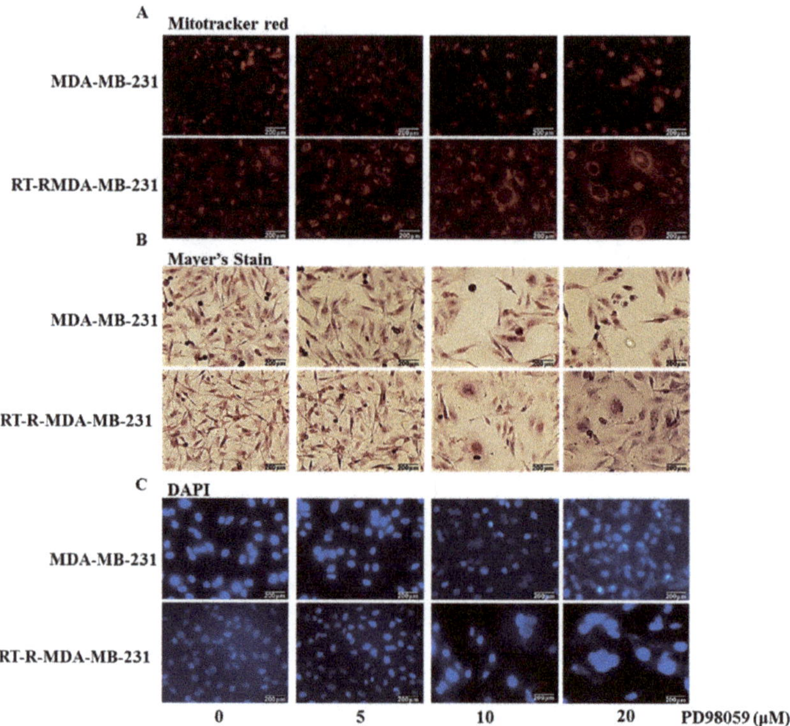

Figure 5. The difference in cell morphology between p-MDA-MB-231 and RT-R-MDA-MB-231 cells during ERK inhibition. Cells were seeded in 12-well plates with a 1×10^5 cell/well density treated with the indicated concentrations of the MEK/ERK inhibitor (PD98059) for 24 h. (**A**) Mitochondrial morphology was analyzed with a fluorescent microscope after staining with MitoTracker (red). (**B**) Light microscopy of hematoxylin-stained cells showed the whole cell morphology of the MEK/ERK inhibitor-induced cell death. (**C**) Nuclear morphology analysis of the MEK/ERK inhibitor-induced cell death with DAPI staining. Results were confirmed by three independent experiments.

2.7. ERK Inhibition Reduced the Expression of Cancer Stem Cell (CSC) Markers (CD44 and Oct $\frac{3}{4}$) and the EMT Phenotype, Which Is Closely Related to the Radio-Resistance of RT-R-MDA-MB-231 Cells

It was reported that CSC markers and EMT phenotypes were highly expressed in RT-R-MDA-MB-231 cells compared to the p-MDA-MB-231 cells, and that their high expression was closely related to radio-resistance [8]. Here, we assessed the effect of ERK inhibition on the expression of CSC markers and the EMT phenotype on both MDA-MB-231 and RT-R-MDA-MB-231 cells. As previously reported, Western blot analysis revealed that RT-R-MDA-MB-231 cells showed a higher expression of CSC markers (CD44 and Oct 3/4) and EMT markers (N-cadherin and β-catenin) compared to the p-MDA-MB-231 cells (Figure 7A,B). ERK inhibition significantly suppressed the expression of CSC markers and the EMT phenotype in both p-MDA-MB-231 and RT-R-MDA-MB-231 cells (Figure 7A,B). In addition, the ERK inhibition more prominently suppressed the EMT phenotype of RT-R-MDA-MB-231 cells than p-MDA-MB-231 cells. These findings indicated that the ERK inhibition clearly suppressed the high expression of CSC markers and the EMT phenotype of RT-R-MDA-MB-231 cells that are reportedly associated with radio-resistance [5].

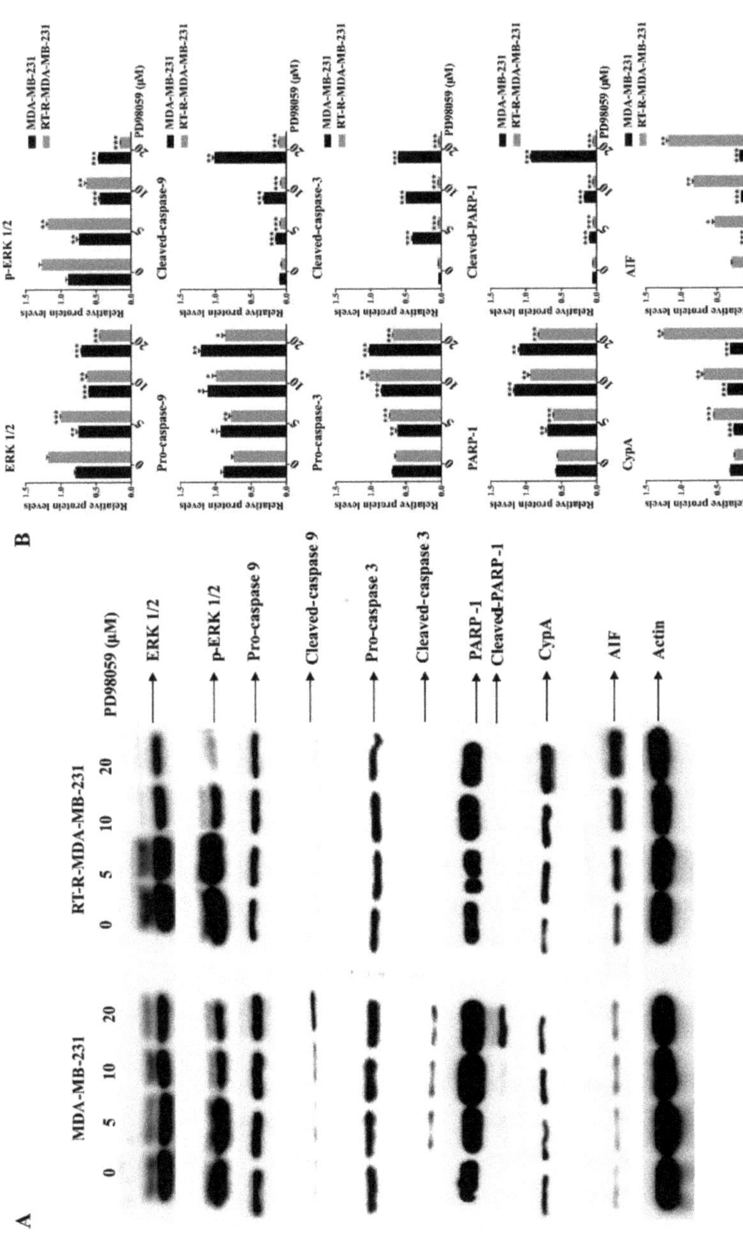

Figure 6. Effects of ERK inhibition on cell death-related proteins in both p-MDA-MB-231 and RT-R-MDA-MB-231 cells. Cells were seeded with a seeding density of 5×10^4 cells and were pretreated with the MEK/ERK inhibitor (PD98059) for 48 h. The control cells were left untreated. The whole cell protein lysate was prepared and 30 µg of proteins was resolved in SDS-polyacrylamide gels. (**A**) Western blot analysis of various cell death-related proteins. (**B**) Densitometry analysis of the data in Western blot analysis by ImageJ software. The values were normalized against β-actin, and they are represented as mean ± standard deviation (SD) ($n = 3$). ** $p < 0.01$; *** $p < 0.005$ vs. the control group.

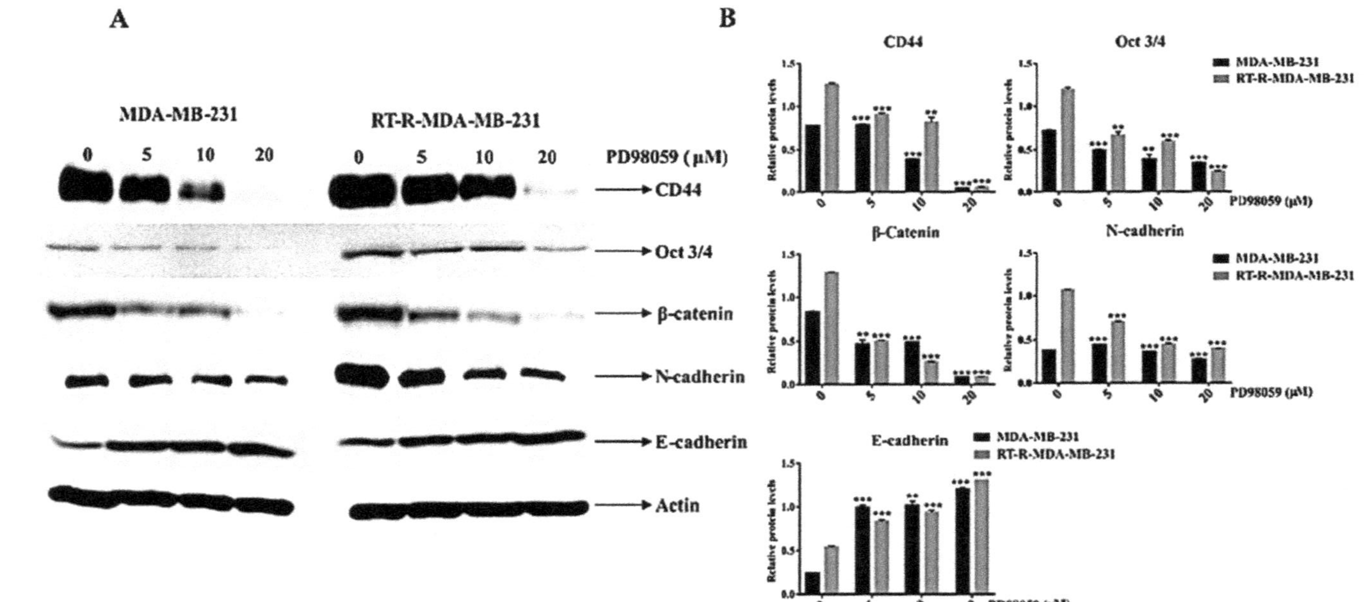

Figure 7. Effects of ERK inhibition on the expression of CSC markers (CD44 and Oct $\frac{3}{4}$), β-catenin, and EMT markers (E-cadherin and N-cadherin) in both p-MDA-MB-231 and RT-R-MDA-MB-231 cells. (**A**) Western blot analysis of CSC markers (CD44 and Oct $\frac{3}{4}$), β-catenin, and EMT markers (E-cadherin and N-cadherin). (**B**) Densitometry analysis of the data in Western blot analysis by ImageJ software. The values were normalized against β-actin, and they are represented as mean ± standard deviation (SD) ($n = 5$). ** $p < 0.01$; *** $p < 0.005$ vs. the control group.

3. Material and Methods

3.1. Protein Array Analysis

The total proteins of p-MDA-MB-231 and RT-R-MDA-MB-231 cells were isolated with a radioimmunoprecipitation assay (RIPA) buffer, which contained 0.1% NP-40 and 0.1% sodium dodecyl sulfate in phosphate-buffered saline (PBS) containing a protease inhibitor cocktail (Sigma Aldrich, St. Louis, MO, USA). The expression profiling of proteins was analyzed by a Signaling Explorer Antibody Array (Ebiogen, Seoul, Korea).

3.2. Bioinformatics Analysis

The obtained proteins from the antibody array analysis were further submitted to DAVID (The Database for Annotation, Visualization, and Integrated Discovery). DAVID is an online tool that provides a biological understanding between two or more data sets of genes, and it can also be used to determine gene ontology (GO) in terms of biological processes and cellular processes. To determine the pathways involved in the identified genes, KEGG (Kyoto Encyclopedia of Genes and Genomes) (https://www.genome.jp/kegg/pathway.html, accessed on 20 October 2018) was employed. The selected genes were investigated for potential protein–protein interactions using STRING (Search Tool for the Retrieval of Interacting Genes) database version: 10.5 (https://string-db.org, accessed on 19 February 2021). For the display of protein interactions, selected proteins were uploaded into the STRING database and assessed using Cytoscape Software version Cytoscape_v3.7.1 (https://www.cytoscape.org, accessed on 19 February 2021). To access the interaction of the experimental data and to provide unambiguous comprehensive coverage, the online tool string was used.

3.3. Cell Culture

RT-MDA-MB-231 cells were established as previously described [8]. Briefly, MDA-MB-231 cells were fractionated with X-ray irradiation until a final dose of 50 Gy was reached. p-MDA-MB-231 and RT-MDA-MB-231 cells were cultured in RPMI-1450 medium supplemented with 10% heat-inactivated FBS and 1% penicillin/streptomycin. The cells were maintained at 37 °C in a 5% CO_2 incubator. The cells were grown with 80% confluence and were treated with a MEK/ERK inhibitor (PD98059) dissolved in DMSO or DMSO alone.

3.4. Cell Viability Assay

We used a calorimetric assay, MTT (3-(4,5-dimethylthiazol-2-yl)-2,5-diphenyltetrazolium bromide), to analyze the cell viability. The cells were seeded in 24-well plates with a confluence of 1×10^5 cells/well, they were treated with the MEK/ERK inhibitor, and they were maintained for 24 and 48 h at 37 °C in a 5% CO_2 incubator. After incubation, 50 μL of MTT (0.5 mg in $1\times$ PBS) was added to each well and incubated for about 2 h at 37 °C in a 5% CO_2 incubator. The media were removed and the formazan crystals that formed in the live cells were dissolved with the 500 μL of DMSO. The solubilized formazan crystals were transferred to 96-well plates and the absorbance was read by an enzyme-linked immunosorbent assay (ELISA) reader at 540 nm. The cell viability was quantified in percentage, while vehicle-treated control cells were set at 100%.

3.5. Colony Formation Assay

P-MDA-MB-231 or RT-R MDA-MB-231 cells (1×10^3 cells/well) were seeded in six-well plates, treated with the indicated doses of the MEK/ERK inhibitor, and maintained at 37 °C in a 5% CO_2 incubator. The cells were irradiated with a given concentration, and the media were discarded after 24 h and replaced with fresh complete media every 2–3 days. After 14 days, the medium was discarded and the cells were washed with $1\times$ PBS thrice. The colonies were fixed with absolute methanol for 10 min, stained with Giemsa staining solution, and then maintained at room temperature. The number of colonies was counted using ImageJ software.

3.6. Cytochemical Staining Methods

3.6.1. Mitotracker Red Analysis

For mitochondrial morphology analysis, Mitotracker Red dye was used. The cells were seeded with a confluence of 1×10^5 cell/well in 12-well plates, they were treated with the MEK/ERK inhibitor for 24 h, and they were washed with $1\times$ PBS and then stained with 0.5 μL of Mitrotracker red in 500 μL of $1\times$ PBS. The cells were incubated for 30 min in a 5% CO_2 incubator at 37 °C. After incubation, the cells were viewed under a fluorescent microscope for the analysis of the live mitochondrial status after the treatment of a MEK/ERK inhibitor.

3.6.2. Hematoxylin Staining

The cells were seeded in 12-well plates with a confluence of 1×10^5 cells/well and were grown for 24 h with the MEK/ERK inhibitor at 37 °C in a CO_2 incubator. After incubation, the cells were washed with $1\times$ PBS and then fixed with 4% para-formaldehyde overnight. The fixed cells were washed thrice with $1\times$ PBS for about 5 min per wash, they were stained with 200 μL of Mayer's hematoxylin staining solution, and they were incubated for 20 min in the dark at room temperature. Then, the cells were washed thoroughly with $1\times$ PBS, followed by 1 mL of 90% glycerol, and they were observed under a phase-contrast microscope.

3.6.3. DAPI (4′,6-diamidino-2-phenylindole) Staining

For the nuclear morphological changes, DAPI staining was performed. The cells were seeded in 12-well plates at a density of 1×10^5 cells/well with the treatment of the MEK/ERK inhibitor for about 24 h, they were washed with $1\times$ PBS, and they were fixed overnight with 4% formaldehyde. After fixation, the cells were washed with $1\times$ PBS thrice for about 5 min per wash. DAPI solution (0.5 μL) was added to the 500 μL of $1\times$ PBS, which was incubated for 30 min at 37 °C with a 5% CO_2 incubator. After incubation, the cells were washed with $1\times$ PBS and were fixed with 90% glycerol in $1\times$ PBS. The cells were viewed under a fluorescent microscope (Leica Microsystems GmbH, Wetzlar, Germany).

3.7. Western Blot Analysis

P-MDA-MB-231 and RT-MDA-MB-231 cells were seeded in 100 mm plates with a cell density of 2×10^6 cells/plate. The cells were treated with the MEK/ERK inhibitor or DMSO as a vehicle control and were maintained for 48 h at 37 °C with a 5% CO_2 incubator. After 48 h, the cells were harvested and transferred to 15 mL falcon tubes, and they were centrifuged for 5 min at 2000 rpm. The supernatant was discarded, and the tubes were centrifuged again to remove the residual supernatant. After complete removal of the supernatant, 500 μL of the 2X sample buffer containing 100 mM of Tris-Cl (pH 6.8), 4% (*w*/*v*) sodium dodecyl sulphate (SDS), 0.2% (*w*/*v*) bromophenol blue, and 200 mM of dithiothreitol was added. The protein lysates were collected in 1.5 mL Eppendorf tubes and kept at 100 °C for 10 min. The protein concentration was determined by the Bradford assay. In addition, 30 μg of the proteins was resolved in 8–12% SDS-PAGE and was transferred to a polyvinylidene difluoride membrane. After transfer, the membranes were blocked with 3% skimmed milk in Tris-buffered saline containing 1% Tween 20 (TBST) buffer for 30 min at room temperature, and they were incubated at 4 °C overnight with antibodies against actin (A5441, 1:10,000, Sigma-Aldrich, St. Louis, MO, USA), ERK (SC-94, 1:2000, Santa Cruz Biotechnology, Dallas, TX, USA), p-ERK (SC-7383, 1:2000, Santa Cruz Biotechnology), CypA (SC-134310, 1:2000, Santa Cruz Biotechnology), pro-caspase 9 (SC-56076, 1:2000, Santa Cruz Biotechnology), pro-caspase 3 (SC-7272, 1:2000, Santa Cruz Biotechnology), AIF (SC-55519, 1:1000, Santa Cruz Biotechnology), PARP-1 (SC-8007, 1:2000, Santa Cruz Biotechnology), CD44 (ab51037, 1:2000, Abcam, Cambridge, UK), β-catenin (SC-7199, 1:2000, Santa Cruz Biotechnology), Oct 3/4 (SC-5279, 1:2000, Santa Cruz Biotechnology), E-cadherin (ab1416, 1:2000, Abcam), and N-cadherin (ab76011, 1:2000, Abcam). After overnight incubation in primary antibodies, the membranes were washed with TBST buffer thrice for about

10 min per wash. Then, the membranes were incubated in horseradish peroxidase (HRP)-conjugated secondary antibody for 2 h at room temperature with 1:2000 dilution. The membranes were later washed with TBST buffer three times (10 min/wash) and were developed with ECL (electrochemiluminescence) solutions (Bio-Rad Laboratory, Hercules, CA, USA).

3.8. Statistical Analysis

All experiments were performed at least in triplicate, and all analyses were performed with the use of GraphPad Prism 7 software (GraphPad Software, San Diego, CA, USA). One-way ANOVA followed by the Newman–Keuls post hoc test was performed to compare various treatment groups. The data were presented as mean ± standard deviation (SD). A *p*-value <0.05 was considered statistically significant.

4. Discussion

Radiation therapy is one of the common and essential parts of breast cancer treatment. Around half of the cancer patients go through radiation therapy at some point in their treatment [23]. Ionizing radiation (IR) induces DNA damage through oxidative stress. The free OH radicals are capable of promoting single-stranded and double-stranded DNA breaks (SSB and DSB, respectively), which, in turn, triggers cell death [24]. Thus, cells develop IR resistance by counteracting the four 'R's,' which is DNA damage by DNA repair, redistribution, repopulation, and reoxygenation through the activation of various pathways [25]. Several studies have described the role of irradiation in breast cancer pathways and the involvement of several proteins in the development of resistance against radiation, which we have listed in Table 2. Only a small number of studies have investigated mechanisms of acquired radio-resistance through the generation of radio-resistant cell lines, which tend to focus on a single specific pathway. Thus, it is paramount to focus on the signaling mechanism through the generation of the radio-resistant breast cancer cell line.

This study was designed to find the hub signaling involved in the RT resistance of RT-R-MDA-MB-231 cells and to investigate the importance of the hub signaling in the reversal of radio-resistance and the regulation of the CSC and EMT phenotype that is highly associated with radio-resistance. We found that ERK signaling was highly activated in RT-R-MDA-MB-231 cells compared to in p-MDA-MB-231 cells and that ERK signaling was essential for the survival of both p-MDA-MB-231 and RT-R-MDA-MB-231 cells. In addition, the RT resistance of RT-R-MDA-MB-231 cells was reversed by the inhibition of ERK signaling (Figure 4). Furthermore, we demonstrated that the activated ERK signaling was associated with cancer stemness and EMT phenotype (Figure 7). Considering all these findings, we can conclude that activated ERK signaling is one of the major hub signaling related to the acquisition of radio-resistant MDA-MB-231 cells.

Table 2. List of studies involving radiation resistance and sensitivity in MDA-MB-231 cells or TNBC, and their possible mechanisms behind it. The reports were mainly categorized into three parts as follows: (i) specific signaling pathways involved in radio-resistant breast cancer cells, (ii) signaling pathways involved in radio-sensitivity, (iii) altered expression of gene/proteins involved in radio-resistant breast cancer cells. (iv) Studies that did not fall into any of the three above categories. Abbreviations: IFIT2—interferon-induced protein with tetratricopeptide repeats 2, TRIB3—tribbles homolog 3, ESM-1—endothelial cell-specific molecule-1, DLX2—vertebrate distal-less homeobox 2.

No.	Author	Key Findings	Year	References
		(i) Specific signaling pathways involved in radio-resistant breast cancer cells		
1.	Gray et al.	The radiation-resistant ER+ breast cancer cell line (MCF-7, ZR-751) showed increased migration and invasion compared to the radiation-resistant ER- breast cancer cell line (MDA-MB-231). ER+ cells also showed a shift from ER to EGFR signaling pathways with increased MAPK and PI3K activity.	2019	[26]
2.	Ediriweera et al.	A phenolic lipid, 10-Gingerol, promotes apoptosis in radiation-resistant MDA-MB-231 cells through the PI3K/Akt signaling pathway.	2020	[27]
3.	Jin et al.	The overexpression of ESM-1 plays a critical role in radiation-resistant MDA-MB-231 cells through the regulation of PDK, PKC, and ERK1/2 pathways, and the subsequent activation of transcription factors HIF-1α, NF-κB, and STAT-3 to regulate adhesion molecules, MMPs, and VEGF.	2020	[28]
4.	lu et al.	The Wnt/β-catenin signaling pathway plays an important role in the development of radioresistance and Niclosamide, an FDA-approved anthelmintic drug that induces radiosensitivity in radiation-resistant MDA-MB-231 via inhibiting STAT3 and Bcl-2.	2018	[29]
5.	Bravatà et al.	Gene expression profiles of the MDA-MB-231 radiation cell fraction show increased TNF signaling, Phagosome, NF-kappa B signaling, Jak-STAT signaling, and PI3K-Akt signaling.	2019	[30]
6.	Choi et al.	DLX2 expression with irradiation incidence causes the increase in the EMT process and CSCs population through the Smad2/3 signaling pathway in MDA-MB-231 and A549 cells.	2016	[31]
		(ii) Signaling pathways involved in radio sensitivity of MDA-MB-231 cells		
7.	Yin et al.	Niclosamide, an antihelminthic drug, inhibited the Wnt/β-catenin signaling pathway and increased the radiation sensitivity to triple-negative breast cancer (TNBC) cells (MDA-MB-231, MDA-MB-468, and Hs578T cells).	2016	[32]
8.	Lin et al.	COX-2 upregulation promotes radioresistance in MDA-MB-231 cells through the p38/MAPK-mediated alteration of apoptosis and metastasis.	2013	[33]
9.	KO et al.	Radiation-resistant MDA-MB-231 cells showed an increased cell proliferation, cell adhesion, EMT process, and increased stem cell population.	2018	[8]
10.	Koh et al.	Baicalein reduced the stem cell-like properties and metastasis in radiation-resistant MDA-MB-231 cells through the upregulation of IFIT2.	2019	[34]

Table 2. *Cont.*

No.	Author	Key Findings	Year	References
		(iii) Altered expression of gene/proteins involved in radio-resistant MDA-MB-231 cells.		
11.	Lammering et al.	Irradiation increased the expression of the EGFR protein in MDA-MB-231 xenograft tumors.	2004	[35]
12	Kim et al.	Proteomic analysis revealed single and the fraction of radiation increased cathepsin D (CTSD), gelsolin (GSN), argininosuccinate synthase 1 (ASS1), and C-type mannose receptor 2 (MRC2) in MDA-MB-231 cells.	2015	[36]
13.	Miao et al.	Radiation-resistant MDA-MB-231 and MCF-7 cells showed an altered expression of several members of the HSP70 and HSP40, subfamilies of HSPs, and an increased level of HSPB8, a target of NF-κB that could be responsible for the development of radioresistance.	2019	[37]
14.	Lee et al.	Increased expression of TRIB3 in radiation-resistant MDA-MB-231 cells causes the resistance and knockdown of TRIB3 sensitized toward radiation.	2019	[38]
15.	HOU et al.	Microarray analysis of radiation-resistant MDA-MB-231 cells showed increased cell adhesion and EMT factors.	2019	[39]
16.	Yang et al.	Overexpression of a small RNA molecule miR-634 decreases the survival rate of radiation-resistant MCF-7 and MDA-MB-231 cells by direct interaction with STAT3.	2020	[40]
		(iv) Studies that did not fall into any of the three above categories.		
17.	Li et al.	A small molecule, ABT-787, induces radiosensitivity in radiation-resistant MDA-MB-231 by targeting Bcl-2 and Bcl-xL.	2012	[41]
18.	Nguyen et al.	A phytochemical phenethyl isothiocyanate reduces the CSC population in radiation-resistant MDA-MB-231 cells through upregulating ROS levels and targeting Metadherin at the post-transcriptional levels.	2020	[42]
19.	Oommen and Prise	A novel benzylidene lactam compound, KNK437, inhibits HIF-1α, HSF1, and AKT in hypoxia-induced MDA-MB-231 and T98G cells, which, in turn, induces radiosensity.	2012	[43]
20.	Kuger et al.	PI3K/mTOR inhibitor NVP-BEZ235 showed a synergistic effect with irradiation (IR) in MCF-7 and MDA-MB-231 cells.	2014	[44]
21.	Holler et al.	The molecular targeting of Akt by Akt inhibitor MK2206 or the knockdown of Akt1 led to a rapamycin-induced radiosensitization of SK-MES-1, HTB-182, or MDA-MB-231 cells by increasing DNA-double-stranded breaks.	2016	[45]
22.	Chen et al.	The estrogen receptor mediates the radiosensitivity of TNBC cells.	2017	[46]
23.	Liu et al.	Hypoxia due to a high cell density downregulated the EGFR expression and increased the sensitivity to ionizing radiation in MCF-7 and MDA-MB-231 cells.	2018	[47]
24.	Arnold et al.	STAT3 inhibition combined with radiation reduces the cellular plasticity in MDA-MB-231 and SUM159PT cells.	2019	[48]

Before concluding, we should discuss some questions. The first question would be whether activated ERK signaling is the main mechanism for the radio-resistance of MDA-MB-231 cells. In Figures 4 and 7, the MEK/ERK inhibition test revealed that MEK/ERK inhibition induced cell death and suppressed the expression of CSC markers and the EMT phenotypes of both p-MDA-MB-231 cells and RT-R-MDA-MB-231 cells. This finding also suggested that ERK signaling is essential for the survival of MDA-MB-231 cells and that it may not be related to RT resistance per se. We also agree with the point. In addition, it was reported that other signaling pathways such as PI3K/Akt, and STAT or other anti-apoptotic proteins are important in the radio-resistance or radio-sensitivity of MDA-MB 231 cells (Table 2). However, there is some evidence to support that the ERK signaling is related to the RT-resistance of RT-R-MDA-MB-231 cells. Recent studies have depicted that the ERK1/2 activation prevails over the cell cycle arrest in the G2/M phase where cancer cells are susceptible to IR, thus inducing radio-resistance [5]. Another is that MDA-MB-231 cells are a triple-negative breast cancer cell line, which is known to have high CSC properties [49]. The third is that in our previous report, RT-R-MDA-MB-231 cells also showed an increased STAT 3 activity, which was reported to be related to cancer stemness and EMT, but that the inhibition of STAT 3 activity by the JNK inhibitor or Janus-activated kinase 2 (JAK2) inhibitor could not suppress the increased expression of CSC markers [50]. To solve this question, we carried out this study with antibody microarray analysis, which revealed that highly increased MAPK1 was enriched in all of the upregulated pathways of RT-R-MDA-MB-231 cells (MAPK signaling, NOD-like receptor signaling, PI3K-Akt signaling, and Pathways in cancer). In addition, the KEGG pathway enrichment analysis showed that all enriched pathways include MAPK1, and the PPI network analysis of differentially expressed proteins showed that MAPK1 could be related to the resistance of cell death. The third is that the inhibition of ERK signaling reversed RT resistance (Figure 4). The inhibition of ERK signaling was reported to increase the anti-cancer efficacy of RT [51,52]. This finding could support the reversal of RT resistance caused by the inhibition of ERK signaling. With all of these findings, we can say that activated ERK signaling is one of the main mechanisms for the radio-resistance of MDA-MB-231 cells.

The second point to discuss would be the relationship between ERK signaling and EMT, as well as the CSC phenotype of RT-R-MDA-MB-231 cells, because it is mentioned that several other signaling pathways such as JAK/STAT, Hedgehog, Wnt, Notch, PI3K/PTEN, and nuclear factor-κB (NF-κB) signaling pathways, compared to ERK signaling, are closely related to CSC properties [53–55], and the blocking of these pathways involved might be an effective way to target CSCs [55]. Even though it is not common, PRMT6-dependent CRAF/ERK signaling was reported to regulate CSC plasticity [56]. In addition, it was reported that the CSC properties-related signal is frequently complexed, and there is cross-talk between and among the mentioned various pathways [57]. In addition, the influence of ERK activity in regulating the CSC phenotype is also reported in gemcitabine-resistant pancreatic cells [58], cisplatin-resistant non-small cell lung cancer cells [59], and docetaxel and carboplatin-resistant ovarian cancer cells [60]. To determine the real cause of the activated ERK signaling of RT-R-MDA-MB-231 cells, we performed whole genome sequencing. In the study, we could not find any mutations on the linear line for the activation of ERK signaling (data not shown), such as EGFR, SOS, B-raf, Ras, or MEK. Regarding this question, we could not give the audience clear evidence. Therefore, further studies are required.

The third point to discuss is why the phenotype of cell death induced by ERK inhibition differed between p-MDA-MB-231 cells and RT-R-MDA-MB-231 cells, while ERK inhibition induced cell death and suppressed the increased expression of CSC markers and the EMT phenotype of both p-MDA-MB-231 cells and RT-R-MDA-MB-231 cells. We speculate that the reason could be that RT-R-MDA-MB-231 cells exhibit a decreased activity of caspase. It has been reported that the cancer cells harboring caspase defects frequently undergo necroptosis or necrosis instead of apoptosis when the death signal appears [61]. Initially, we thought that the defects in caspase 3 activity were the main cause that was associated with increased ERK signaling, which was revealed by string analysis of the PPI network. By whole genome sequencing, we also found that RT-R-MDA-MB-231 cells harbor a nonsynonymous single nucleotide mutation in CASP9 (Arg173His; rs2308950), which is known to be involved in the pathogenesis of various cancers (data not shown) [62,63]. We speculate that this mutation is also responsible for RT-R-MDA-MB-231 cells undergoing necroptosis during the inhibition of ERK signaling.

The fourth point to discuss would be the role of the other upregulated and downregulated proteins in RT-R-MDA-MB-231 cells. Although we could not discuss here all of the 26 proteins, recent studies have suggested that the inhibition of CLK1 also decreases cell proliferation [64]. CLK1 and FGF22 are oncogenes in cancer and their inhibition leads to the inhibition of breast cancer growth in cell culture and xenograft models [65,66]. This supports the possible contribution of upregulated CLK1 to the rapid growth of RT-R-MDA-MB-231 cells. The most downregulated protein, caspase 3, may also contribute to the radio-resistance of RT-R-MDA-MB-231 cells, by avoiding IR-induced apoptosis [67]. All of these findings suggest that the changes in the expression of proteins may be involved in the biological phenotype of RT-R-MDA-MB-231 cells. Regarding these points, further studies are warranted.

The fifth point to discuss would be the mechanisms driving the upregulation of ERK signaling in RT-R-MDA-MB-231 cells. As we know that radiotherapy works by damaging the DNA of cancer cells, our first thought was that the upregulation of ERK signaling would be related to some of the mutations in the Ras-Raf-MEK-ERK pathway. Therefore, we performed whole genome sequencing, but there was no additional mutation of ERFR, PI3K/Akt, Ras, Raf, MEK, or ERK molecules of MDA-MB-231 cells (data not shown). In this paper, we inhibited ERK signaling with PD98059, a non-adenosine triphosphate competitive MAPK (MEK) inhibitor [68]. Therefore, we can speculate that the upregulation of ERK signaling would be caused by some hidden mutations in the upstream of the Ras-Raf-MEK-ERK pathway.

The weakness of this study is that we performed the experiment with only one cell line. It is in question whether the main mechanism of the radio-resistance of RT-R-MDA-MB-231 cells can be applied to all radiation-resistant breast cancer cell lines or can be generalized to triple-negative breast cancer cells. In addition, even regarding the radio-resistance of RT-R-MDA-MB 231 cells, other signaling pathways are also suggested as a key signaling pathway involved in the resistance. Similar to the signaling involved in CSC, the signaling involved in the radio-resistance of RT-R-MDA-MB 231 cells could also be complexed. However, aberrantly upregulated ERK signaling contributes to cancer cell proliferation, survival, and metastasis [14], and many other reports have suggested that ERK signaling is an important signaling pathway in radio-resistance [5,15,52]. Therefore, further research is also warranted regarding ERK signaling on the radio-resistance of breast cancer, especially on TNBC.

5. Conclusions

In summary, we found that ERK signaling was highly activated in RT-R-MDA-MB-231 cells compared to p-MDA-MB-231 cells. The activated ERK signaling was associated with an increased cancer stemness and EMT phenotype. In addition, the RT resistance of RT-R-MDA-MB-231 cells was reversed by the inhibition of ERK signaling. Furthermore, the inhibition of ERK suppressed the CSC marker proteins. With all of these findings, we conclude that activated ERK signaling is one of the major hub signals related to the acquisition of radio-resistant MDA-MB-231 cells. This study suggests a distinct and advantageous therapeutic value of the targeting of the ERK signaling pathway in MDA-MB-231 cells. Further research is also warranted regarding ERK signaling on the radio-resistance of breast cancer, especially on TNBC.

Author Contributions: A.P. performed experiments and data analysis. E.J.J., S.-I.G., and W.S.L. performed data analysis. B.K.J., G.S.K., J.-M.J., S.C.H., and W.S.L. reviewed the manuscript and gave comments on the results. W.S.L. conceived of the hypothesis and directed the project. G.S.K. supported this project. A.P. and W.S.L. prepared the manuscript and confirmed the final manuscript. All authors have read and agreed to the published version of the manuscript.

Funding: This work was supported by the Basic Science Research Program through the National Research Foundation of Korea (NRF) funded by the Ministry of Education (2017R1D1A3B05030971).

Institutional Review Board Statement: Not applicable.

Informed Consent Statement: Not applicable.

Data Availability Statement: Samples are not available from the authors.

Conflicts of Interest: The authors declare no conflict of interest.

Abbreviations

CSCs	Cancer stem cells
CypA	Cyclophilin A
AIF	Apoptosis-inducing factor
EMT	Epithelial-to-mesenchymal transition
ERK	Extracellular-signal-regulated kinase
DDR	Damage response
IR	Irradiation
PI3K	Phosphatidylinositol 3-kinase
MAPK	Mitogen-activated protein kinase
Wnt	Wingless-related integration site
GO	Gene Ontology
KEGG	Kyoto Encyclopedia of Genes and Genomes
STRING	Search Tool for the Retrieval of Interacting Genes
DMSO	Dimethyl sulfoxide is an organosulfur
MTT	3-(4,5-dimethylthiazol-2-yl)-2,5-diphenyltetrazolium bromide
PBS	Phosphate-buffered saline
DAPI	4′,6-diamidino-2-phenylindole
SDS-PAGE	Sodium dodecyl sulfate-polyacrylamide gel electrophoresis
SDS	Sodium dodecyl sulphate
RT	Radiotherapy
RT-R	Radiotherapy-resistant
SD	Standard deviation
CLK1	CDC-like kinases
FGF22	Fibroblast growth factor 22
F2R	Coagulation factor II (thrombin) receptor
Gy	Gray
CCD	Charge-coupled device
JAK2	Janus-activated kinase 2
PI3K-Akt	Phosphatidylinositol 3-kinase (PI3K)-protein kinase B (AKT)
STAT	Signal transducer and activator of transcription
NF-κB	Nuclear factor-κB
EGFR	Epidermal growth factor receptor
SOS	Son of Sevenless
B-raf	B-Rapidly Accelerated Fibrosarcoma
Ras	Rat sarcoma
MEK	Mitogen-activated protein kinase kinase
CASP9	Caspase 9
IFIT2	interferon-induced protein with tetratricopeptide repeats 2
TRIB3	Tribbles homolog 3
ESM-1	Endothelial cell-specific molecule-1
DLX2	Vertebrate distal-less homeobox 2

References

1. Siegel, R.L.; Miller, K.D.; Jemal, A. Cancer statistics, 2019. *CA Cancer J. Clin.* **2019**, *69*, 7–34. [CrossRef]
2. Griffiths, C.L.; Olin, J.L. Triple negative breast cancer: A brief review of its characteristics and treatment options. *J. Pharm. Pract.* **2012**, *25*, 319–323. [CrossRef]
3. Brownlee, Z.; Garg, R.; Listo, M.; Zavitsanos, P.; Wazer, D.E.; Huber, K.E. Late complications of radiation therapy for breast cancer: Evolution in techniques and risk over time. *Gland Surg.* **2018**, *7*, 371–378. [CrossRef] [PubMed]
4. Langlands, F.E.; Horgan, K.; Dodwell, D.D.; Smith, L. Breast cancer subtypes: Response to radiotherapy and potential radiosensitisation. *Br. J. Radiol.* **2013**, *86*, 20120601. [CrossRef]

5.	Marampon, F.; Ciccarelli, C.; Zani, B.M. Biological Rationale for Targeting MEK/ERK Pathways in Anti-Cancer Therapy and to Potentiate Tumour Responses to Radiation. *Int. J. Mol. Sci.* **2019**, *20*, 2530. [CrossRef]
6.	Luo, M.; Ding, L.; Li, Q.; Yao, H. miR-668 enhances the radioresistance of human breast cancer cell by targeting IκBα. *Breast Cancer* **2017**, *24*, 673–682. [CrossRef] [PubMed]
7.	Huang, M.; Wei, H.; Ling, R.; Lv, Y. Opposite effects of TNF-α on proliferation via ceramide in MDA-MB-231 and MCF-7 breast cancer cell lines. *Int. J. Clin. Exp. Med.* **2018**, *11*, 9239–9247.
8.	Ko, Y.S.; Jin, H.; Lee, J.S.; Park, S.W.; Chang, K.C.; Kang, K.M.; Jeong, B.K.; Kim, H.J. Radioresistant breast cancer cells exhibit increased resistance to chemotherapy and enhanced invasive properties due to cancer stem cells. *Oncol. Rep.* **2018**, *40*, 3752–3762. [CrossRef]
9.	Kopf, E.; Zharhary, D. Antibody arrays—An emerging tool in cancer proteomics. *Int. J. Biochem. Cell Biol.* **2007**, *39*, 1305–1317. [CrossRef]
10.	Bernichon, E.; Vallard, A.; Wang, Q.; Attignon, V.; Pissaloux, D.; Bachelot, T.; Heudel, P.E.; Ray-Coquard, I.; Bonnet, E.; de la Fouchardière, A.; et al. Genomic alterations and radioresistance in breast cancer: An analysis of the ProfiLER protocol. *Ann. Oncol.* **2017**, *28*, 2773–2779. [CrossRef] [PubMed]
11.	Strasser-Wozak, E.M.; Hartmann, B.L.; Geley, S.; Sgonc, R.; Böck, G.; Santos, A.J.; Hattmannstorfer, R.; Wolf, H.; Pavelka, M.; Kofler, R. Irradiation induces G2/M cell cycle arrest and apoptosis in p53-deficient lymphoblastic leukemia cells without affecting Bcl-2 and Bax expression. *Cell Death Differ.* **1998**, *5*, 687–693. [CrossRef] [PubMed]
12.	Bernhard, E.J.; Maity, A.; Muschel, R.J.; McKenna, W.G. Effects of ionizing radiation on cell cycle progression. *Radiat. Environ. Biophys.* **1995**, *34*, 79–83. [CrossRef]
13.	Pawlik, T.M.; Keyomarsi, K. Role of cell cycle in mediating sensitivity to radiotherapy. *Int. J. Radiat. Oncol. Biol. Phys.* **2004**, *59*, 928–942. [CrossRef] [PubMed]
14.	Meier, K.E.; Gause, K.C.; Wisehart-Johnson, A.E.; Gore, A.C.S.; Finley, E.L.; Jones, L.G.; Bradshaw, C.D.; McNair, A.F.; Ella, K.M. Effects of Propranolol on Phosphatidate Phosphohydrolase and Mitogen-Activated Protein Kinase Activities in A7r5 Vascular Smooth Muscle Cells. *Cell. Signal.* **1998**, *10*, 415–426. [CrossRef]
15.	Munshi, A.; Ramesh, R. Mitogen-activated protein kinases and their role in radiation response. *Genes Cancer* **2013**, *4*, 401–408. [CrossRef]
16.	Dudley, D.T.; Pang, L.; Decker, S.J.; Bridges, A.J.; Saltiel, A.R. A synthetic inhibitor of the mitogen-activated protein kinase cascade. *Proc. Natl. Acad. Sci. USA* **1995**, *92*, 7686. [CrossRef]
17.	Favata, M.F.; Horiuchi, K.Y.; Manos, E.J.; Daulerio, A.J.; Stradley, D.A.; Feeser, W.S.; Van Dyk, D.E.; Pitts, W.J.; Earl, R.A.; Hobbs, F.; et al. Identification of a Novel Inhibitor of Mitogen-activated Protein Kinase Kinase. *J. Biol. Chem.* **1998**, *273*, 18623–18632. [CrossRef] [PubMed]
18.	Sebolt-Leopold, J.S.; Dudley, D.T.; Herrera, R.; Becelaere, K.V.; Wiland, A.; Gowan, R.C.; Tecle, H.; Barrett, S.D.; Bridges, A.; Przybranowski, S.; et al. Blockade of the MAP kinase pathway suppresses growth of colon tumors in vivo. *Nat. Med.* **1999**, *5*, 810. [CrossRef]
19.	Marampon, F.; Gravina, G.L.; Di Rocco, A.; Bonfili, P.; Di Staso, M.; Fardella, C.; Polidoro, L.; Ciccarelli, C.; Festuccia, C.; Popov, V.M.; et al. MEK/ERK inhibitor U0126 increases the radiosensitivity of rhabdomyosarcoma cells in vitro and in vivo by downregulating growth and DNA repair signals. *Mol. Cancer Ther.* **2011**, *10*, 159–168. [CrossRef] [PubMed]
20.	Santos, N.F.D.; Silva, R.F.; Pinto, M.M.; Silva, E.B.; Tasat, D.R.; Amaral, A. Active caspase-3 expression levels as bioindicator of individual radiosensitivity. *An. Acad. Bras. Cienc.* **2017**, *89*, 649–659. [CrossRef]
21.	Samudio, I.; Konopleva, M.; Hail, N., Jr.; Shi, Y.-X.; McQueen, T.; Hsu, T.; Evans, R.; Honda, T.; Gribble, G.W.; Sporn, M.; et al. 2-Cyano-3,12-dioxooleana-1,9-dien-28-imidazolide (CDDO-Im) Directly Targets Mitochondrial Glutathione to Induce Apoptosis in Pancreatic Cancer. *J. Biol. Chem.* **2005**, *280*, 36273–36282. [CrossRef]
22.	Christofferson, D.E.; Yuan, J. Cyclophilin A release as a biomarker of necrotic cell death. *Cell Death Differ.* **2010**, *17*, 1942–1943. [CrossRef]
23.	Shepard, D.M.; Ferris, M.C.; Olivera, G.H.; Mackie, T.R. Optimizing the Delivery of Radiation Therapy to Cancer Patients. *SIAM Rev.* **1999**, *41*, 721–744. [CrossRef]
24.	Cannan, W.J.; Pederson, D.S. Mechanisms and Consequences of Double-Strand DNA Break Formation in Chromatin. *J.Cell. Physiol.* **2016**, *231*, 3–14. [CrossRef]
25.	Schulz, A.; Meyer, F.; Dubrovska, A. Cancer Stem Cells and Radioresistance: DNA Repair and Beyond. *Cancers* **2019**, *11*, 862. [CrossRef]
26.	Gray, M.; Turnbull, A.K.; Ward, C.; Meehan, J.; Martínez-Pérez, C.; Bonello, M.; Pang, L.Y.; Langdon, S.P.; Kunkler, I.H.; Murray, A.; et al. Development and characterisation of acquired radioresistant breast cancer cell lines. *Radiat. Oncol.* **2019**, *14*, 64. [CrossRef] [PubMed]
27.	Ediriweera, M.K.; Moon, J.Y.; Nguyen, Y.T.; Cho, S.K. 10-Gingerol Targets Lipid Rafts Associated PI3K/Akt Signaling in Radio-Resistant Triple Negative Breast Cancer Cells. *Molecules* **2020**, *25*, 3164. [CrossRef]
28.	Jin, H.; Rugira, T.; Ko, Y.S.; Park, S.W. ESM-1 Overexpression is Involved in Increased Tumorigenesis of Radiotherapy-Resistant Breast Cancer Cells. *Cancers* **2020**, *12*, 1363. [CrossRef]

29. Lu, L.; Dong, J.; Wang, L.; Xia, Q.; Zhang, D.; Kim, H.; Yin, T.; Fan, S.; Shen, Q. Activation of STAT3 and Bcl-2 and reduction of reactive oxygen species (ROS) promote radioresistance in breast cancer and overcome of radioresistance with niclosamide. *Oncogene* **2018**, *37*, 5292–5304. [CrossRef]

30. Bravatà, V.; Cammarata, F.P.; Minafra, L.; Musso, R.; Pucci, G.; Spada, M.; Fazio, I.; Russo, G.; Forte, G.I. Gene Expression Profiles Induced by High-dose Ionizing Radiation in MDA-MB-231 Triple-negative Breast Cancer Cell Line. *Cancer Genom. Proteom.* **2019**, *16*, 257–266. [CrossRef] [PubMed]

31. Choi, Y.J.; Baek, G.Y.; Park, H.R.; Jo, S.K.; Jung, U. Smad2/3-Regulated Expression of DLX2 Is Associated with Radiation-Induced Epithelial-Mesenchymal Transition and Radioresistance of A549 and MDA-MB-231 Human Cancer Cell Lines. *PLoS ONE* **2016**, *11*, e0147343. [CrossRef]

32. Yin, L.; Gao, Y.; Zhang, X.; Wang, J.; Ding, D.; Zhang, Y.; Zhang, J.; Chen, H. Niclosamide sensitizes triple-negative breast cancer cells to ionizing radiation in association with the inhibition of Wnt/β-catenin signaling. *Oncotarget* **2016**, *7*, 42126–42138. [CrossRef]

33. Lin, F.; Luo, J.; Gao, W.; Wu, J.; Shao, Z.; Wang, Z.; Meng, J.; Ou, Z.; Yang, G. COX-2 promotes breast cancer cell radioresistance via p38/MAPK-mediated cellular anti-apoptosis and invasiveness. *Tumour Biol. J. Int. Soc. Oncodev. Biol. Med.* **2013**, *34*, 2817–2826. [CrossRef] [PubMed]

34. Koh, S.Y.; Moon, J.Y.; Unno, T. Baicalein Suppresses Stem Cell-Like Characteristics in Radio- and Chemoresistant MDA-MB-231 Human Breast Cancer Cells through Up-Regulation of IFIT2. *Nutrients* **2019**, *11*, 624. [CrossRef]

35. Lammering, G.; Valerie, K.; Lin, P.S.; Hewit, T.H.; Schmidt-Ullrich, R.K. Radiation-induced activation of a common variant of EGFR confers enhanced radioresistance. *Radiother. Oncol. J. Eur. Soc. Ther. Radiol. Oncol.* **2004**, *72*, 267–273. [CrossRef] [PubMed]

36. Kim, M.H.; Jung, S.Y.; Ahn, J.; Hwang, S.G.; Woo, H.J.; An, S.; Nam, S.Y.; Lim, D.S.; Song, J.Y. Quantitative proteomic analysis of single or fractionated radiation-induced proteins in human breast cancer MDA-MB-231 cells. *Cell Biosci.* **2015**, *5*, 2. [CrossRef] [PubMed]

37. Miao, W.; Fan, M.; Huang, M.; Li, J.J.; Wang, Y. Targeted Profiling of Heat Shock Proteome in Radioresistant Breast Cancer Cells. *Chem. Res. Toxicol.* **2019**, *32*, 326–332. [CrossRef]

38. Lee, Y.C.; Wang, W.L.; Chang, W.C.; Huang, Y.H.; Hong, G.C.; Wang, H.L.; Chou, Y.H. Tribbles Homolog 3 Involved in Radiation Response of Triple Negative Breast Cancer Cells by Regulating Notch1 Activation. *Cancers* **2019**, *11*, 127. [CrossRef]

39. Hou, J.; Li, L.; Zhu, H.; Chen, H.; Wei, N.; Dai, M.; Ni, Q.; Guo, X. Association between breast cancer cell migration and radiosensitivity in vitro. *Oncol. Lett.* **2019**, *18*, 6877–6884. [CrossRef] [PubMed]

40. Yang, B.; Kuai, F.; Chen, Z.; Fu, D.; Liu, J.; Wu, Y.; Zhong, J. miR-634 Decreases the Radioresistance of Human Breast Cancer Cells by Targeting STAT3. *Cancer Biother. Radiopharm.* **2020**, *35*, 241–248. [CrossRef]

41. Li, J.Y.; Li, Y.Y.; Jin, W.; Yang, Q.; Shao, Z.M.; Tian, X.S. ABT-737 reverses the acquired radioresistance of breast cancer cells by targeting Bcl-2 and Bcl-xL. *J. Exp. Clin. Cancer Res.* **2012**, *31*, 102. [CrossRef]

42. Nguyen, Y.T.; Moon, J.Y.; Ediriweera, M.K.; Cho, S.K. Phenethyl Isothiocyanate Suppresses Stemness in the Chemo- and Radio-Resistant Triple-Negative Breast Cancer Cell Line MDA-MB-231/IR Via Downregulation of Metadherin. *Cancers* **2020**, *12*, 268. [CrossRef] [PubMed]

43. Oommen, D.; Prise, K.M. KNK437, abrogates hypoxia-induced radioresistance by dual targeting of the AKT and HIF-1α survival pathways. *Biochem. Biophys. Res. Commun.* **2012**, *421*, 538–543. [CrossRef]

44. Kuger, S.; Cörek, E.; Polat, B.; Kämmerer, U.; Flentje, M.; Djuzenova, C.S. Novel PI3K and mTOR Inhibitor NVP-BEZ235 Radiosensitizes Breast Cancer Cell Lines under Normoxic and Hypoxic Conditions. *Breast Cancer Basic Clin. Res.* **2014**, *8*, 39–49. [CrossRef] [PubMed]

45. Holler, M.; Grottke, A.; Mueck, K.; Manes, J.; Jücker, M.; Rodemann, H.P.; Toulany, M. Dual Targeting of Akt and mTORC1 Impairs Repair of DNA Double-Strand Breaks and Increases Radiation Sensitivity of Human Tumor Cells. *PLoS ONE* **2016**, *11*, e0154745. [CrossRef]

46. Chen, X.; Ma, N.; Zhou, Z.; Wang, Z.; Hu, Q.; Luo, J.; Mei, X.; Yang, Z.; Zhang, L.; Wang, X.; et al. Estrogen Receptor Mediates the Radiosensitivity of Triple-Negative Breast Cancer Cells. *Med Sci. Monit. Int. Med J. Exp. Clin. Res.* **2017**, *23*, 2674–2683. [CrossRef]

47. Liu, B.; Han, D.; Zhang, T.; Cheng, G.; Lu, Y.; Wang, J.; Zhao, H.; Zhao, Z. Hypoxia-induced autophagy promotes EGFR loss in specific cell contexts, which leads to cell death and enhanced radiosensivity. *Int. J. Biochem. Cell. Biol.* **2019**, *111*, 12–18. [CrossRef]

48. Arnold, K.M.; Opdenaker, L.M.; Flynn, N.J.; Appeah, D.K.; Sims-Mourtada, J. Radiation induces an inflammatory response that results in STAT3-dependent changes in cellular plasticity and radioresistance of breast cancer stem-like cells. *Int. J. Radiat. Biol.* **2020**, *96*, 434–447. [CrossRef] [PubMed]

49. Park, S.Y.; Choi, J.H.; Nam, J.S. Targeting Cancer Stem Cells in Triple-Negative Breast Cancer. *Cancers* **2019**, *11*, 965. [CrossRef]

50. Ko, Y.S.; Jung, E.J.; Go, S.-I.; Jeong, B.K.; Kim, G.S.; Jung, J.-M.; Hong, S.C.; Kim, C.W.; Kim, H.J.; Lee, W.S. Polyphenols Extracted from Artemisia annua L. Exhibit Anti-Cancer Effects on Radio-Resistant MDA-MB-231 Human Breast Cancer Cells by Suppressing Stem Cell Phenotype, β-Catenin, and MMP-9. *Molecules* **2020**, *25*, 1916. [CrossRef]

51. Qin, J.; Xin, H.; Nickoloff, B.J. Specifically targeting ERK1 or ERK2 kills melanoma cells. *J. Transl. Med.* **2012**, *10*, 15. [CrossRef]

52. Ciccarelli, C.; Di Rocco, A.; Gravina, G.L.; Mauro, A.; Festuccia, C.; Del Fattore, A.; Berardinelli, P.; De Felice, F.; Musio, D.; Bouché, M.; et al. Disruption of MEK/ERK/c-Myc signaling radiosensitizes prostate cancer cells in vitro and in vivo. *J. Cancer Res. Clin. Oncol.* **2018**, *144*, 1685–1699. [CrossRef] [PubMed]

53. Matsui, W.H. Cancer stem cell signaling pathways. *Medicine* **2016**, *95*, S8–S19. [CrossRef]
54. Tuasha, N.; Petros, B. Heterogeneity of Tumors in Breast Cancer: Implications and Prospects for Prognosis and Therapeutics. *Scientifica* **2020**, *2020*, 4736091. [CrossRef] [PubMed]
55. Moncharmont, C.; Levy, A.; Gilormini, M.; Bertrand, G.; Chargari, C.; Alphonse, G.; Ardail, D.; Rodriguez-Lafrasse, C.; Magné, N. Targeting a cornerstone of radiation resistance: Cancer stem cell. *Cancer Lett.* **2012**, *322*, 139–147. [CrossRef] [PubMed]
56. Chan, L.H.; Zhou, L.; Ng, K.Y.; Wong, T.L.; Lee, T.K.; Sharma, R.; Loong, J.H.; Ching, Y.P.; Yuan, Y.-F.; Xie, D.; et al. PRMT6 Regulates RAS/RAF Binding and MEK/ERK-Mediated Cancer Stemness Activities in Hepatocellular Carcinoma through CRAF Methylation. *Cell Rep.* **2018**, *25*, 690–701. [CrossRef]
57. Skvortsov, S.; Debbage, P.; Lukas, P.; Skvortsova, I. Crosstalk between DNA repair and cancer stem cell (CSC) associated intracellular pathways. *Semin. Cancer Biol.* **2015**, *31*, 36–42. [CrossRef] [PubMed]
58. Chai, X.; Chu, H.; Yang, X.; Meng, Y.; Shi, P.; Gou, S. Metformin Increases Sensitivity of Pancreatic Cancer Cells to Gemcitabine by Reducing CD133+ Cell Populations and Suppressing ERK/P70S6K Signaling. *Sci. Rep.* **2015**, *5*, 14404. [CrossRef]
59. Chen, M.-J.; Wu, D.-W.; Wang, Y.-C.; Chen, C.-Y.; Lee, H. PAK1 confers chemoresistance and poor outcome in non-small cell lung cancer via β-catenin-mediated stemness. *Sci. Rep.* **2016**, *6*, 34933. [CrossRef]
60. Tung, S.L.; Huang, W.C.; Hsu, F.C.; Yang, Z.P.; Jang, T.H.; Chang, J.W.; Chuang, C.M.; Lai, C.R.; Wang, L.H. miRNA-34c-5p inhibits amphiregulin-induced ovarian cancer stemness and drug resistance via downregulation of the AREG-EGFR-ERK pathway. *Oncogenesis* **2017**, *6*, e326. [CrossRef]
61. Su, Z.; Yang, Z.; Xie, L.; DeWitt, J.P.; Chen, Y. Cancer therapy in the necroptosis era. *Cell Death Differ.* **2016**, *23*, 748–756. [CrossRef]
62. Azevedo, A.P.; Silva, S.N.; Reichert, A.; Lima, F.; Júnior, E.; Rueff, J. Effects of polymorphic DNA genes involved in BER and caspase pathways on the clinical outcome of myeloproliferative neoplasms under treatment with hydroxyurea. *Mol. Med. Rep.* **2018**, *18*, 5243–5255. [CrossRef] [PubMed]
63. Zhang, Z.Y.; Xuan, Y.; Jin, X.Y.; Tian, X.; Wu, R. CASP-9 gene functional polymorphisms and cancer risk: A large-scale association study plus meta-analysis. *Genet. Mol. Res. GMR* **2013**, *12*, 3070–3078. [CrossRef]
64. Araki, S.; Dairiki, R.; Nakayama, Y.; Murai, A.; Miyashita, R.; Iwatani, M.; Nomura, T.; Nakanishi, O. Inhibitors of CLK protein kinases suppress cell growth and induce apoptosis by modulating pre-mRNA splicing. *PLoS ONE* **2015**, *10*, e0116929. [CrossRef] [PubMed]
65. Yoshida, T.; Kim, J.H.; Carver, K.; Su, Y.; Weremowicz, S.; Mulvey, L.; Yamamoto, S.; Brennan, C.; Mei, S.; Long, H.; et al. CLK2 Is an Oncogenic Kinase and Splicing Regulator in Breast Cancer. *Cancer Res.* **2015**, *75*, 1516–1526. [CrossRef] [PubMed]
66. Katoh, M. FGFR inhibitors: Effects on cancer cells, tumor microenvironment and whole-body homeostasis (Review). *Int. J. Mol. Med.* **2016**, *38*, 3–15. [CrossRef] [PubMed]
67. Rahmanian, N.; Hosseinimehr, S.J.; Khalaj, A. The paradox role of caspase cascade in ionizing radiation therapy. *J. Biomed. Sci.* **2016**, *23*, 88. [CrossRef]
68. Wu, P.K.; Park, J.I. MEK1/2 Inhibitors: Molecular Activity and Resistance Mechanisms. *Semin. Oncol.* **2015**, *42*, 849–862. [CrossRef]

International Journal of
Molecular Sciences

MDPI

Article

Transcriptomic Profiles of CD47 in Breast Tumors Predict Outcome and Are Associated with Immune Activation

María del Mar Noblejas-López [1,2,†], Mariona Baliu-Piqué [3,†], Cristina Nieto-Jiménez [3], Francisco J. Cimas [1,2], Esther C. Morafraile [3], Atanasio Pandiella [4,5,6,7], Ángel L. Corbi [8], Balázs Győrffy [9,10,11] and Alberto Ocaña [1,2,3,*]

1   Translational Oncology Laboratory, Translational Research Unit, Albacete University Hospital, 02008 Albacete, Spain; mariadelmar.noblejas@uclm.es (M.d.M.N.-L.); franciscojose.cimas@uclm.es (F.J.C.)
2   Centro Regional de Investigaciones Biomédicas, Castilla-La Mancha University (CRIB-UCLM), 02008 Albacete, Spain
3   Experimental Therapeutics Unit, Medical Oncology Department, Hospital Clínico San Carlos (HCSC), Instituto de Investigación Sanitaria (IdISSC) and CIBERONC, 28029 Madrid, Spain; mariona.baliu@salud.madrid.org (M.B.-P.); cnietoj@salud.madrid.org (C.N.-J.); esther.cabanas@salud.madrid.org (E.C.M.)
4   Instituto de Biología Molecular y Celular del Cáncer (IBMCC-CIC), 37007 Salamanca, Spain; atanasio@usal.es
5   Instituto de Investigación Biomédica de Salamanca (IBSAL), 37007 Salamanca, Spain
6   CIBERONC, 37007 Salamanca, Spain
7   Consejo Superior de Investigaciones Científicas (CSIC), 37007 Salamanca, Spain
8   Centro de Investigaciones Biológicas (CIB), Departamento de Biología Celular, Consejo Superior de Investigaciones Científicas (CSIC), 28029 Madrid, Spain; acorbi@cib.csic.es
9   Department of Bioinformatics, Semmelweis University, H-1094 Budapest, Hungary; gyorffy.balazs@ttk.mta.hu
10 2nd Department of Pediatrics, Semmelweis University, H-1094 Budapest, Hungary
11 TTK Lendület Cancer Biomarker Research Group, Institute of Enzymology, H-1117 Budapest, Hungary
*   Correspondence: alberto.ocana@salud.madrid.org
†   Both authors contributed equally to this work.

check for
updates

Citation: Noblejas-López, M.d.M.; Baliu-Piqué, M.; Nieto-Jiménez, C.; Cimas, F.J.; Morafraile, E.C.; Pandiella, A.; Corbi, Á.L.; Győrffy, B.; Ocaña, A. Transcriptomic Profiles of CD47 in Breast Tumors Predict Outcome and Are Associated with Immune Activation. *Int. J. Mol. Sci.* **2021**, *22*, 3836. https://doi.org/10.3390/ijms22083836

Academic Editor: Daniela Grimm

Received: 18 February 2021
Accepted: 4 April 2021
Published: 7 April 2021

Publisher's Note: MDPI stays neutral with regard to jurisdictional claims in published maps and institutional affiliations.

Abstract: Targeting the innate immune system has attracted attention with the development of anti-CD47 antibodies. Anti-CD47 antibodies block the inhibition of the phagocytic activity of macrophages caused by the up-regulation of CD47 on tumor cells. In this study, public genomic data was used to identify genes highly expressed in breast tumors with elevated CD47 expression and analyzed the association between the presence of tumor immune infiltrates and the expression of the selected genes. We found that 142 genes positively correlated with CD47, of which 83 predicted favorable and 32 detrimental relapse-free survival (RFS). From those associated with favorable RFS, we selected the genes with immunologic biological functions and defined a CD47-immune signature composed of PTPRC, HLA-E, TGFBR2, PTGER4, ETS1, and OPTN. In the basal-like and HER2+ breast cancer subtypes, the expression of the CD47-immune signature predicted favorable outcome, correlated with the presence of tumor immune infiltrates, and with gene expression signatures of T cell activation. Moreover, CD47 up-regulated genes associated with favorable survival correlated with pro-tumoral macrophages. In summary, we described a CD47-immune gene signature composed of 6 genes associated with favorable prognosis, T cell activation, and pro-tumoral macrophages in breast cancer tumors expressing high levels of CD47.

Keywords: CD47; immune activation; pro-tumoral macrophages; immunotherapy; breast cancer

1. Introduction

Administration of inhibitors of immunosuppressive signals has become an effective therapeutic strategy for the treatment of different types of cancer [1,2]. This approach has clearly modified the concept of cancer therapeutics, opening the door for the exploitation of the immune system to treat oncogenic processes [3]. In contrast to classical chemotherapy or targeted agents, immunotherapy aims to stimulate the patient's own immune system to attack tumor cells, therefore inducing long-lasting responses [3].

Programmed cell death 1 (PD1) and its ligand, PD-L1, are negative regulators of T cell activation that act as 'checkpoint molecules' [4]. Targeting PD1 and PD-L1 for the treatment of cancer enhances the response of T cells against the tumor [1–3,5]. Unfortunately, not all treated patients respond to checkpoint inhibitors, but immune-activated tumors, including those with high PD-L1 expression, are more prepared to orchestrate an adequate immune response following checkpoint blockade [6,7]. In this context, several studies have explored the potential of genomic signatures to predict outcomes and response to checkpoint inhibitors by identifying immune-activated tumors [8–10].

Exploiting immunotherapy as a therapeutic tool has mainly focused on the adaptive immune system to induce and boost an efficient T cell response [3]. Targeting innate immunity has recently attracted attention as a potential therapeutic option for many types of human cancers [11,12]. Macrophages, one of the components of the innate immune system, can contribute to the elimination of tumor cells by phagocytosis and contact-dependent and independent killing. Of note, two subsets of macrophages have been described: The M1 subtype, which exhibits a pro-inflammatory phenotype and displays anti-tumoral activities and phagocytic functions, and the M2 subtype, closely related to the so-called tumor-associated macrophages (TAMs), which has potent anti-inflammatory and tissue-repair (fibrotic) functions and promotes tumor progression [13].

Inhibitory signals also control macrophages activation. The signal-regulatory protein α (SIRPα) is an inhibitory receptor that presents immunoreceptor tyrosine-based inhibition motifs (ITIMs) [14,15]. SIRPα is expressed not only on macrophages but also on dendritic cells (DC) and neutrophils [16,17]. SIRPα plays an inhibitory role when activated by its ligand CD47. This interaction generates a "do not eat me" signal that prevents phagocytosis. Cancer cells may escape the immune surveillance of macrophages by the upregulation of CD47 expression [17,18].

Strategies to block this inhibitory pathway are under evaluation and aim at mimicking the success achieved with PD1 and PD-L1 inhibitors. Inhibition of the SIRPα-CD47 axis will enable macrophages to phagocytize and eliminate tumor cells in an efficient manner [16,17]. Currently, more than twenty early-stage clinical studies evaluating antibodies against the SIRPα-CD47 axis are ongoing [17]. Following the approach with checkpoint inhibitors evaluating PD-L1 expression, some of these new studies explore the presence of CD47 in relation to clinical efficacy. However, it is expected that the mere expression of this marker, like for PD-L1 expression, would not completely identify responder tumors.

In this study, we aimed to identify genomic correlates associated with the expression of CD47 in breast cancer to get insights into the immunologic characteristics of those tumors.

2. Results

2.1. Identification of Genes Expressed in Breast Tumors with High Expression of CD47

We used public genomic data to identify genes highly associated with CD47 expression at mRNA level in breast cancer tumors ($n = 1764$). Thereafter, we analyzed the transcripts which were positively (Spearman correlation, SC > 0.4 and $p < 0.05$) and negatively (SC < -0.4 and $p < 0.05$) correlated with CD47 expression. We identified 142 genes with a positive and five genes with a negative correlation with CD47 expression (Figure 1a).

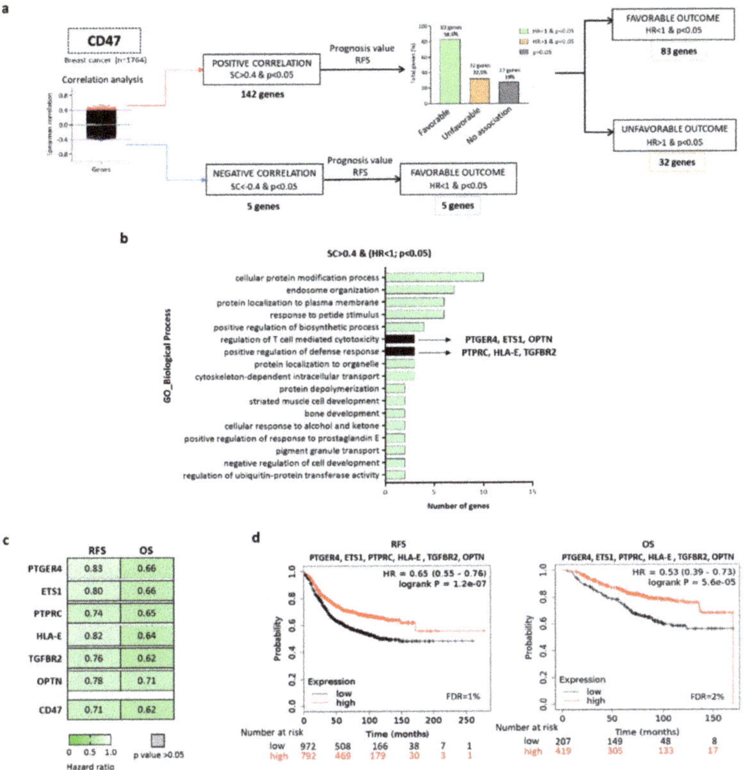

Figure 1. Immune system related transcriptional profiles associated with outcome and with CD47 expression in breast cancer. (**a**) Flow chart of gene selection, describing the tools and selection criteria used. (**b**) Functional analyses of the selected genes described in A using Enrichr Online Tool. For positively correlated genes (Spearman correlation (SC) > 0.4 and Hazard Ratio (HR) < 1), gene ontologies (GO) of biological process with a $p < 0.05$ are shown. The processes related to the immune system are highlighted. (**c**) Heat map displaying HR values extracted from Kaplan–Meier survival plots of the association between PTGER4, ETS1, PTPRC, HLA-E, TGFBR2, OPTN, and CD47 individually expressed and patient prognosis, including relapse-free survival (RFS) ($n = 1764$) and overall survival (OS) ($n = 626$), for all breast subtypes from the exploratory cohort. (**d**) Kaplan–Meier survival plots of the association between PTGER4, ETS1, PTPRC, HLA-E, TGFBR2, and OPTN mean expression levels and patient prognosis, including relapse-free survival (RFS) ($n = 1764$) and overall survival (OS) ($n = 626$) for all breast subtypes from the exploratory cohort.

We next explored the association of the selected gene transcripts with patient outcome, in terms of relapse-free survival (RFS), in 1764 breast cancer patients from all subtypes. From the 142 genes positively correlated with CD47, 83 genes (58.5%) predicted favorable RFS, and 32 genes (22.5%) were predicted to be detrimental to RFS. Of note, no association with survival was observed for 27 genes (19%). The five genes negatively correlated with CD47 were associated with a favorable prognosis (Figure 1a). A complete list of the identified genes is shown in Figure S1.

The get insights into the biological functions of the identified genes, we used the gene set enrichment analysis tool Enrichr (http://amp.pharm.mssm.edu/Enrichr/, accessed on 20 March 2020) [19]. For those genes correlated with detrimental prognosis, no immunologic-related functions were found (Figure S2a), neither for the five genes negatively correlated with CD47 expression (Figure S2b). For the genes positively correlated with CD47 and associated with favorable RFS, 17 biological processes with a $p < 0.01$ value were detected (Figure 1b). As the main goal of this project was to explore immunological correlates associated with the expression of CD47, we focused only on those biological functions related to immunology. From those, we selected only those genes from biological processes related to the immune system: (i) regulation of T cell-mediated cytotoxicity (PTPRC, HLA-E, and TGFBR2) and (ii) positive regulation of defense response (PTGER4, ETS1, and OPTN) (Figure 1b). Other biological functions included 'cell protein modification', 'endosome organization', 'protein localization to plasma membrane', 'response to peptide stimulus', and 'positive regulation of biosynthetic process', among others (Figure 1b). A complete list of the biological functions of the proteins codified by PTPRC, HLA-E, TGFBR2, PTGER4, ETS1, and OPTN genes obtained via UniProt is shown in Table S1.

2.2. CD47-Immune Signature is Associated with Favorable Prognosis in Breast Cancer, Especially for the BASAL-Like and HER2+ Subtypes

Each gene individually, PTGER4, ETS1, PTPRC, HLA-E, TGFBR2, and OPTN was associated with favorable outcomes (RFS and OS) in a statistically significant manner (Figure 1c); although, some did not predict better than the just the expression of CD47. However, given the fact that CD47 expression has been described as a negative regulator of the anti-tumoral action of macrophages [20,21], we aimed to explore the association of the CD47-immune signature composed by PTGER4, ETS1, PTPRC, HLA-E, TGFBR2, and OPTN with clinical outcomes. Using the exploratory dataset which includes more than 1764 patients with RFS data, and 626 patients with OS information, we found that the combination of PTGER4, ETS1, PTPRC, HLA-E, TGFBR2, and OPTN predicted favorable RFS (HR = 0.65; CI = 0.55–0.76; $p = 1.2 \times 10^{-7}$ and OS (HR = 0.53; CI = 0.39–0.73; $p = 5.6 \times 10^{-5}$) in breast cancer (Figure 1d). This prediction was better than single gene prediction, including CD47, and displayed a very low false-discovery rate (FDR).

Next, we explored if clinical outcomes could differ based on different breast cancer subtypes, as immune surveillance in each tumor subtype can be substantially different. In line with this heterogeneity, the prediction capacity of each gene varied, being the basal-like and HER2+ subtypes those in which the majority of genes predicted a favorable outcome, particularly for OS (Figure 2a). A similar correlation was observed for most of the genes in the basal-like, but not for the HER2 subtype, in the Molecular Taxonomy of Breast Cancer International Consortium (METABRIC) study (Figure 2b). For the combined signature, we found that differences in outcome were more evident for the basal-like and HER2+ breast cancer subtypes for both RFS and OS (Figure 2c,d). In the basal-like subgroup, the expression of the CD47-immune signature predicted favorable outcome for RFS (HR = 0.4; CI = 0.29–0.56; $p = 1.2 \times 10^{-8}$) and OS (HR = 0.23; CI = 0.12–0.44; $p = 1 \times 10^{-6}$) (Figure 2c). In the HER2+ subgroup a similar association was identified for RFS (HR = 0.43; CI = 0.27–0.68; $p = 0.00021$) and OS (HR = 0.25; CI = 0.11–0.57; $p = 0.00034$) (Figure 2d).

Given the exploratory nature of this cohort, we next analyzed these results using a confirmatory dataset. To do so, we used the METABRIC study that involved more than 1988 patients (PMID: 22522925). This dataset only provides information about OS. Using the validation cohort, we confirmed that for basal-like (HR = 0.54; CI = 0.4–0.73; $p = 4.4 \times 10^{-5}$) and HER2+ (HR = 0.6; CI = 0.38–0.95; $p = 0.025$) tumors the CD47-immune signature predicted favorable OS (Figure 2e,f). Altogether, this data demonstrates that the prediction observed was particularly strong and reproducible in the basal-like subtype.

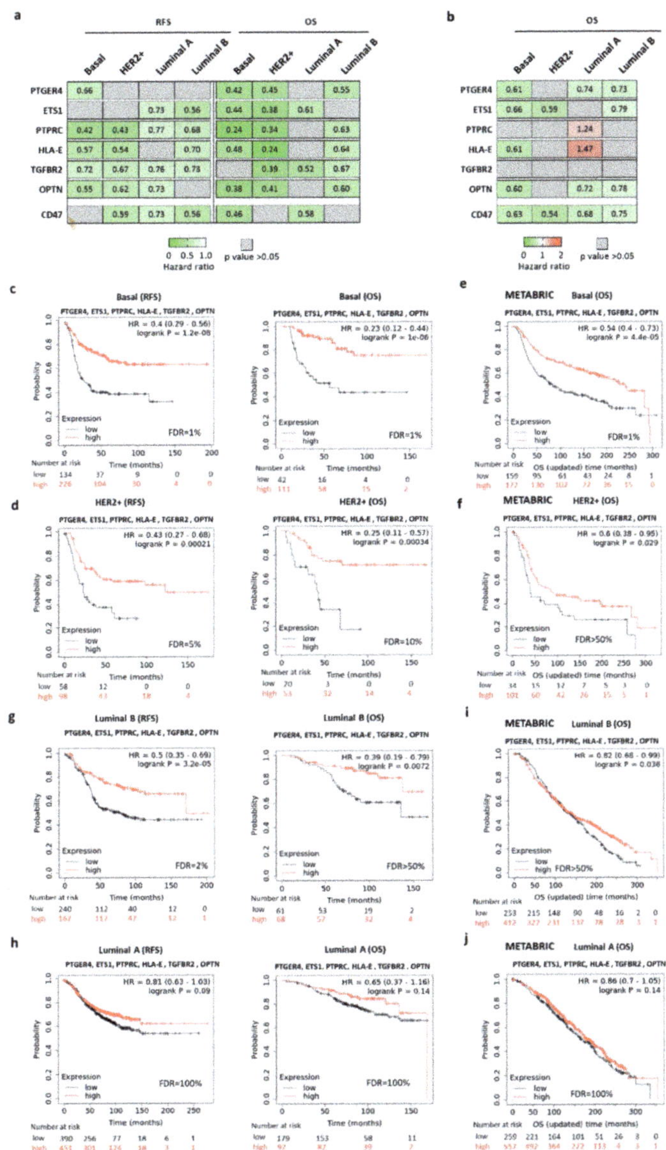

Figure 2. Kaplan–Meier survival curves of the association between transcriptomic expression of PTGER4, ETS1, PTPRC, HLA-E, TGFBR2, and OPTN and clinical outcome in basal-like, HER2+, luminal A, and luminal B breast cancer patients. (**a,b**) Heat map displaying HR values extracted from Kaplan–Meier survival plots for the association between PTGER4, ETS1, PTPRC, HLA-E, TGFBR2, OPTN, and CD47 individually expression levels and patient prognosis, in the exploratory cohort and Molecular Taxonomy of Breast Cancer International Consortium (METABRIC) validation cohort, respectively. (**c,d**) basal-like (relapse-free survival (RFS); *n* = 360 and OS; *n* = 153), and HER2+ (RFS; *n* = 156 and OS; *n* = 53), breast tumors in the exploratory cohort. (**e,f**) basal-like (OS; *n* = 331) and HER2+ (OS; *n* = 135) in the validation cohort (METABIRC project). (**g,h**) luminal B (RFS; *n* = 841 and OS; *n* = 271), and luminal A (RFS; *n* = 407 and OS; *n* = 129) in the exploratory cohort. (**i,j**) luminal B (OS; *n* = 816), and luminal A (OS; *n* = 665) breast tumors in the validation cohort (METABIRC project).

In luminal B tumors, the association between the CD47-immune signature and patient outcome was statistically associated with outcome (RFS: HR = 0.5; CI = 0.35–0.69; $p = 3.2 \times 10^{-5}$; OS: HR = 0.39; CI = 0.19–0.79; $p = 0.0072$) (Figure 2g). This result was confirmed using the validation cohort (OS: HR = 0.82; CI = 0.68–0.99; $p = 0.038$) (Figure 2i). However, for the more frequent breast cancer subtype, the luminal A subgroup, no association was observed in either of the two cohorts: for the exploratory cohort RFS: HR = 0.81; CI = 0.63–1.03; $p = 0.09$; OS: HR = 0.65; CI = 0.37–1.16; $p = 0.14$ (Figure 2h) and for the validation cohort: OS: HR = 0.86; CI = 0.7–1.05; $p = 0.14$ (Figure 2j).

2.3. CD47-Immune Signature Correlated with the Presence of Immune Infiltrates in Basal-Like and HER2+ Breast Tumors

The expression of the six genes composing the CD47-immune signature was associated with low tumor purity in all breast cancer subtypes analyzed (Figure 3), suggesting a high infiltration of non-tumor cells. For the basal-like subtype a positive correlation (partial correlation (pc) > 0.4) was observed for PTGER4, ETS1, OPTN, PTRC, and HLA-E for DCs, neutrophils, and CD4+ T cells. PTPRC, HLA-E, and TGFBR2 were also highly correlated with the presence of CD8+ T cells (pc > 0.5). No association was observed with macrophages, with the exception of TGFBR2 (pc > 0.5) (Figure 3b). For the HER2+ subtype, a positive correlation (pc > 0.4) was observed between ETS1, PTPRC, and HLA-E and CD8+ T cells, CD4+ T cells, neutrophils, and DCs, with ETS1 and PTPRC showing a pc > 0.7. Expression of TGFBR2 was again linked with a high presence of macrophages (pc > 0.6), while no association with macrophages was identified for the other genes (Figure 3c). For the luminal subtype, a positive association was observed for most genes, but more significantly for PTPRC and DC, neutrophils, CD4+ T cells, and CD8+ T cells (Figure 3d).

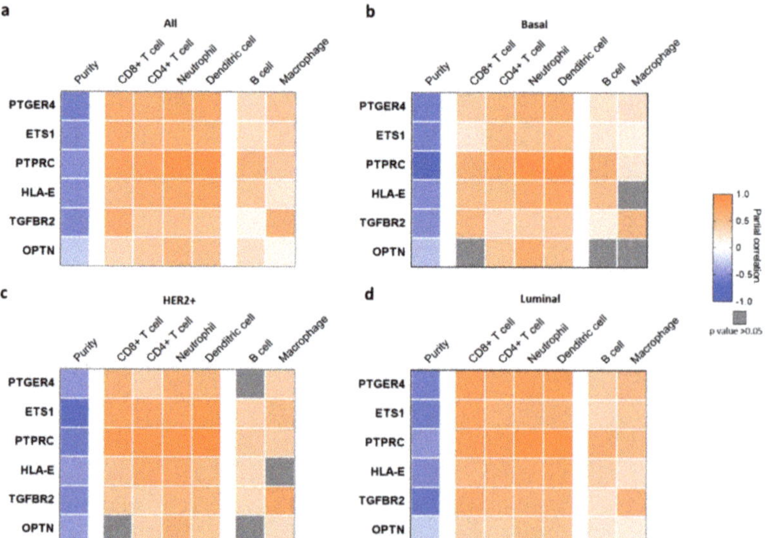

Figure 3. Association of the expression of the selected genes with immune infiltrates in breast cancer. Heat map depicting the Pearson correlation coefficient (R) between gene expression (PTGER4, ETS1, PTPRC, HLA-E, TGFBR2, and OPTN), tumor purity, and the presence of tumor immune infiltrates in (**a**) all, (**b**) basal-like, (**c**) HER2+, and (**d**) luminal breast cancer tumors using TIMER. Tumor immune infiltrates were separated into two groups: first, CD8+ T cells, CD4+ T cells, neutrophils, and dendritic cells (DCs); second, B cells and macrophages.

2.4. CD47-Immune Signature is Associated with Markers of T Cell Activation and Antigen Presentation

Next, we explored the association between the genes included within the CD47-immune signature and genes that encode for markers of T cell activation and antigen presentation. We found a strong positive correlation between the expression of CD69 and HLA-DRA, markers of T cell activation, with the expression of all the genes except for OPTN and ETS1 in all breast cancer subtypes (Figure 4a). Similar findings were observed when markers of antigen presentation, namely CD40, CD86, and CD83, were evaluated (Figure 4b). Figure S3 shows the data obtained using the cohort from the TCGA project.

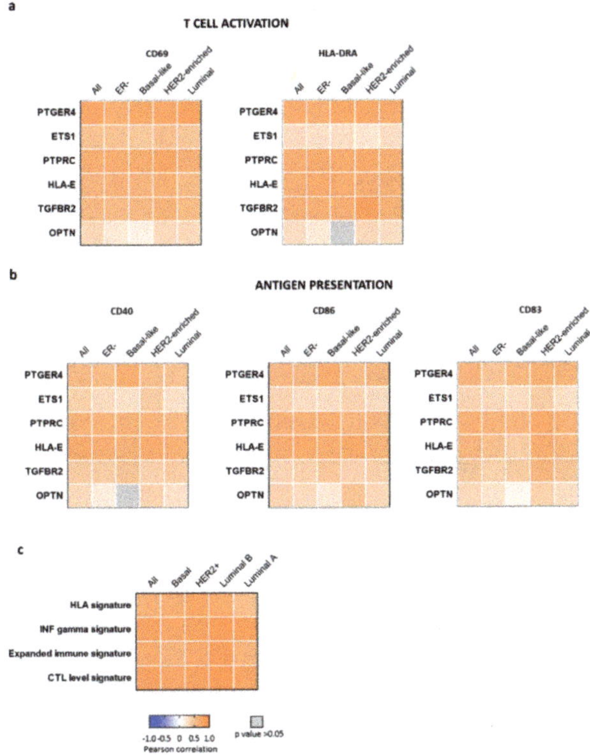

Figure 4. Relationship between gene expression and genomic signatures of immune activation. Heat map depicting the Pearson correlation coefficient (R) of the association between (**a**) markers of T cell activation (CD69 and HLA-DRA) or (**b**) antigen presentation (CD40, CD86, and CD83) and the expression of the selected genes using CANCERTOOL and the METABRIC cohort. (**c**) Heat map of Pearson correlation coefficient (R) of the expression of the CD47-immune signature and the HLA signature, IFN gamma signature, expanded immune gene signature, and CTL level signature in all (n = 1988), basal (n = 334), HER2+ (n = 137), luminal B (n = 680), and luminal A (n = 837) breast cancer.

2.5. CD47-Immune Signature Correlated with Gene Signatures of T Cell Activation

The findings described before suggested that the tumors expressing CD47 were enriched with T cells, DCs, and neutrophils. To further assess immune activation, we explored the correlation of our six-gene signature with already described genomic profiles of immune activation, including the HLA-A/B signature, the IFN gamma signature, the expanded immune gene signature, and the CTL signature [10–12]. We found that the CD47-immune signature was positively associated with these four transcriptomic profiles (Figure 4c), suggesting that this signature identifies tumors with a high presence of activated T cells and DCs. Given the fact that PTPRC or CD45 is a gene that codes for a marker globally expressed in different immune populations, being considered as a pan-leukocyte antigen, we repeated the analysis excluding this gene. As can be seen in Supplementary Figure S4a, the correlation was present even with the absence of this gene.

2.6. Gene-Set Enrichment Analysis (GSEA) Confirm the Association of the CD47-Immune Signature with Pro-Tumoral Macrophages

We next used gene-set enrichment analysis (GSEA) to test whether the CD47-immune signature composed by PTPRC, HLA-E, TGFBR2, PTGER4, ETS1, and OPTN was preferentially associated with a specific macrophage polarization state. The CD47-immune signature was not significantly enriched in the transcriptome of either anti-inflammatory macrophages (M-MØ) or monocyte-derived pro-inflammatory (GM-MØ), suggesting that the expression of this gene set was independent of the macrophage polarization state (Figure 5a). However, the signature was found to be significantly enriched (FDR q value = 0.018) in the transcriptome of IL-10-treated adherent peripheral blood mononuclear cells (monocytes) (Figure 5b). Since IL-10 is a major factor that determines the pro-tumoral action of tumor-associated macrophages (TAM) [20,22], this result suggests that the 6-gene CD47-immune signature might be regulated by factors promoting TAMs. In a similar manner, we performed the same analysis but excluded PTPRC, observing that the results were in the same direction (Figure S4b,c).

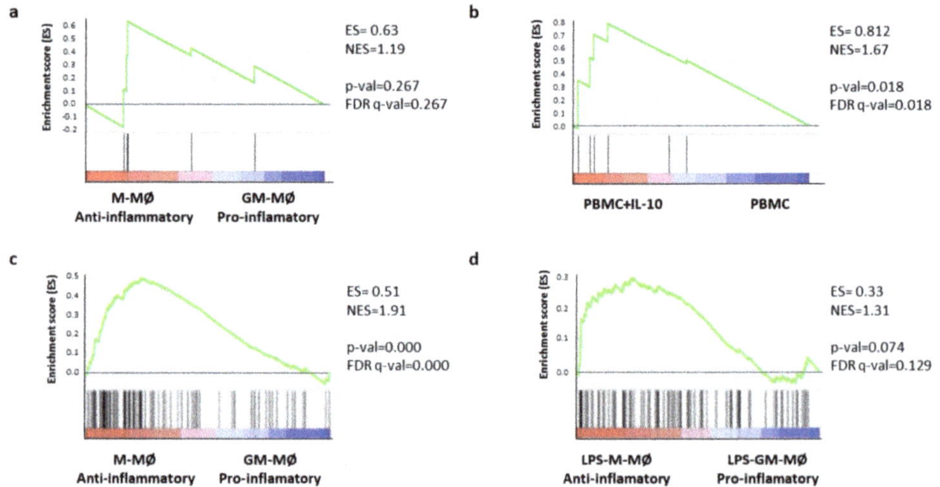

Figure 5. Gene-set enrichment analysis (GSEA) between gene expression and macrophage signatures. Gene-set enrichment analysis (GSEA) of the six-gene CD47-immune signature on (**a**) the ranked comparison of anti-inflammatory macrophages (M-MØ) or M2 and monocyte-derived pro-inflammatory (GM-MØ) or M1 whole transcriptomes, previously described in [20,22] (GSE27792 and GSE68061) or (**b**) the transcriptomes of adherent human peripheral blood mononuclear cells either untreated (PBMC) or treated with 10 ng/mL IL-10 for 24 h (PBMC + IL-10) that have been previously described. GSEA of the genes that positively correlate with CD47 and are associated with good prognosis (Figure S1) on (**c**) the ranked comparison of M-MØ or M2 and GM-MØ or M1 whole transcriptomes, previously described in [21,23] (GSE27792 and GSE68061) or (**d**) the ranked comparison of the transcriptome of lipopolysaccharide (LPS)-treated M-MØ or M2 and LPS-treated GM-MØ or M1 transcriptomes, previously described in [24] (GSE99056).

Last, we hypothesized that the genes that positively correlate with CD47 and that were associated with a good prognosis (see Figure 1 and Figure S1) might be associated with a specific type of macrophage polarization. To test this hypothesis, we analyzed the expression of this gene set on the ranked comparison of M-MØ and GM-MØ transcriptomes [21,23]. GSEA revealed a very significant enrichment of this set of genes in the transcriptome of pro-tumoral M-MØ (ES = 0.51; NES = 1.91; *p*-value = 0.000; FDR q-value = 0.000) (Figure 5c). Indeed, and using the available information on the transcriptomes of activated M-MØ and GM-MØ [24], a lower but significant enrichment was also observed in activated pro-tumoral M-MØ (ES = 0,33; NES = 1.31; *p*-value = 0.074; FDR q-value = 0.129) (Figure 5d).

Altogether, these results indicate that the expression of genes that positively correlate with CD47 and are associated with a good prognosis are preferentially expressed by macrophages with anti-inflammatory capacity.

3. Discussion

In the present article, we described a transcriptomic immune signature formed by six genes that were expressed in breast tumors with high expression of CD47 and were associated with favorable outcomes. The CD47 ligand is present on the surface of tumoral cells, and by binding to its receptor SIRPα, inhibits the induction of phagocytosis by macrophages [18–20]. CD47 is a perfect target to stimulate the activation of macrophages as well as other innate immunity cells, and several therapeutic strategies blocking the CD47-SIRPα axis are under evaluation in clinical studies.

When analyzing the genes positively correlated to CD47 expression in breast cancer, we observed that only a very limited number of these genes, 33 (22.5%) predicted detrimental outcomes, while 83 genes (58.5%) predicted good outcomes. Among the identified functions of the genes linked to favorable prognosis, our attention was attracted to the two functions related to the immune system; the genes included within these functions were PTGER4, ETS1, PTPRC, HLA-E, TGFBR2, and OPTN. We termed the combination of these genes as the "CD47-immune signature", and predicted favorable prognosis in all breast tumors, but particularly in the HER2+ and basal-like subtype. These results obtained from the exploratory dataset including 1764 patients was confirmed by an independent cohort that included 1947 patients.

The CD47-immune signature includes a variety of genes that code for proteins with a wide range of functions that have been rarely described in relation to the immune response in cancer. PTGER4 which is a member of the G-protein coupled receptor family and can activate T cell factor signaling [25]; ETS1 which codes for a member of the ETS family of transcription factors that are involved in stem cell development, cell senescence and death, and tumorigenesis [26]; PTPRC a member of the protein tyrosine phosphatase (PTP) family that regulates a variety of cellular processes including cell growth, differentiation, mitosis, and oncogenic transformation [27]; HLA-E which belongs to the HLA class I heavy chain paralogues and functions as a ligand for natural killer (NK) cell inhibitory receptor KLRD1-KLRC1, enabling NK cells to monitor the expression of other MHC class I molecules in healthy cells [28]; TGFBR2 a transmembrane protein that has a protein kinase domain, forms a heterodimeric complex with TGF-beta receptor type-1, and binds TGF-beta [29]; and finally, OPTN that encodes the coiled-coil containing protein optineurin that interacts with adenovirus E3-14.7K protein and may utilize tumor necrosis factor-alpha or Fas-ligand pathways to mediate apoptosis, inflammation or vasoconstriction [30].

To study whether CD47 positive tumors are linked to an active T cell response, we correlated the CD47-immune signature with immune populations using bioinformatic approaches. Data demonstrated that there was a positive correlation at a single gene level with the presence of infiltrating T cells, DCs, and neutrophils. It is relevant to mention that although the strongest effect was observed in the basal-like and HER2+ population, such associations were also observed for all breast cancer patients. A limitation of our analysis was that we were not able to dissect the presence of immune infiltrates in the two different luminal breast cancer subtypes, luminal A and B.

Finally, we aimed to confirm this association by exploring the correlation of each gene contained in the CD47-immune signature with well-known markers of T cell activation and of antigen presentation. We observed a positive correlation for all genes except for OPTN and ETS1. Moreover, a strong positive correlation was identified between the whole expression of the signature and that of described signatures of T cell activation, including HLA-A/B, the CTL signature, the expanded immune gene signature, and the IFN gamma signature [8–10,31]. Finally, a surprising finding was that no increase in the macrophage population was observed, but those identified were pro-tumoral. This finding was further confirmed when exploring the whole transcriptomic signature of each of the different subtypes of macrophages, in line with the inhibitory action of CD47.

Our study had several limitations. This was an in silico study which needs further clinical validation using human samples. We do acknowledge that our study is based on associations and correlations among biological parameters. However, we tried to use different datasets as well as associations between genes, biomarkers, and immune populations to avoid findings produced by casualty.

We believe it is relevant to explore if this signature could help identify patients that would respond to anti-CD47 agents. In line with this, the data observed here suggest the combined administration of anti-PD-L1 inhibitors with anti-CD47/ SIRPα agents to boost the T cell response and the activation of macrophages. In this regard, ongoing studies are exploring the activity of targeting both pathways [16,17]. Given the fact that strategies are under evaluation targeting CD47 and SIRPα, it might be interesting to evaluate the transcriptomic profile of tumors with high expression of SIRPα.

4. Materials and Methods

4.1. Exploratory Cohort

Samples included in the KM Plotter Online Tool (http://www.kmplot.com, accessed on 20 March 2020) [32] were used as an exploratory cohort. This publicly available database shows the relationship between gene expression and patient outcome in different breast cancer subtypes, including relapse-free survival (RFS) and overall survival (OS). Patients were distributed according to the best cutoff values of the gene expression (lowest p-value) into "high" vs. "low". RFS was defined as the time from diagnosis to the first recurrence, and OS as the time from diagnosis to patient death. The number of breast cancer patients included in each subtype for RFS was: all: $n = 1764$; basal-like: $n = 360$; HER2+: $n = 156$; luminal B: $n = 407$; luminal A: $n = 841$. For OS: all: $n = 626$; basal-like: $n = 156$; HER2+: $n = 73$; luminal B: $n = 129$; luminal A: $n = 271$ (HGU133 array 2.0) (available data April 2020).

To identify genes whose expression correlated to CD47 expression, the probe set 226016 was correlated using all samples from the exploratory cohort. For each gene, Spearman rank correlation was computed to compare its normalized gene expression and CD47 expression. Then, the genes were ranked based on the achieved Spearman correlation coefficients.

4.2. Validation Cohort

Survival analysis was performed in basal-like ($n = 331$), HER2+ ($n = 135$), luminal B ($n = 668$), and luminal A ($n = 825$) patient samples from the METABRIC (Molecular Taxonomy of Breast Cancer International Consortium) project (PMID: 22522925). Gene expression values were quantile normalized in R. The molecular subtype designation was based on St. Gallen criteria [33], and the expression of ESR1, ERBB2, and MKI67 on the arrays was used to define the patient cohorts. In this, basal breast cancer was defined by those negative for ESR1 and ERBB2, luminal A was defined as those ESR1 positive ERBB2 negative MKI67 negative, HER2+enriched was defined as those ERBB2 positive ESR1 negative, and luminal B comprises all remaining samples.

4.3. Gene Function Analysis

Genes positive-correlated with CD47 and associated with good or detrimental outcomes were analyzed using the biological function enrichment analyses tool Enrichr (http://www.amp.pharm.mssm.edu/Enrichr/, accessed on 20 March 2020) [34]. Biological process with a p-value < 0.01 were selected for CD47 positively correlated genes and with a p-value < 0.05 CD47 negatively correlated genes. Biological processes related to the immune system were grouped. Immune system processes contained the following gene ontologies: positive regulation of defense response (GO:0031349), regulation of T cell-mediated cytotoxicity (GO:0001914), and positive regulation of alpha-beta T cell proliferation (GO:0046641).

4.4. Protein Functional Analyses

We used UniProt Online Tool (http://www.uniprot.org/, accessed on 20 March 2020) [35] for protein functional analysis. UniProt provides a comprehensive, high-quality, and freely accessible resource of protein sequence and functional information. All data is freely available on the web. Proteins codified by genes related to immune system processes were studied. For complete protein functional analyses, we collected all biological function gene ontologies included in UniProt.

4.5. Association between Tumor Immune Infiltrates and Gene Expression

Tumor Immune Estimation Resource (TIMER) platform (https://cistrome.shinyapps. io/timer/, accessed on 2 April 2020) [36] was used to analyze tumor purity, and the association between the presence of tumor immune infiltrates, namely CD4+ T cells, CD8+ T cells, DCs, macrophages, neutrophils, B cell, and macrophages and the expression of the selected genes. TIMER contains 10,897 samples from diverse cancer types from the TCGA (The Cancer Genome Atlas) project. We explored the tumor immune infiltrates in breast cancer subtypes: basal-like, HER2+, and luminal. TIMER does not allow for the analysis of the luminal A and luminal B subtypes separately.

4.6. Correlation between Gene Expression and T Cell Activation and Antigen Presentation

CANCERTOOL (http://web.bioinformatics.cicbiogune.es/CANCERTOOL/index. html, accessed on 2 April 2020) [37] was used to explore the relationship between the expression of CD47-immune signature genes and the expression of T cell activation (CD69 and HLA-DRA) and antigen presentation (CD40, CD86, and CD83) markers in all, basal-like, HER2+, luminal B, and luminal A breast cancer. This open-access resource for the analysis of gene expression provides the Pearson correlation coefficients of every pair of genes analyzed. The datasets used for the analysis included METABRIC as the primary cohort, and TCGA, as a validation cohort.

The complete METABRIC cohort (total: n = 1988; basal: n = 331; HER2+: n = 135; luminal B: n = 668; luminal A: n = 825 breast cancer) was used to explore the correlation between the identified CD47-immune signature and previously described signatures, including: the HLA signature (HLA-A and HLA-B) [10] the interferon (IFN) gamma signature (IDO1, CXCL10, CXCL9, HLA-DRA, ISGF-3, and IFNG) [12], the expanded immune gene signature (CD30, IDO1, CIITA, CD3E, CCL5, GZMK, CD2, HLA-DRA, CXCL13, IL3RG, NKG7, HLA-E, CXCR6, LAG3, TAGAP, CXCL10, STAT1, and GZMB) [29] and the cytotoxic T lymphocyte (CTL) signature (CD8A, CD8B, GZMA, GZMB, and PRF1) [11].

4.7. Correlation between Gene Expression and Macrophage Signatures

Gene set enrichment analysis (GSEA) (http://software.broadinstitute.org/gsea/index. jsp, accessed on 5 May 2020) [38] was done to assess enrichment of the indicated gene-sets in the transcriptomes of human monocyte-derived pro-inflammatory (GM-MØ) and anti-inflammatory macrophages (M-MØ) (GSE27792 and GSE68061) [20,22].

4.8. Statistical Analysis

The Kaplan–Meier (KM) plots are presented with the hazard ratio (HR), the 95% confidence interval (CI), the log-rank p-value (p), and the false discovery rate (FDR). The FDR was computed after performing the Cox regression analysis across all cutoff values between the lower and upper quartiles of expression, and only results with an FDR below 10% were accepted as significant. For the METABRIC dataset, only the median was used to define high and low cohorts, and therefore FDR values were not calculated. In addition to the HR and FDR cutoff values described above, statistical significance was defined as $p < 0.05$. Genes that had an HR < 1 and a $p < 0.05$ were considered predictors of a favorable outcome, while genes that had an HR > 1 and $p < 0.05$ were considered predictors of detrimental outcome.

5. Conclusions

Here we described an immune gene signature associated with elevated levels of CD47 that predicts favorable outcomes in breast cancer tumors. In addition, the described signature was linked with the presence of T cell, DC, and neutrophil infiltrates, T cell activation and antigen presentation, and correlated with pro-tumoral macrophages. Further studies should confirm the predictive capacity of this signature in ongoing clinical studies.

Supplementary Materials: The following are available online at https://www.mdpi.com/article/10.3390/ijms22083836/s1. Figure S1. Tables depicting genes 83 positively correlated with CD47 (SC > 0.4 and p-value < 0.05) associated with good prognosis (HR < 1 and p-value < 0.05); 32 genes positively correlated with CD47 (SC > 0.4 and p-value < 0.05) associated with worse prognosis (HR > 1 and p-value < 0.05); 27 genes positively correlated with CD47 (SC > 0.4 and p-value < 0.05) no associated with prognosis (p-value > 0.05) and 5 negatively correlated with CD47 (SC < −0.4 and p-value < 0.05) associated with good prognosis (HR < 1 and p-value < 0.05). Figure S2. Functional analyses using Enrichr Online Tool. For positively correlated genes (SC.0.4 and HR < 1) Gene ontologies (GO) of biological process with a $p < 0.05$ are shown for (A) genes positively correlated with CD47 expression (SC > 0.4 and $p < 0.05$) associated with unfavourable prognosis (HR > 1) and (B) genes negatively correlated with CD47 expression (SC < −0.4 and $p < 0.05$) associated with favourable prognosis (HR < 1). Figure S3. Relationship between markers of T cell activation (CD69 and HLA-DRA) and antigen presentation (CD40, CD86 and CD83) expression and the expression of the selected genes using CANCERTOOL in TCGA cohort. Figure S4. Heat map of Pearson correlation coefficient (R) of the expression of the CD47-immune signature (excluding PTPRC) and the HLA signature, IFN gamma signature, expanded immune gene signature, and CTL level signature in all ($n = 1988$), basal ($n = 334$), HER2+ ($n = 137$), luminal B ($n = 680$), and luminal A ($n = 837$) breast cancer subgroups. Gene-set enrichment analysis (GSEA) of the CD47-immune signature excluding PTPRC on (b) the ranked comparison of M-MØ or M2 and GM-MØ or M1 whole transcriptomes, previously described in (GSE27792, GSE68061) or (c) the transcriptomes of adherent human peripheral blood mononuclear cells either untreated (PBMC) or treated with 10 ng/mL IL-10 for 24 h (PBMC+IL-10) that have been previously described. Table S1. Complete list of the biological functions of the proteins codified by PTPRC, HLA-E, TGFBR2, PTGER4, ETS1, and OPTN genes obtained via UniProt Online Tool.

Author Contributions: Conceptualization, M.B.-P. and A.O.; Data curation, M.d.M.N.-L., M.B.-P., C.N.-J., F.J.C., E.C.M., and B.G.; Formal analysis, M.d.M.N.-L., M.B.-P., C.N.-J., E.C.M., and Á.L.C.; Funding acquisition, A.P. and A.O.; Investigation, M.d.M.N.-L., M.B.-P., C.N.-J., and Á.L.C.; Methodology, M.d.M.N.-L., M.B.-P., C.N.-J., and B.G.; Project administration, A.P. and A.O.; Resources, B.G.; Software, B.G.; Supervision, A.P.; Validation, M.d.M.N.-L. and F.J.C.; Visualization, F.J.C., E.C.M., and Á.L.C.; Writing—original draft, M.d.M.N.-L., M.B.-P., Á.L.C., and A.O.; Writing—review and editing, F.J.C., E.C.M., A.P., and B.G. All authors have read and agreed to the published version of the manuscript.

Funding: This work was supported by Instituto de Salud Carlos III (PI19/00808), ACEPAIN, Diputación de Albacete, CIBERONC and CRIS Cancer Foundation (to A.O.). Ministry of Economy and Competitiveness of Spain (BFU2015-71371-R), the Instituto de Salud Carlos III through the Spanish Cancer Centers Network Program (RD12/0036/0003) and CIBERONC, the scientific foundation of the AECC and the CRIS Foundation (to A.P.). M.d.M.N.-L. was supported by the Spanish Ministry of Education (FPU grant; Ref.: FPU18/01319). The work carried out in our laboratories received support from the European Community through the regional development funding program (FEDER). B.G. was supported by the grants NVKP_16-1-2016-0037, 2018-1.3.1-VKE-2018-00032, and KH-129581.

Institutional Review Board Statement: Not applicable.

Informed Consent Statement: Not applicable.

Data Availability Statement: Data are available upon reasonable request to the corresponding author. Part of the data that support the findings of this study are publicly available in the Gene Expression Omnibus (GEO) repository; data set accession numbers: GSE27792, GSE68061, GSE99056, and GSE84622.

Conflicts of Interest: A.O. receives research funding from Entrechem, travel expenses from Merck and advisory board fees from Daiichi Sankyo. A.P. receives consultancy fees from Daiichi Sankyo.

References

1. Hargadon, K.M.; Johnson, C.E.; Williams, C.J. Immune checkpoint blockade therapy for cancer: An overview of FDA-approved immune checkpoint inhibitors. *Int. Immunopharmacol.* **2018**, *62*, 29–39. [CrossRef]
2. Luke, J.J.; Flaherty, K.T.; Ribas, A.; Long, G.V. Targeted agents and immunotherapies: Optimizing outcomes in melanoma. *Nat. Rev. Clin. Oncol.* **2017**, *14*, 463–482. [CrossRef] [PubMed]
3. Waldman, A.D.; Fritz, J.M.; Lenardo, M.J. A guide to cancer immunotherapy: From T cell basic science to clinical practice. *Nat. Rev. Immunol.* **2020**, *20*, 651–668. [CrossRef]
4. Dong, Y.; Sun, Q.; Zhang, X. PD-1 and its ligands are important immune checkpoints in cancer. *Oncotarget* **2017**, *8*, 2171–2186. [CrossRef]
5. Ribas, A.; Wolchok, J.D. Cancer immunotherapy using checkpoint blockade. *Science* **2018**, *359*, 1350–1355. [CrossRef]
6. Gettinger, S.N.; Horn, L.; Gandhi, L.; Spigel, D.R.; Antonia, S.J.; Rizvi, N.A.; Powderly, J.D.; Heist, R.S.; Carvajal, R.D.; Jackman, D.M.; et al. Overall Survival and Long-Term Safety of Nivolumab (Anti–Programmed Death 1 Antibody, BMS-936558, ONO-4538) in Patients with Previously Treated Advanced Non–Small-Cell Lung Cancer. *J. Clin. Oncol.* **2015**, *33*, 2004–2012. [CrossRef]
7. Zappasodi, R.; Merghoub, T.; Wolchok, J.D. Emerging Concepts for Immune Checkpoint Blockade-Based Combination Therapies. *Cancer Cell* **2018**, *33*, 581–598. [CrossRef]
8. Noblejas-López, M.d.M.; Nieto-Jiménez, C.; Morcillo García, S.; Pérez-Peña, J.; Nuncia-Cantarero, M.; Andrés-Pretel, F. Expression of MHC class I, HLA-A and HLA-B identifies immune-activated breast tumors with favorable outcome. *Oncoimmunology* **2019**, *8*, e1629780. [CrossRef]
9. Jiang, P.; Gu, S.; Pan, D.; Fu, J.; Sahu, A.; Hu, X. Signatures of T cell dysfunction and exclusion predict cancer immunotherapy response. *Nat. Med.* **2018**, *24*, 1550–1558. [CrossRef]
10. Ayers, M.; Lunceford, J.; Nebozhyn, M.; Murphy, E.; Loboda, A.; Kaufman, D.R.; Andrew, A.; Jonathan, D.; Cheng, S.; Peter, K.; et al. IFN-γ–related mRNA profile predicts clinical response to PD-1 blockade. *J. Clin. Investig.* **2017**, *127*, 2930–2940. [CrossRef]
11. Liu, Z.; Han, C.; Fu, Y.-X. Targeting innate sensing in the tumor microenvironment to improve immunotherapy. *Cell. Mol. Immunol.* **2020**, *17*, 13–26. [CrossRef] [PubMed]
12. Ponzoni, M.; Pastorino, F.; Di Paolo, D.; Perri, P.; Brignole, C. Targeting Macrophages as a Potential Therapeutic Intervention: Impact on Inflammatory Diseases and Cancer. *Int. J. Mol. Sci.* **2018**, *19*, 1953. [CrossRef] [PubMed]
13. Atri, C.; Guerfali, F.; Laouini, D. Role of Human Macrophage Polarization in Inflammation during Infectious Diseases. *Int. J. Mol. Sci.* **2018**, *19*, 1801. [CrossRef]
14. Zhang, W.; Huang, Q.; Xiao, W.; Zhao, Y.; Pi, J.; Xu, H. Advances in Anti-Tumor Treatments Targeting the CD47/SIRPα Axis. *Front. Immunol.* **2020**, *11*, 18. [CrossRef]
15. Jalil, A.R.; Andrechak, J.C.; Discher, D.E. Macrophage checkpoint blockade: Results from initial clinical trials, binding analyses, and CD47-SIRPα structure–function. *Antib. Ther.* **2020**, *3*, 80–94. [CrossRef]
16. Brown, E. Integrin-associated protein (CD47) and its ligands. *Trends Cell Biol.* **2001**, *11*, 130–135. [CrossRef]
17. Adams, S.; van der Laan, L.J.; Vernon-Wilson, E.; Renardel de Lavalette, C.; Döpp, E.A.; Dijkstra, C.D. Signal-regulatory protein is selectively expressed by myeloid and neuronal cells. *J. Immunol.* **1998**, *161*, 1853–1859.
18. Chao, M.P.; Weissman, I.L.; Majeti, R. The CD47–SIRPα pathway in cancer immune evasion and potential therapeutic implications. *Curr. Opin. Immunol.* **2012**, *24*, 225–232. [CrossRef]
19. Kuleshov, M.V.; Jones, M.R.; Rouillard, A.D.; Fernandez, N.F.; Duan, Q.; Wang, Z. Enrichr: A comprehensive gene set enrichment analysis web server 2016 update. *Nucleic Acids Res.* **2016**, *44*, W90–W97. [CrossRef]
20. Chuang, Y.; Hung, M.E.; Cangelose, B.K.; Leonard, J.N. Regulation of the IL-10-driven macrophage phenotype under incoherent stimuli. *Innate Immun.* **2016**, *22*, 647–657. [CrossRef]
21. Sierra-Filardi, E.; Puig-Kröger, A.; Blanco, F.J.; Nieto, C.; Bragado, R.; Palomero, M.I. Activin A skews macrophage polarization by promoting a proinflammatory phenotype and inhibiting the acquisition of anti-inflammatory macrophage markers. *Blood* **2011**, *117*, 5092–5101. [CrossRef]
22. Zhou, J.; Tang, Z.; Gao, S.; Li, C.; Feng, Y.; Zhou, X. Tumor-Associated Macrophages: Recent Insights and Therapies. *Front. Oncol.* **2020**, *10*, 188. [CrossRef]
23. Gonzalez-Dominguez, E.; Dominguez-Soto, A.; Nieto, C.; Flores-Sevilla, J.L.; Pacheco-Blanco, M.; Campos-Pena, V. Atypical Activin A and IL-10 Production Impairs Human CD16+ Monocyte Differentiation into Anti-Inflammatory Macrophages. *J. Immunol.* **2016**, *196*, 1327–1337. [CrossRef] [PubMed]
24. Riera-Borrull, M.; Cuevas, V.D.; Alonso, B.; Vega, M.A.; Joven, J.; Izquierdo, E. Palmitate Conditions Macrophages for Enhanced Responses toward Inflammatory Stimuli via JNK Activation. *J. Immunol.* **2017**, *199*, 3858–3869. [CrossRef] [PubMed]

25. Mori, K.; Tanaka, I.; Kotani, M.; Miyaoka, F.; Sando, T.; Muro, S. Gene expression of the human prostaglandin E receptor EP4 subtype: Differential regulation in monocytoid and lymphoid lineage cells by phorbol ester. *J. Mol. Med.* **1996**, *74*, 333–336. [CrossRef]

26. Bhat, N.K.; Thompson, C.B.; Lindsten, T.; June, C.H.; Fujiwara, S.; Koizumi, S. Reciprocal expression of human ETS1 and ETS2 genes during T-cell activation: Regulatory role for the protooncogene ETS1. *Proc. Natl. Acad. Sci. USA* **1990**, *87*, 3723–3727. [CrossRef] [PubMed]

27. Felberg, J.; Lefebvre, D.C.; Lam, M.; Wang, Y.; Ng, D.H.W.; Birkenhead, D. Subdomain X of the Kinase Domain of Lck Binds CD45 and Facilitates Dephosphorylation. *J. Biol. Chem.* **2004**, *279*, 3455–3462. [CrossRef]

28. Llano, M.; Lee, N.; Navarro, F.; García, P.; Albar, J.P.; Geraghty, D.E. HLA-E-bound peptides influence recognition by inhibitory and triggering CD94/NKG2 receptors: Preferential response to an HLA-G-derived nonamer. *Eur. J. Immunol.* **1998**, *28*, 2854–2863. [CrossRef]

29. Tzachanis, D.; Li, L.; Lafuente, E.M.; Berezovskaya, A.; Freeman, G.J.; Boussiotis, V.A. Twisted gastrulation (Tsg) is regulated by Tob and enhances TGF-β signaling in activated T lymphocytes. *Blood* **2007**, *109*, 2944–2952. [CrossRef]

30. Sahlender, D.A.; Roberts, R.C.; Arden, S.D.; Spudich, G.; Taylor, M.J.; Luzio, J.P. Optineurin links myosin VI to the Golgi complex and is involved in Golgi organization and exocytosis. *J. Cell Biol.* **2005**, *169*, 285–295. [CrossRef]

31. Rooney, M.S.; Shukla, S.A.; Wu, C.J.; Getz, G.; Hacohen, N. Molecular and Genetic Properties of Tumors Associated with Local Immune Cytolytic Activity. *Cell* **2015**, *160*, 48–61. [CrossRef]

32. Győrffy, B.; Lanczky, A.; Eklund, A.C.; Denkert, C.; Budczies, J.; Li, Q. An online survival analysis tool to rapidly assess the effect of 22,277 genes on breast cancer prognosis using microarray data of 1809 patients. *Breast Cancer Res. Treat.* **2010**, *123*, 725–731. [CrossRef] [PubMed]

33. Goldhirsch, A.; Winer, E.P.; Coates, A.S.; Gelber, R.D.; Piccart-Gebhart, M.; Thürlimann, B. Personalizing the treatment of women with early breast cancer: Highlights of the St Gallen International Expert Consensus on the Primary Therapy of Early Breast Cancer 2013. *Ann. Oncol.* **2013**, *24*, 2206–2223. [CrossRef] [PubMed]

34. Chen, E.Y.; Tan, C.M.; Kou, Y.; Duan, Q.; Wang, Z.; Meirelles, G. Enrichr: Interactive and collaborative HTML5 gene list enrichment analysis tool. *BMC Bioinform.* **2013**, *14*, 128. [CrossRef]

35. Bateman, A. UniProt: A worldwide hub of protein knowledge. *Nucleic Acids Res.* **2019**, *47*, D506–D515.

36. Li, T.; Fan, J.; Wang, B.; Traugh, N.; Chen, Q.; Liu, J.S. TIMER: A Web Server for Comprehensive Analysis of Tumor-Infiltrating Immune Cells. *Cancer Res.* **2017**, *77*, e108–e110. [CrossRef]

37. Cortazar, A.R.; Torrano, V.; Martín-Martín, N.; Caro-Maldonado, A.; Camacho, L.; Hermanova, I. CANCERTOOL: A Visualization and Representation Interface to Exploit Cancer Datasets. *Cancer Res.* **2018**, *78*, 6320–6328. [CrossRef] [PubMed]

38. Subramanian, A.; Tamayo, P.; Mootha, V.K.; Mukherjee, S.; Ebert, B.L.; Gillette, M.A. Gene set enrichment analysis: A knowledge-based approach for interpreting genome-wide expression profiles. *Proc. Natl. Acad. Sci. USA* **2005**, *102*, 15545–15550. [CrossRef]

International Journal of
Molecular Sciences

Article

Combination of Niraparib, Cisplatin and Twist Knockdown in Cisplatin-Resistant Ovarian Cancer Cells Potentially Enhances Synthetic Lethality through ER-Stress Mediated Mitochondrial Apoptosis Pathway

Entaz Bahar [1,†], Ji-Ye Kim [2,†], Dong-Chul Kim [3], Hyun-Soo Kim [4,*] and Hyonok Yoon [1,*]

[1] College of Pharmacy, Research Institute of Pharmaceutical Sciences, Gyeongsang National University, Jinju 52828, Korea; entazbahar@gnu.ac.kr

[2] Department of Pathology, Ilsan Paik Hospital, Inje University, Goyang 10380, Korea; alucion@gmail.com

[3] Department of Pathology, Gyeongsang National University School of Medicine and Gyeongsang National University Hospital, Jinju 52828, Korea; kdcjes@gmail.com

[4] Samsung Medical Center, Department of Pathology and Translational Genomics, Sungkyunkwan University School of Medicine, Seoul 06351, Korea

* Correspondence: hyun-soo.kim@samsung.com (H.-S.K.); hoyoon@gnu.ac.kr (H.Y.); Tel.: +82-2-3410-1243 (H.-S.K.); +82-55-772-2422 (H.Y.)

† These authors contributed equally to this work.

Citation: Bahar, E.; Kim, J.-Y.; Kim, D.-C.; Kim, H.-S.; Yoon, H. Combination of Niraparib, Cisplatin and Twist Knockdown in Cisplatin-Resistant Ovarian Cancer Cells Potentially Enhances Synthetic Lethality through ER-Stress Mediated Mitochondrial Apoptosis Pathway. *Int. J. Mol. Sci.* **2021**, *22*, 3916. https://doi.org/10.3390/ijms22083916

Academic Editor: Valentina De Falco

Received: 10 February 2021
Accepted: 8 April 2021
Published: 10 April 2021

Publisher's Note: MDPI stays neutral with regard to jurisdictional claims in published maps and institutional affiliations.

Abstract: Poly (ADP-ribose) polymerase 1 inhibitors (PARPi) are used to treat recurrent ovarian cancer (OC) patients due to greater survival benefits and minimal side effects, especially in those patients with complete or partial response to platinum-based chemotherapy. However, acquired resistance of platinum-based chemotherapy leads to the limited efficacy of PARPi monotherapy in most patients. Twist is recognized as a possible oncogene and contributes to acquired cisplatin resistance in OC cells. In this study, we show how Twist knockdown cisplatin-resistant (CisR) OC cells blocked DNA damage response (DDR) to sensitize these cells to a concurrent treatment of cisplatin as a platinum-based chemotherapy agent and niraparib as a PARPi on in vitro two-dimensional (2D) and three-dimensional (3D) cell culture. To investigate the lethality of PARPi and cisplatin on Twist knockdown CisR OC cells, two CisR cell lines (OV90 and SKOV3) were established using step-wise dose escalation method. In addition, in vitro 3D spheroidal cell model was generated using modified hanging drop and hydrogel scaffolds techniques on poly-2-hydroxylethly methacrylate (poly-HEMA) coated plates. Twist expression was strongly correlated with the expression of DDR proteins, PARP1 and XRCC1 and overexpression of both proteins was associated with cisplatin resistance in OC cells. Moreover, combination of cisplatin (Cis) and niraparib (Nira) produced lethality on Twist-knockdown CisR OC cells, according to combination index (CI). We found that Cis alone, Nira alone, or a combination of Cis+Nira therapy increased cell death by suppressing DDR proteins in 2D monolayer cell culture. Notably, the combination of Nira and Cis was considerably effective against 3D-cultures of Twist knockdown CisR OC cells in which Endoplasmic reticulum (ER) stress is upregulated, leading to initiation of mitochondrial-mediated cell death. In addition, immunohistochemically, Cis alone, Nira alone or Cis+Nira showed lower ki-67 (cell proliferative marker) expression and higher cleaved caspase-3 (apoptotic marker) immuno-reactivity. Hence, lethality of PARPi with the combination of Cis on Twist knockdown CisR OC cells may provide an effective way to expand the therapeutic potential to overcome platinum-based chemotherapy resistance and PARPi cross resistance in OC.

Keywords: ovarian cancer; cisplatin; cisplatin resistance; PARPi; niraparib; Twist; lethality

1. Introduction

Ovarian cancer (OC) is the deadliest gynecologic malignancy, responsible for over 50 percent of mortality among all gynecologic malignancies worldwide [1,2]. Currently, cytoreductive operation preceded by combination therapy based on platinum and taxane is used as a typical OC treatment regimen for OC [3]. Despite the high response rate of most patients receiving platinum-based chemotherapy, the majority of patients with advanced OC die from recurrent diseases and resistance to platinum-based chemotherapy. Therefore, an alternative therapeutic strategy to treat patients with recurrent OC is urgently needed.

Recently, several novel strategies for potential personalized therapy have been established, including molecular-targeted medicine (small-molecule inhibitors or antibodies), clinical immunotherapeutic applications and the recognition of synthetic lethal partners [4]. Genetic screening techniques, including computer methods, drug screening, genetic modification with shRNA/siRNA or CRISPR, or a combination of these methods, can recognize synthetic lethal partners [5–7]. The alteration of the DNA damage response (DDR) pathway is a predictive biomarker of platinum-based sensitivity in various cancers, including OC [8]. The synthetic lethality interfaces between altered genes and molecules involved in DDR that could be therapeutically exploited to preferentially eradicate cancer cells [9]. Therefore, synthetic lethality could be an alternative therapeutic approach to overcome platinum based chemotherapy resistance.

Poly (ADP-ribose) polymerase 1 (PARP1) is a DNA repair protein that regulates the growth and differentiation of cells by repairing single-strand break (SSB) and double-strand breaks (DSB) of DNA [10]. Inhibition of PARP1 is considered to be the most active and exciting new personalized target therapy for the treatment of OC, especially in those patients with relapsed platinum-sensitive OC [2,11]. PARP1 inhibitors (PARPi), including olaparib (LYNPARZA), niraparib (ZEJULA) and rucaparib (RUBRACA), are recommended for the maintenance of care in patients with recurrent ovarian cancer due to improved benefits and fewer adverse reactions in patients with a full or partial platinum-based chemotherapy response [12–16]. However, PARPi agents also become resistant to other chemotherapy agents, as almost half of BRCA mutated OC patients fai to benefit from PARPi [17,18].

Twist is a member of the core transcription factor helix-loop-helix (bHLH), known to be an epithelial–mesenchymal transition (EMT) master regulator associated with tumor recurrence and chemo-resistance [19–26]. Functionally, Twist was identified as a potential oncogene and our previous studies have identified that Twist also contributed to acquire cisplatin resistance in OC cells [27–29]. In several studies, Twist has been linked to the development of resistance to platinum-based chemotherapy and metastasis in cancer [20,30–35].

We hypothesized that Twist knockdown cisplatin-resistant (CisR) OC cells sup-pressed DDR and exerted synthetic lethal effect by sensitizing these cells to concurrent treatment of cisplatin and niraparib.

2. Results

2.1. Twist Expression Is Strongly Correlated with Expression of PARP1 Involved in DNA Damage Response and Repair in CisR OC Cells

To investigate the expression pattern of Twist during the development of cisplatin resistance, we have established two CisR OC cell lines (OV90-CisR and SKOV3-CisR) using intermittent incremental dosing method, starting from a low dose of 10, 20, 40, 80 to 100 µM (10 doses for each term) of cisplatin. Beside this, we generated another 4 subline with the level of platinum resistance in between parental (P) and CisR cell, named G1, G2, G3 and G4 for cisplatin 10, 20, 40 and 80 µm respectively (Figure 1A). Our result revealed that Twist expression gradually increased with increasing intermittent cisplatin doses at 10 to 100 µm which may contribute to the development of cisplatin resistance. In both cancer cell lines, Twist amplification coincided with an increased expression of DNA damage response (DDR) genes, including PARP1 and XRCC1 (Figure 1B and Figure S1). Next, we examined all sublines treated with cisplatin (Cis), niraparib (Nira) and olaparib (Ola) with 20 µM doses, each respectively, for 72 h of treatment followed by 72 h of recovery for clonogenic assay (Figure 1C). The cisplatin and PARP inhibitor (PARPi) sensitivity gradually decreased during the development of resistance (Figure 1C and Figure S2). Both cell lines demonstrated decreased sensitivity toward cisplatin and PARPi (Nira and Ola) in their respective resistant cells (Figure 1D); in other words, CisR OC cells were cross-resistant to PARPi.

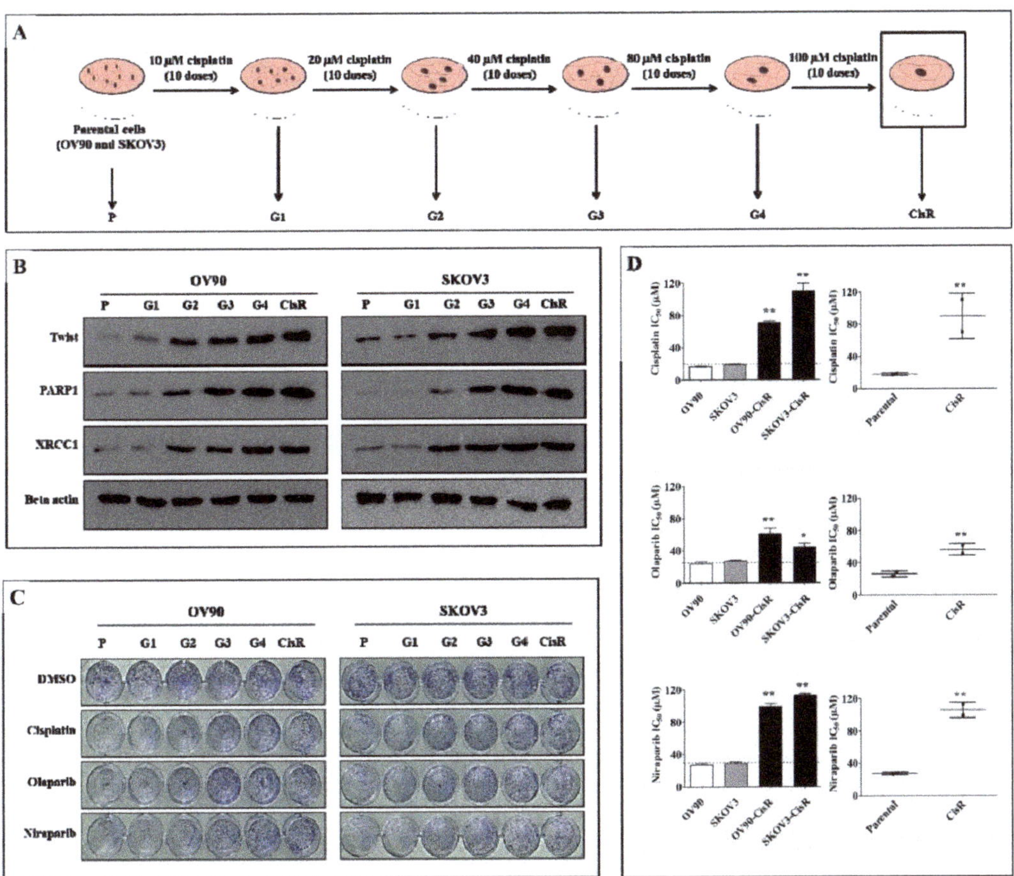

Figure 1. Twist expression regulates DNA damage response (DDR) proteins, PARP1 and XRCC1 during the development of cisplatin resistance in ovarian cancer (OC) cells. (**A**) Schematic repre-setantion of how four intermediate sublines were generated between parental and CisR OC cells. (**B**) Western blot analysis of Twist and DDR proteins, PARP1 and XRCC1 in each subline of OV90-CisR and SKOV3-CisR. (**C**) Clonogenic cell growth assay for cisplatin, olaparib and niraparib with the doses of 20 µM, each respectively, for 72 h of treatment followed by 72 h recovery in each subline of OV90-CisR and SKOV3-CisR. (**D**) Measurement of 50% inhibitory concentration (IC_{50}) values of cisplatin, olaparib and niraparib on CisR OC cells compared with their respective parental cells. P: parental cell; CisR: cisplatin-resistant cells; G: sublines of each respective generation. Values were represented as mean ± SD. * $p < 0.05$, ** $p < 0.01$, compared with their respective parental cells.

2.2. The Twist Deficient Cisplatin-Resistant OC Cells Were More Susceptible to Cisplatin and PARPi

To investigate whether Twist regulates DNA repair pathway in CisR OC, we transfected cells with Twist siRNA (siTwist) and siRNA negative control (siNC). Twist knockdown effectively downregulated PARP1 and XRCC1 expression in OV90-CisR and SKOV3-CisR OC cells (Figure 2A).

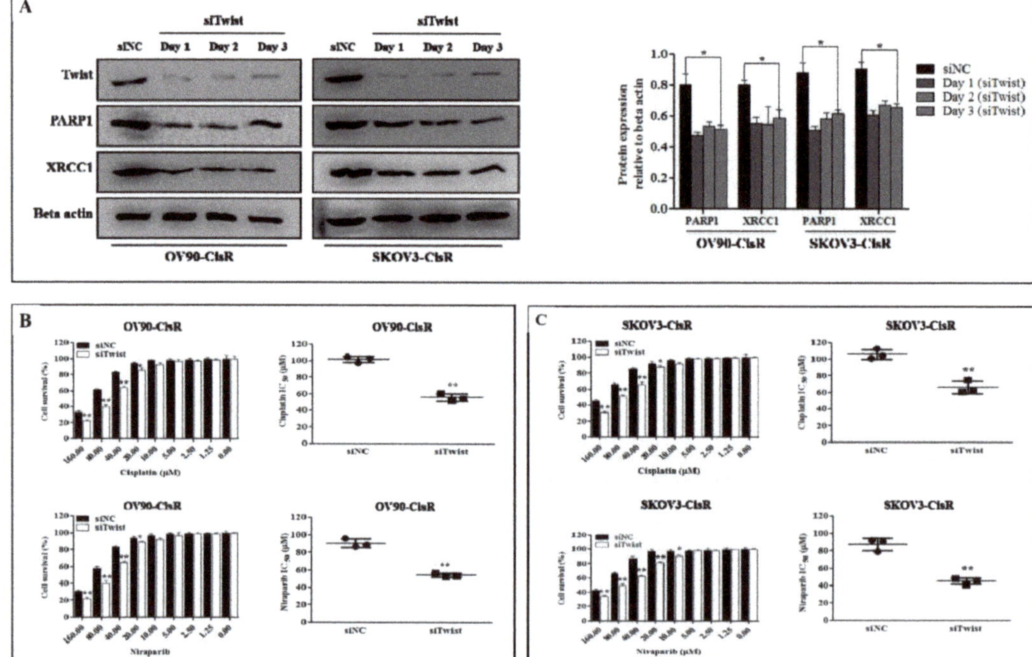

Figure 2. Twist regulates DDR proteins to sensitize cisplatin and PARPi in CisR OC cells. (**A**) Western blot analysis of Twist-knockdown with PARP1 and XRRRC1 proteins expression at day 1, 2 and 3 in OV90-CisR and SKOV3-CisR compared with the siNC group. (**B,C**) The treatment effect expressed as cell survival and IC$_{50}$ of cisplatin and PARPi (niraparib) on CisR OV90 and SKOV3 OC cells depending on Twist expression. The both drus were treated at a various concentrations ranging from 0 to 160 μM for 72 h of incubation. CisR: cisplatin-resistant; siRNA: small interfering RNA; siNC: non-targeting negative control siRNA; siTwist: Twist siRNA. Values were represented as mean ± SD. * $p < 0.05$, ** $p < 0.01$ compared with the siNC group.

To examine if knockdown of Twist alters the sensitivity of CisR OC cells to Cis and Nira, we treated both drugs at a various concentrations ranging from 0 to 160 μM for 72 h. Twist-deficient cells demonstrated more susceptibility to Cis and Nira compared with Twist-proficient cells at doses starting from 20 μM and up (Figure 2B,C). Twist-knockdown CisR cells showed a decreased IC$_{50}$ in response to Cis and Nira as compared with Twist-proficient CisR OC cells.

2.3. Combination Therapy Exhibits Synergetic Effect with Suppression of DNA Damage Response Capacity and Inducing Apoptosis on Twist Deficient Cisplatin-Resistant OC Cells

Combination drug therapies induced synergistic, additive or antagonistic effects depending on the dosage ratio [36–38]. We investigated the dose response of both drugs (Cis and Nira) in three combination ratios of 1:1, 2:1 and 1:2 in the OV90-P and SKOV3-P, siNC OV90-CisR and SKOV3-CisR and Twist-deficient OV90-CisR and SKOV3-CisR cells (Figure 3A). The IC$_{50}$ measurements of three combination ratio after 72 h of incubation with original parental, siNC CisR and Twist-deficient CisR cell lines are summarized in Supplementary Table S1. In Twist knockdown CisR cells, the IC$_{50}$ values of Cis+Nira at ratios of 1:1 was significantly smaller than that of the 2:1 and 1:2 ratio in the cell lines studied ($p < 0.05$).

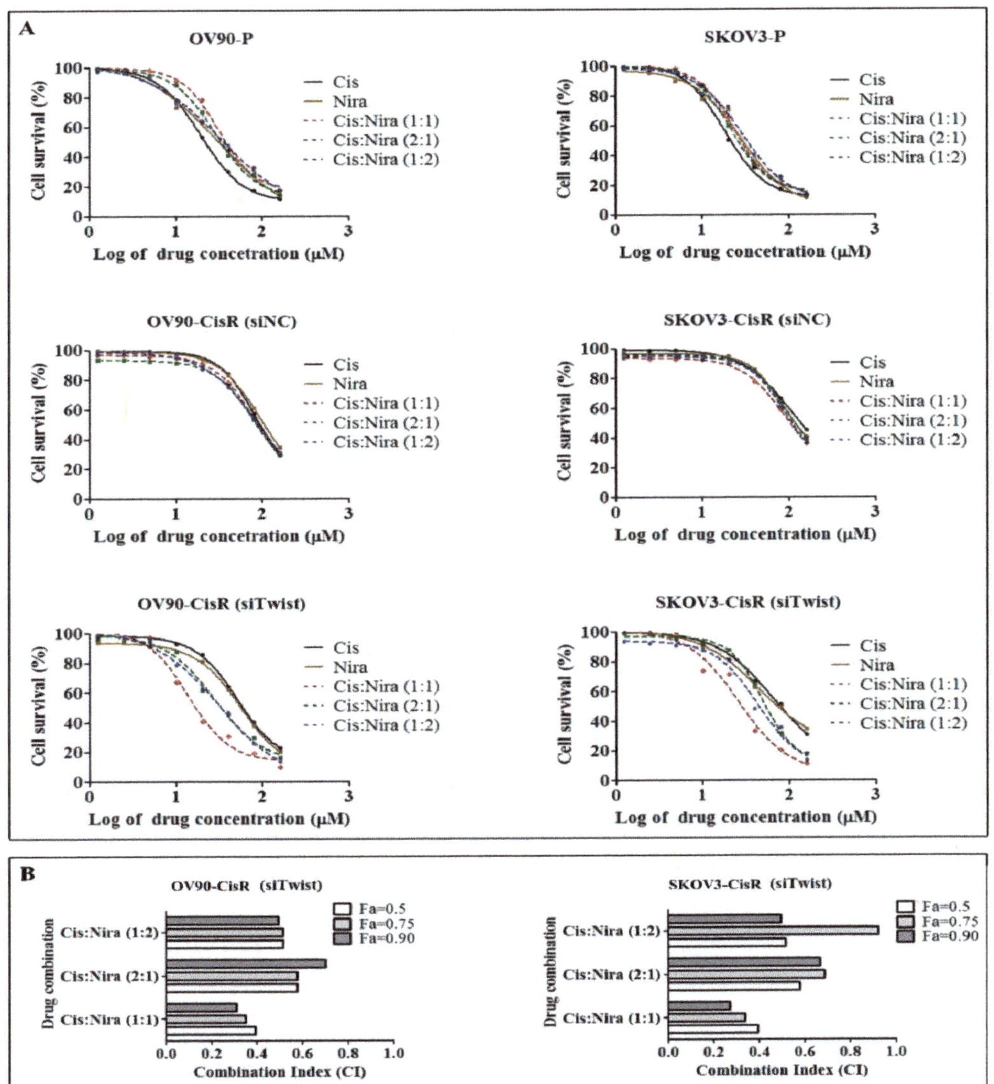

Figure 3. Twist knockdown sensitizes CisR OC cells to Cis and niraparib (Nira) therapy. (**A**) The dose response curve of Cis and Nira at 1:1, 2:1 and 1:2 combination ratio on parental, siNC CisR and siTwist CisR OV90 and SKOV3 cells. The cells were co-treated with increase dose of Cis and Nira (0–160 μM) for 72 h and percentage of cell survival was determined by EZ-cytox cell viability kit. (**B**) Measurement of combination index (CI) using compusyn software. P: parental CisR: cisplatin-resistant; siRNA: small interfering RNA; siNC: non-targeting negative control, siRNA; siTwist: Twist siRNA; DMSO: dimethyl sulfoxide, Cis: cisplatin; Nira; niraparib; Fa: fraction of affected cells.

Furthermore, using the Compusyn software program, cell cytotoxicity and quantitative values of the drug interaction combination index (CI) were calculated for Twist knockdown CisR OC cells, where CI < 1, CI = 1, and CI > 1 indicates synergism, additive effect, and antagonism, each respectively [39]. The corresponding CI values have been determined to interpret the value of combinations. The combination of Cis and Nira at the ratio of 1:1 had the highest synergistic value compared to other two ratios of 2:1 and 1:2

at fraction of affected cells value of 0.5, 0.75 and 0.90 in the both Twist-deficient CisR OC cells (Figure 3B). For further experiments, the 1:1 mixture in which the IC_{50} (IC_{25} dosage of Cis plus IC_{25} dosage of Nira) dose of respective drug was used as it showed the lowest CI value in both Twist-deficient CisR OC cells (Table S2).

Moreover, we investigated the effectiveness of Cis and Nira on cell survival and DDR proteins, XRCC1 and PARP1 in original parental, siNC CisR and Twist Knockdown CisR OC. The Cis and Nira effectively reduced cell survival capacity and suppressed PARP1 and XRCC1 expression in both parental OC cells (Figure 4A,B). However, the cell survival capacity and DDR potential of siNC CisR remained constant even after Cis and Nira treatment (Figure 4C,D).

Established that Twist knockdown reduced cell survival capacity by suppressing DDR proteins, we determined whether a combination of Cis and Nira (Cis+Nira therapy) enhanced the DDR blocking capacity and cell death of Twist-knockdown CisR cells compared to single agents; the Twist knockdown CisR were treated with Cis alone, Nira alone and Cis+Nira therapy. As expected, Cis+Nira markedly decreased cell clonogenicity, where the DMSO group had greater clonal proliferation after 72 h of drug incubation followed by 72 h of recovery time ($p < 0.05$, Figure 5A). Indeed, the Cis+Nira effectively blocked DDR capacity by decreased PARP1 and XRCC1 protein expression level, while the DMSO group had significantly greater expression of DDR genes ($p < 0.05$, Figure 5B).

We also investigated the cell death potential of Cis+Nira by measuring relative cell viability and detecting apoptotic proteins, Bax, cleaved caspase-9 and cleaved caspase-3. Additionally, we did anti-apoptotic proteins, Bcl-2. The Cis+Nira significantly reduced relative cell viability compared to the DMSO group ($p < 0.05$, Figure 5C). The Western blot data revealed that Cis+Nira markedly increased cell death by inducing Bax, cleaved caspase-9 and cleaved caspase-3, and reducing Bcl-2 proteins expression level ($p < 0.05$, Figure 5D).

2.4. Combination Therapy Is Considerably Effective against Three-Dimensional Cultured of Twist Knockdown CisR OC Cells

The monolayer culture conditions usually do not replicate the in vivo environment [40,41]. Thus, we employed a 3D spheroids culture to recapitulate the organoid traits of the in vivo environment (Figure 6A). The DMSO treated siNC transfected microspheroids progressively increased in size with tight aggregate spheres formation, while DMSO treated siTwist transfected microspheroids increased in size with loosely aggregate sphere formation (Figure 6B). The Cis+Nira therapy remarkably reduced sphere formation demonstrating significantly lower cell viability ($p < 0.05$, Figure 5C). These findings suggest that the Cis+Nira therapy could potentially eliminate recurred CisR OC cells in vivo.

2.5. Cancer Cell-Death by Cisplatin and Niraparib Accomplished through Boosting ER Stress-Mediated Mitochondrial-Depended Apoptosis on Three-Dimensional Cultured of Twist Knockdown Cisplatin-Resistant OC Cells

To determine how Endoplasmic reticulum (ER) stress effects on apoptosis following Cis+Nira therapy in 3D cell culture, the protein expression of several ER stress markers was examined by Western blotting. The release of ER chaperons and activation of ER resident proteins enhancing the expression of CHOP are usually thought to be associated with activation of ER stress [42–44]. In the present study, the protein levels of GRP78, calnexin, cleaved ATF6, and CHOP were markedly elevated after the treatment of Cis alone, Nira alone or Cis+Nira therapy (Figure 7A,B).

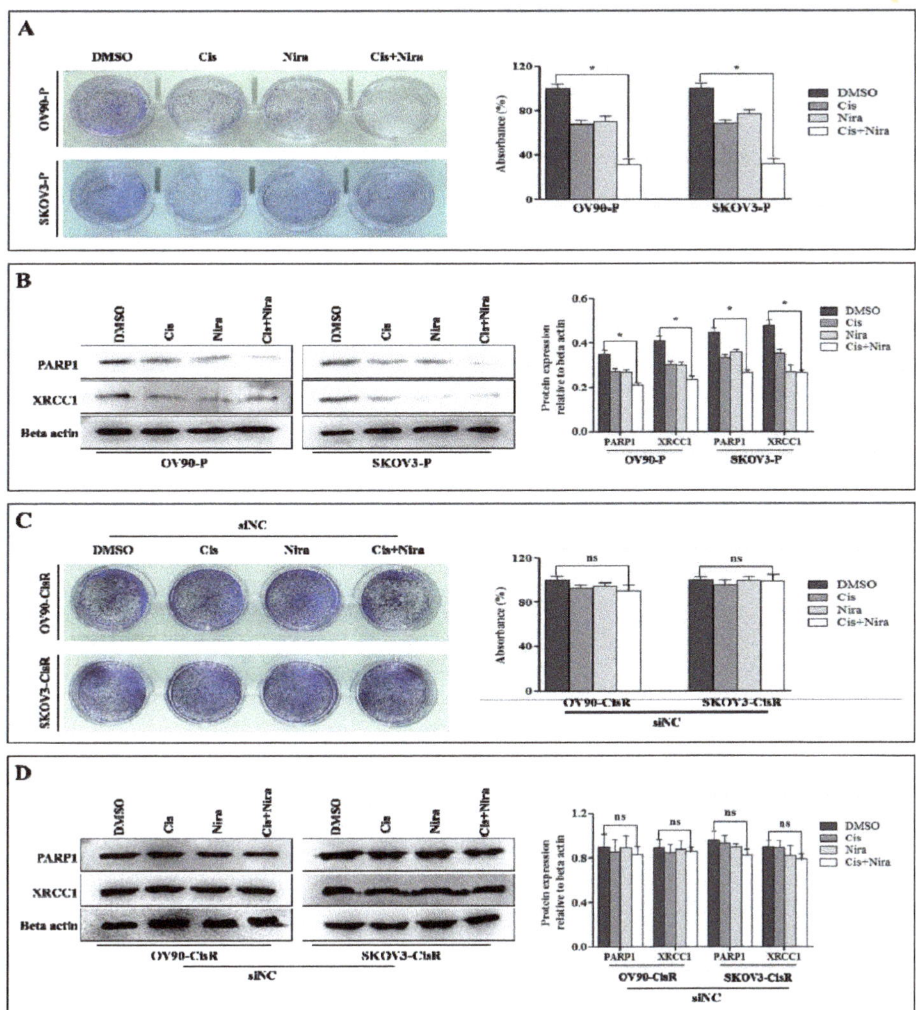

Figure 4. The effect of Cis and Nira on DDR proteins, PARP1 and XRCC1 in parental and non-silenced CisR OC cells. (**A**) Measurement of cell growth rate using clonogenic assay in parental cells administered with DMSO, Cis, Nira, and Cis+Nira. The cells were cultured in absence or presence of Cis (18 μm, OV90—P; 21 μm, SKOV3-P) and Nira (17.5 μm, OV90—P; 15 μm, SKOV3-P) for 72 h followed by 72 h of recovery period. (**B**) Western blot analysis of DDR proteins, PARP1 and XRCC1 expression levels to determine the effect on parental and siNC cells treated with DMSO, Cis, Nira and Cis+Nira. The transfected cells were cultured in absence or presence of Cis (18 μm, OV90—P; 21 μm, SKOV3-P) and Nira (17.5 μm, OV90—P; 15 μm, SKOV3-P) for 72 h. (**C**) Measurement of cell growth rate using clonogenic assay in siNC transfected CisR cells administered with DMSO, Cis, Nira, and Cis+Nira. The cells were cultured in absence or presence of Cis (18 μm, OV90—CisR; 21 μm, SKOV3-CisR) and Nira (17.5 μm, OV90—CisR; 15 μm, SKOV3-CisR) for 72 h followed by 72 h of recovery period. (**D**) Western blot analysis of DDR proteins, PARP1 and XRCC1 expression levels to determine the effect on parental and siNC cells treated with DMSO, Cis, Nira and Cis+Nira. The transfected cells were cultured in absence or presence of Cis (18 μm, OV90—CisR; 21 μm, SKOV3-CisR) and Nira (17.5 μm, OV90—CisR; 15 μm, SKOV3-CisR) for 72 h. P: parental; CisR: cisplatin-resistant; siRNA: small interfering RNA; siNC: non-targeting negative control siRNA; siTwist: Twist siRNA. Values were represented as mean ± SD. * $p < 0.05$, compared with the DMSO group and [ns] $p > 0.05$, compared with the DMSO group.

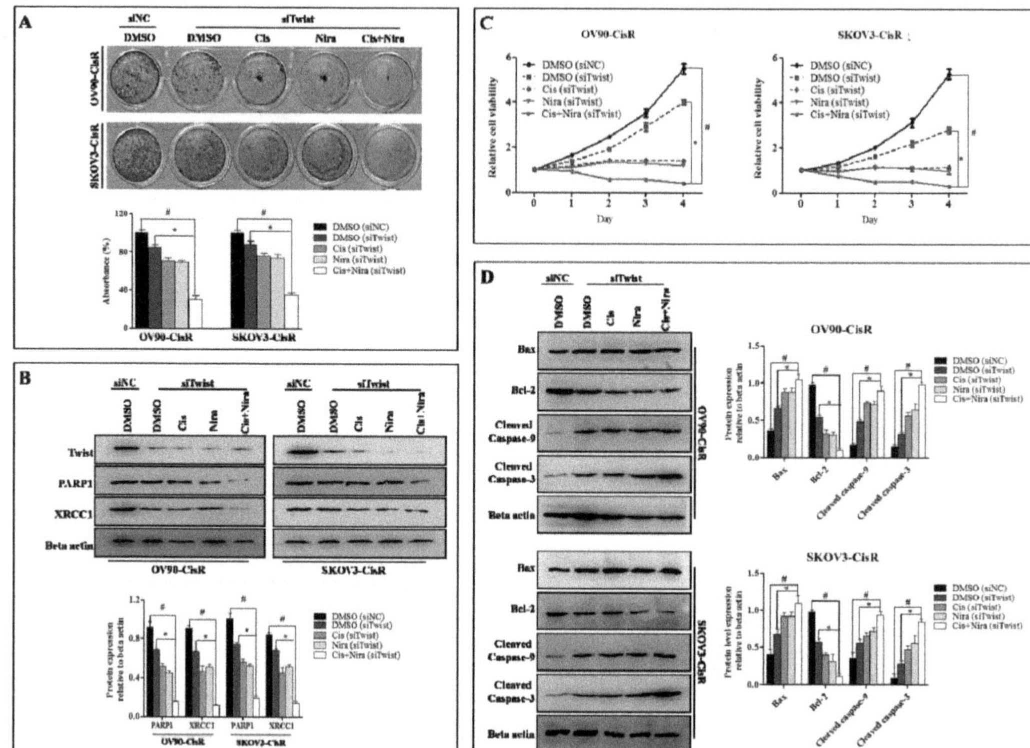

Figure 5. The Cis+Nira therapy induces cell death with blocking DDR capacity on Twist-knockdown CisR OC cells. (**A**) Measurement of cell growth rate using clonogenic assay in siTwist transfected cells administered with DMSO, Cis, Nira, and Cis+Nira. The siTwist transfected cells were cultured in absence or presence of Cis (18 μm, OV90—CisR; 21 μm, SKOV3-CisR) and Nira (17.5 μm, OV90—CisR; 15 μm, SKOV3-CisR) for 72 h followed by 72 h of recovery period. (**B**) Western blot analysis of DDR proteins, PARP1 and XRCC1 expression levels to determine the effect of siNC and siTwist transfected cells treated with DMSO, Cis, Nira and Cis+Nira. The transfected cells were cultured in absence or presence of Cis (18 μm, OV90—CisR; 21 μm, SKOV3-CisR) and Nira (17.5 μm, OV90—CisR; 15 μm, SKOV3-CisR) for 72 h. (**C**) Measurement of relative cell viability in siNC and siTwist transfected cells administered with DMSO, Cis, Nira, and Cis+Nira in OV90-CisR (18 μm of Cis and 17.5 μm of Nira) and SKOV3-CisR (21 μm of Cis and 15 μm of Nira) OC cells for 4 days. (**D**) Western blot analysis of cell death proteins, including Bax, cleaved caspase-9 and cleaved caspase-3 and Bcl-2 proteins and their expression levels to determine the effect of siNC and siTwist transfected cells treated with DMSO, Cis, Nira and Cis+Nira. The transfected cells were cultured in absence or presence of Cis (18 μm, OV90—CisR; 21 μm, SKOV3-CisR) and Nira (17.5 μm, OV90—CisR; 15 μm, SKOV3-CisR) for 72 h. CisR: cisplatin-resistant; siRNA: small interfering RNA; siNC: non-targeting negative control siRNA; siTwist: Twist siRNA. Values were represented as mean ± SD. # $p < 0.05$, compared with siNC group; * $p < 0.05$, compared with the siTwist (DMSO) group.

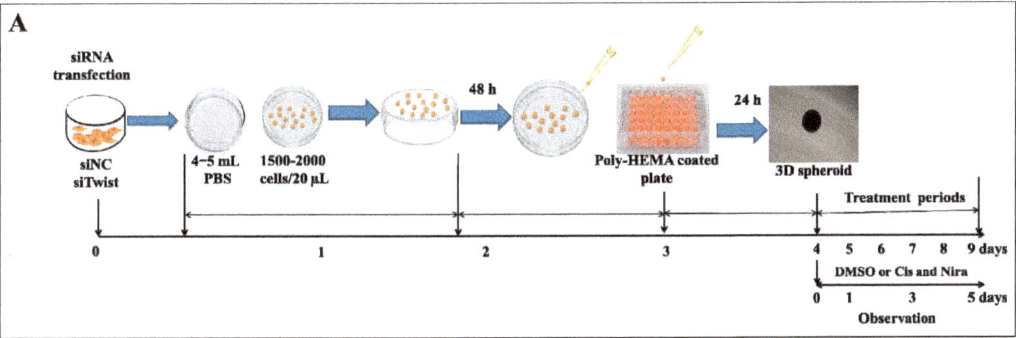

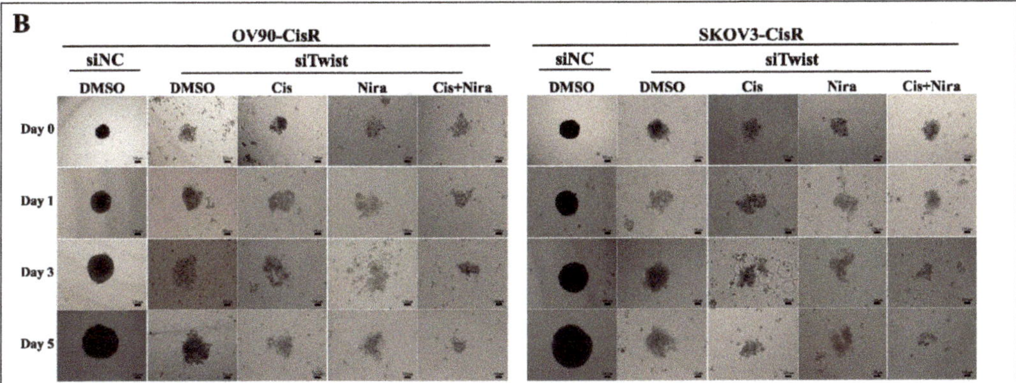

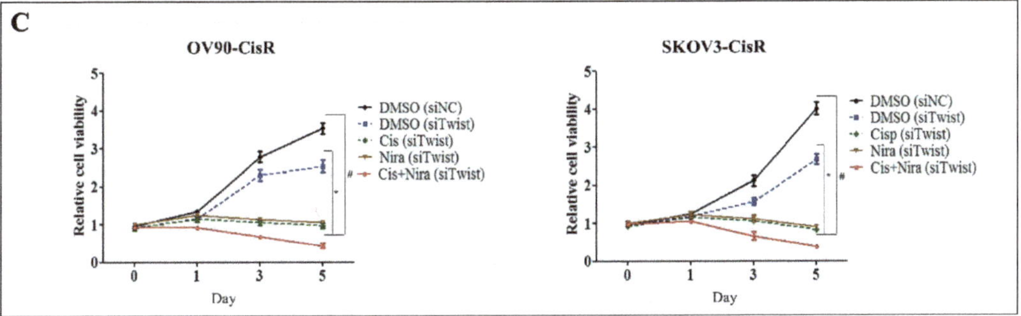

Figure 6. The Cis+Nira therapy is effective against three-dimensional (3D) cultures of Twist-knockdown CisR OC cells. (**A**) The schematic diagram showing the experimental protocol for the formation 3D spheroid and treatment plan of DMSO, Cis, and Nira. (**B**) The images representing 3D spheroids formation of siNC and siTwist transfected cells at 0, 1, 3 and 5 days on OV90-CisR and SKOV3-CisR (Magnification, 10×, scale bar 100 μm). The cells were cultured in absence or presence of Cis (18 μm, OV90—CisR; 21 μm, SKOV3-CisR) and Nira (17.5 μm, OV90—CisR; 15 μm, SKOV3-CisR) for 120 h. (Magnification, 10×, scale bar 100 μm) (**C**) Relative cell viability of siNC and siTwist transfected 3D culture cells administered with DMSO, Cis, Nira, and Cis+Nira in OV90-CisR (18 μm of Cis and 17.5 μm of Nira) and SKOV3-CisR (21 μm of Cis and 15 μm of Nira) OC cells. CisR: cisplatin-resistant; siRNA: small interfering RNA; siNC: non-targeting negative control, siRNA; siTwist: Twist siRNA; DMSO: dimethyl sulfoxide, Cis: cisplatin; Nira: niraparib. Values were represented as mean ± SD. # $p < 0.05$, compared with siNC group; * $p < 0.05$, compared with the siTwist (DMSO) group.

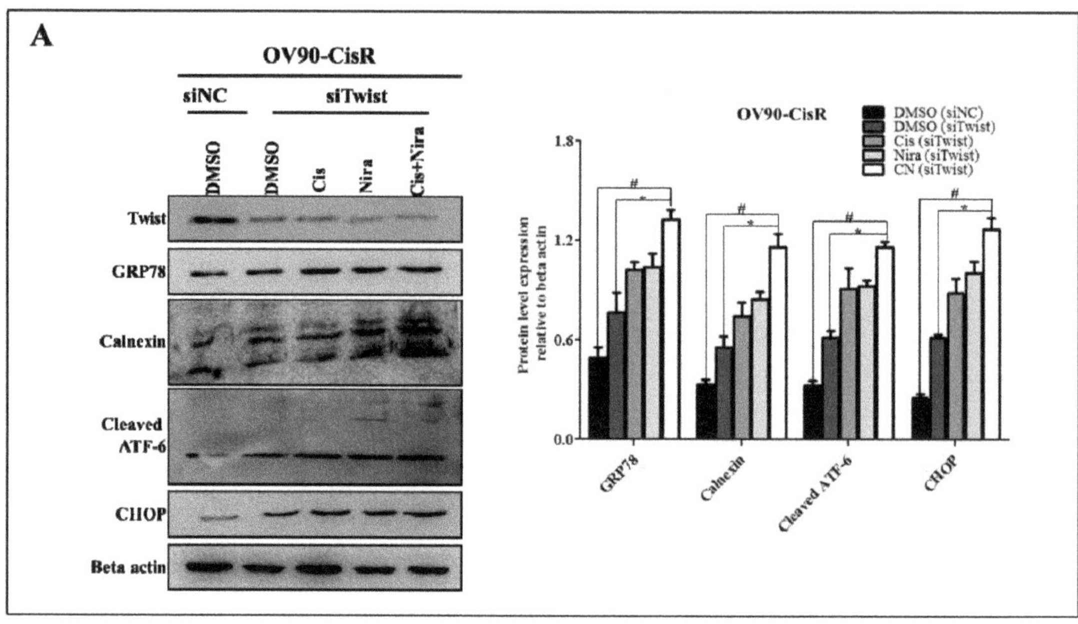

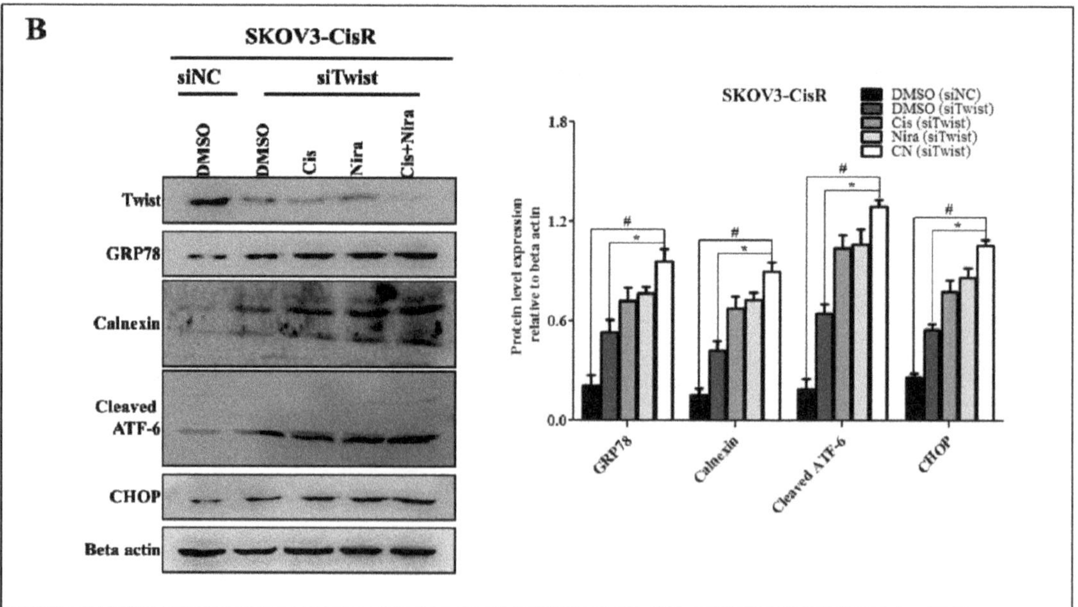

Figure 7. The Cis+Nira therapy enhances ER stress in 3D cultures of Twist-knockdown CisR OC cells. Western blot analysis of ER stress proteins, GRP78, calnexin, cleaved ATF6 and CHOP in 3D cultures of Twist-knockdown (**A**) OV90-CisR and (**B**) SKOV3-CisR OC cells. CisR: cisplatin-resistant; siRNA: small interfering RNA; siNC: non-targeting negative control, siRNA; siTwist: Twist siRNA; DMSO: dimethyl sulfoxide, Cis: cisplatin; Nira; niraparib. Values were represented as mean ± SD. # $p < 0.05$, compared with siNC group; * $p < 0.05$, compared with the siTwist (DMSO) group.

Mitochondrial changes are critical for apoptosis where disturbance in mitochondrial transmembrane potential is one of the earliest intracellular actions to occur following induction of apoptosis [45–48]. It is identified that mitochondrial outer membrane proteins, Bax (the pro-apoptotic proteins) and Bcl-2 (anti-apoptotic) and inner protein, cytochrome C play a vital role in triggering caspase-dependent apoptosis [47,48]. The mitochondrial apoptosis kit employs Mitostain, a cationic dye that produce fluorescence to differentiate healthy and apoptotic cells. Our result demonstrated that red/green ratio reduced by DMSO (siNC), DMSO (siTwist), Cis (siTwist), Nira (siTwist), and Cis+Nira (siTwist) in Twist-knockdown OV90-CisR cells were 8.25 ± 0.94, 6.28 ± 0.76, 4.42 ± 0.47, 3.61 ± 0.40 and 1.63 ± 0.186, respectively (Figure 8A). The values were identical to those observed in SKOV3-CisR (9.85 ± 0.73, 7.01 ± 0.72, 4.63 ± 0.81, 3.99 ± 0.43, 2.24 ± 0.42) (Figure 8B). By comparison, triple combination of Twist-knockdown, Cis, Nira treatment showed the greatest decrease of mitochondrial membrane potential—four to five-times lower than that of the DMSO (siNC) group. Triple combination treatment caused greater mitochondrial damage in both OV90-CisR and SKOV3-CisR cell lines, as indicated by the decrease in mitochondrial membrane potential. Most importantly, triple combination therapy significantly upregulated Bax in 9the mitochondrial fraction (Figure S3) and upregulated cytochrome C in the cytosolic fraction (Figure S4) in Twist-knockdown CisR OC cells. In contrast, downregulation of Bcl-2 and cytochrome C in the mitochondrial fraction was observed with triple combination therapy (Figure 9 A and Figure S3). The ratio of cytochrome C in cytosol to that in mitochondria was significantly higher in the group of combination therapy due to the release of cytochrome C from mitochondria to cytosol (Figure 9B). These findings suggest that the efficacy of Cis and Nira on Twist-Knockdown CisR OC cells is associated with decreasing the mitochondrial potential, followed by the translocation of apoptotic protein Bax from cytoplasm to mitochondrial outer membrane with diminishing Bcl-2 activity to facilitate the cytochrome C release from mitochondria to cytoplasm.

2.6. The Synthetic Lethality of Cisplatin and Niraparib Consists of Expressing Apoptotic Marker and Suppressing Cell Proliferative Marker on Three-Dimensional Cultured of Twist Knockdown Cisplatin-Resistant OC Cells

We applied hematoxylin and eosin (H&E) staining and immunohistochemistry to investigate histological features together with markers of cell proliferation and apoptosis (Figure 9C). Proliferative activity was examined by anti-Ki-67 and apoptosis potential was represented by anti-cleaved caspase-3 immunohistochemistry. The H&E staining of 3D spheroids after generation exhibited tumor cell characteristics of high nucleus to cytoplasm ratios. The siNC transfected spheroids were densely arranged in tight balls. The siTwist transfected spheroids treated with Cis alone, Nira alone or both combinations of Cis+Nira showed developed vascular structures with loose arrangements of discohesive OC cells. Immunohistochemically, Cis alone, Nira alone or combinations of Cis+Nira showed lower ki-67 activity along with higher cleaved caspase-3 immunoreactivity.

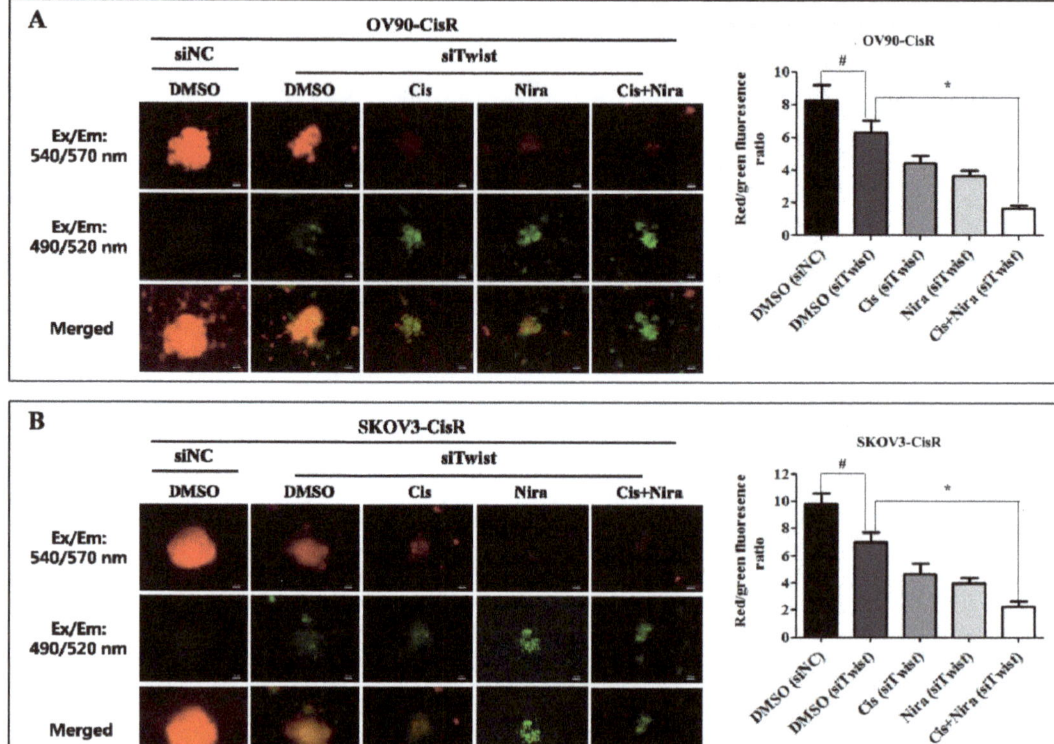

Figure 8. The Cis+Nira therapy significantly reduced mitochondrial membrane potential in Twist-knockdown CisR OC cells. Microscopic visualization and quantification of mitochondrial membrane potential in 3D cultures of Twist-knockdown (**A**) OV90-CisR and (**B**) SKOV3-CisR OC cells. Magnification, 10×, scale bar 100 μm. CisR: cisplatin-resistant; siRNA: small interfering RNA; siNC: non-targeting negative control, siRNA; siTwist: Twist siRNA; DMSO: dimethyl sulfoxide, Cis: cisplatin; Nira; niraparib. Values were represented as mean ± SD. # $p < 0.05$, compared with siNC group; * $p < 0.05$, compared with the siTwist (DMSO) group.

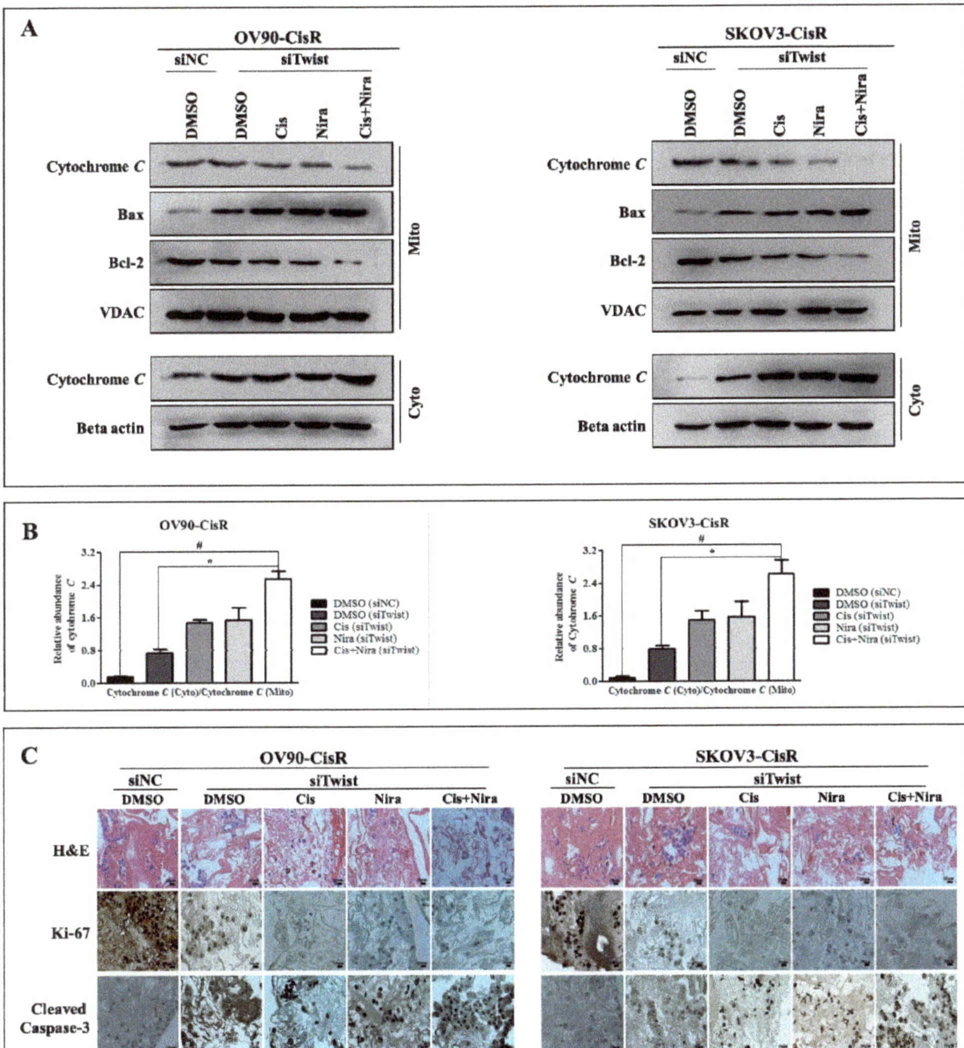

Figure 9. The combination therapy efficiently activated mitochondrial-dependent apoptotic pathway in Twist-knockdown CisR OC cells. (**A**) Western blot analysis of Bax, Bcl-2 and cytochrome C in 3D cultured of Twist knockdown OC cells. (**B**) The ratio of cytochrome C in cytosolic fraction to that in mitochondria fraction. (**C**) H&E staining for morphology evaluation and immunohistochemistry of the cell proliferation marker, ki-67 and apoptosis marker, cleaved caspase-3 immunoreactivity prediction (Magnification, 40×, scale bar 20 μm). CisR: cisplatin-resistant; siRNA: small interfering RNA; siNC: non-targeting negative control, siRNA; siTwist: Twist siRNA; DMSO: dimethyl sulfoxide, Cis: cisplatin; Nira; Niraparib. Values were represented as mean ± SD. # $p < 0.05$, compared with siNC group; * $p < 0.05$, compared with the siTwist (DMSO) group.

3. Discussion

In this study, we demonstrated that the Nira and Cis proficiently killed Twist-knockdown CisR OC cell via induction of apoptosis under both in 2D and 3D cell culture conditions. Additionally, through in-detail in vitro experiments, we investigated the potential mechanisms underlying the synthetic lethality of Nira and Cis on Twist-deficient CisR OC cells and provided a schematic representation of the mechanisms and interactions in Figure 10. We found that one of the mechanisms of synthetic lethality is through blocking DNA repair, i.e., suppression of PARP1 and XRCC1 activation. Another important mechanism is the ER stress-mediated mitochondrial apoptosis through boosting release of ER chaperons, GRP78 and calnexin, and activation of ER resident protein, cleaved ATF6 with enhanced expression of CHOP. Mitochondrial outer membrane proteins, Bax and Bcl-2, and inner protein, cytochrome C play a vital role in triggering caspase-dependent apoptosis [47–49]. Hence, Twist-knockdown of CisR OC cells greatly improved the translocation of Bax to the outer mitochondrial membrane with reduced activity of Bcl-2, and thus the release of cytochrome C from mitochondria to cytoplasm, leading to irreversible cell death.

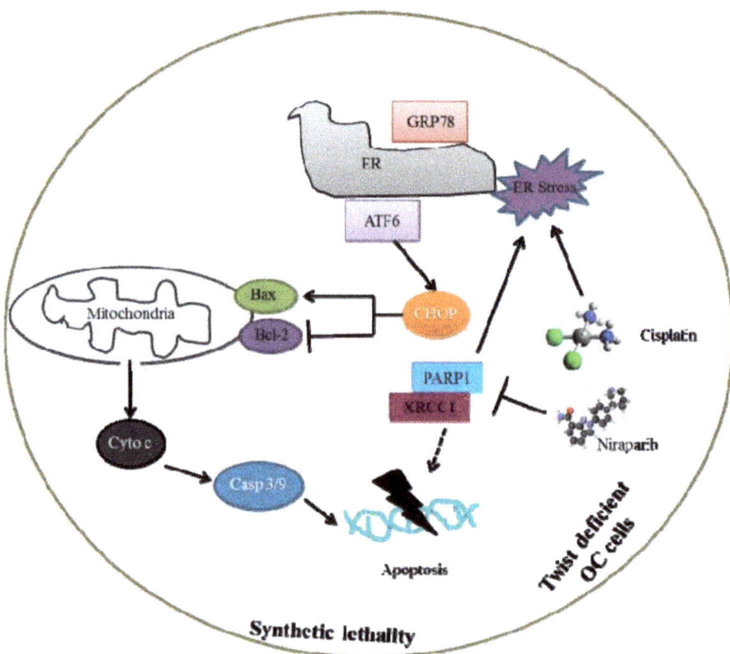

Figure 10. The pathway involved in the synthetic lethality of PARPi with combination of cisplatin on Twist knockdown CisR OC cells.

Cancer cells are often considered to have abnormalities in DDR including defects in cell cycle checkpoints and/or DNA repair [18,50]. Cancers with DDR impairment demonstrated vulnerability to genome instability by "synthetic lethality" targeting chemotherapy [18,51,52]. Synthetic lethality between DDR pathways has provided a standard for cancer therapy by targeting DDR. PARP1 is an essential cofactor in DDR by incorporating DNA repair effector proteins, such as XRCC1, into the cisplatin-induced DNA injury site and coordinating their action of DNA single-strand break repair [53–56]. The most notable event of DDR directing mediators is the capability of PARP1 repression as a regulated, synthetic lethal approach, demonstrated in both in vitro and in vivo studies [57–62]. Although FDA approved PARPi includes rucaparib, olaparib, talazoparib and niraparib were hypothesized to be effective in a wide range of patients, only small percentage of cancer patients receiving benefit from it due to cross-resistance to other chemotherapeutic drugs [63–66].

In patients with OC, Twist expression forecasts poor prognostic outcomes, indicating that Twist can play a critical role in OC development [67,68]. Several studies have indicated that Twist expression could be a useful predictor of unfavorable prognosis for OC [69,70]. However, little information is available as to the role of Twist in platinum resistance of OC.

In this study, we demonstrated that CisR OC cells exhibit higher expression of Twist with amplification of DDR proteins, PARP1 and XRCC1. We confirmed that both Cis and PARPi sensitivity gradually decreased during the development of resistant cell lines, indicating CisR cells were also cross-resistance to PARPi. Although the knockdown of Twist in CisR OC cells reportedly blocks DDR genes activity, we confirmed that Twist-knockdown sensitizes OC cells to Cis and Nira confirmed by the significantly lowered IC_{50} value.

More importantly, we observed synergetic drug interactions between Cis and Nira in both Twist-knockdown CisR OC monolayer cell culture. Treatment of Nira causes DNA injury and cellular damage by restricting PARP enzymatic action and enhancing the development of PARP-DNA clusters [71,72]. Similarly, Cis induces DNA damage by generating monoadducts and interstrand crosslinks leading to inhibition of DNA and ultimately to cell death [56,73]. The Cis+Nira therapy significantly reduced cell growth rate, and increased the effects of DNA damage and cell death compared with either monotherapy. Based on these results, we believe that knockdown of the Twist with PARPi could be a highly effective strategy in overcoming CisR in OC.

The 3D cell culture aims at recapitulating in vivo tissue architectures to provide physiologically relevant models to study normal development and disease [74,75]. Although the 3D culture model cannot fully recapitulate all aspects of the in vivo environment, it offers more benefits than a 2D culture. Our study revealed that Cis+Nira therapy with Twist-knockdown considerably decreased the formation of the micr-organoids in 3D cell culture.

ER stress, measured by the magnitude and length of ER stress, is important for both pro-survival and pro-apoptotic functions. It has been a possible target for developing potential drug candidates that regulate specific signaling pathways to either suppress tumor or overcome chemotherapy resistance [76]. Recently several chemotherapy candidates have been investigated in connection to ER stress, which could directly or indirectly affect cancers [77]. Among them, ER stress inducing agents that activated CHOP-GADD34 axis is a promising anti-cancer approach [78–81]. Our study showed that after Cis and Nira treatment, potential levels of GRP78, calnexin, cleaved ATF6 and CHOP were simultaneously elevated in 3D spheroid cell culture. Although the underlying basis influencing ER stress- mediated apoptosis have not been fully elucidated, growing evidence indicates that mitochondria and ER work together to trigger cell damage signal [82]. The protein levels of pro-apoptotic Bax elevated, whereas anti-apoptotic Bcl-2 dropped after Cis+Nira therapy in Twist-knockdown CisR OC cells. These two proteins cause the external mitochondrial membrane to be permeabilized and consequently release cytochrome C to the cytosol and eventually activating caspases-3, as reflected in our research.

In addition, the IHC examination of Ki-67 revealed downregulated proliferation and increased cleaved caspase-3 expression after combination therapy in Twist knockdown 3D spheroid OC cell culture.

4. Materials and Methods

4.1. Cell Lines

Two human ovarian cancer (OC) cell lines, OV90 and SKOV3 were obtained from Korean Biotech Co., Ltd., Seoul, Korea, the domestic distributor of American Type Culture Collection (ATCC). The cells were cultured in their respective ATCC recommended growth supplemented with fetal bovine serum (FBS). Two CisR OC cells, OV90/CisR and SKOV3/CisR were generated by the stepwise increment of cisplatin dosing method described in our previous publication [27].

4.2. siRNA Transfection

Small interfering RNA (siRNA) transfections were performed according to the manufacturer's instructions using Lipofectamine 2000 RNAiMAX transfection reagent (Invitrogen). Twist siRNA (#sc38604, Santa Cruz Biotechnology, Dallas, TX, USA) and control siRNA (#sc37007, Santa Cruz Biotechnology, Dallas, TX, USA) were used to generate Twist knockdown (siTwist) and negative control (siNC) cells. The lyophilized siRNA duplex was reconstituted in RNase-free water to create 10 μM stock solutions. The transfected cells were confirmed by Western blot analysis.

4.3. Measurement of Cell Viability and IC$_{50}$

Cell viability and IC$_{50}$ were determined using EZ-cytox cell viability kits (#EZ-1000, DLS-1906, DoGenBio Co., Ltd., Seoul, Korea). The cells (1×10^4 cells/well) were plated in 96-well plates and incubated for desired period of time. The cells were treated with different concentration of drugs. After incubation, EZ-cytox solution was added to each well and incubated for 2 h. Absorbance was then recorded on a microplate reader (Synergy H1; BioTek Instruments, Inc., Winooski, VT, USA) at the wavelength of 490 nm. The IC$_{50}$ values were analyzed using GraphPad Prism software (version 5.0, GraphPad Software Inc., San Diego, CA, USA).

4.4. Colony Formation Assay

The clonogenic assay was employed for the determination of cell growth using previously described method [83]. In brief, the cells were cultured plates at low density (~500 cells per well) in 12-well for 24 h and then treated with drugs for 72 h followed by 72 h recovery. The plates were then washed with PBS and stained with 0.1% crystal violet solution. The cells were washed until no stain was visible, air dried, and photographed. The dye was extracted using 1% sodium dodecyl sulfate (SDS) solution by continuous shaking and then quantified using a spectrophotometer at 570 nm.

4.5. Generation of Three Dimensional (3D) Spheroidal Model

To generate in vitro three–dimensional (3D) spheroidal model utilized combination of hanging drop and hydrogel scaffolds modified methods [27,84]. The details of generation of three dimensional (3D) spheroidal model are described in the supporting information, Supplementary materials and methods, Section S1.1.

4.6. Mitochondrial Apoptosis Staining

Mitochondrial transmembrane potential was evaluated by using mitochondrial apoptosis staining kit (#PK-CA577-K250-25, Promo Kine, Heidelberg, Germany), according to manufacturer instruction. The details of mitochondrial apoptosis staining are described in the supporting information, Supplementary materials and methods, Section S1.2.

4.7. Mitochondrial and Cytosolic Separation

Mitochondrial and cytosolic fractions were isolated as described earlier [43]. The details of mitochondrial and cytosolic separation are described in the supporting information, Supplementary materials and methods, Section S1.3.

4.8. Western Blotting

The Western blotting was performed as described previously [27]. The details of Western blotting are described in the supporting information, Supplementary materials and methods, Section S1.4.

4.9. Histology and Immunohistochemistry

The hematoxylin and eosin (H&E) and immunohistochemistry (IHC) staining was performed according to routine protocols. The details of staining are described in the supporting information, Supplementary materials and methods, Section S1.5.

4.10. Statistical Analysis

Data were statistically expressed using mean $\pm$ SD and one-way analysis of variance (ANOVA) followed by Tukey's multiple comparison test was used for the statistical analysis of various three groups. $p < 0.05$ was considered statistically significant.

5. Conclusions

The present study links Twist and DDR proteins (PARP1 and XRCC1) with cisplatin resistance in OC. We demonstrated that Twist knockdown CisR OC cells are highly sensitive to cisplatin and PARPi niraparib in both in vitro 2D and 3D cell culture models (Figure 11). Our study also suggests that targeting Twist and PARPi (a synthetic lethal partner) could potentially be a beneficial approach to overcoming cisplatin-based resistance and a promising option for the treatment of OC.

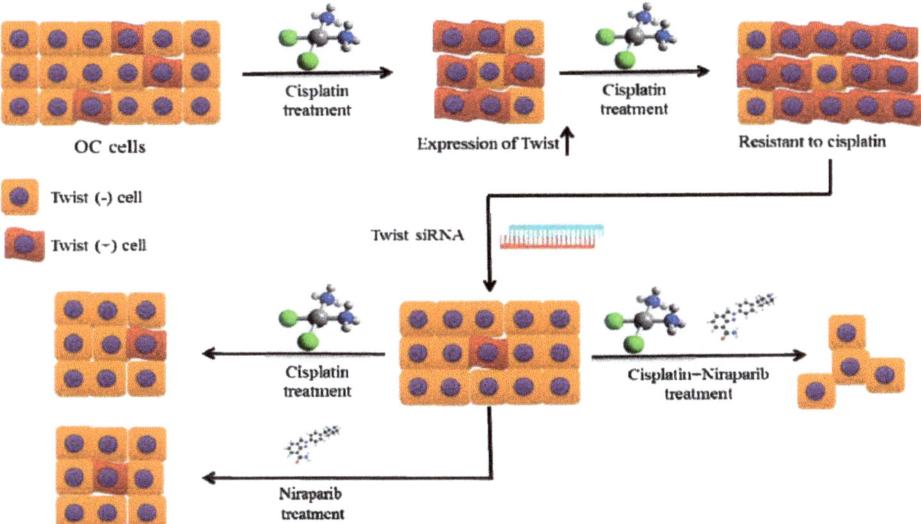

Figure 11. The proposed therapeutic strategy to overcome cisplatin resistance in OC.

Supplementary Materials: The following are available online at https://www.mdpi.com/article/10.3390/ijms22083916/s1.

Author Contributions: E.B., conceptualization, methodology, data curation, software and formal analysis, original draft preparation; J.-Y.K., conceptualization, methodology, data curation, software, review and editing manuscript; H.Y., validation, investigation, resources, supervision, project administration, funding acquisition and writing—review and editing manuscript; H.-S.K., validation, investigation, resources, supervision, project administration, funding acquisition, writing—review and editing manuscript; D.-C.K., histopathological interpretation, writing—review and editing manuscript. All authors have read and agreed to the published version of the manuscript.

Funding: This research was supported by the Basic Science Research Program through the National Research Foundation of Korea (NRF) funded by the Ministry of Education (2016R1D1A1B03935584: 01 November 2016) and by the Ministry of Science and ICT (2019R1FA1059148: 01 June 2019).

Institutional Review Board Statement: Not applicable for this study.

Informed Consent Statement: Not applicable for this study.

Data Availability Statement: The data presented in this study are available in this article Int. J. Mol. Sci. and its Supplementary material and methods.

Conflicts of Interest: The authors declare no conflict of interest.

References

1. Reid, B.M.; Permuth, J.B.; Sellers, T.A. Epidemiology of ovarian cancer: A review. *Cancer Biol. Med.* **2017**, *14*, 9–32. [CrossRef]
2. Kawahara, N.; Ogawa, K.; Nagayasu, M.; Kimura, M.; Sasaki, Y.; Kobayashi, H. Candidate synthetic lethality partners to PARP inhibitors in the treatment of ovarian clear cell cancer. *Biomed. Rep.* **2017**, *7*, 391–399. [CrossRef] [PubMed]

3. Kim, A.; Ueda, Y.; Naka, T.; Enomoto, T. Therapeutic strategies in epithelial ovarian cancer. *J. Exp. Clin. Cancer Res.* **2012**, *31*, 14. [CrossRef] [PubMed]

4. Thompson, N.; Adams, D.J.; Ranzani, M. Synthetic lethality: Emerging targets and opportunities in melanoma. *Pigment Cell Melanoma Res.* **2017**, *30*, 183–193. [CrossRef] [PubMed]

5. Lu, X.; Megchelenbrink, W.; Notebaart, R.A.; Huynen, M.A. Predicting human genetic interactions from cancer genome evolution. *PLoS ONE* **2015**, *10*, e0125795. [CrossRef]

6. Iorio, F.; Knijnenburg, T.A.; Vis, D.J.; Bignell, G.R.; Menden, M.P.; Schubert, M.; Aben, N.; Goncalves, E.; Barthorpe, S.; Lightfoot, H.; et al. A Landscape of Pharmacogenomic Interactions in Cancer. *Cell* **2016**, *166*, 740–754. [CrossRef]

7. Possik, P.A.; Muller, J.; Gerlach, C.; Kenski, J.C.; Huang, X.; Shahrabi, A.; Krijgsman, O.; Song, J.Y.; Smit, M.A.; Gerritsen, B.; et al. Parallel in vivo and in vitro melanoma RNAi dropout screens reveal synthetic lethality between hypoxia and DNA damage response inhibition. *Cell Rep.* **2014**, *9*, 1375–1386. [CrossRef]

8. Bonanno, L. Predictive models for customizing chemotherapy in advanced non-small cell lung cancer (NSCLC). *Transl. Lung Cancer Res.* **2013**, *2*, 160–171. [CrossRef]

9. Reinhardt, H.C.; Jiang, H.; Hemann, M.T.; Yaffe, M.B. Exploiting synthetic lethal interactions for targeted cancer therapy. *Cell Cycle* **2009**, *8*, 3112–3119. [CrossRef]

10. Ray Chaudhuri, A.; Nussenzweig, A. The multifaceted roles of PARP1 in DNA repair and chromatin remodelling. *Nat. Rev. Mol. Cell Biol.* **2017**, *18*, 610–621. [CrossRef]

11. Li, M.; Li, H.; Liu, F.; Bi, R.; Tu, X.; Chen, L.; Ye, S.; Cheng, X. Characterization of ovarian clear cell carcinoma using target drug-based molecular biomarkers: Implications for personalized cancer therapy. *J. Ovarian Res.* **2017**, *10*, 9. [CrossRef]

12. Friedlander, M.; Gebski, V.; Gibbs, E.; Davies, L.; Bloomfield, R.; Hilpert, F.; Wenzel, L.B.; Eek, D.; Rodrigues, M.; Clamp, A.; et al. Health-related quality of life and patient-centred outcomes with olaparib maintenance after chemotherapy in patients with platinum-sensitive, relapsed ovarian cancer and a BRCA1/2 mutation (SOLO2/ENGOT Ov-21): A placebo-controlled, phase 3 randomised trial. *Lancet Oncol.* **2018**, *19*, 1126–1134. [CrossRef]

13. Ledermann, J.A.; Harter, P.; Gourley, C.; Friedlander, M.; Vergote, I.; Rustin, G.; Scott, C.; Meier, W.; Shapira-Frommer, R.; Safra, T.; et al. Overall survival in patients with platinum-sensitive recurrent serous ovarian cancer receiving olaparib maintenance monotherapy: An updated analysis from a randomised, placebo-controlled, double-blind, phase 2 trial. *Lancet Oncol.* **2016**, *17*, 1579–1589. [CrossRef]

14. Coleman, R.L.; Oza, A.M.; Lorusso, D.; Aghajanian, C.; Oaknin, A.; Dean, A.; Colombo, N.; Weberpals, J.I.; Clamp, A.; Scambia, G.; et al. Rucaparib maintenance treatment for recurrent ovarian carcinoma after response to platinum therapy (ARIEL3): A randomised, double-blind, placebo-controlled, phase 3 trial. *Lancet* **2017**, *390*, 1949–1961. [CrossRef]

15. Mirza, M.R.; Monk, B.J.; Herrstedt, J.; Oza, A.M.; Mahner, S.; Redondo, A.; Fabbro, M.; Ledermann, J.A.; Lorusso, D.; Vergote, I.; et al. Niraparib Maintenance Therapy in Platinum-Sensitive, Recurrent Ovarian Cancer. *N. Engl. J. Med.* **2016**, *375*, 2154–2164. [CrossRef] [PubMed]

16. Robson, M.; Im, S.A.; Senkus, E.; Xu, B.; Domchek, S.M.; Masuda, N.; Delaloge, S.; Li, W.; Tung, N.; Armstrong, A.; et al. Olaparib for Metastatic Breast Cancer in Patients with a Germline BRCA Mutation. *N. Engl. J. Med.* **2017**, *377*, 523–533. [CrossRef]

17. Fong, P.C.; Yap, T.A.; Boss, D.S.; Carden, C.P.; Mergui-Roelvink, M.; Gourley, C.; De Greve, J.; Lubinski, J.; Shanley, S.; Messiou, C.; et al. Poly(ADP)-ribose polymerase inhibition: Frequent durable responses in BRCA carrier ovarian cancer correlating with platinum-free interval. *J. Clin. Oncol. Off. J. Am. Soc. Clin. Oncol.* **2010**, *28*, 2512–2519. [CrossRef]

18. Audeh, M.W.; Carmichael, J.; Penson, R.T.; Friedlander, M.; Powell, B.; Bell-McGuinn, K.M.; Scott, C.; Weitzel, J.N.; Oaknin, A.; Loman, N.; et al. Oral poly(ADP-ribose) polymerase inhibitor olaparib in patients with BRCA1 or BRCA2 mutations and recurrent ovarian cancer: A proof-of-concept trial. *Lancet* **2010**, *376*, 245–251. [CrossRef]

19. Ru, G.Q.; Wang, H.J.; Xu, W.J.; Zhao, Z.S. Upregulation of Twist in gastric carcinoma associated with tumor invasion and poor prognosis. *Pathol. Oncol. Res.* **2011**, *17*, 341–347. [CrossRef] [PubMed]

20. Yang, J.; Mani, S.A.; Donaher, J.L.; Ramaswamy, S.; Itzykson, R.A.; Come, C.; Savagner, P.; Gitelman, I.; Richardson, A.; Weinberg, R.A. Twist, a master regulator of morphogenesis, plays an essential role in tumor metastasis. *Cell* **2004**, *117*, 927–939. [CrossRef]

21. Ohuchida, K.; Mizumoto, K.; Ohhashi, S.; Yamaguchi, H.; Konomi, H.; Nagai, E.; Yamaguchi, K.; Tsuneyoshi, M.; Tanaka, M. Twist, a novel oncogene, is upregulated in pancreatic cancer: Clinical implication of Twist expression in pancreatic juice. *Int. J. Cancer* **2007**, *120*, 1634–1640. [CrossRef]

22. Kang, Y.; Massague, J. Epithelial-mesenchymal transitions: Twist in development and metastasis. *Cell* **2004**, *118*, 277–279. [CrossRef] [PubMed]

23. Yu, J.; Xie, F.; Bao, X.; Chen, W.; Xu, Q. miR-300 inhibits epithelial to mesenchymal transition and metastasis by targeting Twist in human epithelial cancer. *Mol. Cancer* **2014**, *13*, 121. [CrossRef] [PubMed]

24. Kwon, C.H.; Park, H.J.; Choi, Y.; Won, Y.J.; Lee, S.J.; Park, D.Y. TWIST mediates resistance to paclitaxel by regulating Akt and Bcl-2 expression in gastric cancer cells. *Tumour Biol. J. Int. Soc. Oncodev. Biol. Med.* **2017**, *39*, 1010428317722070. [CrossRef]

25. Ji, H.; Lu, H.W.; Li, Y.M.; Lu, L.; Wang, J.L.; Zhang, Y.F.; Shang, H. Twist promotes invasion and cisplatin resistance in pancreatic cancer cells through growth differentiation factor 15. *Mol. Med. Rep.* **2015**, *12*, 3841–3848. [CrossRef] [PubMed]

26. Wang, X.; Ling, M.T.; Guan, X.Y.; Tsao, S.W.; Cheung, H.W.; Lee, D.T.; Wong, Y.C. Identification of a novel function of TWIST, a bHLH protein, in the development of acquired taxol resistance in human cancer cells. *Oncogene* **2004**, *23*, 474–482. [CrossRef] [PubMed]

27. Bahar, E.; Kim, J.Y.; Kim, H.S.; Yoon, H. Establishment of Acquired Cisplatin Resistance in Ovarian Cancer Cell Lines Characterized by Enriched Metastatic Properties with Increased Twist Expression. *Int. J. Mol. Sci.* **2020**, *21*, 7613. [CrossRef]

28. Maestro, R.; Dei Tos, A.P.; Hamamori, Y.; Krasnokutsky, S.; Sartorelli, V.; Kedes, L.; Doglioni, C.; Beach, D.H.; Hannon, G.J. Twist is a potential oncogene that inhibits apoptosis. *Genes Dev.* **1999**, *13*, 2207–2217. [CrossRef]

29. Valsesia-Wittmann, S.; Magdeleine, M.; Dupasquier, S.; Garin, E.; Jallas, A.C.; Combaret, V.; Krause, A.; Leissner, P.; Puisieux, A. Oncogenic cooperation between H-Twist and N-Myc overrides failsafe programs in cancer cells. *Cancer Cell* **2004**, *6*, 625–630. [CrossRef]

30. Roberts, C.M.; Tran, M.A.; Pitruzzello, M.C.; Wen, W.; Loeza, J.; Dellinger, T.H.; Mor, G.; Glackin, C.A. TWIST1 drives cisplatin resistance and cell survival in an ovarian cancer model, via upregulation of GAS6, L1CAM, and Akt signalling. *Sci. Rep.* **2016**, *6*, 37652. [CrossRef] [PubMed]

31. Wu, J.; Liao, Q.; He, H.; Zhong, D.; Yin, K. TWIST interacts with beta-catenin signaling on osteosarcoma cell survival against cisplatin. *Mol. Carcinog.* **2014**, *53*, 440–446. [CrossRef]

32. Cheng, G.Z.; Chan, J.; Wang, Q.; Zhang, W.; Sun, C.D.; Wang, L.H. Twist transcriptionally up-regulates AKT2 in breast cancer cells leading to increased migration, invasion, and resistance to paclitaxel. *Cancer Res.* **2007**, *67*, 1979–1987. [CrossRef]

33. Song, Y.H.; Shiota, M.; Yokomizo, A.; Uchiumi, T.; Kiyoshima, K.; Kuroiwa, K.; Oda, Y.; Naito, S. Twist1 and Y-box-binding protein-1 are potential prognostic factors in bladder cancer. *Urol. Oncol.* **2014**, *32*, 31.e31–31.e37. [CrossRef]

34. Shiota, M.; Song, Y.; Yokomizo, A.; Kiyoshima, K.; Tada, Y.; Uchino, H.; Uchiumi, T.; Inokuchi, J.; Oda, Y.; Kuroiwa, K.; et al. Foxo3a suppression of urothelial cancer invasiveness through Twist1, Y-box-binding protein 1, and E-cadherin regulation. *Clin. Cancer Res. Off. J. Am. Assoc. Cancer Res.* **2010**, *16*, 5654–5663. [CrossRef]

35. Hung, J.J.; Yang, M.H.; Hsu, H.S.; Hsu, W.H.; Liu, J.S.; Wu, K.J. Prognostic significance of hypoxia-inducible factor-1alpha, TWIST1 and Snail expression in resectable non-small cell lung cancer. *Thorax* **2009**, *64*, 1082–1089. [CrossRef] [PubMed]

36. Liu, Y.; Fang, J.; Kim, Y.J.; Wong, M.K.; Wang, P. Codelivery of doxorubicin and paclitaxel by cross-linked multilamellar liposome enables synergistic antitumor activity. *Mol. Pharm.* **2014**, *11*, 1651–1661. [CrossRef] [PubMed]

37. Tardi, P.; Johnstone, S.; Harasym, N.; Xie, S.; Harasym, T.; Zisman, N.; Harvie, P.; Bermudes, D.; Mayer, L. In vivo maintenance of synergistic cytarabine:daunorubicin ratios greatly enhances therapeutic efficacy. *Leuk. Res.* **2009**, *33*, 129–139. [CrossRef] [PubMed]

38. Mayer, L.D.; Harasym, T.O.; Tardi, P.G.; Harasym, N.L.; Shew, C.R.; Johnstone, S.A.; Ramsay, E.C.; Bally, M.B.; Janoff, A.S. Ratiometric dosing of anticancer drug combinations: Controlling drug ratios after systemic administration regulates therapeutic activity in tumor-bearing mice. *Mol. Cancer Ther.* **2006**, *5*, 1854–1863. [CrossRef] [PubMed]

39. Chou, T.C. Theoretical basis, experimental design, and computerized simulation of synergism and antagonism in drug combination studies. *Pharmacol. Rev.* **2006**, *58*, 621–681. [CrossRef]

40. Lee, G.Y.; Kenny, P.A.; Lee, E.H.; Bissell, M.J. Three-dimensional culture models of normal and malignant breast epithelial cells. *Nat. Methods* **2007**, *4*, 359–365. [CrossRef]

41. Shibue, T.; Weinberg, R.A. Integrin beta1-focal adhesion kinase signaling directs the proliferation of metastatic cancer cells disseminated in the lungs. *Proc. Natl. Acad. Sci. USA* **2009**, *106*, 10290–10295. [CrossRef] [PubMed]

42. Bahar, E.; Kim, H.; Yoon, H. ER Stress-Mediated Signaling: Action Potential and Ca(2+) as Key Players. *Int. J. Mol. Sci.* **2016**, *17*, 1558. [CrossRef] [PubMed]

43. Shi, J.; Jiang, Q.; Ding, X.; Xu, W.; Wang, D.W.; Chen, M. The ER stress-mediated mitochondrial apoptotic pathway and MAPKs modulate tachypacing-induced apoptosis in HL-1 atrial myocytes. *PLoS ONE* **2015**, *10*, e0117567. [CrossRef]

44. Bahar, E.; Kim, J.Y.; Yoon, H. Chemotherapy Resistance Explained through Endoplasmic Reticulum Stress-Dependent Signaling. *Cancers* **2019**, *11*, 338. [CrossRef] [PubMed]

45. Denning, M.F.; Wang, Y.; Tibudan, S.; Alkan, S.; Nickoloff, B.J.; Qin, J.Z. Caspase activation and disruption of mitochondrial membrane potential during UV radiation-induced apoptosis of human keratinocytes requires activation of protein kinase C. *Cell Death Differ.* **2002**, *9*, 40–52. [CrossRef]

46. Reers, M.; Smiley, S.T.; Mottola-Hartshorn, C.; Chen, A.; Lin, M.; Chen, L.B. Mitochondrial membrane potential monitored by JC-1 dye. *Methods Enzymol.* **1995**, *260*, 406–417. [CrossRef] [PubMed]

47. Chipuk, J.E.; Green, D.R. How do BCL-2 proteins induce mitochondrial outer membrane permeabilization? *Trends Cell Biol.* **2008**, *18*, 157–164. [CrossRef] [PubMed]

48. Wang, K.; Zhan, Y.; Chen, B.; Lu, Y.; Yin, T.; Zhou, S.; Zhang, W.; Liu, X.; Du, B.; Wei, X.; et al. Tubeimoside I-induced lung cancer cell death and the underlying crosstalk between lysosomes and mitochondria. *Cell Death Dis.* **2020**, *11*, 708. [CrossRef] [PubMed]

49. Ma, M.; Lin, X.H.; Liu, H.H.; Zhang, R.; Chen, R.X. Suppression of DRP1mediated mitophagy increases the apoptosis of hepatocellular carcinoma cells in the setting of chemotherapy. *Oncol. Rep.* **2020**, *43*, 1010–1018. [CrossRef]

50. Hu, Y.; Guo, M. Synthetic lethality strategies: Beyond BRCA1/2 mutations in pancreatic cancer. *Cancer Sci.* **2020**, *111*, 3111–3121. [CrossRef]

51. Lord, R.V.; Brabender, J.; Gandara, D.; Alberola, V.; Camps, C.; Domine, M.; Cardenal, F.; Sanchez, J.M.; Gumerlock, P.H.; Taron, M.; et al. Low ERCC1 expression correlates with prolonged survival after cisplatin plus gemcitabine chemotherapy in non-small cell lung cancer. *Clin. Cancer Res. Off. J. Am. Assoc. Cancer Res.* **2002**, *8*, 2286–2291.

52. Cabelof, D.C.; Guo, Z.; Raffoul, J.J.; Sobol, R.W.; Wilson, S.H.; Richardson, A.; Heydari, A.R. Base excision repair deficiency caused by polymerase beta haploinsufficiency: Accelerated DNA damage and increased mutational response to carcinogens. *Cancer Res.* **2003**, *63*, 5799–5807.

53. Weaver, A.N.; Yang, E.S. Beyond DNA Repair: Additional Functions of PARP-1 in Cancer. *Front. Oncol.* **2013**, *3*, 290. [CrossRef] [PubMed]

54. Satoh, M.S.; Lindahl, T. Role of poly(ADP-ribose) formation in DNA repair. *Nature* **1992**, *356*, 356–358. [CrossRef]

55. Hilton, J.F.; Hadfield, M.J.; Tran, M.T.; Shapiro, G.I. Poly(ADP-ribose) polymerase inhibitors as cancer therapy. *Front. Biosci.* **2013**, *18*, 1392–1406. [CrossRef] [PubMed]

56. Rocha, C.R.R.; Silva, M.M.; Quinet, A.; Cabral-Neto, J.B.; Menck, C.F.M. DNA repair pathways and cisplatin resistance: An intimate relationship. *Clinics* **2018**, *73*, e478s. [CrossRef]

57. Lord, C.J.; Ashworth, A. PARP inhibitors: Synthetic lethality in the clinic. *Science* **2017**, *355*, 1152–1158. [CrossRef]

58. Johnson, N.; Johnson, S.F.; Yao, W.; Li, Y.C.; Choi, Y.E.; Bernhardy, A.J.; Wang, Y.; Capelletti, M.; Sarosiek, K.A.; Moreau, L.A.; et al. Stabilization of mutant BRCA1 protein confers PARP inhibitor and platinum resistance. *Proc. Natl. Acad. Sci. USA* **2013**, *110*, 17041–17046. [CrossRef]

59. Ashworth, A. A synthetic lethal therapeutic approach: Poly (ADP) ribose polymerase inhibitors for the treatment of cancers deficient in DNA double-strand break repair. *J. Clin. Oncol. Off. J. Am. Soc. Clin. Oncol.* **2008**, *26*, 3785–3790. [CrossRef] [PubMed]

60. Gourley, C.; Balmana, J.; Ledermann, J.A.; Serra, V.; Dent, R.; Loibl, S.; Pujade-Lauraine, E.; Boulton, S.J. Moving from Poly (ADP-Ribose) Polymerase Inhibition to Targeting DNA Repair and DNA Damage Response in Cancer Therapy. *J. Clin. Oncol. Off. J. Am. Soc. Clin. Oncol.* **2019**, *37*, 2257–2269. [CrossRef]

61. Farmer, H.; McCabe, N.; Lord, C.J.; Tutt, A.N.; Johnson, D.A.; Richardson, T.B.; Santarosa, M.; Dillon, K.J.; Hickson, I.; Knights, C.; et al. Targeting the DNA repair defect in BRCA mutant cells as a therapeutic strategy. *Nature* **2005**, *434*, 917–921. [CrossRef]

62. Pettitt, S.J.; Krastev, D.B.; Brandsma, I.; Drean, A.; Song, F.; Aleksandrov, R.; Harrell, M.I.; Menon, M.; Brough, R.; Campbell, J.; et al. Genome-wide and high-density CRISPR-Cas9 screens identify point mutations in PARP1 causing PARP inhibitor resistance. *Nat. Commun.* **2018**, *9*, 1849. [CrossRef]

63. Carey, J.P.W.; Karakas, C.; Bui, T.; Chen, X.; Vijayaraghavan, S.; Zhao, Y.; Wang, J.; Mikule, K.; Litton, J.K.; Hunt, K.K.; et al. Synthetic Lethality of PARP Inhibitors in Combination with MYC Blockade Is Independent of BRCA Status in Triple-Negative Breast Cancer. *Cancer Res.* **2018**, *78*, 742–757. [CrossRef]

64. Kim, G.; Ison, G.; McKee, A.E.; Zhang, H.; Tang, S.; Gwise, T.; Sridhara, R.; Lee, E.; Tzou, A.; Philip, R.; et al. FDA Approval Summary: Olaparib Monotherapy in Patients with Deleterious Germline BRCA-Mutated Advanced Ovarian Cancer Treated with Three or More Lines of Chemotherapy. *Clin. Cancer Res. Off. J. Am. Assoc. Cancer Res.* **2015**, *21*, 4257–4261. [CrossRef]

65. Balasubramaniam, S.; Beaver, J.A.; Horton, S.; Fernandes, L.L.; Tang, S.; Horne, H.N.; Liu, J.; Liu, C.; Schrieber, S.J.; Yu, J.; et al. FDA Approval Summary: Rucaparib for the Treatment of Patients with Deleterious BRCA Mutation-Associated Advanced Ovarian Cancer. *Clin. Cancer Res. Off. J. Am. Assoc. Cancer Res.* **2017**, *23*, 7165–7170. [CrossRef]

66. Hoy, S.M. Talazoparib: First Global Approval. *Drugs* **2018**, *78*, 1939–1946. [CrossRef] [PubMed]

67. Grzegrzolka, J.; Biala, M.; Wojtyra, P.; Kobierzycki, C.; Olbromski, M.; Gomulkiewicz, A.; Piotrowska, A.; Rys, J.; Podhorska-Okolow, M.; Dziegiel, P. Expression of EMT Markers SLUG and TWIST in Breast Cancer. *Anticancer Res.* **2015**, *35*, 3961–3968. [PubMed]

68. Nuti, S.V.; Mor, G.; Li, P.; Yin, G. TWIST and ovarian cancer stem cells: Implications for chemoresistance and metastasis. *Oncotarget* **2014**, *5*, 7260–7271. [CrossRef] [PubMed]

69. Kajiyama, H.; Hosono, S.; Terauchi, M.; Shibata, K.; Ino, K.; Yamamoto, E.; Nomura, S.; Nawa, A.; Kikkawa, F. Twist expression predicts poor clinical outcome of patients with clear cell carcinoma of the ovary. *Oncology* **2006**, *71*, 394–401. [CrossRef]

70. Elloul, S.; Vaksman, O.; Stavnes, H.T.; Trope, C.G.; Davidson, B.; Reich, R. Mesenchymal-to-epithelial transition determinants as characteristics of ovarian carcinoma effusions. *Clin. Exp. Metastasis* **2010**, *27*, 161–172. [CrossRef]

71. Caruso, D.; Papa, A.; Tomao, S.; Vici, P.; Panici, P.B.; Tomao, F. Niraparib in ovarian cancer: Results to date and clinical potential. *Ther. Adv. Med. Oncol.* **2017**, *9*, 579–588. [CrossRef] [PubMed]

72. Moore, K.N.; Mirza, M.R.; Matulonis, U.A. The poly (ADP ribose) polymerase inhibitor niraparib: Management of toxicities. *Gynecol. Oncol.* **2018**, *149*, 214–220. [CrossRef] [PubMed]

73. Dasari, S.; Tchounwou, P.B. Cisplatin in cancer therapy: Molecular mechanisms of action. *Eur. J. Pharmacol.* **2014**, *740*, 364–378. [CrossRef]

74. Vidi, P.A.; Bissell, M.J.; Lelievre, S.A. Three-dimensional culture of human breast epithelial cells: The how and the why. *Methods Mol. Biol.* **2013**, *945*, 193–219. [CrossRef] [PubMed]

75. Sumi, T.; Hirai, S.; Yamaguchi, M.; Tanaka, Y.; Tada, M.; Yamada, G.; Hasegawa, T.; Miyagi, Y.; Niki, T.; Watanabe, A.; et al. Survivin knockdown induces senescence in TTF1-expressing, KRAS-mutant lung adenocarcinomas. *Int. J. Oncol.* **2018**, *53*, 33–46. [CrossRef] [PubMed]

76. Kraskiewicz, H.; FitzGerald, U. InterfERing with endoplasmic reticulum stress. *Trends Pharmacol. Sci.* **2012**, *33*, 53–63. [CrossRef]

77. Schleicher, S.M.; Moretti, L.; Varki, V.; Lu, B. Progress in the unraveling of the endoplasmic reticulum stress/autophagy pathway and cancer: Implications for future therapeutic approaches. *Drug Resist. Updat.* **2010**, *13*, 79–86. [CrossRef]

78. Rosati, E.; Sabatini, R.; Rampino, G.; De Falco, F.; Di Ianni, M.; Falzetti, F.; Fettucciari, K.; Bartoli, A.; Screpanti, I.; Marconi, P. Novel targets for endoplasmic reticulum stress-induced apoptosis in B-CLL. *Blood* **2010**, *116*, 2713–2723. [CrossRef]

79. Kim, I.; Xu, W.; Reed, J.C. Cell death and endoplasmic reticulum stress: Disease relevance and therapeutic opportunities. *Nat. Rev. Drug Discov.* **2008**, *7*, 1013–1030. [CrossRef]
80. Dalton, L.E.; Clarke, H.J.; Knight, J.; Lawson, M.H.; Wason, J.; Lomas, D.A.; Howat, W.J.; Rintoul, R.C.; Rassl, D.M.; Marciniak, S.J. The endoplasmic reticulum stress marker CHOP predicts survival in malignant mesothelioma. *Br. J. Cancer* **2013**, *108*, 1340–1347. [CrossRef]
81. Huber, A.L.; Lebeau, J.; Guillaumot, P.; Petrilli, V.; Malek, M.; Chilloux, J.; Fauvet, F.; Payen, L.; Kfoury, A.; Renno, T.; et al. p58(IPK)-mediated attenuation of the proapoptotic PERK-CHOP pathway allows malignant progression upon low glucose. *Mol. Cell* **2013**, *49*, 1049–1059. [CrossRef] [PubMed]
82. Malhotra, J.D.; Kaufman, R.J. ER stress and its functional link to mitochondria: Role in cell survival and death. *Cold Spring Harbor Perspect. Biol.* **2011**, *3*, a004424. [CrossRef] [PubMed]
83. Franken, N.A.; Rodermond, H.M.; Stap, J.; Haveman, J.; van Bree, C. Clonogenic assay of cells in vitro. *Nat. Protoc.* **2006**, *1*, 2315–2319. [CrossRef] [PubMed]
84. Foty, R. A simple hanging drop cell culture protocol for generation of 3D spheroids. *J. Vis. Exp.* **2011**. [CrossRef]

International Journal of
Molecular Sciences

Article

Molecular Mechanisms and Tumor Biological Aspects of 5-Fluorouracil Resistance in HCT116 Human Colorectal Cancer Cells

Chinatsu Kurasaka [1], Yoko Ogino [1,2] and Akira Sato [1,*]

1 Department of Biochemistry and Molecular Biology, Faculty of Pharmaceutical Sciences, Tokyo University of Science, 2641 Yamazaki, Noda, Chiba 278-8510, Japan; 3a16042@ed.tus.ac.jp (C.K.); ogino@rs.tus.ac.jp (Y.O.)
2 Department of Gene Regulation, Faculty of Pharmaceutical Sciences, Tokyo University of Science, 2641 Yamazaki, Noda, Chiba 278-8510, Japan
* Correspondence: akirasat@rs.tus.ac.jp; Tel.: +81-4-7121-3620

Abstract: 5-Fluorouracil (5-FU) is a cornerstone drug used in the treatment of colorectal cancer (CRC). However, the development of resistance to 5-FU and its analogs remain an unsolved problem in CRC treatment. In this study, we investigated the molecular mechanisms and tumor biological aspects of 5-FU resistance in CRC HCT116 cells. We established an acquired 5-FU-resistant cell line, HCT116R^{F10}. HCT116R^{F10} cells were cross-resistant to the 5-FU analog, fluorodeoxyuridine. In contrast, HCT116R^{F10} cells were collaterally sensitive to SN-38 and CDDP compared with the parental HCT16 cells. Whole-exome sequencing revealed that a cluster of genes associated with the 5-FU metabolic pathway were not significantly mutated in HCT116 or HCT116R^{F10} cells. Interestingly, HCT116R^{F10} cells were regulated by the function of thymidylate synthase (TS), a 5-FU active metabolite 5-fluorodeoxyuridine monophosphate (FdUMP) inhibiting enzyme. Half of the TS was in an active form, whereas the other half was in an inactive form. This finding indicates that 5-FU-resistant cells exhibited increased TS expression, and the TS enzyme is used to trap FdUMP, resulting in resistance to 5-FU and its analogs.

Keywords: colorectal cancer cells; drug resistance; 5-Fluorouracil; thymidylate synthase; exome sequencing

1. Introduction

Colorectal cancer (CRC) is the third-most common cancer in the world [1], and 5-Fluorouracil (5-FU) is the most important chemotherapeutic agent used in its treatment [2,3]. 5-FU is also widely used to treat other cancers, such as gastric, pancreatic, breast, ovarian, and head and neck cancers [2,3]. 5-FU is converted to 5-fluorodeoxyuridine monophosphate (FdUMP), which is a potent inhibitor of thymidylate synthase (TS) [3–5]. FdUMP forms a covalent complex with TS in the presence of 5,10-methylenetetrahydrofolate (CH$_2$–THF) [2,3,5]. The inhibition of TS depletes the intracellular dTTP pool and subsequently inhibits DNA synthesis [2–5]. Another effect by which 5-FU can exert its cytotoxic action is its incorporation as fluorodeoxyuridine triphosphate (FdUTP) and fluorouridine triphosphate (FUTP) into DNA and RNA, respectively [2–4]. Experimental and clinical studies indicate that continuous exposure of CRC cells to 5-FU results in acquired resistance to 5-FU and its derivatives. This is often caused by common cancer resistance mechanisms, such as drug inactivation, drug efflux, drug target alterations, bypass pathway activation, DNA damage repair, and cell death [2,3]. 5-FU resistance is correlated with the level of TS protein and enzymatic activity in cancer cells [2,3,6–8]. In addition, high TS protein and RNA expression levels in tumor tissue is also a useful biomarker for poor prognosis for 5-FU-based chemotherapy in CRC patients [2,3,9]. Furthermore, 5-FU sensitivity is influenced by the expression levels of dihydropyrimidine dehydrogenase (DPD) [9,10], which converts 5-FU to dihydrofluorouracil (DHFU) during the catabolic process [2–4,9,10]. However, 5-FU resistance has not yet been circumvented clinically.

In this study, we established a 5-FU-resistant HCT116 CRC cell line (HCT116R^{F3} and HCT116R^{F10}) and analyzed its biological features. HCT116R^{F10} cells, which are cross-resistant to the 5-FU analog fluorodeoxyuridine (FUdR), were collaterally sensitive to SN-38 and CDDP compared with the parental HCT16 cells. In addition, HCT116R^{F10} cells exhibited a lower ability to form tumor spheres compared with parental HCT116 cells. Notably, HCT116R^{F10} cells maintained the tumor sphere formation ability compared with HCT116 cells under 5-FU exposure conditions. Furthermore, a gene cluster associated with 5-FU metabolic pathway was not significantly mutated in HCT116 and HCT116R^{F10} cells as determined by whole-exome sequencing. We found that HCT116R^{F10} cells regulate intracellular TS states in which half of the TS enzyme is in a functional form and the other half exists as an FdUMP-covalent complex (inactive form). These findings provide a better understanding of resistance to anticancer 5-FU and its analogs.

2. Results

2.1. Establishment of the 5-Fluorouracil-Resistant HCT116 Cells

To elucidate the mechanisms underlying resistance to 5-FU, we generated a variant of the HCT116 human colorectal cancer cell line that was resistant to 5-FU, an important anticancer drug used for CRC treatment [2,3]. We established 5-FU-resistant HCT116R^{F3} or HCT116R^{F10} cells by repeated exposure of parental HCT116 cells to stepwise increasing concentrations of 5-FU over a period of approximately 12 weeks at 3 μM and 14 weeks at 10 μM, respectively (Figure 1a). The EC$_{50}$ of 5-FU in HCT116R^{F3} (intermediate variant) and HCT116R^{F10} cells were determined by a WST-8 assay after continuous exposure for 72 h. As shown in Table 1 and Figure 1b, the EC$_{50}$ value of the 5-FU-resistant HCT11 6 cells was higher (1.5 × 10^{-5} M in HCT116R^{F3} and 2.9 × 10^{-5} M in HCT116R^{F10} cells) than that of sensitive, parental HCT116 cells (5.1 × 10^{-6} M). The RI was approximately 2.9 for HCT116R^{F3} cells and 5.7 for HCT116R^{F10} cells (Table 1). In addition, similar results were obtained by colony formation assay (Figure 1c,d). The EC$_{50}$ value of 5-FU-resistant HCT116 cells was significantly higher (1.6 × 10^{-5} M in HCT116R^{F3} and 3.8 × 10^{-5} M in HCT116R^{F10} cells) than that of the parental HCT116 cells (5.5 × 10^{-6} M) (Table 1). The RI of HCT116R^{F3} and HCT116R^{F10} cells was approximately 2.9 and 6.9, respectively (Table 1). Furthermore, parental HCT116, HCT116R^{F3}, and HCT116R^{F10} cells exhibited nearly similar morphological features (Figure 1e).

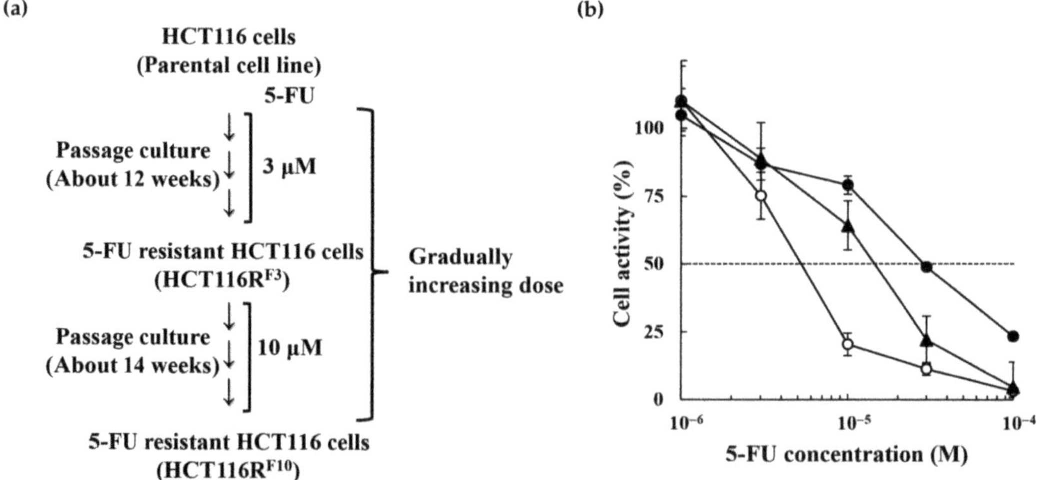

Figure 1. *Cont.*

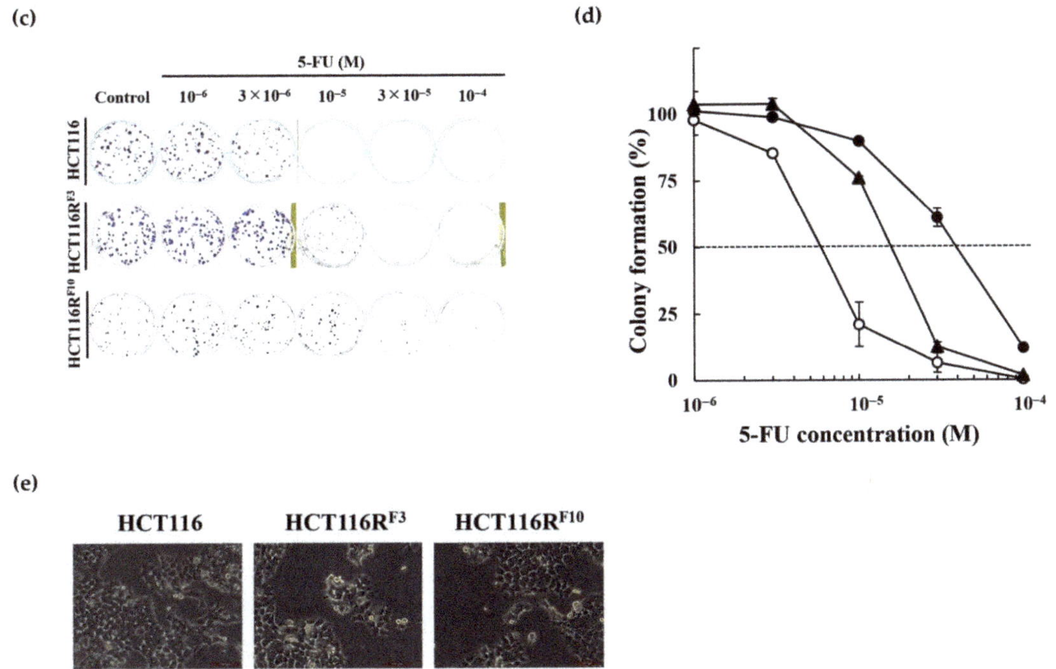

Figure 1. Establishment of HCT116R^{F10}, a 5-FU-resistant derivative of the human colorectal cancer cell line HCT116. (**a**) Scheme for the establishment of the 5-FU-resistant HCT116 cells (HCT116R^{F10}). (**b**) HCT116R^{F10} and parental HCT116 cells were tested for cell viability after a 72 h treatment with 5-FU. Results represent the averages of three independent experiments with error bars showing ±SE from triplicates. (**c**) Drug sensitivities of HCT116R^{F10} and HCT116 using the colony formation assay. HCT116R^{F10} and HCT116 cells were treated with the indicated concentration of 5-FU and incubated for 10 days. (**d**) HCT116 cells: Colony formation (%) represents the average of two independent experiments, each performed in duplicate, with error bars showing ±SE of four measurement values. HCT116R^{F3} and R^{F10} cells: Colony formation (%) represents the average of three independent experiments, each performed in triplicate, with error bars showing ±SE of nine measurement values. White circle, HCT116 cells; black triangle, HCT116R^{F3} cells; black circle, HCT116R^{F10} cells. (**e**) Morphological features were analyzed using a Leica DMi1 microscope with LAS V4.12 at 200× magnification. Scale bar = 100 µm.

Table 1. Sensitivity of 5-fluorouracil in the parental HCT116, HCT116R^{F3} and HCT116R^{F10} cells.

Cell line	EC$_{50}$ (M)		RI	
	WST-8 Assay	**CFA**	**WST-8 Assay**	**CFA**
HCT116	5.1×10^{-6}	5.5×10^{-6}	1	1
HCT116R^{F3}	1.5×10^{-5}	1.6×10^{-5}	2.9	2.9
HCT116R^{F10}	2.9×10^{-5}	3.8×10^{-5}	5.7	6.9

Note. EC$_{50}$, 50% effective concentration; R, resistant; F3, fluorouracil 3×10^{-6} M; F10, fluorouracil 10×10^{-6} M; RI, resistance index.

2.2. Anticancer Drug Response of the 5-FU-Resistant HCT116 Cells

We examined the effects of several anticancer drugs, including the 5-FU analog FUdR, SN-38, an active metabolite of irinotecan, and CDDP, on the proliferation of parental HCT116 and 5-FU-resistant HCT116R^{F10} cells by WST-8 (Figure 2) and colony formation assays (Figure 3). As shown in Table 2 and Figure 2a, HCT116R^{F10} cells were 80.0-fold (EC$_{50}$ = 1.2×10^{-4} M) more resistant to FUdR than parental HCT116 cells (EC$_{50}$ = 1.5×10^{-6} M). In contrast, the resistant index for SN-38 and CDDP was 2.1-fold (EC$_{50}$ = 6.6×10^{-9} M in HCT116R^{F10} cells; 3.1×10^{-9} M in HCT116 cells) and 1.4-fold

(EC_{50} = 1.4 × 10^{-5} M in HCT116R^{F10} cells; 1.0 × 10^{-5} M in HCT116 cells), respectively (Table 2, Figure 2b,c). Similarly, for the colony-forming assay, HCT116R^{F10} cells were 9.7-fold (EC_{50} = 3.3 × 10^{-5} M) more resistant to FUdR than the parental HCT116 cells (EC_{50} = 3.4 × 10^{-6} M) (Figure 3a). In addition, the RI of SN-38 and CDDP was 0.7-fold (EC_{50} = 3.0 × 10^{-9} M in HCT116R^{F10} cells; 4.2 × 10^{-9} M in HCT116 cells) and 0.9-fold (EC_{50} = 4.5 × 10^{-6} M in HCT116R^{F10} cells; 5.2 × 10^{-6} M in HCT116 cells), respectively (Table 2, Figure 3b,c). These results indicate that 5-FU-resistant HCT116R^{F10} cells exhibit cross-resistance to FUdR but collateral sensitivity to the anticancer drugs SN-38 and CDDP. This finding suggests that the HCT116R^{F10} cells are resistant not only to 5-FU but also to other 5-FU deoxyribose analogs such as FUdR.

Table 2. Sensitivities of several anticancer agents in the parental HCT116 and HCT116R^{F10} cells.

	EC_{50} (M, WST-8)			EC_{50} (M, CFA)		
	HCT116	**HCT116R^{F10}**	**RI**	**HCT116**	**HCT116R^{F10}**	**RI**
FUdR	1.5 × 10^{-6}	1.2 × 10^{-4}	80.0	3.4 × 10^{-6}	3.3 × 10^{-5}	9.7
SN-38	3.1 × 10^{-9}	6.6 × 10^{-9}	2.1	4.2 × 10^{-9}	3.0 × 10^{-9}	0.7
CDDP	1.0 × 10^{-5}	1.4 × 10^{-5}	1.4	5.2 × 10^{-6}	4.5 × 10^{-6}	0.9

Note: EC_{50}, 50% effective concentration; R^{F10}, resistant to fluorouracil 10 × 10^{-6} M; RI, resistance index.

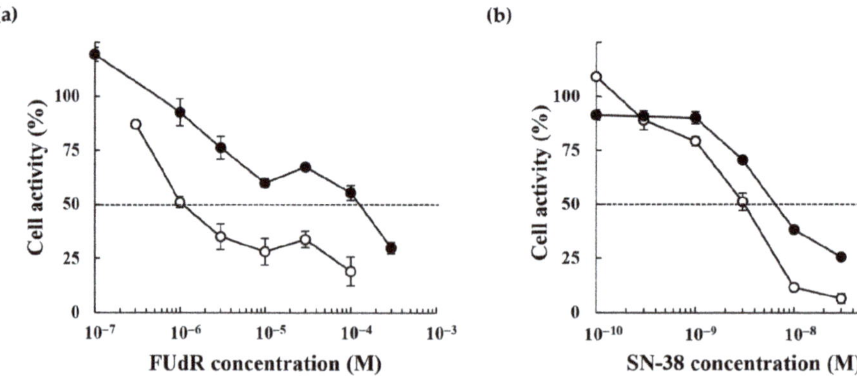

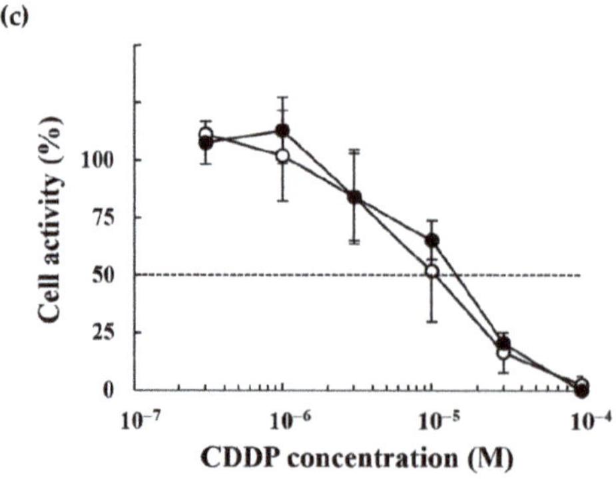

Figure 2. Sensitivity of 5-FU-resistant HCT116R^{F10} and parental HCT116 cells to FUdR, SN-38, and CDDP. The cell proliferation WST-8 assay of HCT116R^{F10} and parental HCT116 cells after a 72 h treatment with (**a**) FUdR, (**b**) SN-38, and (**c**) CDDP. Results represent the averages of two independent experiments, with error bars showing ±SE of triplicates. White circle, HCT116 cells; black circle, HCT116R^{F10} cells.

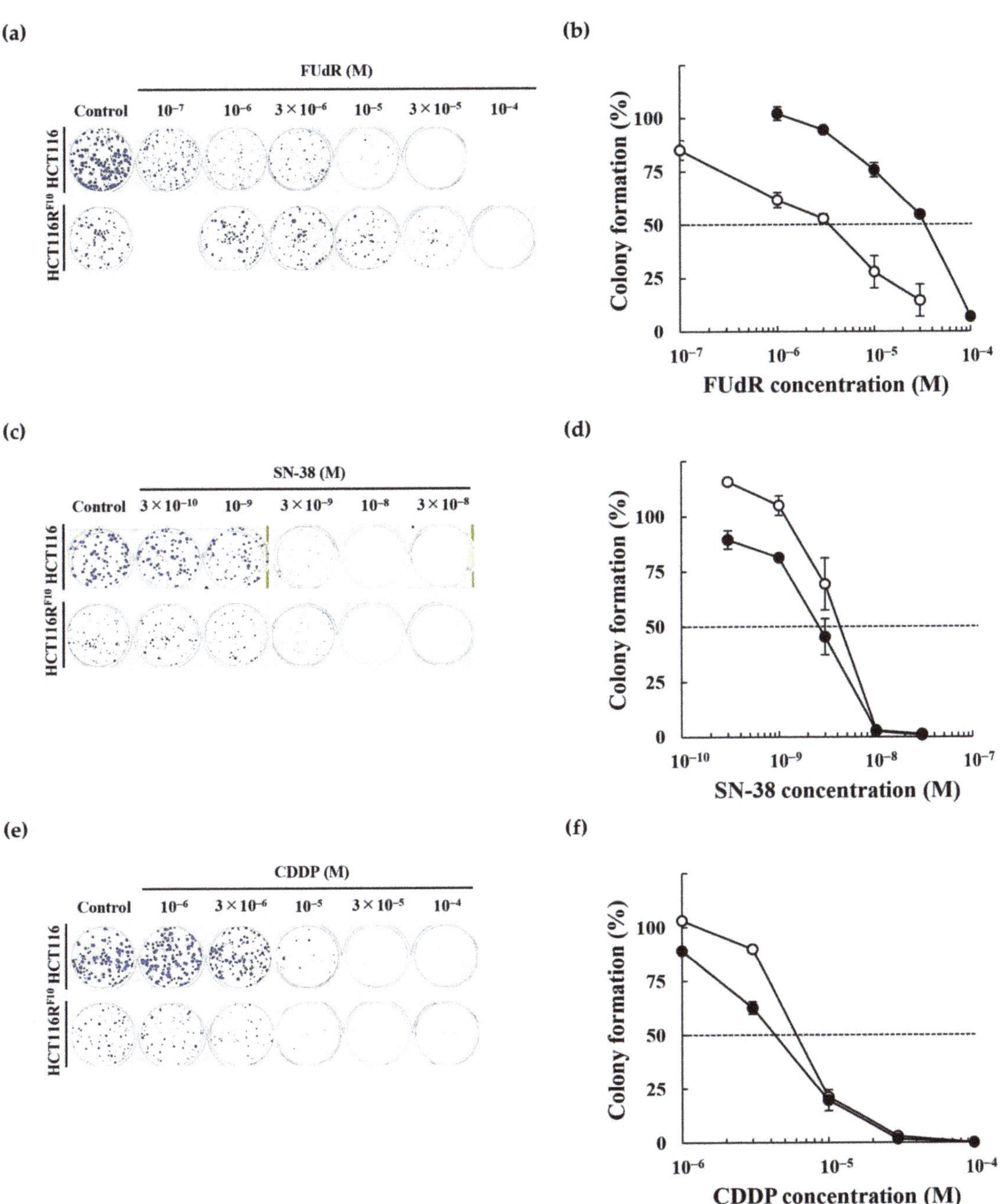

Figure 3. Sensitivity of 5-FU-resistant HCT116R^{F10} and parental HCT116 cells to FUdR, SN-38, and CDDP. Colony formation by HCT116R^{F10} and parental HCT116 cells after 10 days of treatment with (**a,b**) FUdR, (**c,d**) SN-38, and (**e,f**) CDDP. Colony formation (%) represents the averages of two independent experiments each performed in duplicate (**b**) or triplicate (**d,f**), with error bars showing ±SE of four (**b**) or six (**d,f**) measurement values. White circle, HCT116 cells; black circle, HCT116R^{F10} cells.

2.3. Biological Features of the 5-FU-Resistant HCT116 Cells

We analyzed the tumor sphere formation ability of HCT116R^{F10} cells and parental HCT116 cells in three-dimensional cell culture experiments (Figure 4). HCT116R^{F10} cells exhibited a lower ability to form tumor spheres compared with parental HCT116 cells under untreated conditions (Figure 4a left panel and b). Interestingly, HCT116R^{F10} cells maintained a tumor sphere formation ability compared with parent HCT116 cells during 5-FU treatment conditions (Figure 4a,c). We next examined the sensitivity of parental HCT116 and HCT116R^{F10} tumor sphere cells to 5-FU. As shown in Figure 4d, HCT116R^{F10} cells were 18.7-fold (EC$_{50}$ = 2.8 × 10^{-5} M) more resistant to 5-FU than parental HCT116 cells (EC$_{50}$ = 1.5 × 10^{-6} M). These data indicate that 5-FU-resistant HCT116R^{F10} cells are less prone to tumorigenesis than sensitive, parental HCT116 cells, but formed tumor spheres that retained a higher 5-FU resistance.

(a)

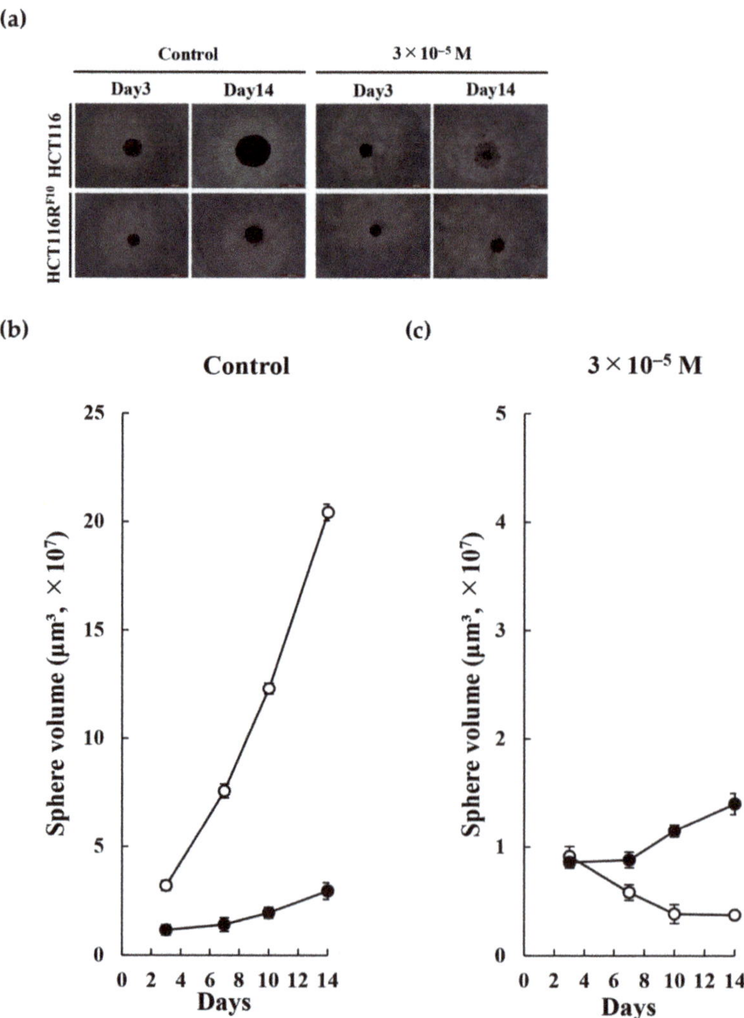

Figure 4. *Cont.*

(d)

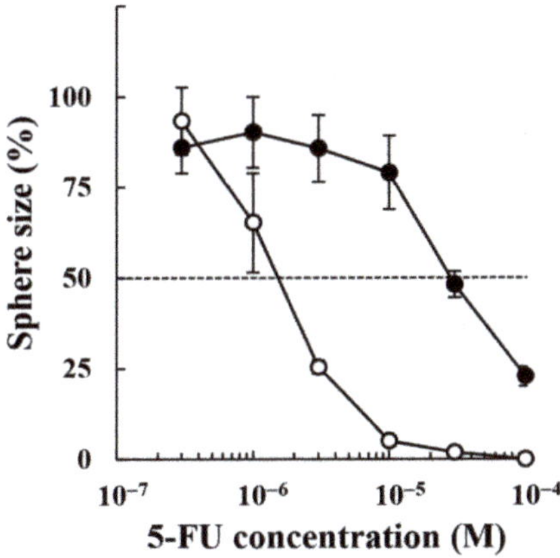

Figure 4. Tumor sphere formation of 5-FU-resistant HCT116R^{F10} and parental HCT116 cells. (**a**) Tumor sphere formation was analyzed using a Leica DMi1 microscope with 50× magnification. Scale bar = 500 μm. HCT116R^{F10} and parental HCT116 cells were treated with or without 3×10^{-5} M 5-FU for 3 or 14 days. Control, no 5-FU, solvent (DMSO) alone. To assess the ability of HCT116R^{F10} and parental HCT116 cells to form tumor spheres, the cells were treated with solvent alone (**b**) or 3×10^{-5} M 5-FU (**c**) for 14 days. Tumor sphere size was calculated as described in the Materials and Methods. White circle, HCT116 cells; black circle, HCT116R^{F10} cells. (**d**) Drug sensitivity of 5-FU in HCT116 and HCT116R^{F10} tumor spheres. Tumor sphere formation by HCT116R^{F10} and parental HCT116 cells after a 14-day treatment with 5-FU at the indicated concentrations. Results are the averages for groups of three tumor spheres each with error bars showing SE. White circle, HCT116 cells; black circle, HCT116R^{F10} cells.

2.4. Exome Sequencing Analysis of HCT116 Parent Cells and 5-FU-Resistant HCT116R^{F10} Cells

We analyzed variants of 5-FU metabolic pathway-related enzyme genes, including *TYMS*, which encodes for TS, and *DPYD*, which encodes for DPD in HCT116 and HCT116R^{F10} cells. TS is a major intracellular target of 5-FU, whereas DPD catalyzes the rate-limiting step in the catabolism of 5-FU [2,3,11]. The pathways involved in the metabolism of 5-FU and its analog FUdR are shown in Figure 5. The genetic alteration status of nearly all of the 5-FU metabolic pathway-related genes was of similar status in both cells. Importantly, the variants of *TYMS* and *DPYD* in HCT116 and HCT116R^{F10} cells contained heterozygous mutations or intron variants. We identified two *TYMS* intron variants, 454+197_454+202delTTTTTT and 454+199_454+202delTTTT, in HCT116R^{F10} cells. In contrast, only one *TYMS* intron variant, 454+200_454+202delTTT, was present in sensitive parental HCT116 cells. Similarly, the three *DPYD* variants, 2999A>T, 2623-59T>G, and 2442+78delA, were present in the HCT116R^{F10} cells. In addition, three *DPYD* variants, 2442+77_2442+delAA, 40-461delT, and -113T>C, were present in HCT116 cells. Herein, we show that one of the *DPYD* heterozygous variants, 2999A>T, is a missense mutation (Asp1000Val) in DPD of 5-FU-resistant HCT116R^{F10} cells.

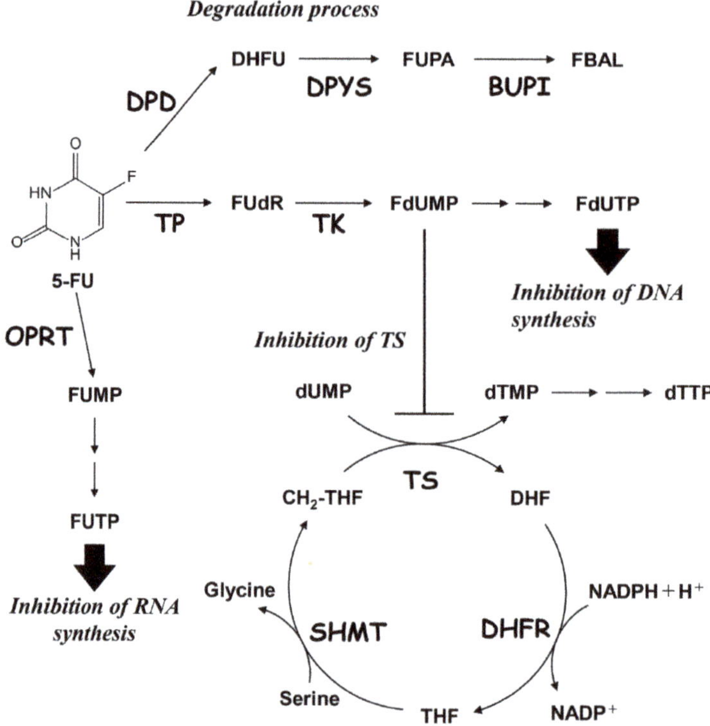

Figure 5. Metabolic pathways associated with 5-FU. 5-FU, 5-fluorouracil; DHFU, dihydrofluorouracil; FUPA, fluoroureidopropionate; FBAL, fluoroalanine; FUdR, fluorodeoxyuridine; FdUMP, fluorodeoxyuridine monophosphate; FdUTP, fluorodeoxyuridine triphosphate; FUMP, fluorouridine monophosphate; FUTP, fluorouridine triphosphate; dUMP, deoxyuridine monophosphate; dTMP, deoxythymidine monophosphate; dTTP, deoxythymidine triphosphate; CH₂-THF, 5,10-methylenetetrahydrofolate; DHF, dihydrofolate; THF; tetrahydrofolate; TS, thymidylate synthase; DPD, dihydropyrimidine dehydrogenase; DPYS, dihydro pyrimidase; DHFR, dihydrofolate reductase; BUPI, β-ureido propionase; TP, thymidine phosphorylase; TK, thymidine kinase; SHMT, serine hydroxymethyltransferase; and OPRT, orotate phosphoribosyltransferase 1.

2.5. Regulation of TS and DPD in HCT116 Parent Cell and 5-FU-Resistant HCT116R^F10 Cells

To elucidate the association of TS and DPD expression with 5-FU resistance, we analyzed TS and DPD expression levels in parental HCT116 and 5-FU-resistant HCT116R^F10 cells by Western blot analysis (Figure 6a). Interestingly, as shown in Figure 6a (top panel) and 6b, free-TS protein levels were almost identical in HCT116R^F10 and HCT116 cells. Conversely, the FdUMP–TS covalent complex was 1.8-fold higher in HCT116R^F10 cells than in HCT116 cells (Figure 6a top panel and 6c). Importantly, it should be noted that total TS, the free form, the FdUMP-covalent form, and total TS was overexpressed in HCT116R^F10 cells rather than in HCT116 cells (Figure 6a top panel and Figure 6d). The upper band of TS, indicated FdUMP-covalent form, which represents TS in ternary complexes and is correlated with the intracellular concentration of FdUMP [12–14]. In addition, DPD protein levels were slightly decreased in HCT116R^F10 cells than in parental HCT116 cells (Figure 6a second panel and Figure 6e). GAPDH and beta-actin were used as an internal controls (Figure 6a third and bottom panels). In parental HCT116 cells and HCT116R^F10 cells, both internal control proteins, GAPDH and beta-actin, had similar levels. After treatment with 1×10^{-4} M 5-FU for 24 h, the protein levels of free TS, FdUMP–TS covalent complex, and total TS were individually about 1.5-fold higher in HCT116R^F10 cells than in parental HCT116 cells (Figure 7a–d). Intriguingly, these data indicated that the proportion of active free TS in the intracellular total TS was highly regulated in the 5-FU resistant HCT116R^{F10} cells. These findings suggested that the regulation of TS status, which includes the balance of active free TS or the inactive FdUMP–TS covalent complex, may confer resistance to 5-FU.

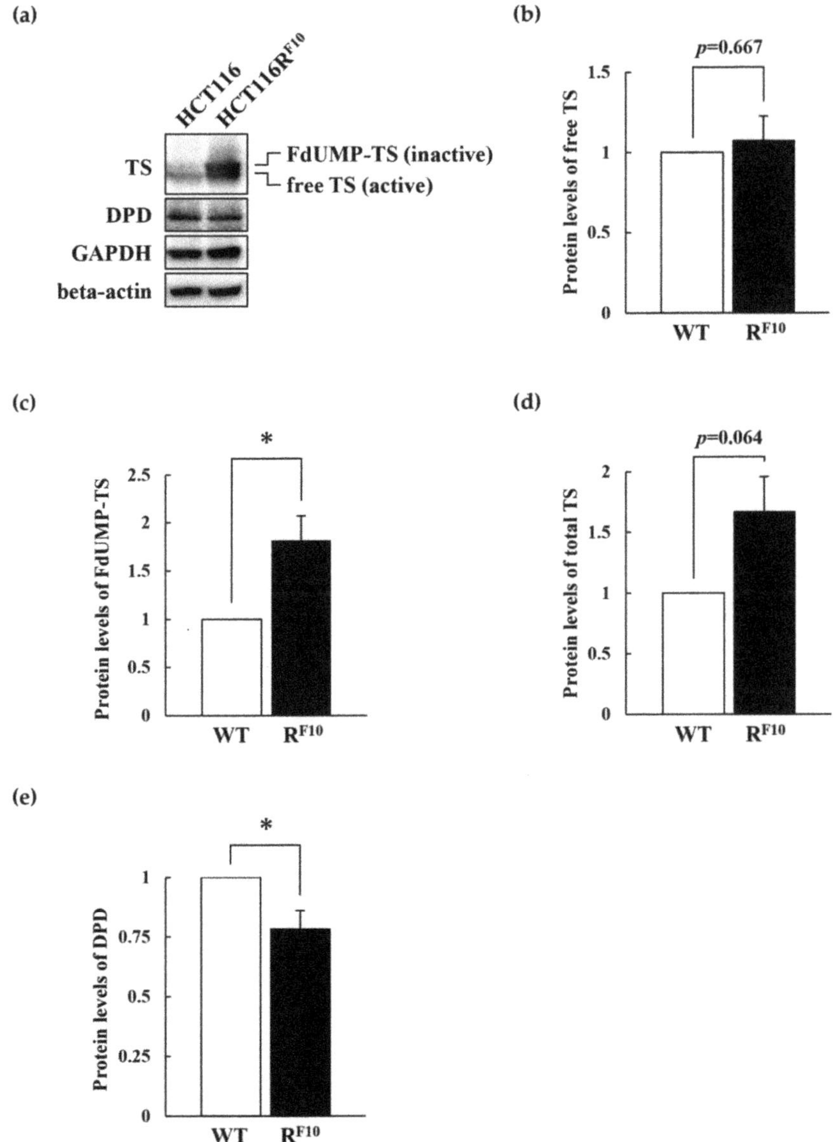

Figure 6. Protein levels of TS and DPD in 5-FU-resistant HCT116R^{F10} and parental HCT116 cells. (**a**) Whole-cell lysates were prepared from parental HCT116 and HCT116R^{F10} cells. Protein levels of TS, DPD, GAPDH, and beta-actin were measured by Western blot analysis. The expression levels of GAPDH and beta-actin were used as an internal control. Data are representative of at least three independent experiments. Protein levels of (**b**) free TS, (**c**) FdUMP-TS, (**d**) total TS, and (**e**) DPD in parental HCT116 and HCT116R^{F10} cells. Levels of TS and DPD protein in HCT116R^{F10} cells are represented by the ratio of TS or DPD density to GAPDH density relative to the value for parental HCT116 cells. Results represent the averages of three independent experiments with error bars showing ±SE of triplicates. * $p < 0.05$.

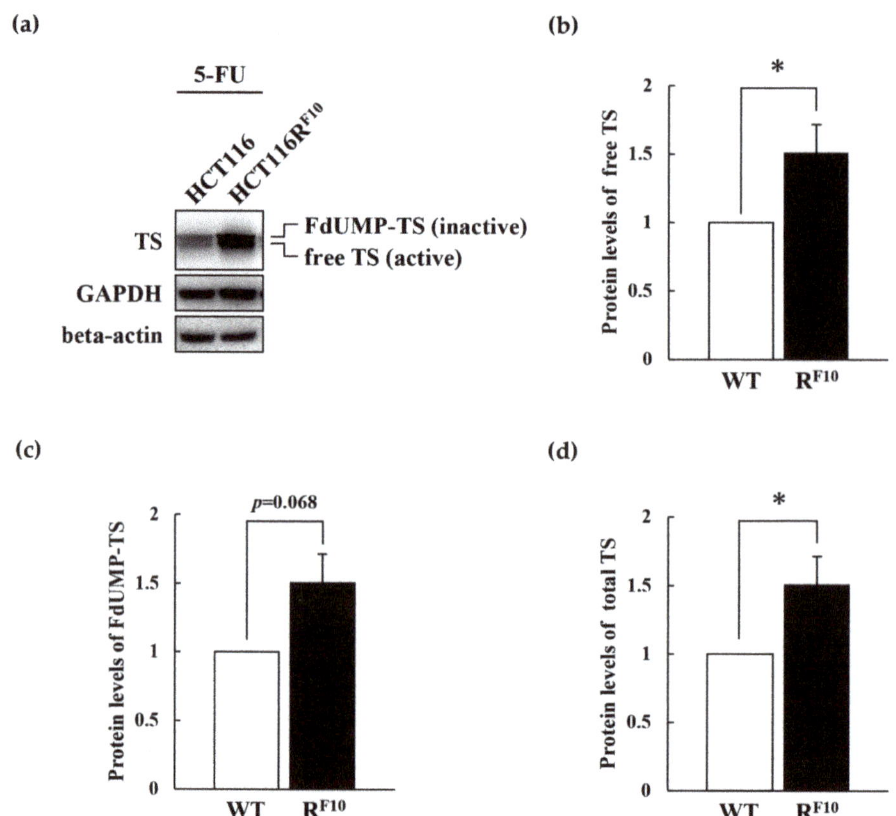

Figure 7. Protein levels of TS in 5-FU-resistant HCT116R^{F10} and parental HCT116 cells after treatment with 5-FU. (**a**) Whole-cell lysates were prepared from parental HCT116 and HCT116R^{F10} cells after 24 h treatment with 1×10^{-4} M 5-FU. Protein levels of TS, GAPDH, and beta-actin were measured by Western blot analysis. Data are representative of at least three independent experiments. Protein levels of (**b**) free TS, (**c**) FdUMP-TS, and (**d**) total TS in parental HCT116 and HCT116R^{F10} cells. Levels of TS protein in HCT116R^{F10} cells are represented by the ratio of TS density to GAPDH density relative to the value for parental HCT116 cells. Results represent the averages of three independent experiments with error bars showing ±SE of triplicates. * $p < 0.05$.

3. Discussion

5-FU and its derivatives are widely used in anticancer chemotherapy [2,3]. Studies to date have shown that cancer cells develop resistance to 5-FU through complex mechanisms [2,3]. Of note, the TS enzyme and other enzymes involved in 5-FU anabolism or catabolism are often altered in expression or function to promote 5-FU resistance [2,3]. In addition, altered cell death and autophagy, expression/functional changes in drug transporters, epigenetic changes, and non-coding RNA (i.e., microRNA and long non-coding RNA) dysfunction represent putative 5-FU-resistant mechanisms [2,3]. It has been widely believed that TS is the main molecular mechanism that influences 5-FU sensitivity and targeting TS is a major strategy for reversing 5-FU resistance. Importantly, there are currently no specific therapies to overcome 5-FU resistance.

We established a 5-FU-resistant cell line, HCT116R^{F10}, and analyzed its characteristics. Importantly, HCT116R^{F10} cells were cross-resistant to the 5-FU analog, FUdR (Figures 2 and 3). In contrast, HCT116R^{F10} cells did not exhibit cross-resistance to the anticancer drugs, SN-38 and CDDP (Figures 2 and 3). Similarly, Boyer et al. also reported that 5-FU-resistant HCT116 cells were not cross-resistant to oxaliplatin or irinotecan [15]. In addition, the sensitivities to 5-FU and FUdR were similar to the levels observed individually in parental HCT116 cells. Of note, previous reports indicated that FUdR is more potent than 5-FU and that the inhibition of cell proliferation was approximately 10- to 100-fold higher than that of 5-FU in multiple cancer cell lines [16–18]. These findings suggest that the common target or mechanism of action of 5-FU and FUdR is the key to 5-FU resistance in this resistant cell model. Furthermore, our results revealed that HCT116R^{F10} cells are resistant to 5-FU and its derivatives, but are not multidrug resistant.

To elucidate the underlying cause of 5-FU resistance, we investigated 5-FU metabolism-related genes, including *TYMS* and *DPYD*, in HCT116R^{F10} and parent HCT116 cells by using whole-exome sequencing. The results revealed that the genetic alteration of almost all of the 5-FU metabolic pathway-related genes was similar in status, intron variants, and heterozygous mutation in both cells (Table 3). Interestingly, we found that the one functional DPD mutation, Asp1000Val, is present in HCT116R^{F10} cells. However, the effects of *DPYD* missense mutation on 5-FU resistance are not well understood.

Next, to evaluate TS and DPD in HCT116R^{F10} and parent HCT116 cells, we analyzed the expression of these genes by Western blot analysis (Figure 6). 5-FU and FUdR are converted to FdUMP, and it has been shown to form a covalent complex with TS in the presence of CH$_2$-THF [2,3,5]. Our results indicated that the free-TS protein (active form) levels were similar in HCT116R^{F10} and HCT116 cells. Interestingly, the FdUMP–TS covalent complex (inactive form) was higher in HCT116R^{F10} cells than in HCT116 cells. Notably, this result indicates that TS is not overexpressed, but rather there are two types of TS in HCT116R^{F10} cells: free TS and FdUMP-coupled TS. We observed that 5-FU-resistant HCT116R^{F10} cells exhibit upregulated *TYMS* expression and use a fraction of TS to trap FdUMP, resulting in resistance to 5-FU and its analogs. In addition, our data suggest that the regulation of the TS complex, which refers to the balance of the active free-TS form and the inactive FdUMP–TS covalent complex, may confer to 5-FU resistance.

Numerous studies have shown that *TYMS* gene amplification, leading to mRNA and enzyme overproduction, is a major mechanism of resistance to fluoropyrimidines 5-FU and FUdR and their derivatives [19]. Also, free TS binds to its own mRNA, resulting in translational repression, that is, translational autoregulation [12,20–23]. Indeed, TS ligands, including 5-FU, disrupt the interaction of the TS enzyme with TS mRNA, leading to translational derepression and enzyme upregulation [12,22,23]. Additionally, to translational derepression, enzyme stabilization has been indicated as the primary mechanism of TS induction by fluoropyrimidines in human colon and ovarian cancer cell lines [24–26]. Furthermore, it is proposed that fluoropyrimidine-mediated increases in TS levels occur through an effect on enzyme stability with no effect on its mRNA [25,27]. It is also suggested that TS stabilization could be the result of conformational changes that may occur upon the formation of a ternary complex, reducing the susceptibility of the TS enzyme to proteolysis [28]. These findings indicated that understanding translational derepression and enzyme stabilization as the process of TS induction has significance for elucidating the mechanism of resistance acquisition. Further investigation is needed on the functions of the FdUMP–TS covalent complex and free TS in both translational regulation and enzyme stabilization for fluoropyrimidine resistance mechanisms using 5-FU-resistance and 5-FU-sensitive parental HCT116 cell lines. Collectively, our findings provide a better understanding of the anticancer drugs, 5-FU and its fluoropyrimidine derivatives, with respect to resistance mechanisms and anticancer treatment strategies.

Table 3. Mutations of 5-FU metabolic enzyme genes in the parental HCT116 and HCT116R^{F10} cells.

Gene Symbol	HCT116	HCT116R^{F10}
	wt	**mt(c.2999A>T)**het
	mt(c.2908-58G>C)het	mt(c.2908-58G>C)het
	mt(.2907+55C>T)hom	mt(.2907+55C>T)hom
	wt	**mt(c.2623-59T>G)**het
	wt	**mt(c.2442+78delA)**het
	mt(c.2442+77_2442+78delAA)het	*wt*
DPYD	mt(c.2059-94G>T)het	mt(c.2059-94G>T)het
	mt(c.1740+40A>G)hom	mt(c.1740+40A>G)hom
	mt(c.1740+39C>T)het	mt(c.1740+39C>T)het
	mt(c.1627A>G)het	mt(c.1627A>G)het
	mt(c.234-123G>C)het	mt(c.234-123G>C)het
	mt(c.40-461delT)het	*wt*
	mt(c.-113T>C)het	*wt*
	mt(c.1444-145C>T)hom	mt(c.1444-145C>T)hom
	mt(c.951-113T>C)hom	mt(c.951-113T>C)hom
DPYS	mt(c.424-62G>T)hom	mt(c.424-62G>T)hom
	mt(c.265-58T>C)hom	mt(c.265-58T>C)hom
	mt(c.216C>T)hom	mt(c.216C>T)hom
	mt(c.-1T>C)hom	mt(c.-1T>C)hom
BUPI	n.d.	n.d.
TP	n.d.	n.d.
	mt(c.393+168C>T)het	*wt*
	mt(c.393+1G>A)het	*wt*
TK1	mt(c.225A>G)het	mt(c.225A>G)het
	mt(c.98+97_98+101delCCCCT)het	mt(c.98+97_98+101delCCCCT)het
	mt(c.33T>C)het	mt(c.33T>C)het
	mt(c.619-53A>G)het	mt(c.619-53A>G)het
	mt(c.619-63C>G)het	mt(c.619-63C>G)het
	mt(c.156+836G>A)het	mt(c.156+836G>A)het
TK2	mt(c.156+742G>A)het	mt(c.156+742G>A)het
	mt(c.125-116G>A)het	mt(c.125-116G>A)het
	mt(c.-30C>G)het	mt(c.-30C>G)het
	mt(c.-38A>G)het	mt(c.-38A>G)het
	mt(c.97T>C)het	mt(c.97T>C)het
	mt(c.280-43G>A)hom	mt(c.280-43G>A)hom
	mt(c.454+50T>C)hom	mt(c.454+50T>C)hom
	wt	**mt(c.454+197_454+202delTTTTTT)**het
TYMS	*wt*	**mt(c.454+199_454+202delTTTT)**het
	mt(c.454+200_454+202delTTT)hom	*wt*
	mt(c.556+123_556+126delATTG)hom	mt(c.556+123_556+126delATTG)hom
	mt(c.*19C>T)hom	mt(c.*19C>T)hom
	mt(c.*89A>G)het	mt(c.*89A>G)het
	wt	**mt(c.137-25T>G)**het
DHFR1	*wt*	**mt(c.137-43T>C)**het
	mt(c.-204T>C)het	mt(c.-204T>C)het
DHFR2	mt(c.247C>G)hom	mt(c.247C>G)hom
	mt(c.*66C>T)het	mt(c.*66C>T)het
	mt(c.*47C>G)het	mt(c.*47C>G)het
	mt(c.1420C>T)het	mt(c.1420C>T)het
	mt(c.1171+59A>G)het	mt(c.1171+59A>G)het
	mt(c.1054+141C>T)het	mt(c.1054+141C>T)het
SHMT1	mt(c.815-23C>T)het	mt(c.815-23C>T)het
	mt(c.601+174C>T)het	mt(c.601+174C>T)het
	mt(c.601+173G>A)het	mt(c.601+173G>A)het
	mt(c.243-256A>G)het	mt(c.243-256A>G)het
	mt(c.-19-101T>C)hom	mt(c.-19-101T>C)hom
	mt(c.595-6G>A)het	mt(c.595-6G>A)het
SHMT2	mt(c.717+14dupG)het	mt(c.717+14dupG)het
	mt(c.1279+30G>A)het	mt(c.1279+30G>A)het

Note. *wt*, wild-type; *mt*, mutation-type; n.d., not detected; *hom*, homozygous; *het*, heterozygous.

4. Materials and Methods

4.1. Reagents

The anticancer drugs 5-FU, FUdR, CDDP, and SN-38 were obtained from FUJIFILM Wako Pure Chemical Corporation (Osaka, Japan). 5-FU, CDDP, and SN-38 were stored as 100 mM stocks in dimethyl sulfoxide (DMSO, Sigma-Aldrich; Merck KGaA, Darmstadt, Germany) at $-20\,°C$. FUdR was stored as a 20 mM stock solution in ultrapure water at $-20\,°C$.

4.2. Cell Culture

The human colon cancer cell line HCT116 was obtained from the American Type Culture Collection. Parental and 5-FU-resistant HCT116 cell lines were cultured in DMEM medium containing 10% heat-inactivated fetal bovine serum, 100 units/mL penicillin, and 100 µg/mL streptomycin in a 37 °C incubator under an atmosphere containing 5% CO_2 and 100% relative humidity.

4.3. Generation of the 5-FU-Resistant HCT116 Cell Line

5-FU-resistant HCT116 cells were obtained by continuous exposure of cells to 3 µM 5-FU for approximately 12 weeks and following at 10 µM for an approximate 14-week period. A derivative of HCT116 was isolated and named HCT116R^{F3} or HCT116R^{F10}. The HCT116R^{F10} cells were maintained in culture in the presence of 10 µM 5-FU.

4.4. Cell Viability by WST-8 Assay

Cell viability assays were performed as previously described [29]. Cell viability was determined using the WST-8 (Cell Counting Kit-8) cell proliferation assay (Dojindo, Tokyo, Japan). Briefly, cells were seeded into 96-well plates (1000 cells per well) in triplicate and then treated with various concentrations of anticancer drugs or DMSO and water (as a negative control). Following incubation for 72 h, WST-8 reagent was added to each well and the plate was placed in a 5% CO_2 incubator at 37 °C for an additional 1 h. Optical density was measured at 450 nm on a Tecan microplate reader (Mannedorf, Switzerland). The EC$_{50}$ value was defined as the concentration of drug producing 50% inhibition of cell proliferation. The resistance index (RI) was defined as the ratio of EC$_{50}$ values between the resistant and parental cell lines. Experiments were repeated at least three times.

4.5. Colony Formation Assay

Colony formation assay was performed as previously described [29–32]. HCT116 and HCT116R^{F10} cells were dissociated with Accutase, suspended in medium, inoculated into 6-well plates (200 cells per well) in triplicate, and then incubated overnight. The cells were treated with various concentrations of drugs or with solvent (DMSO or water) as a negative control. After incubation for 10 days, cells were fixed with 4% formaldehyde solution and stained with 0.1% (w/v) crystal violet, and the number of colonies in each well was counted.

4.6. Tumor Sphere Assay

HCT116 and HCT116R^{F10} cells were seeded into 96-well PrimeSurface® plate 96U (Sumitomo Bakelite Co., Ltd., Tokyo, Japan) plates (1000 cells per well) in triplicate and then treated with various concentrations of 5-FU or DMSO (as a negative control). Following incubation for 14 days, tumor sphere size was monitored once every 3–4 days. Tumor sphere volume (V) was calculated using the following formula: $V = ab^2/2$ (a and b are the long and short diameters of the tumor sphere, respectively).

4.7. Exome Sequencing Analysis

DNA extraction was performed as previously described [29]. Genomic DNA was extracted from cells (5×10^6 cells) by using a DNeasy Tissue Kit (QIAGEN, Venlo, Netherlands), according to the manufacturer's instructions. Exome sequencing of parental HCT116 and

HCT116R^{F10} cells was performed by APRO Life Science Institute, Inc. (Tokushima, Japan) and Macrogen Global Headquarters (Seoul, Korea).

4.8. Western Blot Analysis

Western blot analysis was performed as previously described [29,32–34]. The antibodies used were rabbit anti-thymidylate synthase (D5B3) monoclonal antibody (9045S, 1:1000, Cell Signaling Technologies, Danvers, MA, USA), mouse anti-DPYD (A-5) monoclonal antibody (sc-376712, 1:1000, Santa Cruz Biotechnology, Dallas, TX, USA), rabbit anti-GAPDH antibody (2275-PC-100, 1:20,000, Trevigen, Gaithersburg, MD, USA), mouse anti-beta-actin monoclonal antibody (A19178-200UL, 1:20,000, Sigma-Aldrich), horseradish peroxidase-linked anti-rabbit IgG (1:20,000, GE Healthcare, Pittsburgh, PA, USA), and horseradish peroxidase-linked whole antibody anti-mouse IgG (1:20,000, GE Healthcare).

4.9. Statistical Analysis

The data are presented as means $\pm$ standard deviation. The significance of differences among groups was evaluated using a Student's *t*-test; $p < 0.05$ was considered statistically significant.

Author Contributions: Conceptualization, A.S.; investigation, C.K., Y.O. and A.S.; writing—original draft preparation, A.S.; writing—review and editing, C.K., Y.O. and A.S.; visualization, C.K., Y.O. and A.S.; project administration, A.S. All authors have read and agreed to the published version of the manuscript.

Funding: This research received no external funding.

Institutional Review Board Statement: Not applicable.

Informed Consent Statement: Not applicable.

Data Availability Statement: Not applicable.

Acknowledgments: We thank Yusuke Wataya (Okayama University) for their helpful discussions.

Conflicts of Interest: The authors declare no conflict of interest.

References

1. Bray, F.; Ferlay, J.; Soerjomataram, I.; Siegel, R.L.; Torre, L.A.; Jemal, A. Global cancer statistics 2018: GLOBOCAN estimates of incidence and mortality worldwide for 36 cancers in 185 countries. *CA Cancer J. Clin.* **2018**, *68*, 394–424. [CrossRef] [PubMed]
2. Longley, D.B.; Harkin, D.P.; Johnston, P.G. 5-fluorouracil: Mechanisms of action and clinical strategies. *Nat. Rev. Cancer* **2003**, *3*, 330–338. [CrossRef] [PubMed]
3. Blondy, S.; David, V.; Verdier, M.; Mathonnet, M.; Perraud, A.; Christou, N. 5-Fluorouracil resistance mechanisms in colorectal cancer: From classical pathways to promising processes. *Cancer Sci.* **2020**, *111*, 3142–3154. [CrossRef] [PubMed]
4. Heidelberger, C. Fluorinated pyrimidines. *Prog. Nucleic Acid. Res. Mol. Biol.* **1965**, *4*, 1–50.
5. Santi, D.V. Perspective on the design and biochemical pharmacology of inhibitors of thymidylate synthetase. *J. Med. Chem.* **1980**, *23*, 103–111. [CrossRef]
6. Copur, S.; Aiba, K.; Drake, J.C.; Allegra, C.J.; Chu, E. Thymidylate synthase gene amplification in human colon cancer cell lines resistant to 5-fluorouracil. *Biochem. Pharmacol.* **1995**, *49*, 1419–1426. [CrossRef]
7. Johnston, P.G.; Lenz, H.J.; Leichman, C.G.; Danenberg, K.D.; Allegra, C.J.; Danenberg, P.V.; Leichman, L. Thymidylate synthase gene and protein expression correlate and are associated with response to 5-fluorouracil in human colorectal and gastric tumors. *Cancer Res.* **1995**, *55*, 1407–1412.
8. Wang, W.; Marsh, S.; Cassidy, J.; McLeod, H.L. Pharmacogenomic dissection of resistance to thymidylate synthase inhibitors. *Cancer Res.* **2001**, *61*, 5505–5510.
9. Popat, S.; Matakidou, A.; Houlston, R.S. Thymidylate synthase expression and prognosis in colorectal cancer: A systematic review and meta-analysis. *J. Clin. Oncol.* **2004**, *22*, 529–536. [CrossRef]
10. Ishibiki, Y.; Kitajima, M.; Sakamoto, K.; Tomiki, Y.; Sakamoto, S.; Kamano, T. Intratumoural thymidylate synthase and dihydropyrimidine dehydrogenase activities are good predictors of 5-fluorouracil sensitivity in colorectal cancer. *J. Int. Med. Res.* **2003**, *31*, 181–187. [CrossRef]
11. Zhang, N.; Yin, Y.; Xu, S.J.; Chen, W.S. 5-Fluorouracil: Mechanisms of resistance and reversal strategies. *Molecules* **2008**, *13*, 1551–1569. [CrossRef] [PubMed]

12. Chu, E.; Koeller, D.M.; Johnston, P.G.; Zinn, S.; Allegra, C.J. Regulation of thymidylate synthase in human colon cancer cells treated with 5-fluorouracil and interferon-gamma. *Mol. Pharmacol.* **1993**, *43*, 527–533. [PubMed]
13. Drake, J.C.; Allegra, C.J.; Johnston, P.G. Immunological quantitation of thymidylate synthase-FdUMP-5,10-methylenetetrahydrofolate ternary complex with the monoclonal antibody TS 106. *Anticancer Drugs* **1993**, *4*, 431–435. [CrossRef] [PubMed]
14. Mori, R.; Futamura, M.; Tanahashi, T.; Tanaka, Y.; Matsuhashi, N.; Yamaguchi, K.; Yoshida, K. 5FU resistance caused by reduced fluoro-deoxyuridine monophosphate and its reversal using deoxyuridine. *Oncol. Lett.* **2017**, *14*, 3162–3168. [CrossRef]
15. Boyer, J.; McLean, E.G.; Aroori, S.; Wilson, P.; McCulla, A.; Carey, P.D.; Longley, D.B.; Johnston, P.G. Characterization of p53 wild-type and null isogenic colorectal cancer cell lines resistant to 5-fluorouracil, oxaliplatin, and irinotecan. *Clin. Cancer Res.* **2004**, *10*, 2158–2167. [CrossRef] [PubMed]
16. Laskin, J.D.; Evans, R.M.; Slocum, H.K.; Burke, D.; Hakala, M.T. Basis for natural variation in sensitivity to 5-fluorouracil in mouse and human cells in culture. *Cancer Res.* **1979**, *39*, 383–390.
17. Yoshioka, A.; Tanaka, S.; Hiraoka, O.; Koyama, Y.; Hirota, Y.; Ayusawa, D.; Seno, T.; Garrett, C.; Wataya, Y. Deoxyribonucleoside triphosphate imbalance. 5-Fluorodeoxyuridine-induced DNA double strand breaks in mouse FM3A cells and the mechanism of cell death. *J. Biol. Chem.* **1987**, *262*, 8235–8241. [CrossRef]
18. Yamada, M.; Nakagawa, H.; Fukushima, M.; Shimizu, K.; Hayakawa, T.; Ikenaka, K. In vitro study on intrathecal use of 5-fluoro-2′-deoxyuridine (FdUrd) for meningeal dissemination of malignant brain tumors. *J. Neurooncol.* **1998**, *37*, 115–121. [CrossRef]
19. Peters, G.J.; Backus, H.H.; Freemantle, S.; van Triest, B.; Codacci-Pisanelli, G.; van der Wilt, C.L.; Smid, K.; Lunec, J.; Calvert, A.H.; Marsh, S.; et al. Induction of thymidylate synthase as a 5-fluorouracil resistance mechanism. *Biochim. Biophys. Acta* **2002**, *1587*, 194–205. [CrossRef]
20. Chu, E.; Koeller, D.M.; Casey, J.L.; Drake, J.C.; Chabner, B.A.; Elwood, P.C.; Zinn, S.; Allegra, C.J. Autoregulation of human thymidylate synthase messenger RNA translation by thymidylate synthase. *Proc. Natl. Acad. Sci. USA* **1991**, *88*, 8977–8981. [CrossRef]
21. Chu, E.; Voeller, D.; Koeller, D.M.; Drake, J.C.; Takimoto, C.H.; Maley, G.F.; Maley, F.; Allegra, C.J. Identification of an RNA binding site for human thymidylate synthase. *Proc. Natl. Acad. Sci. USA* **1993**, *90*, 517–521. [CrossRef]
22. Keyomarsi, K.; Samet, J.; Molnar, G.; Pardee, A.B. The thymidylate synthase inhibitor, ICI D1694, overcomes translational detainment of the enzyme. *J. Biol. Chem.* **1993**, *268*, 15142–15149. [CrossRef]
23. Chu, E.; Allegra, C.J. The role of thymidylate synthase as an RNA binding protein. *Bioessays* **1996**, *18*, 191–198. [CrossRef]
24. Kitchens, M.E.; Forsthoefel, A.M.; Barbour, K.W.; Spencer, H.T.; Berger, F.G. Mechanisms of acquired resistance to thymidylate synthase inhibitors: The role of enzyme stability. *Mol. Pharmacol.* **1999**, *56*, 1063–1070. [CrossRef] [PubMed]
25. Kitchens, M.E.; Forsthoefel, A.M.; Rafique, Z.; Spencer, H.T.; Berger, F.G. Ligand-mediated induction of thymidylate synthase occurs by enzyme stabilization. Implications for autoregulation of translation. *J. Biol. Chem.* **1999**, *274*, 12544–12547. [CrossRef]
26. Marverti, G.; Ligabue, A.; Paglietti, G.; Corona, P.; Piras, S.; Vitale, G.; Guerrieri, D.; Luciani, R.; Costi, M.P.; Frassineti, C.; et al. Collateral sensitivity to novel thymidylate synthase inhibitors correlates with folate cycle enzymes impairment in cisplatin-resistant human ovarian cancer cells. *Eur. J. Pharmacol.* **2009**, *615*, 17–26. [CrossRef]
27. Washtien, W.L. Increased levels of thymidylate synthetase in cells exposed to 5-fluorouracil. *Mol. Pharmacol.* **1984**, *25*, 171–177. [PubMed]
28. Mohsen, A.W.; Aull, J.L.; Payne, D.M.; Daron, H.H. Ligand-induced conformational changes of thymidylate synthase detected by limited proteolysis. *Biochemistry* **1995**, *34*, 1669–1677. [CrossRef] [PubMed]
29. Ogino, Y.; Sato, A.; Uchiumi, F.; Tanuma, S.I. Cross resistance to diverse anticancer nicotinamide phosphoribosyltransferase inhibitors induced by FK866 treatment. *Oncotarget* **2018**, *9*, 16451–16461. [CrossRef] [PubMed]
30. Ogino, Y.; Sato, A.; Uchiumi, F.; Tanuma, S.I. Genomic and tumor biological aspects of the anticancer nicotinamide phosphoribosyltransferase inhibitor FK866 in resistant human colorectal cancer cells. *Genomics* **2019**, *111*, 1889–1895. [CrossRef] [PubMed]
31. Inada, M.; Shindo, M.; Kobayashi, K.; Sato, A.; Yamamoto, Y.; Akasaki, Y.; Ichimura, K.; Tanuma, S.I. Anticancer effects of a non-narcotic opium alkaloid medicine, papaverine, in human glioblastoma cells. *PLoS ONE* **2019**, *14*, e0216358. [CrossRef] [PubMed]
32. Ogino, Y.; Sato, A.; Kawano, Y.; Aoyama, T.; Uchiumi, F.; Tanuma, S.I. Association of ABC Transporter With Resistance to FK866, a NAMPT Inhibitor, in Human Colorectal Cancer Cells. *Anticancer Res.* **2019**, *39*, 6457–6462. [CrossRef] [PubMed]
33. Sato, A.; Satake, A.; Hiramoto, A.; Wataya, Y.; Kim, H.S. Protein expression profiles of necrosis and apoptosis induced by 5-fluoro-2′-deoxyuridine in mouse cancer cells. *J. Proteome Res.* **2010**, *9*, 2329–2338. [CrossRef] [PubMed]
34. Sato, A.; Nakama, K.; Watanabe, H.; Satake, A.; Yamamoto, A.; Omi, T.; Hiramoto, A.; Masutani, M.; Wataya, Y.; Kim, H.S. Role of activating transcription factor 3 protein ATF3 in necrosis and apoptosis induced by 5-fluoro-2′-deoxyuridine. *FEBS J.* **2014**, *281*, 1892–1900. [CrossRef] [PubMed]

International Journal of
Molecular Sciences

Article

Hyperthermia Induced by Gold Nanoparticles and Visible Light Photothermy Combined with Chemotherapy to Tackle Doxorubicin Sensitive and Resistant Colorectal Tumor 3D Spheroids

Catarina Roma-Rodrigues [†], Inês Pombo [†], Alexandra R. Fernandes [*,‡] and Pedro V. Baptista [*,‡]

UCIBIO, Department of Life Sciences, Faculdade de Ciências e Tecnologia, Universidade NOVA de Lisboa, 2829-516 Caparica, Portugal; catromar@fct.unl.pt (C.R.-R.); id.pombo@campus.fct.unl.pt (I.P.)
* Correspondence: ma.fernandes@fct.unl.pt (A.R.F.); pmvb@fct.unl.pt (P.V.B.); Tel.: +351-21-2948530 (P.V.B.)
† These authors contributed equally to this work.
‡ These authors share the co-last authorship.

Received: 7 October 2020; Accepted: 26 October 2020; Published: 28 October 2020

Abstract: Current cancer therapies are frequently ineffective and associated with severe side effects and with acquired cancer drug resistance. The development of effective therapies has been hampered by poor correlations between pre-clinical and clinical outcomes. Cancer cell-derived spheroids are three-dimensional (3D) structures that mimic layers of tumors in terms of oxygen and nutrient and drug resistance gradients. Gold nanoparticles (AuNP) are promising therapeutic agents which permit diminishing the emergence of secondary effects and increase therapeutic efficacy. In this work, 3D spheroids of Doxorubicin (Dox)-sensitive and -resistant colorectal carcinoma cell lines (HCT116 and HCT116-DoxR, respectively) were used to infer the potential of the combination of chemotherapy and Au-nanoparticle photothermy in the visible (green laser of 532 nm) to tackle drug resistance in cancer cells. Cell viability analysis of 3D tumor spheroids suggested that AuNPs induce cell death in the deeper layers of spheroids, further potentiated by laser irradiation. The penetration of Dox and earlier spheroid disaggregation is potentiated in combinatorial therapy with Dox, AuNP functionalized with polyethylene glycol (AuNP@PEG) and irradiation. The time point of Dox administration and irradiation showed to be important for spheroids destabilization. In HCT116-sensitive spheroids, pre-irradiation induced earlier disintegration of the 3D structure, while in HCT116 Dox-resistant spheroids, the loss of spheroid stability occurred almost instantly in post-irradiated spheroids, even with lower Dox concentrations. These results point towards the application of new strategies for cancer therapeutics, reducing side effects and resistance acquisition.

Keywords: 3D spheroids; photothermy; gold nanoparticles; doxorubicin resistance; colorectal cancer

---

1. Introduction

Effective cancer eradication is being hampered by tumor cells' acquired resistance to a wide spectrum of unrelated drugs, resulting in the so-called multidrug resistance (MDR) [1]. Despite the existence of different strategies for cancer treatment, including surgery, immune-, endocrine, gene, radio- or chemotherapies, the latter remains the method of choice [1]. However, exposure for a prolonged time to drugs, together with genetic alterations in tumor cells, hinder increased resistance to chemotherapeutics [1,2]. The acquired MDR, together with tumor heterogeneity, which is often correlated to characteristic tumor microenvironments, is forcing the development of new strategies to tackle cancer [1,3,4]. These strategies mainly rely on the application of combined therapies aiming to target multiple pathways in the tumor, circumventing MDR and lowering secondary effects [1,3,5,6].

Nanomedicine puts forward a multitude of conceptual tools to tackle cancer cells directly and/or to improve the delivery of therapeutic moieties, such as drugs, small interfering RNA (siRNA), antibodies, etc. [7–11]. Among these, nanoparticles able to convey different therapeutic cargos and with unique physico-chemical properties have allowed the design of novel combinatory strategies to tackle cancer [7–11]. Gold nanoparticles (AuNPs) possess remarkable optical properties that depend on their easy tunable size and shape and a large surface area to volume ratio, optimal for ease of functionalization with different agents [12]. Among these optical properties, the localized surface plasmon resonance (SPR) of AuNPs makes them suitable as photothermal agents in non-invasive photothermal therapy since they are extremely effective at converting light into heat [12–20]. Importantly, if the size is right, the irradiation may be in the visible region of the spectrum, allowing for visual control of the target subjected to therapy while improving the photothermal conversion ratio by using more energetic radiation than the conventional rear-infrared. What is more, this photothermy may be attained by means of green lasers commonly used in medical surgery for photocoagulation [16,21].

Thus far, most of the pre-clinical research and development of new cancer chemotherapeutics has been focused on two-dimensional (2D) cell cultures and murine animal models, whose limitations have been critical aspects for failure of clinical translation [22–24]. In fact, solid tumors are three-dimensional (3D) entities composed of tumor cells with variable proliferation, amidst oxygen and nutrient gradients, and communicating with a complex network of stromal, immune and endothelial cells in a characteristic extracellular matrix (ECM)—the tumor microenvironment [3,7]. Anticancer drugs are found in gradients in tumors, associated to deferential cell response. However, these intra-tumoral gradients are absent in 2D monolayers, which, coupled to differentially activated signaling pathways, might introduce a misleading bias when interpreting data originating from these in vitro 2D systems [25]. Another handicap of these 2D systems is the lack of ECM, which is preponderant for the mechanical properties of the tumor and improve intra- and inter-cellular communications [23]. Murine models, while substantially more complex to handle, still fall short of accurately representing the human tissues and organs [23]. To bridge this gap, 3D cultures, such as spheroids and organoids, have been introduced as relevant models for in vitro testing that combine the simplicity of 2D cell cultures with the three-dimensional, more complex tumor microenvironment (TME) [22,23]. Spheroids, consisting of 3D cell aggregates formed in ultra-low attachment coating microplates, mimic some important characteristics of tumors, such as the existence of ECM and the typical layering of solid tumors for spheroids with more than 200 µM [26–29]. The outermost proliferative zone with more oxygen supply and less interstitial pressure; the middle layer is the region with senescence cells (quiescent zone) characterized by higher interstitial pressure and middle level of oxygen supply; the innermost layer, or necrotic core, is characterized by low oxygen supply and high interstitial pressure with few to none cell proliferation [27,28]. The hypoxic and acidic environments found in the quiescent zone and necrotic core have been associated with low penetrability of drugs into tumors, and acquisition of radiotherapy and chemotherapy resistance [7,27,30,31]. The growth kinetics and the genetic expression of in vivo tumors are also mimicked in spheroids [26–28].

Herein, we evaluated the combination of photothermal irradiation using AuNPs with the chemotherapeutic agent Doxorubicin (Dox) to tackle a 3D model of colorectal cancer. Results suggested that spheroid treatment with AuNPs functionalized with polyethylene glycol (AuNP@PEG) combined with Dox and irradiation are promising therapeutic strategies to increase the therapeutic efficacy and decrease the dosage of the agent, even in drug-resistant cells.

2. Results and Discussion

We used 3D tumor spheroids to assess the drug gradient uptake and the effect of localized photothermy mediated by AuNP functionalized with polyethylene glycol (AuNP@PEG) alone or in combination with Dox. With that purpose, spheroids of colorectal cancer cells sensitive and resistant to

Dox (HCT116 and HCT116-DoxR, respectively) were produced using ultra-low attachment plate wells. The HCT116-DoxR cell line was derived from a standard sensitive HCT116 cell line by culturing with increasing concentrations of Dox up to 3.6 µM [21].

2.1. Localized Irradiation of Spheroids

Firstly, the spheroids' cell viability was inferred after a 24 h incubation with AuNP@PEG, followed by 30 min, 1 h 30 min or 24 h incubation with a fresh medium supplemented with CellTox green dye. Results suggest a decreased viability of cells at the periphery of the spheroid after treatment with AuNP@PEG (Figure 1A,B) when compared to untreated spheroids (Figure 1C,D). The fluorescent peripherical corona increases over time, suggesting increased cell death induced by AuNP@PEG, which is corroborated by the disintegration of the spheroid observed in Brightfield images (Figure 1, Figure S1 and Movie S1–S4). A decrease in cell viability in the presence of AuNP@PEG has previously been observed and associated to reactive oxygen species (ROS)-mediated apoptosis [32–35]. After 24 h, there is an increase of fluorescence in the central area of the spheroids correlating to widespread cell death at the core (Figure S1). This effect might be associated to the diminished supply of nutrients and oxygen, indicating that a necrotic region was formed in HCT116 spheroids, mimicking those of tumors growing in vivo [28]. It is of particular relevance to observe the dynamics of fluorescence within the spheroid, which provides valuable information on the evolution of the overall systema rather than the momentary intensity of a particular observation field. The dynamics of fluorescence signal between core and periphery of the spheroid observed over time as function of challenging the structure with AuNP@PEG provide a more precise approximation to that of solid tumors (Figure S1).

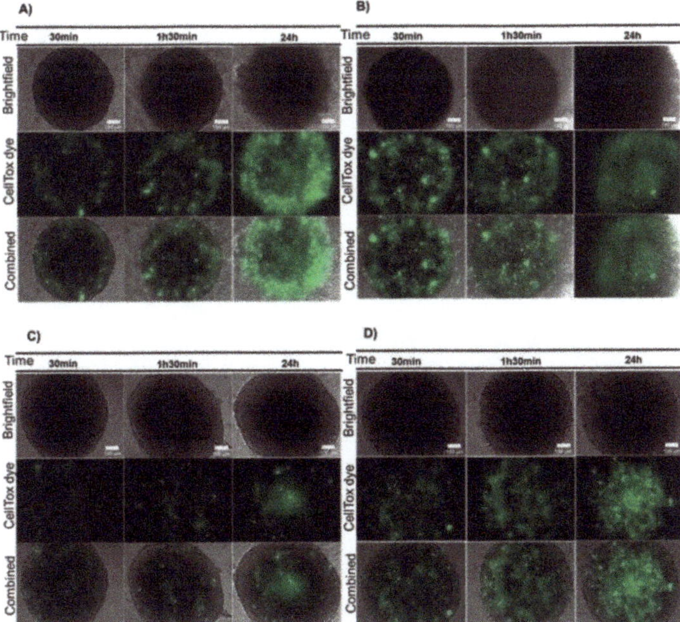

Figure 1. AuNP@PEG effect on cell viability measured by CellTox green dye. (**A**) HCT116 spheroids incubated for 24 h with 8 nM AuNP@PEG; (**B**) HCT116-DoxR spheroids incubated for 24 h with 8 nM AuNP@PEG; (**C**) HCT116 spheroids alone; (**D**) HCT116-DoxR spheroids alone. Microscopy images were acquired in Brightfield or with a green fluorescence filter to evaluate CellTox green dye fluorescence, after 30 min, 1 h 30 min or 24 h incubation. The combined images result from the overlap between Brightfield and green filter images. Scale bar corresponds to 100 µm.

These spheroids were then subjected to localized photothermy mediated by AuNP@PEG [16,21]. HCT116 and HCT116-DoxR spheroids were incubated with AuNP@PEG as described above, washed three times to ensure removal of AuNP@PEG in suspension and irradiated with a 532 nm laser for 1 min, and cell viability was assessed with CellTox after 30 min, 1 h 30 min or 24 h. As observed for the AuNP@PEG alone, there is a more pronounced loss of viability of cells located at the periphery (Figure 2A–D). Analyzing this process with further detail shows that for HCT116 spheroids, the fluorescent signal increases over time, allowing the distinction of three layers (Figure 2A): (1) the peripheral layer that presents a high fluorescent signal that is not observed in irradiated spheroids (Figure 2B and Figure S2), suggesting that AuNP@PEG are responsible for cell death at the periphery; (2) a middle layer with low fluorescence; (3) the innermost core with increased fluorescence observed also in control spheroids (Figure 2B and Figure S2). Interestingly, HCT116-DoxR spheroids incubated for 24 h with AuNP@PEG and irradiated show scattered fluorescence throughout the spheroid, indicating extensive cell death culminating with the disintegration of the whole 3D structure (Figure 2C). The observed effect can be, at least partially, attributed to the photothermal effect, as previously reported [16]. In fact, the increase in temperature (ΔT) following 60 s irradiation is higher for spheroids pre-incubated with AuNP@PEG (ΔT = 10.2 °C), relative to irradiated non-treated spheroids (ΔT = 6.4 °C) or control, i.e., medium alone incubated with AuNP@PEG, washed three times with phosphate buffer saline (PBS) and then irradiated (ΔT = 6.6 °C).

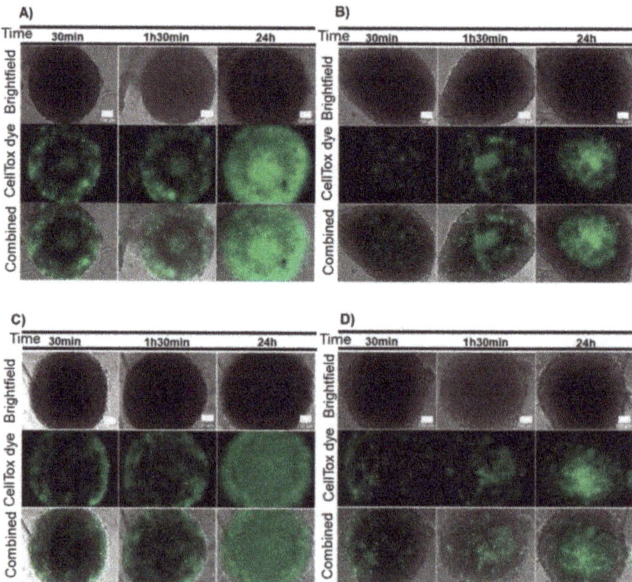

Figure 2. Combined effect of AuNP@PEG and irradiation on cell viability measured by CellTox green. (**A**) HCT116 spheroids incubated for 24 h with 8 nM AuNP@PEG and then irradiated with a 532-nm green laser for 1 min; (**B**) HCT116 spheroids irradiated with a 532-nm green laser for 1 min; (**C**) Doxorubicin-resistant HCT116 (HCT116-DoxR) spheroids incubated for 24 h with 8 nM AuNP@PEG and then irradiated with a 532-nm green laser for 1 min; (**D**) HCT116-DoxR spheroids irradiated with a 532-nm green laser for 1 min. Microscopy images were acquired in Brightfield or with a green fluorescence filter to evaluate CellTox green dye fluorescence, after 30 min, 1 h 30 min or 24 h incubation. The combined images result from the overlap between Brightfield and green filter images. Scale bar corresponds to 100 μm.

When cells are irradiated without AuNPs, a pattern of cell death over time is also observed, with more fluorescent tumor cells in the inner core after 24 h (Figure 2B,D and Figure S2). This might be a side-effect of the light absorption by cytochromes present in cells, which are able to focus the irradiated light creating minute focal heating spots, but with considerably lower photothermal conversion efficiency when compared to AuNP@PEG. It should be noted that cells at the core of the spheroid are under stress due to limitations to the supply of nutrient and oxygen, which cumulatively contribute to the loss of viability in growing 3D cell structures [27–29]. The integrity of spheroids upon incubation with AuNP@PEG and irradiation was also followed in real time (Movie S5 and S6). Upon irradiation, HCT116 spheroids show a localized cell bursting effect at the periphery that follows the trend observed for AuNP@PEG gradient into the 3D structure. This localized photothermal effect continues for at least 48 h. Together, these results suggest that AuNP@PEG induce cell death in spheroid layers that is potentiated by irradiation.

2.2. Combination Therapy in 3D Models–Dox and Localized Photothermy Mediated by AuNP@PEG

Irradiation of AuNPs has been shown to promote cell permeabilization [18–20]. Hence, we hypothesized that AuNPs, with or without laser irradiation, might be able to increase drug penetration into spheroids, thus improving anti-tumor efficacy. A synergistic effect between photothermy mediated by AuNP@PEG and visible irradiation and Dox against breast cancer cells has been reported [16].

Firstly, we analyzed Dox diffusion in HCT116 and HCT116-DoxR spheroids via its intrinsic fluorescence (λ_{exc}=470 nm; λ_{em}=599 nm) [36]. Spheroids were incubated with 8 µM of Dox, corresponding to 20x the IC_{50} of Dox in a standard 2D monolayer of HCT116 cells (0.4 µM), and HCT116-DoxR spheroids were incubated also with an additional concentration of Dox (120 µM) [21,37] for 30 min, 1 h 30 min and 24 h (Figure 3). Dox showed a small diffusion into the central core of the spheroid and preferential accumulation at the periphery of HCT116 spheroids (Figure S3). In HCT116-DoxR spheroids, there was a low amount of fluorescence for the initial time points (30 min and 1 h 30 min) and equally distributed within the spheroid for both Dox concentrations (Figure S3). After 24 h of incubation with 120 µM of Dox, there was a noticeable increase in fluorescence at the periphery and intermediate areas of the spheroid, suggesting that Dox was accumulating in these sections (Figure 3C and Figure S3). In fact, previous reports have described a time-dependent penetration of Dox into HCT116 spheroids with the same pattern of distribution [38,39].

Cell viability evaluation via the CellTox green dye in these models seems to suggest that pre-incubation of sensitive and resistant HCT116 spheroids with Dox did not have an effect on the overall cell viability within the spheroids for the lower concentration (Figures S4 and S5).

We then followed the integrity of the spheroids in real time while monitoring the internalization of Dox by acquiring images every 15 min over a 48 h period. After 17 h incubation with Dox, disintegration of the 3D structure of HCT116 spheroid could be observed (Movie S7), as well as disintegration of HCT116-DoxR spheroid incubated with even lower concentrations of Dox after 32 h (Movie S8).

To assess the combinatory effect on 3D spheroids of chemo- and localized photothermy, we first pre-incubated the spheroids with AuNP@PEG and subsequently challenged these with Dox. As a result, an overall increase in red fluorescence in all spheroids was observed, suggesting that Dox had diffused faster and/or more easily within the 3D structures (Figure 4 and Figure S6). Curiously, there was a slight difference in observed accumulation of Dox between HCT116 and HCT1116-DoxR. The former showed higher accumulation of Dox in the intermediary region of the spheroid, whereas the latter showed a lower extent of penetration for the lower concentration of Dox (Figure 4 and Figure S6). Together, these data seem to suggest that the lower cell viability at the periphery (Figure 1) might allow deeper penetration of the drug into the spheroids due to disturbance of the 3D cell structure. This hypothesis is further supported by the observation that following pre-incubation with AuNP@PEG, the simultaneous incubation with Dox and the CellTox dye showed a general increase in green fluorescence at the intermediary section of these spheroids (Figure S7).

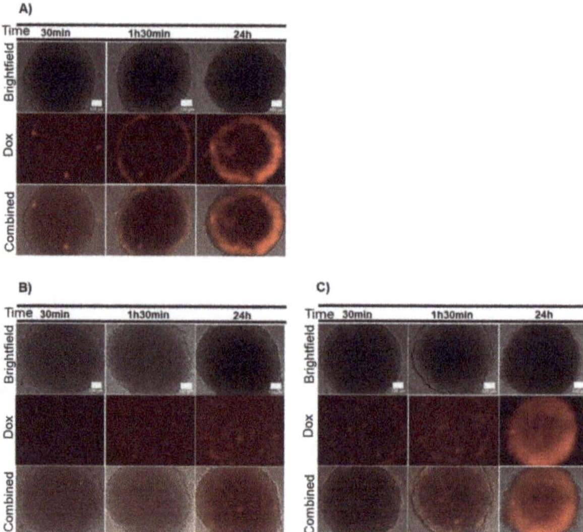

Figure 3. Dox diffusion in sensitive and resistant spheroids. (**A**) HCT116 spheroids incubated with 8 μM of Dox; (**B**) Dox-resistant HCT116 (HCT116-DoxR) spheroids incubated with 8 μM of Dox; (**C**) HCT116-DoxR spheroids incubated with 120 μM of Dox. Microscopy images were acquired in Brightfield or with a red fluorescence filter after 30 min, 1 h 30 min or 24 h incubation. The combined images result from the overlap between Brightfield and red filter images. Scale bar corresponds to 100 μm.

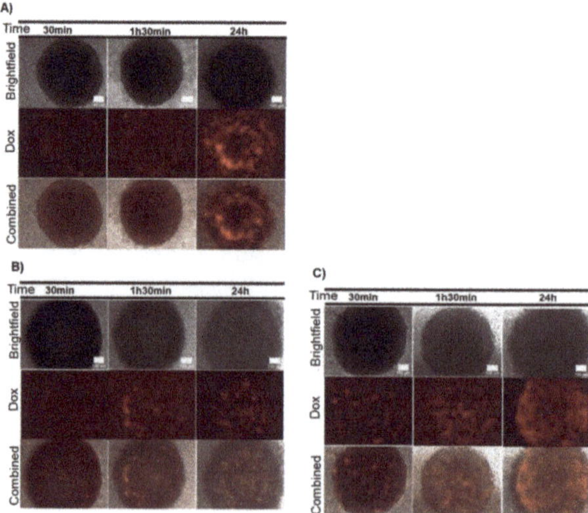

Figure 4. Doxorubicin (Dox) diffusion in spheroids after incubation with AuNP@PEG. (**A**) HCT116 spheroids incubated with 8 nM AuNP@PEG and then with 8 μM of Dox; (**B**) Dox-resistant HCT116 (HCT116-DoxR) spheroids incubated with 8 nM AuNP@PEG and then with 8 μM of Dox; (**C**) HCT116-DoxR spheroids incubated with 8 nM AuNP@PEG and then with 120 μM of Dox. Microscopy images were acquired in Brightfield or with a red fluorescence filter after 30 min, 1 h 30 min or 24 h incubation. The combined images result from the overlap between Brightfield and red filter images. Scale bar corresponds to 100 μm.

Figures 5 and 6 show the effect of irradiation in the penetration of Dox into the spheroids. HCT116 spheroids incubated with Dox showed a similar penetration profile of the drug in pre-irradiated and non-irradiated cells (Figure 3A, Figure 5A and Figures S3 and S8). Noteworthily, spheroid disintegration occurs earlier for non-irradiated spheroids treated with Dox (after 17 h) while pre-irradiated spheroids treated with Dox disintegrate only after 24 h (Movie S7 and S9). Interestingly, irradiation after incubation with Dox resulted in the swelling of the spheroid with no visible disintegration until at least 48 h (Movie S10).

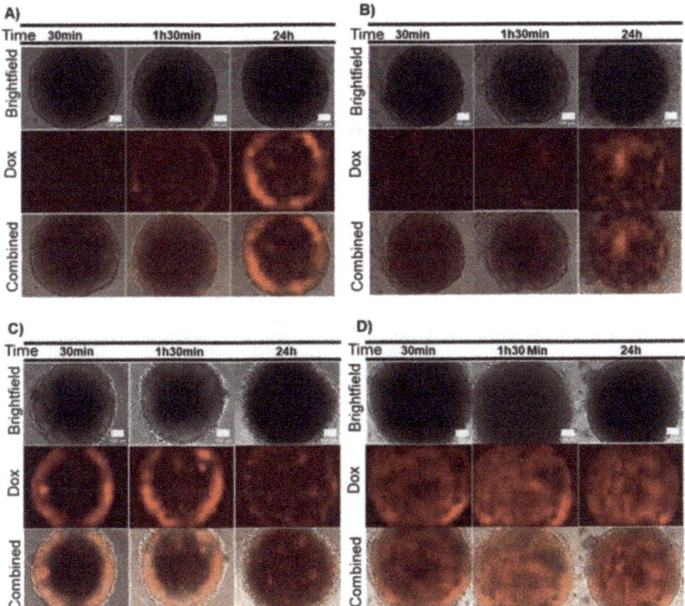

Figure 5. Effect of AuNPs and irradiation on Doxorubicin (Dox) diffusion in HCT116 spheroids. (**A**) HCT116 spheroids irradiated with a 532-nm laser for 1 min and then incubated with 8 μM Dox; (**B**) HCT116 spheroids incubated with 8 nM AuNP@PEG, irradiated with a 532-nm laser for 1 min and then incubated with 8 μM Dox; (**C**) HCT116 spheroids incubated for 6 h with 8 μM Dox and then irradiated with a 532-nm laser for 1 min; (**D**) HCT116 spheroids incubated with 8 nM AuNPs, incubated for 6 h with 8 μM Dox and then irradiated with a 532-nm laser for 1 min. Microscopy images were acquired in Brightfield or with a red fluorescence filter after 30 min, 1 h 30 min or 24 h incubation. The combined images result from the overlap between Brightfield and red filter images. Scale bar corresponds to 100 μm.

These results clearly indicate that combination of Dox and AuNP@PEG photothermal irradiation might be an excellent strategy to tackle 3D tumor structures. What is more, the harmonization of Dox accumulation and moment of irradiation is critical for spheroid destabilization, paving the way for the establishment of therapeutic strategies deferred in time to maximize efficacy.

Concerning the Dox-resistant spheroids (HCT116-DoxR), a similar penetration profile in pre-irradiated, post-irradiated and non-irradiated cells is observed (Figure 6A,B, Figure 3B and Figure S10), which is associated to lower cell death (Figure S11). The real-time analysis showed a similar behavior of pre-irradiated and non-irradiated spheroids incubated with 8 μM Dox, with disintegration observed after 30 h and 32 h, respectively (Movie S8 and S13). However, spheroid disintegration of post-irradiated spheroids occurs earlier—after 10 h of time of irradiation (Movie S14). Pre-incubation with AuNP@PEG showed an increase in drug diffusion within the spheroid, which is more evident when the spheroid is irradiated following incubation with AuNP@PEG and Dox

(Figure 6C,D). Disintegration of the pre-irradiated HCT116-DoxR spheroid occurred at 30h, while that of the HCT116-DoxR spheroid incubated with AuNP@PEG, irradiated and incubated with Dox was observed after 6h (Movie S15). Once again, time of irradiation is critical to maximize disintegration and tumor ablation since irradiation after pre-incubation with AuNP@PEG and Dox resulted in an almost immediate loss of spheroid integrity (Movie S16). Once again, disintegration of the 3D structure leads to lower values of fluorescence intensity than expected (Figure S10D), pointing out the importance of real-time imaging and monitoring of spheroid dynamics.

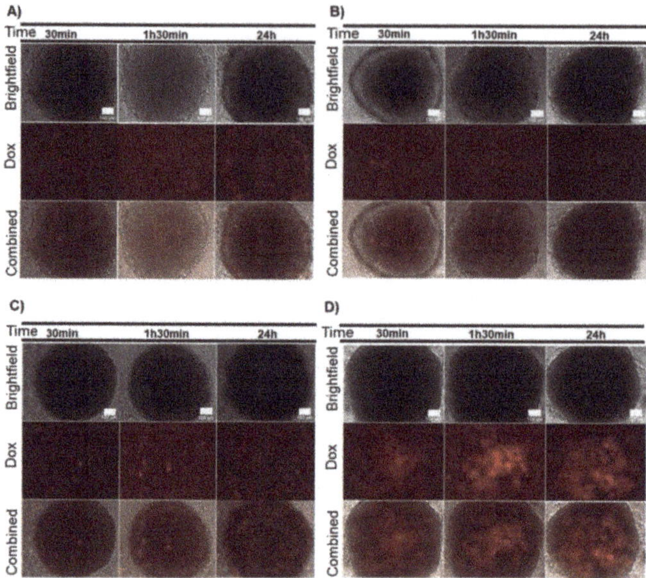

Figure 6. Effect of AuNPs and irradiation in Doxorubicin (Dox) diffusion in Dox-resistant HCT116 (HCT116-DoxR) spheroids. (**A**) HCT116-DoxR spheroids irradiated with a 532-nm laser for 1 min and then incubated with 8 μM Dox; (**B**) HCT116-DoxR spheroids incubated for 6 h with 8 μM Dox and then irradiated with a 532-nm laser for 1 min; (**C**) HCT116-DoxR spheroids incubated with 8 nM AuNP@PEG, irradiated with a 532-nm laser for 1 min and then incubated with 8 μM Dox; (**D**) HCT116-DoxR spheroids incubated with 8 nM AuNPs, incubated for 6 h with 8 μM Dox and then irradiated with a 532-nm laser for 1 min. Microscopy images were acquired in Brightfield or with a red fluorescence filter after 30 min, 1h 30 min or 24h incubation. The combined images result from the overlap between Brightfield and red filter images. Scale bar corresponds to 100 μm.

An instantaneous increase to cell death was observed in the spheroid intermediary area when HCT116-DoxR spheroids were irradiated and then incubated with 120 μM Dox (Figure S12C). Once again, improved penetration of the drug occurs after pre-incubation with AuNP@PEG followed by irradiation and incubation with 120 μM of Dox, or vice-versa (Figure S12).

The cumulative effect of AuNP@PEG and irradiation might allow the use of lower doses of chemotherapeutic agents for the effective destabilization of tumor cells.

3. Materials and Methods

3.1. Gold Nanoparticles Synthesis, Functionalization and Characterization

AuNPs with a diameter of ~18 nm (18.4 ± 0.3 nm) were synthesized via the citrate methods and subsequently functionalized with polyethylene glycol (PEG, MW 350 g/mol, Sigma-Aldrich, St. Louis, MO, USA) to obtain 100% PEG coverage and characterized by dynamic light scattering for hydrodynamic size and zeta potential with a SZ-100 equipment from Horiba (Kyoto, Japan), and by transmission electron microscopy (TEM) in a JEOL 1200EX electron microscope (Tokyo, Japan) as previously described (Figure S13) [16,21,40].

3.2. Cell Cultures and Cell Cultures Maintenance

Colorectal carcinoma cell line HCT116 (CCL-247) was obtained from the American Type Culture Collection (ATCC®, Manassas, VA, USA). HCT116 Doxorubicin-resistant cell line (HCT116-DoxR) was derived from the sensitive HCT116 cells as previously described [21]. The HCT116 and HCT116-DoxR cultures were cultured as previously described [21]. Briefly, cells were maintained in Dulbecco's Modified Eagle Medium (DMEM) (Thermo Fisher Scientific, Waltham, MA, USA) supplemented with 10 % (*v/v*) fetal bovine serum (FBS, Thermo Fisher Scientific, Waltham, MA, USA), and a mixture of Penicillin 100 U/mL and Streptomycin 100 µg/mL (Pen/Strep, Thermo Fisher Scientific, Waltham, MA, USA) at 37 °C with 99% (*v/v*) humidity and 5% (*v/v*) CO_2. The culture medium for the HCT116-DoxR cell line was additionally supplemented with 3.6 µM Dox (for easier reading, supplemented media will be termed simply as DMEM).

3.3. Spheroids Preparation and Handling

HCT116 and HCT116-DoxR spheroids were prepared according to Baek et al. (2016), with few modifications [29]. Cultures were seeded at a density of 5×10^3 cells per well in a super-low attachment 96-well culture plate (NunclonTM SpheraTM Microplate, Thermo Fisher Scientific, Waltham, MA, USA), shacked in an orbital direction and incubated for 7 days, when spheroids with a 750-µm diameter are obtained (Figure S14, HCT116, Movie S17, HCT116 and Movie S18, HCT116-DoxR).

After 7 days of growth, the medium was replaced by a fresh medium supplemented with 8 nM AuNP@PEG and incubated for 24 h. Spheroids were then washed three times for 1 min with PBS to remove non-internalized nanoparticles. HCT116 spheroids were incubated with DMEM (without phenol red) supplemented with 8 µM Dox (Merck, Darmstadt, Germany), and HCT116-DoxR spheroids were incubated with DMEM (without phenol red) supplemented with 8 µM or 120 µM Dox. As a control, spheroids were also incubated with Dox vehicle dimethyl sulfoxide (DMSO, Sigma-Aldrich, St. Louis, MO, USA), under the same conditions.

3.4. Spheroids Irradiation

The spheroids were irradiated with a 532- nm green diode-pumped solid-state laser (dpss) (Changchun New Industries Optoelectronics Tech. Co., Ltd., Changchun, China) coupled to a 1-mm diameter optical fiber with the tip placed 2 cm above the bottom of the well and a power set to 3.78 w.cm^{-2} for 1 min. The medium temperature before and after irradiation was measured using a multilogger thermometer hh806u (Omega Engineering, Norwalk, CT, USA).

3.5. Cell Viability

The CellTox™ Green cytotoxicity assay (Promega Corporation, Madison, WI, USA) was used, following the manufacturer's recommendations, for evaluating cell viability. Briefly, following spheroid incubation, the culture medium was removed and replaced by DMEM, without phenol red, supplemented with CellTox™ Green dye 1x. The following settings were used: C1) spheroids incubated with CellTox green dye for 30 min, 1 h 30 min or 24 h; C2) spheroids were irradiated and then incubated with CellTox green dye for 30 min, 1 h 30 min or 24 h; C3) spheroids were incubated for 24 h with AuNP@PEG and then with CellTox green dye for 30 min, 1 h 30 min or 24 h; C4) spheroids were incubated for 24 h with AuNP@PEG, irradiated and then incubated with CellTox green dye for 30 min, 1 h 30 min or 24 h. Spheroids were observed and images were acquired with a Ti-U eclipse inverted microscope (Nikon, Tokyo, Japan) and respective software NIS Elements Basic software vs 3.1 (Nikon, Tokyo, Japan), using the Fluorescein isothiocyanate (FITC) filter (excitation at 480/30 nm and emission at 535/45 nm). The CellTox Green dye is a cyanine dye that enters cells with compromised membrane integrity and become fluorescent (excitation at 485–500 nm and emission at 520–530 nm) after binding to DNA, allowing the distinction between healthy cells (low fluorescence) from dead cells (high fluorescence). The cytotoxicity of the treatment could be inferred through the fluorescence intensity, as the fluorescent signal is proportional to the binding of the dye to the DNA of cells with compromised membranes [41,42]. ImageJ software (https://imagej.net/, version 1.53a) was used to measure the green fluorescence intensity (mean fluorescence x area) of the periphery and core (denser zone of the spheroid) of each spheroid. Bars in graphs represent the average of three different z-stacks of the peripheral and core zone of the spheroid (corresponding to middle layer and innermost core).

3.6. Effect of Doxorubicin

For the analysis of Dox diffusion, the following settings were used: D1) spheroids were incubated with DMSO for 30 min, 1 h 30 min or 24 h; D2) spheroids were incubated with Dox for 30 min, 1 h 30 min or 24 h; D3) spheroids incubated with Dox for 6 h, irradiated and incubated for 30 min, 1 h 30 min or 24 h; D4) spheroids incubated first with AuNP@PEG for 24 h, and then with Dox for 30 min, 1 h 30 min or 24 h; D5) spheroids incubated first with AuNP@PEG for 24 h, then with Dox for 6 h, irradiated and incubated for 30 min, 1 h 30 min or 24 h. Images of Dox diffusion were acquired in a Ti-U eclipse inverted microscope (Nikon, Tokyo, Japan) and respective software NIS Elements Basic software vs 3.1 (Nikon, Tokyo, Japan), using a red filter (G2A filter—excitation at 535/50 nm and emission > 580 nm). ImageJ software (https://imagej.net/, vs 1.53a) was used to measure the red fluorescence intensity (mean fluorescence x area) of the periphery and core (denser zone of the spheroid) of each spheroid. Bars in graphs represent the average of three different z-stacks of the peripheral and core zone of the spheroid.

The analysis of cell viability was also evaluated via incubation with CellTox Green dye after spheroids' treatment (as described above).

Spheroids' morphology was visualized via a cell culture video monitoring Cytosmart Lux2 (Cytosmart technologies, Eindhoven, The Netherlands) with Brightfield with digital contrast phase. The monitorization of Dox internalization in spheroids was performed for 2 days, with the exact time stated in movie caption, in a Fluorescent Lux2 (Cytosmart technologies, Eindhoven, The Netherlands), and red fluorescence with maximum exposure and gain 4. The focus of the microscope was set to be at the bottom of the spheroid. The loss of spheroid integrity was considered to begin when cell debris and/or many cells started to detach from the surface of the spheroid.

4. Conclusions

The combinatory effect of different therapeutic approaches has been put forward as an effective strategy to tackle drug resistance in cancer. Here, we have demonstrated the combinatory effect of Dox and localized photothermy mediated by spherical AuNPs against colorectal tumor cells in 3D spheroid models. Using these spheroid models, we highlight the effect of diffusion through the 3D structures in cell viability and structural organization of the tumor cells. Data suggest that AuNP@PEG are effective against cancer cells and induce severe loss of integrity in HCT116 and HCT116-DoxR spheroids. This is more so when combined with photo-induced hyperthermia. A combination of these approaches enhanced the penetration of the chemotherapeutic agent into the 3D structures. The moment of irradiation is critical to potentiate the loss of spheroid stability. These results seem to indicate that such a combinatory strategy may reduce the amount of drug needed to attain the same efficacy, thus reducing the possibility of side effects. In cancer cells already showing resistance to chemotherapy, this combinatory strategy may allow to increase the stress imposed to these cells and tackle resistance and relapse.

Our study clearly emphasizes the importance of real-time assessment of spheroid dynamics over time. Often, the effect of agents on spheroids is reported by evaluation of single time points, i.e., data retrieved from a still image. More important than the fluorescence intensity, per se, is the dynamic of fluorescence variation, both in time and space, namely between the core and the periphery of the spheroid. This is of particular relevance when combinatory strategies are being studied and evaluated, since each one of these ought to tackle a distinct pathway and/or chain of events, leading to ablation of the tumor. What is more, only dynamic acquisition of data allows to follow the structural alteration of the 3D spheroids (e.g., cell death and consequent disintegration) and the stages leading to this ultimate fate. Further studies will be needed to demonstrate whether this strategy might be broadly applicable to other chemotherapeutics used in clinical oncology.

Supplementary Materials: Supplementary materials can be found at http://www.mdpi.com/1422-0067/21/21/8017/s1.

Author Contributions: Conceptualization, A.R.F. and P.V.B.; Methodology, C.R.-R. and I.P.; Investigation, C.R.-R. and I.P.; Resources, A.R.F. and P.V.B.; Writing—Original Draft Preparation, I.P. and C.R.-R.; Writing—Review and Editing, A.R.F. and P.V.B.; Visualization, I.P., C.R.-R., A.R.F. and P.V.B.; Supervision, A.R.F. and P.V.B.; Project Administration, A.R.F. and P.V.B.; Funding Acquisition, A.R.F. and P.V.B. All authors have read and agreed to the published version of the manuscript.

Funding: This work was supported by the Applied Molecular Biosciences Unit, UCIBIO, which is financed by national funds from FCT (UIDB/04378/2020). C.R.-R. was funded by FCT/MCTES, grant number SFRH/BPD/124612/2016.

Conflicts of Interest: The authors declare no conflict of interest.

References

1. Bukowski, K.; Kciuk, M.; Kontek, R. Mechanisms of Multidrug Resistance in Cancer Chemotherapy. *Int. J. Mol. Sci.* **2020**, *21*, 3233. [CrossRef] [PubMed]

2. Singh, M.S.; Tammam, S.N.; Boushehri, M.A.S.; Lamprecht, A. MDR in cancer: Addressing the underlying cellular alterations with the use of nanocarriers. *Pharmacol. Res.* **2017**, *126*, 2–30. [CrossRef] [PubMed]

3. Roma-Rodrigues, C.; Mendes, R.; Baptista, P.V.; Fernandes, A.R. Targeting Tumor Microenvironment for Cancer Therapy. *Int. J. Mol. Sci.* **2019**, *20*, 840. [CrossRef] [PubMed]

4. Palmer, A.C.; Sorger, P.K. Combination Cancer Therapy Can Confer Benefit via Patient-to-Patient Variability without Drug Additivity or Synergy. *Cell* **2017**, *171*, 1678–1691. [CrossRef]

5. Yao, Y.; Zhou, Y.; Liu, L.; Xu, Y.; Chen, Q.; Wang, Y.; Wu, S.; Deng, Y.; Zhang, J.; Shao, A. Nanoparticle-Based Drug Delivery in Cancer Therapy and Its Role in Overcoming Drug Resistance. *Front. Mol. Biosci.* **2020**, *7*, 193. [CrossRef]

6. Wang, Y.; Deng, W.; Li, N.; Neri, S.; Sharma, A.; Jiang, W.; Lin, S.H. Combining Immunotherapy and Radiotherapy for Cancer Treatment: Current Challenges and Future Directions. *Front. Pharmacol.* **2018**, *9*, 185. [CrossRef]

7. Roma-Rodrigues, C.; Pombo, I.; Raposo, L.; Pedrosa, P.; Fernandes, A.R.; Baptista, P.V. Nanotheranostics Targeting the Tumor Microenvironment. *Front. Bioeng. Biotechnol.* **2019**, *7*, 197. [CrossRef]

8. Baeza, A. Tumor Targeted Nanocarriers for Immunotherapy. *Molecules* **2020**, *25*, 1508. [CrossRef]

9. Gao, D.; Guo, X.; Zhang, X.; Chen, S.; Wang, Y.; Chen, T.; Huang, G.; Gao, Y.; Yang, Z.; Yang, Z. Multifunctional phototheranostic nanomedicine for cancer imaging and treatment. *Mater. Today Bio.* **2020**, *5*, 100035. [CrossRef] [PubMed]

10. Ovais, M.; Mukherjee, S.; Pramanik, A.; Das, D.; Mukherjee, A.; Raza, A.; Chen, C. Designing Stimuli-Responsive Upconversion Nanoparticles that Exploit the Tumor Microenvironment. *Adv. Mater.* **2020**, *32*, e2000055. [CrossRef]

11. Roma-Rodrigues, C.; Rivas-García, L.; Baptista, P.V.; Fernandes, A.R. Gene Therapy in Cancer Treatment: Why Go Nano? *Pharmaceutics* **2020**, *12*, 233. [CrossRef] [PubMed]

12. Singh, P.; Pandit, S.; Mokkapati, V.; Garg, A.; Ravikumar, V.; Mijakovic, I. Gold Nanoparticles in Diagnostics and Therapeutics for Human Cancer. *Int. J. Mol. Sci.* **2018**, *19*, 1979. [CrossRef] [PubMed]

13. Mendes, R.; Fernandes, A.R.; Baptista, P.V. Gold Nanoparticle Approach to the Selective Delivery of Gene Silencing in Cancer—The Case for Combined Delivery? *Genes* **2017**, *8*, 94. [CrossRef]

14. Adnan, N.N.M.; Cheng, Y.Y.; Ong, N.M.N.; Kamaruddin, T.T.; Rozlan, E.; Schmidt, T.W.; Duong, H.; Boyer, C. Effect of gold nanoparticle shapes for phototherapy and drug delivery. *Polym. Chem.* **2016**, *7*, 2888–2903. [CrossRef]

15. Hwang, S.; Nam, J.; Jung, S.; Song, J.; Doh, H.; Kim, S. Gold nanoparticle-mediated photothermal therapy: Current status and future perspective. *Nanomedicine* **2014**, *9*, 2003–2022. [CrossRef] [PubMed]

16. Mendes, R.; Pedrosa, P.; Lima, J.C.; Fernandes, A.R.; Baptista, P.V. Photothermal enhancement of chemotherapy in breast cancer by visible irradiation of Gold Nanoparticles. *Sci. Rep.* **2017**, *7*, 1–9. [CrossRef]

17. D'Acunto, M. Detection of Intracellular Gold Nanoparticles: An Overview. *Materials* **2018**, *11*, 882.

18. Yao, C.; Rudnitzki, F.; Hüttmann, G.; Zhang, Z.; Rahmanzadeh, R. Important factors for cell-membrane permeabilization by gold nanoparticles activated by nanosecond-laser irradiation. *Int. J. Nanomed.* **2017**, *12*, 5659–5672. [CrossRef]

19. Heinemann, D.; Schomaker, M.; Kalies, S.; Schieck, M.; Carlson, R.; Escobar, H.M.; Ripken, T.; Meyer, H.; Heisterkamp, A. Gold Nanoparticle Mediated Laser Transfection for Efficient siRNA Mediated Gene Knock Down. *PLoS ONE* **2013**, *8*, e58604. [CrossRef]

20. Heinemann, D.; Kalies, S.; Schomaker, M.; Ertmer, W.; Escobar, H.M.; Meyer, H.; Ripken, T. Delivery of proteins to mammalian cells via gold nanoparticle mediated laser transfection. *Nanotechnology* **2014**, *25*, 245101. [CrossRef]

21. Pedrosa, P.; Mendes, R.; Cabral, R.; Martins, L.M.D.R.S.; Baptista, P.V.; Fernandes, A.R. Combination of chemotherapy and Au-nanoparticle photothermy in the visible light to tackle doxorubicin resistance in cancer cells. *Sci. Rep.* **2018**, *8*, 11429. [CrossRef]

22. Sun, W.; Luo, Z.; Lee, J.; Kim, H.-J.; Lee, K.; Tebon, P.; Feng, Y.; Dokmeci, M.R.; Sengupta, S.; Khademhosseini, A. Organ-on-a-Chip for Cancer and Immune Organs Modeling. *Adv. Healthc. Mater.* **2019**, *8*, e1801363. [CrossRef]

23. Saglam-Metiner, P.; Gulce-Iz, S.; Biray-Avci, C. Bioengineering-inspired three-dimensional culture systems: Organoids to create tumor microenvironment. *Gene* **2019**, *686*, 203–212. [CrossRef] [PubMed]

24. Ben-David, U.; Ha, G.; Tseng, Y.-Y.; Greenwald, N.F.; Oh, C.; Shih, J.; McFarland, J.M.; Wong, B.; Boehm, J.S.; Beroukhim, R.; et al. Patient-derived xenografts undergo mouse-specific tumor evolution. *Nat. Genet.* **2017**, *49*, 1567–1575. [CrossRef] [PubMed]

25. Lee, J.M.; Mhawech-Fauceglia, P.; Lee, N.; Parsanian, L.C.; Lin, Y.G.; Gayther, S.A.; Lawrenson, K. A three-dimensional microenvironment alters protein expression and chemosensitivity of epithelial ovarian cancer cells in vitro. *Lab. Investig.* **2013**, *93*, 528–542. [PubMed]

26. Jensen, C.; Teng, Y. Is It Time to Start Transitioning From 2D to 3D Cell Culture? *Front. Mol. Biosci.* **2020**, *7*, 33. [CrossRef]

27. Brancato, V.; Oliveira, J.M.; Correlo, V.M.; Reis, R.L.; Kundu, S.C. Could 3D models of cancer enhance drug screening? *Biomaterials* **2020**, *232*, 119744. [CrossRef]

28. Costa, E.C.; Moreira, A.F.; De Melo-Diogo, D.; Gaspar, V.M.; Carvalho, M.P.; Correia, I.J. 3D tumor spheroids: An overview on the tools and techniques used for their analysis. *Biotechnol. Adv.* **2016**, *34*, 1427–1441. [CrossRef]

29. Baek, N.; Seo, O.W.; Kim, M.; Hulme, J.; An, S.S.A. Monitoring the effects of doxorubicin on 3D-spheroid tumor cells in real-time. *OncoTargets Ther.* **2016**, *9*, 7207–7218. [CrossRef]

30. Shi, H.; Wang, Z.; Huang, C.; Gu, X.; Jia, T.; Zhang, A.; Wu, Z.; Zhu, L.; Luo, X.; Zhao, X.; et al. A Functional CT Contrast Agent for In Vivo Imaging of Tumor Hypoxia. *Small* **2016**, *12*, 3995–4006. [CrossRef]

31. Shannon, A.M.; Bouchier-Hayes, D.J.; Condron, C.M.; Toomey, D. Tumour hypoxia, chemotherapeutic resistance and hypoxia-related therapies. *Cancer Treat. Rev.* **2003**, *29*, 297–307. [CrossRef]

32. Soenen, S.J.; Manshian, B.B.; Abdelmonem, A.; Montenegro, J.-M.; Tan, S.; Balcaen, L.; Vanhaecke, F.; Brisson, A.R.; Parak, W.J.; De Smedt, S.C.; et al. The Cellular Interactions of PEGylated Gold Nanoparticles: Effect of PEGylation on Cellular Uptake and Cytotoxicity. *Part. Part. Syst. Charact.* **2014**, *31*, 794–800. [CrossRef]

33. Huang, Y.-C.; Yang, Y.-C.; Yang, K.-C.; Shieh, H.-R.; Wang, T.-Y.; Hwu, Y.; Chen, Y.-J. Pegylated Gold Nanoparticles Induce Apoptosis in Human Chronic Myeloid Leukemia Cells. *BioMed Res. Int.* **2014**, *2014*, 1–9. [CrossRef] [PubMed]

34. Bhamidipati, M.; Fabris, L. Multiparametric Assessment of Gold Nanoparticle Cytotoxicity in Cancerous and Healthy Cells: The Role of Size, Shape, and Surface Chemistry. *Bioconjugate Chem.* **2017**, *28*, 449–460. [CrossRef]

35. Leite, P.E.C.; Pereira, M.R.; Harris, G.; Pamies, D.; Dos Santos, L.M.G.; Granjeiro, J.M.; Hogberg, H.T.; Hartung, T.; Smirnova, L. Suitability of 3D human brain spheroid models to distinguish toxic effects of gold and poly-lactic acid nanoparticles to assess biocompatibility for brain drug delivery. *Part. Fibre Toxicol.* **2019**, *16*, 1–20. [CrossRef] [PubMed]

36. Shah, S.; Chandra, A.; Kaur, A.; Sabnis, N.; Lacko, A.; Gryczynski, Z.; Fudala, R.; Gryczynski, I. Fluorescence properties of doxorubicin in PBS buffer and PVA films. *J. Photochem. Photobiol. B Biol.* **2017**, *170*, 65–69. [CrossRef] [PubMed]

37. Silva, J.; Luís, D.; Santos, S.; Mendo, A.S.; Coito, L.; Silva, T.F.; Da Silva, M.F.C.G.; Martins, L.M.; Pombeiro, A.J.; Borralho, P.M.; et al. Biological characterization of the antiproliferative potential of Co(II) and Sn(IV) coordination compounds in human cancer cell lines: A comparative proteomic approach. *Drug Metab. Drug Interact.* **2013**, *28*, 167–176. [CrossRef]

38. Mellor, H.R.; Callaghan, R. Accumulation and distribution of doxorubicin in tumour spheroids: The influence of acidity and expression of P-glycoprotein. *Cancer Chemother. Pharmacol.* **2011**, *68*, 1179–1190. [CrossRef]

39. Tchoryk, A.; Taresco, V.; Argent, R.H.; Ashford, M.B.; Gellert, P.R.; Stolnik, S.; Grabowska, A.M.; Garnett, M.C. Penetration and Uptake of Nanoparticles in 3D Tumor Spheroids. *Bioconjugate Chem.* **2019**, *30*, 1371–1384. [CrossRef]

40. Turkevich, J.; Stevenson, P.C.; Hillier, J. A study of the nucleation and growth processes in the synthesis of colloidal gold. *Discuss. Faraday Soc.* **1951**, *11*, 55–75. [CrossRef]

41. Chiaraviglio, L.; Kirby, J.E. Evaluation of Impermeant, DNA-Binding Dye Fluorescence as a Real-Time Readout of Eukaryotic Cell Toxicity in a High Throughput Screening Format. *ASSAY Drug Dev. Technol.* **2014**, *12*, 219–228. [CrossRef]

42. Bittremieux, M.; Gerasimenko, J.V.; Schuermans, M.; Luyten, T.; Stapleton, E.; Alzayady, K.J.; De Smedt, H.; Yule, D.I.; Mikoshiba, K.; Vangheluwe, P.; et al. DPB162-AE, an inhibitor of store-operated Ca2+ entry, can deplete the endoplasmic reticulum Ca2+ store. *Cell Calcium* **2017**, *62*, 60–70. [CrossRef]

Publisher's Note: MDPI stays neutral with regard to jurisdictional claims in published maps and institutional affiliations.

International Journal of
Molecular Sciences

MDPI

Article

Personalised Medicine for Colorectal Cancer Using Mechanism-Based Machine Learning Models

Annabelle Nwaokorie and Dirk Fey *

Systems Biology Ireland, School of Medicine, University College Dublin, Belfield, Dublin 4, Ireland;
annabelle.nwaokorie@ucdconnect.ie
* Correspondence: dirk.fey@ucd.ie

Abstract: Gaining insight into the mechanisms of signal transduction networks (STNs) by using critical features from patient-specific mathematical models can improve patient stratification and help to identify potential drug targets. To achieve this, these models should focus on the critical STNs for each cancer, include prognostic genes and proteins, and correctly predict patient-specific differences in STN activity. Focussing on colorectal cancer and the WNT STN, we used mechanism-based machine learning models to identify genes and proteins with significant associations to event-free patient survival and predictive power for explaining patient-specific differences of STN activity. First, we identified the WNT pathway as the most significant pathway associated with event-free survival. Second, we built linear-regression models that incorporated both genes and proteins from established mechanistic models in the literature and novel genes with significant associations to event-free patient survival. Data from The Cancer Genome Atlas and Clinical Proteomic Tumour Analysis Consortium were used, and patient-specific STN activity scores were computed using PROGENy. Three linear regression models were built, based on; (1) the gene-set of a state-of-the-art mechanistic model in the literature, (2) novel genes identified, and (3) novel proteins identified. The novel genes and proteins were genes and proteins of the extant WNT pathway whose expression was significantly associated with event-free survival. The results show that the predictive power of a model that incorporated novel event-free associated genes is better compared to a model focussing on the genes of a current state-of-the-art mechanistic model. Several significant genes that should be integrated into future mechanistic models of the WNT pathway are DVL3, FZD5, RAC1, ROCK2, GSK3B, CTB2, CBT1, and PRKCA. Thus, the study demonstrates that using mechanistic information in combination with machine learning can identify novel features (genes and proteins) that are important for explaining the STN heterogeneity between patients and their association to clinical outcomes.

Keywords: cancer; colorectal cancer; signal transduction networks; pathways; event-free survival; biomarkers; WNT pathway; targeted therapy

check for
updates

Citation: Nwaokorie, A.; Fey, D. Personalised Medicine for Colorectal Cancer Using Mechanism-Based Machine Learning Models. *Int. J. Mol. Sci.* **2021**, *22*, 9970. https://doi.org/10.3390/ijms22189970

Academic Editor: Valentina De Falco

Received: 2 August 2021
Accepted: 10 September 2021
Published: 15 September 2021

Publisher's Note: MDPI stays neutral with regard to jurisdictional claims in published maps and institutional affiliations.

1. Introduction

Colorectal cancer (CRC) is the second leading cause of cancer-related death worldwide [1–3]. From the four global consensus molecular subtypes (CMS) of CRC, CMS1–4, no specific drugs have been identified that target a specific CMS [4]. However, there are multiple studies showing that some CMS subtypes benefit from specific therapeutic regimens [5–7]. For CMS2 and 3, both of which are poorly immunogenic in comparison to CMS1 and 4, the only associated treatment is toxic chemotherapy, which results in poor patient prognosis [2,8]. Biomarkers are foundations of clinical and personalised medicine; despite being vital tools in the area of clinical diagnostics, many are based on single time-point measurements and lack the dynamic information that is needed to follow diseases and therapy [9]. Therefore, we hypothesised that by integrating tumour profiling data with dynamic information about signal transduction networks (STNs), mechanism-based machine learning models could aid in the prediction of patient survival and response to therapy, and overall provide insight into the disease and drug-response mechanisms and reveal potential drug targets and novel biomarkers.

STNs are extremely adjustable and dynamic [10]. To better understand these STNs in the context of CRC, knowledge about how active these STNs are and how this activity varies between patients and relates to patient outcomes is desired. To date, numerous CRC STNs have been identified, including TGF-β (transforming growth factor beta), PI3K/Akt (phosphatidylinositol 3-kinase/protein kinase B), TP53 (tumour protein P53), MAPK (microtubule-associated protein kinase), Apoptosis, Cell Cycle (cell-division cycle), mTOR (mammalian target of rapamycin), Notch, and WNT (Wingless-related integration site) [3,11]. Table 1 portrays these critical CRC STNs and their associated functions. This study analyses only five of these critical pathways, including TGF-β, PI3K/Akt, TP53, MAPK, and WNT, because PROGENy (pathway responsive genes for activity inference) is not available for Notch, Apoptosis, mTOR, and Cell Cycle.

Table 1. Critical CRC STNs and the associated key functions. Nine major CRC STNs, WNT, Apoptosis, PI3K-Akt, TP53, MAPK, TGF-Beta, Cell Cycle, mTOR, and Notch are located in the first column. The "features" column represents the number of features within each STN, i.e., the number of genes in each pathway. The correlated key functions of each pathway in CRC are briefly explained in the third column [12–23].

CRC Signalling Pathways	Features	Key Functions in Colorectal Cancer
WNT	119	**Function:** Normal activation leads to tumour growth in advanced CRC. This function depends on the amount of B-catenin in the cytoplasm **Cellular Activities:** Cell fate specification, proliferation, migration, and asymmetric cell division **Activated:** WNT signal or APC mutation
PI3K/Akt	340	**Function:** Oncogenic role in the initiation and progression of CRC **Inhibition:** Reduction in CRC cell growth and increase in apoptosis **Cellular Activities:** Cell growth, proliferation, differentiation, and migration **Activated:** EGFR (epidermal growth factor receptor) Signalling • Akt is a downstream effector of PI3K, mediating effects on tumour growth and progression
MAPK	252	**Function:** Oncogenic role in CRC associated with tumour growth and disease progression **Cellular Activities:** Cell growth, differentiation, and survival
TGFß	135	**Function:** Reduces colon epithelial cells proliferation and induces apoptosis and differentiation **Cellular Activities:** Cell proliferation, differentiation, migration, apoptosis, and adhesion. **Activated:** Binding of TGF-β ligands to type II TGF-β receptors • Tumour suppressor in the normal epithelium • Tumour promotor in the last stage of CRC
TP53	38	**Function:** Regulation of the cell cycle, DNA replication and apoptosis
Apoptosis	87	**Function:** Apoptotic cell death induction by two main pathways, intrinsic and extrinsic signalling Resulting in the formation of a death-inducing signalling complex and an apoptosome
Cell Cycle	122	**Function:** Controls cell division
mTOR	51	**Function:** Regulation of cell growth and division **Cellular Activities:** Cell growth, proliferation, and survival
Notch	48	**Function:** Promotes CRC through regulating the cell cycle and cell apoptosis by regulation of p21 and PUMA (p53 upregulated modulator of apoptosis) genes **Cellular Activities:** Normal cell development, differentiation, proliferation, and apoptosis

STNs can be mapped to their most frequent mutations. Interestingly, APC (adenomatous polyposis coli) negatively regulates WNT activity since it is an integral part of the destruction complex, which targets ß-catenin for degradation. Mutations in APC thus lead to hyperactive WNT signalling [18]. Across all CRC STNs, the role of WNT signalling in carcinogenesis has most notably been described for CRC [24]. Roughly all CRCs present hyperactivation of the WNT pathway, which in many cases is believed to be the initiating and driving event [3,25,26]. Despite this, most mechanistic models of the canonical WNT signalling pathway (β-catenin-dependent) are based on solely the intracellular steps, neglecting the extracellular steps [27–32]. Classically, the term "canonical" for this pathway refers to the pathway components, which lead to the stabilisation of β-catenin in response to WNT ligands. Consequently, the analysis for this study was restricted to this specific WNT sub-pathway, i.e., canonical. In contrast to most studies, Kogan et al. developed a flexible framework for identifying potential drug targets in the WNT STN. This mechanistic model was based on the initial steps of the activation solely and analysed the effects of two extracellular inhibitors, DKK (Dickkopf) and sFRP (secreted frizzled-related protein) [27]. This validated mathematical model, which predicts effective combinational therapy by sFRP and DKK, is the most recent WNT STN model to date and, interestingly has never been validated in human cell lines. Overall, it is not instinctively apparent which components of the WNT STN are best to target for therapeutic intervention or how such interventions should be designed in order to achieve the best clinical outcomes [27]. However, it has emerged that in recent years, using patient-specific mathematical models can improve patient-stratification and help to identify potential drug targets and largely improve patient response to therapy [9,10,33]. Additionally, combining mechanism-based machine learning models of disease dynamics has been proven to enhance the development of novel disease interventions [34]. Concretely, mechanism-based machine learning models are firmly based upon the behaviour and critical features of mechanistic signal transduction networks of interest, whilst formed by classic linear regression. Adapting this, a schematic of the approach is shown in Figure 1.

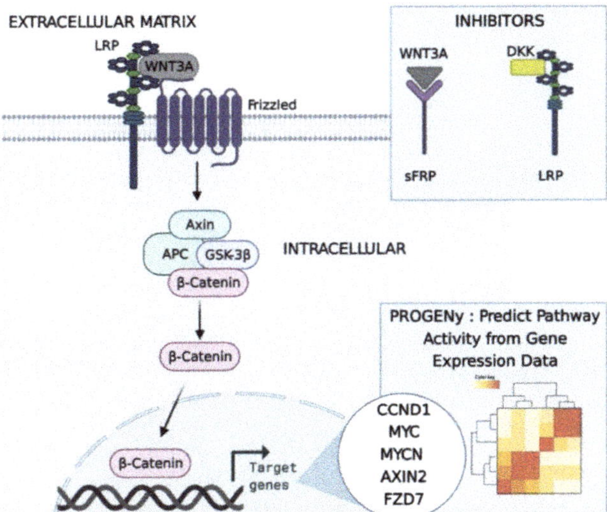

Figure 1. Cascade of events in the WNT Signalling Pathway. Our approach is to combine critical features (genes forming the model) from the most recent mathematical model of the WNT STN, by Kogan et al., with adjustable machine learning models [27]. The schematic represents the initial steps of activation of the WNT pathway. WNT binds to FZD (frizzled), and the co-receptor LRP5/6 (low-density lipoprotein receptor-related protein 5/6), which is recruited in the vicinity of FZD receptors upon WNT binding. This binding prevents the formation of the whole destruction complex consisting of AXIN, APC, and GSK3β (Glycogen Synthase Kinase 3 Beta), thereby causing rising levels of β-catenin, β-catenin translocation into the nucleus, and expression of WNT target genes, CCND1 (Cyclin D1), MYC, MYCN, AXIN2, and FZD7 [27]. Using machine learning tools, such as PROGENy, the WNT pathway activity can be calculated for each individual patient by analysing their gene expression [35,36]. Created with BioRender.com.

In the present study, we developed adjustable linear regression machine learning models of the activation of the canonical WNT signalling pathway and incorporated genes and proteins that were significantly associated with event-free patient survival in CRC. The term "adjustable" in such models represents the capability to vary (include/remove) features with ease to determine which features are most important. The results revealed that the machine learning models can not only help to understand the behaviour of the initial steps of this complex network by identifying novel features not yet included in mechanistic models but also interpret its behaviour by relating pathway activity to clinical outcomes on the level of individual patients.

2. Results

2.1. Determining Significant Genes and Proteins in CRC: A Kaplan–Meier Survival Analysis

In line with the literature, five signalling pathways critical in CRC were selected to analyse, from several network databases, including WikiPathways, Gene Set Enrichment Analysis (GSEA), and KEGG [14,21,22,37,38]. These five signalling pathways comprise WNT, PI3K-Akt, TP53, MAPK, and TGF-Beta. As mentioned above, the motivation behind choosing these five pathways is due to the limited data available to perform a PROGENy analysis. The five STNs analysed are the only CRC STNs available on PROGENy. To determine which genes and proteins are significant in CRC, components of each pathway were used to integrate data from the listed resources. In total, 1156 features were analysed across all pathways. The correlation between patient event-free survival and the expression of proteins and genes from a Kaplan–Meier and log-rank test resulted in 464 significant features. In total, 389 genes and 75 proteins were found to be significant across the eight major CRC signalling pathways. A concrete list of all significant genes and proteins is available in the supplemental data. The associated hazard ratios and *p*-values of each significant feature are denoted below in Tables 2 and 3. All significant genes were obtained from the transcriptomic TCGA (The Cancer Genome Atlas), Firehose legacy dataset, with a cohort of 329/636 patients. The 329/636 TCGA patients selected were solely from the transcriptomic, mRNA (messenger ribonucleic acid) expression dataset; all patients had the necessary clinical data available. Similarly, all significant proteins were obtained from the proteomics TCGA Firehose legacy dataset, a cohort of 73 patients [39,40]. The associated clinical information can be found in the supplemental data.

2.1.1. Significant Genes from RNA Sequencing

The total number of genes whose RNA expression correlated with event-free patient survival in this cohort was 389 of 872 genes across all CRC signalling pathways considered. For most pathways, including the WNT pathway, approximately half of the genes were significant using a fold change and *p*-value cut-off of 0.5 and 0.05, respectively.

Next, we wanted to obtain a more comprehensive picture of the significant genes within a pathway. Using the WNT pathway as an exemplar, the volcano plot in Figure 2 shows the *p*-values of the significant genes in the *y*-axis against their $\log_2$ fold change on the *x*-axis. Genes above the dashed threshold line, to the right, DVL3 (dishevelled segment polarity protein 3), VANGL2 (vang-like protein 2), CER1 (cerberus 1), and TCF7L1 (transcription factor 7-like 1), etc., are found to be significant, with the lowest *p*-values and greatest positive hazard-ratio fold-changes. These genes were associated with increased risk. Conversely, genes to the left above the threshold line, including MAPK8, have a negative hazard-ratio fold change and were thus associated with decreased risk. Overall, 15 genes were found to be significant negative markers, and the remaining 39 genes were significant positive markers determined for the WNT signalling pathway.

Table 2. The top two significant genes across five CRC relevant signalling pathways from the TCGA Legacy dataset. A Kaplan–Meier estimate and log-rank test were used to compute the association between patient event-free survival and gene expressions. The associated hazard ratio, 95% confidence interval, *p*-value, mafdr (estimate positive false discovery rate for multiple hypothesis testing), and standard error are apparent. "Patient" indicates the number of patients for which data were available. The last column, "significant", indicates the number of significant genes out of the total number of genes for this pathway. The fold change and *p*-value cut-off used were 0.5 and 0.05, respectively.

Genes	Hazard Ratio	95% Confidence Interval	*p*-Value	mafdr	Standard Error	Patient	Significant
(1) WNT							53/119
DVL3	2.8158	(1.6159–4.9066)	2.58×10^{-4}	1.37×10^{-2}	0.2833	329	
VANGL2	2.7692	(1.5039–5.0992)	1.08×10^{-3}	1.40×10^{-2}	0.3115	329	
(2) PI3K-Akt							159/340
EFNA1	3.2860	(1.8496–5.8377)	4.96×10^{-5}	9.92×10^{-5}	0.2932	329	
KRAS	0.2853	(0.1502–0.5418)	1.27×10^{-4}	1.26×10^{-4}	0.3272	329	
(3) TP53							11/38
BCL2	0.4610	(0.2960–0.7179)	6.12×10^{-4}	0.0012	0.2260	329	
CDKN2A	1.8156	(1.1931–2.7629)	5.36×10^{-3}	0.0054	0.2142	329	
(4) MAPK							113/252
KRAS	0.2853	(0.1502–0.5418)	1.27×10^{-4}	2.09×10^{-4}	0.3272	329	
CACNA1I	2.7685	(1.6159–4.7432)	2.10×10^{-4}	2.09×10^{-4}	0.2747	313	
(5) TGF-Beta							53/135
TERT	2.8734	(1.6639–4.96201)	1.53×10^{-4}	3.05×10^{-4}	0.2787	328	
TGFB1I1	2.2381	(1.3807–3.6281)	1.08×10^{-3}	0.0011	0.2465	329	

Table 3. The top two significant proteins across five CRC relevant signalling pathways from the TCGA Legacy dataset. A Kaplan–Meier estimate and log-rank test were used to compute the Pearson correlation between patient event-free survival and protein expressions. The associated hazard ratio, 95% confidence interval, *p*-value, mafdr (estimate positive false discovery rate for multiple hypothesis testing), and standard error are apparent. "Patient" indicates the number of patients the corresponding protein was determined in. The last column, "significant", in the format X/Y/Z lists the number of significant proteins in each STN (X), the number of genes within the pathway with available protein information (Y), and the total number of genes in the pathway (Z).

Genes	Hazard Ratio	95% Confidence Intervals	*p*-Value	mafdr	Standard Error	Patient	Significant
(1) WNT							6/27/119
PRKCA	0.1557	(0.0490–0.4947)	1.61×10^{-3}	0.0036	0.5898	63	
ROCK2	5.7810	(1.9229–17.3789)	1.78×10^{-3}	0.0036	0.5616	71	
(2) PI3K-Akt							29/100/340
PPP2R1B	6.8050	(2.2113–20.9407)	8.26×10^{-4}	8.32×10^{-4}	0.5735	73	
EIF4E	7.6285	(2.3168–25.1176)	8.32×10^{-4}	8.32×10^{-4}	0.6080	73	
(3) TP53							2/5/38
BAX	10.8648	(3.3684–35.0445)	6.54×10^{-5}	1.30×10^{-4}	0.5975	71	
BID	0.2650	(0.0814–0.8616)	2.73×10^{-2}	0.0273	0.6016	67	
(4) MAPK							20/80/252
STK3	8.6543	(2.4297–30.8245)	8.69×10^{-4}	0.0092	0.6481	65	
PRKCA	0.1557	(0.0490–0.4947)	1.61×10^{-3}	0.0092	0.5898	63	
(5) TGF-Beta							15/71/135
SMAD2	10.4343	(2.7196–40.0330)	6.30×10^{-4}	0.0013	0.6860	67	
SPTBN1	5.8318	(1.9510–17.4318)	1.60×10^{-3}	0.0016	0.5587	73	

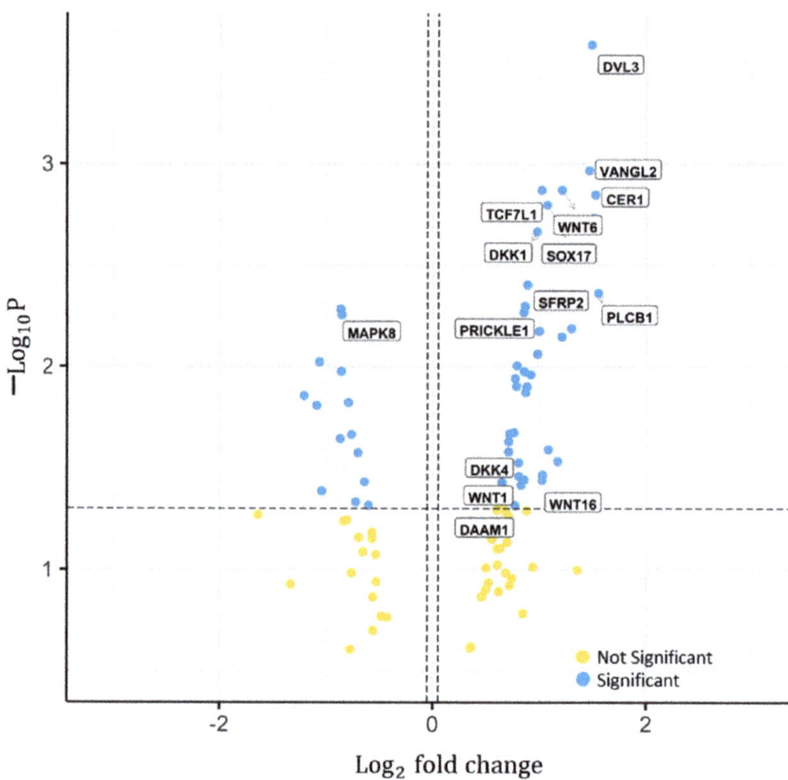

Figure 2. Significant genes in the WNT signalling pathway. Distribution of the *p*-value ($-\log_{10}$P) over the hazard ratio (Log_2 fold change) from the genes analysed from the RNA Sequencing TCGA legacy dataset. Features were divided in two; significant and not significant, represented by the blue and yellow dots, respectively. Labels represent the gene names of some significant features. *p*-value cut-off used = 0.05. Log_2 fold change cut-off used = 0.5. Total number of features in WNT signalling pathway = 119. Total number of significant genes determined = 53.

Plotting the event-free survival Kaplan–Meier curve for DVL3, the most significant gene in the WNT signalling pathway, resulted in a $\log_2$ hazard ratio of 1.49 and a 95% confidence interval between 0.69 and 2.29 (Figure 3). Kaplan–Meier scanning to determine the optimal cut-off divided the patients into high ($n = 26$) and low ($n = 303$) groups by scanning the group size from 10–90 to 90–10 percent splits, where 10–90 means that 10% of the patients were in the low group and 90% of the patients in the high group and calculating the *p*-value for the overall event-free survival difference between the groups using a log-rank test with Yates' correction. A high expression of DVL3 resulted in poor patient survival, where a 2-year event-free survival equals 38% in the high group ($n = 26$). The Kaplan–Meier curves for the most significant gene found for each CRC STN is shown in Figure 3.

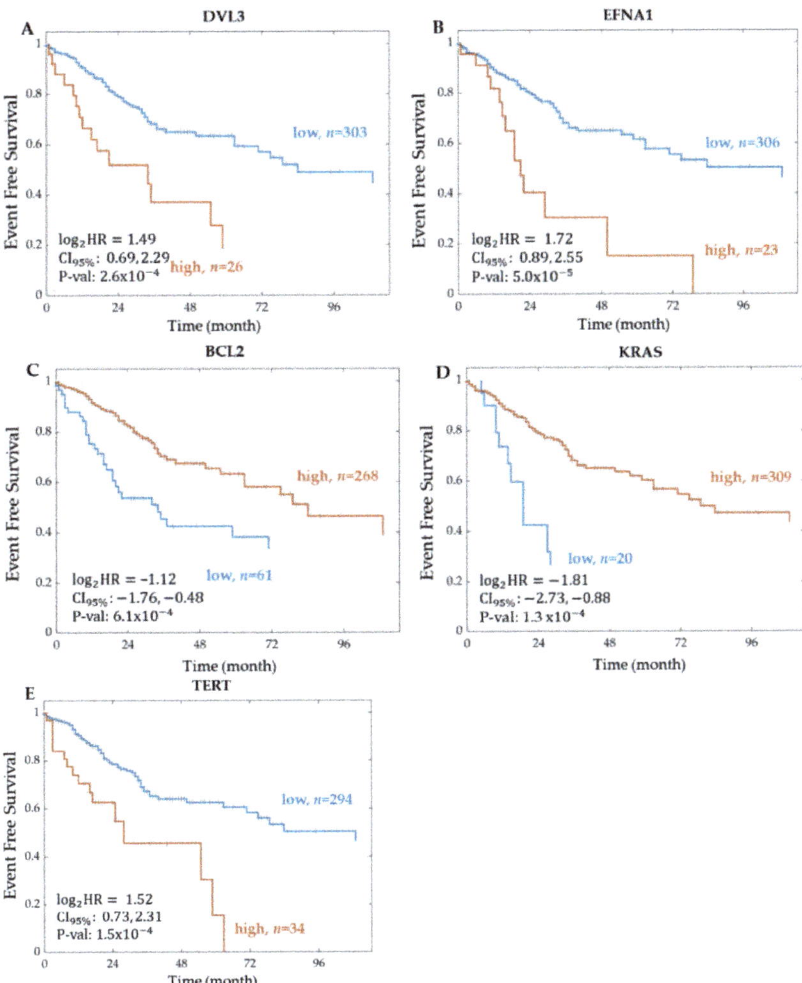

Figure 3. Kaplan–Meier curve of event-free survival for the most significant gene of each CRC STN. The most significant gene in each signalling pathway from the RNA Sequencing TCGA legacy dataset was (**A**) DVL3 for WNT STN, (**B**) EFNA1 for PI3K-Akt STN, (**C**) BCL2 for TP53 STN, (**D**) KRAS for MAPK STN, and (**E**) TERT for TGF-Beta STN. The patients were stratified into two groups according to the expression level of each feature. The optimal cut-off was determined using Kaplan–Meier scanning (see Methods). The groups are represented as high (orange line), and low (blue line), where n indicates the total number of patients in each group. For each group, the Kaplan–Meier curve of event-free survival was tested for statistically significant differences using a log-rank test. $\mathrm{Log_2HR}$ = log base 2 of the hazard ratio. $\mathrm{CI_{95\%}}$ = 95% Confidence Interval. P-val = p-value.

2.1.2. Significant Proteins from Proteomics

Across all of the CRC signalling pathways considered, 75 of 284 proteins exhibited expressions that correlated with event-free patient survival. For each pathway, about 4–10% of the pathway proteins exhibited an association with event-free survival. The top two significant proteins across each CRC signalling pathway are shown in Table 3. PRKCA (protein kinase c alpha) and ROCK2 (rho associated coiled-coil containing protein kinase 2) were found to be the most correlated to the patient's survival in the WNT signalling pathway. PRKCA was found to be the most significant protein with an associated hazard ratio of 0.1557 and a *p*-value of 1.61×10^{-3}. The signalling pathway PI3K-Akt has the greatest number of significant proteins associated, 29 proteins in total.

Six significant proteins within the WNT signalling pathway were found from the proteomics dataset, including PRKCA, ROCK2, CSNK2A1 (casein kinase 2 alpha 1), LRP1 (low-density lipoprotein receptor-related protein 1), CSNK1A1 (casein kinase 1 alpha 1), and GPC4 (glypican 4). This is represented by the volcano plot in Figure 4. In total, 92 proteins in this pathway could not be analysed because of missing values. Significant proteins, ROCK2 and CSNK2A1, have a positive hazard ratio with a $\log_2$ fold change above 1.5. All other significant proteins have a low-risk association with hazard ratios below -1.5. In summary, six proteins were found to be significant, two with an increased risk and four with decreased risk.

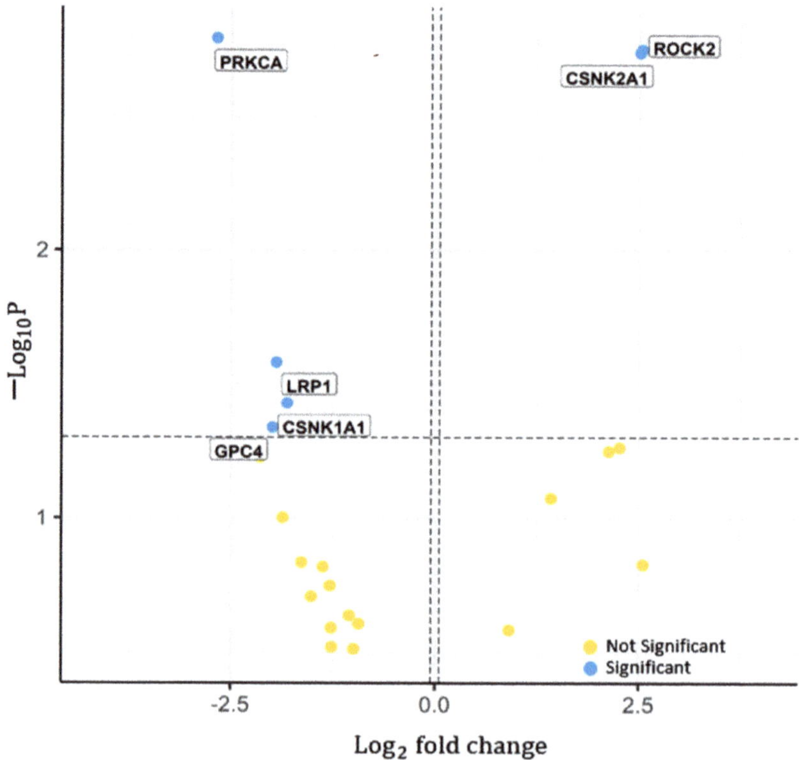

Figure 4. Significant proteins in the WNT signalling pathway. Distribution of the *p*-value ($-\log_{10}\mathrm{P}$) across the hazard ratio ($\log_2$ fold change) from the proteins analysed from proteomics TCGA legacy dataset. Features were divided into two; significant and not significant, represented by the blue and yellow dots, respectively. Labels represent the gene names of some significant features. *p*-value cut-off used = 0.05. $\log_2$ fold change cut-off used = 1. Total number of features in WNT signalling pathway = 119. Total number of significant proteomics determined = 6.

Plotting the event-free survival Kaplan–Meier curve for PRKCA, the most significant protein in the WNT signalling pathway, resulted in a $\log_2$ hazard ratio of -2.68, a confidence interval between -4.35 and -1.02, and a *p*-value of 1.61×10^{-3}. From Kaplan–Meier scanning, 55 patients were in the high expression group, and 8 were in the low expression group. High expression of PRKCA resulted in improved patient survival, with 2-year event-free survival of about 90% in the high group ($n = 55$). The Kaplan–Meier curves for the most significant proteins found for each CRC STN are shown in Figure 5.

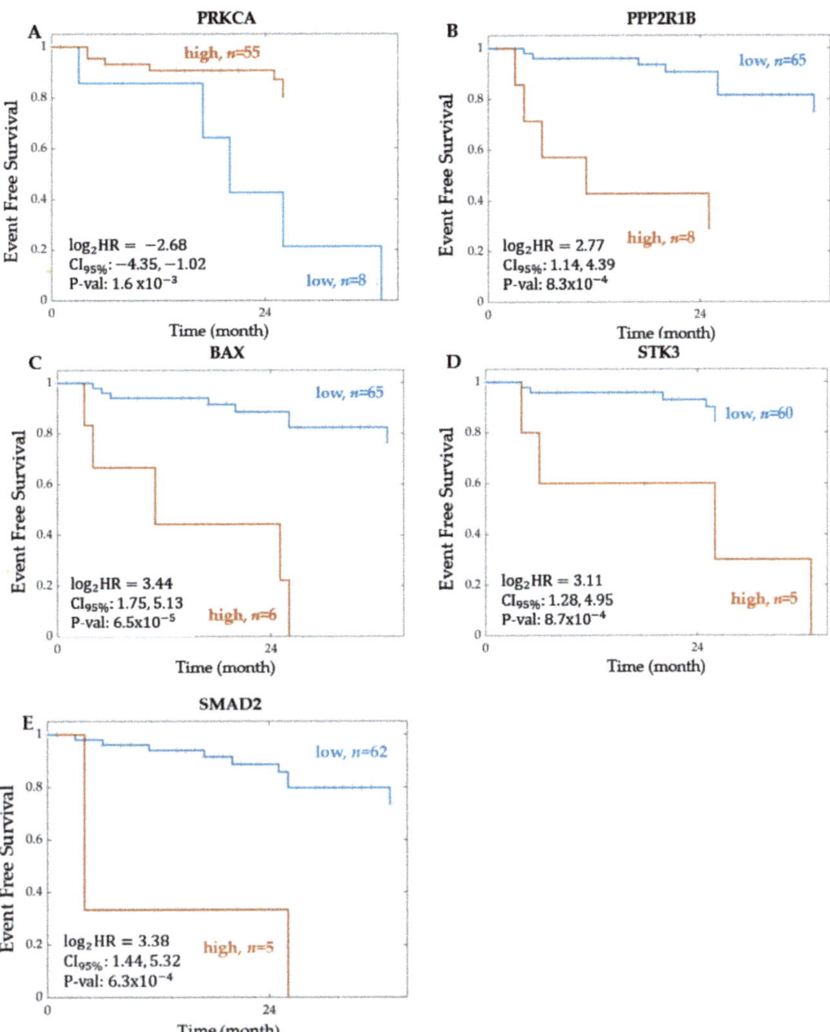

Figure 5. Kaplan–Meier curve of event-free survival for the most significant protein, of each CRC STN. The most significant protein in each signalling pathway from the proteomics TCGA Firehose legacy dataset was (**A**) PRKCA for WNT STN, (**B**) PPP2R1B for PI3K-Akt STN, (**C**) BAX for TP53 STN, (**D**) STK3 for MAPK STN, and (**E**) SMAD2 for TGF-Beta STN. The patients were stratified into two groups according to the expression level of each feature. The optimal cut-off was determined using Kaplan–Meier scanning (see Methods). The groups are represented as high (orange line), and low (blue line), where n indicates the total number of patients in each group. For each group, the Kaplan–Meier curve of event-free survival was tested for statistically significant differences using a log-rank test. Log₂HR = log base 2 of the hazard ratio. CI₉₅% = 95% Confidence Interval. P-val = *p*-value.

2.1.3. Summary of the Event-Free Survival Analysis Results

Across five critical CRC signalling pathways, the results from the WNT signalling pathway introduced several significant genes and proteins that have not yet been accounted for in current WNT mechanistic models. Therefore, the results of this section reinforce the need to propose a strategy to incorporate them into current WNT mechanistic models. To summarise, Table 4 is composed of all 53 significant genes and 6 significant proteins found in the WNT signalling pathway. These results were from the TCGA Legacy, the transcriptomic dataset of 329 CRC patients, and the proteomic from mass spectroscopy dataset of 73 patients. Interestingly, some significant genes (DVL3, DKK1, SFPR2, WNT3A, WNT3, SFPR1, and GSK3B) are found in published mechanistic models of the WNT pathway [27–32]. The remaining 52 significant features found in this study are not found in these mechanistic models. Thus, we have identified many genes that were associated with event-free survival and that constitute prime candidates to improve current WNT mechanistic models. These improved models could then be used for patient stratification and predicting response to therapy [9,33].

Table 4. Significant genes and proteins in the WNT signalling pathway from TCGA legacy datasets. (**A**) List of significant genes and their associated hazard ratio, *p*-value, and mafdr (estimate positive false discovery rate for multiple hypothesis testing). Hazard ratios smaller than one (between 0 and 1) indicate a negative association (decreased risk), hazard ratios of >1 indicate a positive association (increased risk). The last column, proteomics, indicates whether proteomics data for this gene were available. If yes, a Y is depicted. (**B**) List of significant proteins and their associated hazard ratio, *p*-value, and mafdr. Significant features highlighted in blue were found in published mechanistic models of the WNT pathway [27–32].

A	Genes	Hazard Ratio	*p*-Value	mafdr	Patient	Proteomics
1	DVL3	2.8158	2.58×10^{-4}	1.37×10^{-2}	329	
2	VANGL2	2.7693	1.08×10^{-3}	1.40×10^{-2}	329	
3	WNT6	2.3175	1.34×10^{-3}	1.40×10^{-2}	299	
4	TCF7L1	2.0324	1.35×10^{-3}	1.40×10^{-2}	329	
5	CER1	2.8871	1.42×10^{-3}	1.40×10^{-2}	147	
6	SOX17	2.1048	1.59×10^{-3}	1.40×10^{-2}	329	
7	NKD2	2.8598	1.85×10^{-3}	1.40×10^{-2}	329	
8	DKK1	1.9732	2.16×10^{-3}	1.43×10^{-2}	309	
9	SFRP2	1.8539	3.94×10^{-3}	2.09×10^{-2}	327	Y
10	PLCB1	2.9436	4.32×10^{-3}	2.09×10^{-2}	329	
11	PRICKLE1	1.8229	5.01×10^{-3}	2.09×10^{-2}	329	
12	MAPK8	0.5491	5.19×10^{-3}	2.09×10^{-2}	329	Y
13	PRICKLE2	1.8104	5.38×10^{-3}	2.09×10^{-2}	329	
14	MYC	0.5535	5.53×10^{-3}	2.09×10^{-2}	329	
15	WIF1	2.4639	6.46×10^{-3}	2.20×10^{-2}	246	
16	CAMK2B	1.9981	6.66×10^{-3}	2.20×10^{-2}	302	
17	WNT3A	2.3137	7.09×10^{-3}	2.21×10^{-2}	243	
18	LEF1	1.9751	8.65×10^{-3}	2.52×10^{-2}	329	
19	RHOA	0.4792	9.47×10^{-3}	2.52×10^{-2}	329	Y
20	MAPK10	1.7294	9.88×10^{-3}	2.52×10^{-2}	329	

Table 4. *Cont.*

A	Genes	Hazard Ratio	p-Value	mafdr	Patient	Proteomics
21	CAMK2D	0.5514	1.05×10^{-2}	2.52×10^{-2}	329	Y
22	WNT3	1.8141	1.06×10^{-2}	2.52×10^{-2}	328	
23	FZD8	1.8913	1.09×10^{-2}	2.52×10^{-2}	329	
24	SERPINF1	1.7116	1.14×10^{-2}	2.52×10^{-2}	329	Y
25	FZD1	1.7251	1.24×10^{-2}	2.55×10^{-2}	329	
26	NKD1	1.8480	1.25×10^{-2}	2.55×10^{-2}	329	
27	CXXC4	1.8290	2.62×10^{-2}	1.34×10^{-2}	318	
28	PPP3CA	0.4335	2.62×10^{-2}	1.39×10^{-2}	329	Y
29	PLCB3	0.5772	2.74×10^{-2}	1.50×10^{-2}	329	
30	CSNK1A1L	0.4710	2.74×10^{-2}	1.55×10^{-2}	321	
31	PLCB2	1.6955	3.46×10^{-2}	2.10×10^{-2}	329	
32	ROR2	1.6511	3.46×10^{-2}	2.14×10^{-2}	329	
33	PRKCB	0.5888	3.46×10^{-2}	2.15×10^{-2}	329	
34	WNT2	0.5479	3.52×10^{-2}	2.26×10^{-2}	329	
35	SFRP1	1.6386	3.53×10^{-2}	2.33×10^{-2}	322	
36	WNT10A	2.1170	3.69×10^{-2}	2.56×10^{-2}	328	
37	GSK3B	1.6397	3.69×10^{-2}	2.61×10^{-2}	329	Y
38	FRAT1	0.6144	3.69×10^{-2}	2.65×10^{-2}	329	
39	SENP2	2.2535	3.84×10^{-2}	2.92×10^{-2}	329	
40	DKK4	1.7497	3.84×10^{-2}	2.96×10^{-2}	301	
41	CTNNBIP1	1.5838	3.84×10^{-2}	2.97×10^{-2}	329	
42	FZD7	2.0424	4.18×10^{-2}	3.40×10^{-2}	329	
43	WNT1	1.7533	4.18×10^{-2}	3.46×10^{-2}	244	
44	DKK2	1.8152	4.18×10^{-2}	3.59×10^{-2}	326	
45	WNT16	2.0367	4.18×10^{-2}	3.62×10^{-2}	276	
46	WNT2B	0.6417	4.18×10^{-2}	3.68×10^{-2}	329	
47	NFATC4	1.5715	4.18×10^{-2}	3.70×10^{-2}	329	
48	NLK	1.7784	4.23×10^{-2}	3.83×10^{-2}	329	
49	NOTUM	1.5754	4.34×10^{-2}	4.08×10^{-2}	329	
50	CSNK2A3	0.4868	4.34×10^{-2}	4.09×10^{-2}	329	
51	WNT4	0.6060	4.82×10^{-2}	4.64×10^{-2}	329	
52	FZD5	0.6579	4.82×10^{-2}	4.81×10^{-2}	329	
53	DAAM1	1.7076	4.82×10^{-2}	4.82×10^{-2}	329	
B	Proteins	Hazard Ratio	p-Value	mafdr	Patient	
1	PRKCA	0.1557	1.62×10^{-3}	3.66×10^{-3}	63	
2	ROCK2	5.7810	1.78×10^{-3}	3.66×10^{-3}	71	
3	CSNK2A1	5.6824	1.83×10^{-3}	3.66×10^{-3}	73	
4	LRP1	0.2594	2.62×10^{-2}	3.93×10^{-2}	72	
5	CSNK1A1	0.2846	3.71×10^{-2}	4.46×10^{-2}	72	
6	GPC4	0.2512	4.58×10^{-2}	4.58×10^{-2}	50	

2.2. Predicting Pathway Responsive Genes for Activity Interference from Gene Expression: A PROGENy Analysis

Having identified the significant genes and proteins within the pathways, we now sought to obtain activity scores for each pathway. The purpose was two-fold. Firstly, to identify pathway activities that correlate with event-free patient survival; and secondly, to relate the genes and proteins expression of a pathway to its activity score. The RStudio package function PROGENy was used to obtain pathway scores from the RNA sequencing data of the TCGA legacy dataset. Albeit PROGENy's composition of 14-cancer relevant pathways, specifically for this study, only the CRC pathways were analysed in depth. The only available CRC pathways in PROGENy were PI3K, MAPK, TGF-Beta, WNT, and p53.

The PROGENy activity scores for each CRC signalling pathway are shown in Figure 6. Although the TGF-Beta and PI3K appear to have an inverse relationship, most pathways exhibit a cluster of high activity scores in a small subset of patients. However, specific activity patterns are difficult to discern.

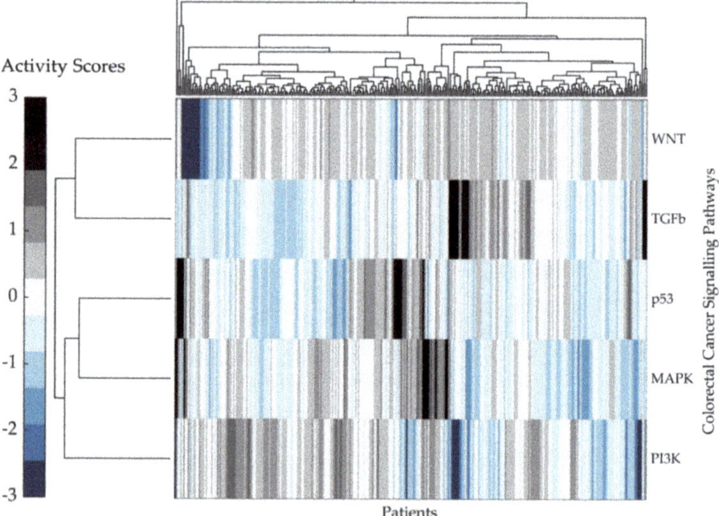

Figure 6. PROGENy pathway activity scores from the RNA Sequencing expression dataset. Heatmap visualising the z-coefficients matrix for all features in each CRC signalling pathway for all 329 patients within the RNA sequencing TCGA legacy dataset cohort. The colourmap on the left indicates the activity score values ranging from 3–−3. The Euclidean distance metric was used to pass the pairwise distance between observations between both rows and columns. The complete linkage method was used to create the hierarchical cluster tree. Data were clustered along the columns of data, then along the rows of row-clustered data.

Kaplan–Meier Survival Analysis on PROGENy Activity Scores

To investigate the correlation between the PROGENy pathway activity scores and patient event-free survival, a Kaplan–Meier analysis and log-rank test were used. The activities of the WNT, PI3K, TGF-Beta, and MAPK pathways were all significantly correlated with event-free survival (Table 5). In particular, we found that activity of the WNT STN exhibited the strongest association with event-free patient survival (hazard ratio and p-value of 1.9731 and 0.0013, respectively).

Table 5. The association between PROGENy pathway activity and patient event-free survival. PROGENy scores were analysed using Kaplan–Meier analysis and the log-rank test to obtain hazard ratios and *p*-values.

Pathway	Hazard Ratio	*p*-Value
WNT	1.9731	0.0013
PI3K	0.5775	0.0276
TGFβ	1.5792	0.0477
MAPK	1.8058	0.0489
p53	0.5972	0.0687

The event-free survival Kaplan–Meier curves for the CRC STNs MAPK, TGF-Beta, PI3K, p53, and WNT, associated with patient survival through PROGENy activity scores, are shown in Figure 7. The STNs MAPK, TGF-Beta, and WNT followed a similar trend, where a high pathway activation was associated with shorter event-free survival. From the literature, this is expected considering that both pathways, MAPK and WNT, drive cell proliferation. Activation of the WNT pathway in CRC increases the levels of β-catenin within the cytosol, causing it to further travel into the nucleus and express WNT target genes, including genes that control the cell cycle [41]. For the WNT signalling pathway, the $\log_2$ hazard ratio was 0.98 with a 95% confidence interval between 0.38 and 1.58 and a *p*-value of 1.3×10^{-3}. The high group characterised by high WNT activity consisted of 108 patients and the low group of 221 patients. Similarly, the results for TGF-Beta are in line with its immunosuppressive function. Conversely, for the other STNs, p53 and PI3K, a high pathway activity was associated with longer event-free survival. The p53 result is in line with its known functions as a tumour-suppressor. The PI3K result is counterintuitive considering that the PI3K/AKT pathway is a classical survival signalling pathway but might be a confounding factor explained by "side-effect" activation due to crosstalk.

2.3. Developing Linear Regression Machine Learning Models

Having ascertained the significant genes, proteins, and pathways, we next sought to determine how well the expression of genes and proteins of a pathway can predict its activity scores. The results should give valuable insight into which candidate genes and proteins should be prioritised in follow-up studies focusing on constructing patient-specific mechanistic models with significant value. The WNT STN was the most significant pathway (Table 5, Figure 7); thus, we focused solely on it. Because there was a limited overlap between the mRNA and proteomics data in the TCGA legacy dataset (only five patients had data for both), we focused the analysis on the CPTAC (Clinical Proteomic Tumor Analysis Consortium) dataset. Both mRNA and proteomics data were available for the entire CPTAC cohort of 79 patients.

In total, three machine learning models were developed to predict the PROGENY pathway scores using different features as inputs. The features were the genes and proteins of the WNT pathway and their associated gene or protein expression values. To minimise the risk of overfitting, 10-fold cross-validation was used to build each model. The predictive power of each model was judged based on the root mean squared error (RMSE), the correlation coefficient (R), and the *p*-value. A *p*-value less than 0.05 was taken as significant for this field of study. The RMSE is an appropriate measure of predictive power for this study because model complexity was not an issue, and RMSE measures the differences between predicted and actual PROGENy values. There is no absolute threshold; however, lower values indicate a better overall fit of how accurately the model predicts the response. An acceptable range for the RSME values is difficult to define, but Model 1, consisting of the genes of the current mechanistic model, can act as a baseline benchmark. Notably, the predicted values of Model 2 show a much better correlation with the true values and a lower RMSE value compared to Model 1. (Figure 8B); thus, demonstrating a much better fit. We focus our analysis on three models:

- Model 1: Features are eleven genes taken from the most recent WNT mechanistic model developed by Kogan et al.: APC, AXIN1, CTNNB1, DKK2, DVL3, GSK3B, LRP6, SFRP1, SFRP2, and WNT3.
- Model 2: Features are the mRNA expression of nine genes: seven features selected using LASSO and the two most significant genes (DVL3, VANGL2): DKK3, FZD5, NKD1, NOTUM (notum, palmitoleoyl-protein carboxylesterase), WNT11, PRKCA, and ROCK2.
- Model 3: Features are the protein expression of seven proteins; five identified using LASSO and the two most significant proteins (PRKCA, ROCK2): CTBP1 (c-terminal binding protein 1), CTBP2 (c-terminal binding protein 2), GPC4, PLCB4 (phospholipase c beta 4), and RAC1 (ras-related C3 botulinum toxin substrate 1).

Model 1 was developed based on the mechanistic WNT signalling pathway. The selected features were the eleven genes of the mechanistic models in the literature [27–32], using the mRNA expression data from the CPTAC dataset (Table 6). The gene DKK1 is typically found in the mechanistic WNT pathway; however, it was not available in the RNA CPTAC dataset; therefore, its paralogue, DKK2, replaced it [42]. Models 2 and 3 were developed using systematic feature selection based on the entire set of 119 WNT pathway genes. LASSO regression was used to identify the features based on the mRNA expression data (Model 2) or protein expression data (Model 3). To complete the model, the two most significant genes (Model 2) or proteins (Model 3) were also included as features. LASSO regression on the mRNA data of the 119 WNT network genes selected seven genes as features. Thus, Model 2 consisted of nine features: seven identified by LASSO plus the two most significant genes. LASSO regression on the protein data of the 27 proteins of the WNT network for which data were available selected five proteins as features. Thus, Model 3 consisted of seven features: five identified by LASSO plus the two most significant proteins.

How well can the mechanism-based, mRNA-based, and protein-based models predict the pathway scores? Model 1, based on the genes of the mechanistic models, serves as a benchmark and achieved a reasonable predictive power with a Spearman correlation coefficient of 0.5241, a p-value of 7.12×10^{-7}, and an RMSE of 1.0167. Model 2, based on LASSO identified genes from the canonical WNT pathway, resulting in a greater predictive power, compared to Model 1, with a Spearman correlation coefficient of 0.8367, a p-value of 7.88×10^{-22}, and an RMSE of 0.6416. Despite the smaller number of input features for Model 2, compared to Model 1, Model 2 evidentially has greater predictive power, shown by the higher p-value, correlation coefficient, and lower RMSE. Similarly to Model 2, Model 3 used LASSO to identify proteins from the canonical WNT pathway. In total, data for only 27 proteins were available due to missing data and imputation (see materials and methods). Model 3 is the least predictive model. The associated Spearman correlation coefficient, RMSE, and p-value were 0.4095, 1.0509, and 1.78×10^{-4}, respectively. Most features used in these machine learning models have nonzero regression coefficients (Figure 8D–F).

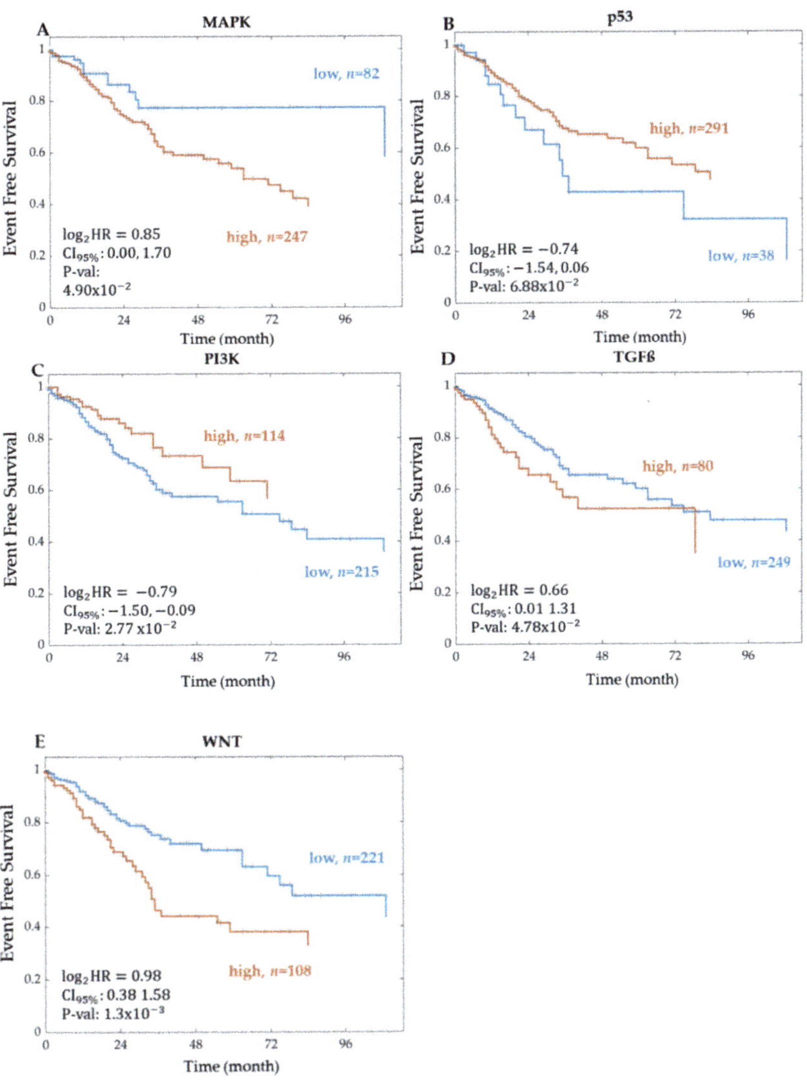

Figure 7. Kaplan–Meier curves of event-free survival associated with CRC PROGENy pathway activity scores. Kaplan–Meier curves associated with PROGENy pathway activity scores for the signalling pathways (**A**) MAPK, (**B**) TGFß, (**C**) PI3K, (**D**) p53, and (**E**) WNT. The patients were stratified into two groups according to the expression level of each feature. The groups are represented as high (orange line) and low (blue line), where *n* indicates the total number of patients in each group. For each group, the Kaplan–Meier curve of event-free survival was tested for statistically significant differences using a log-rank test. Log_2HR = log base 2 of the hazard ratio. $CI_{95\%}$ = 95% Confidence Interval. P-val = *p*-value. RNA Sequencing data is from the TCGA legacy dataset.

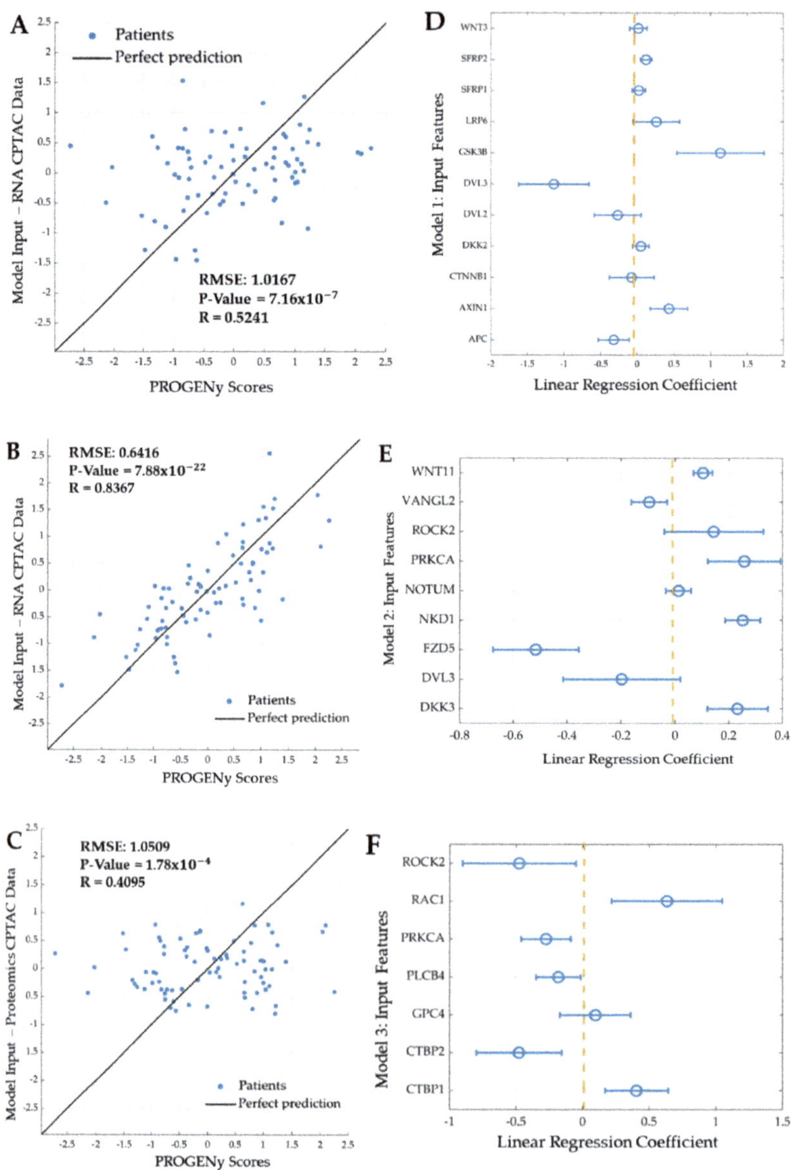

Figure 8. Linear regression machine learning models of the mechanistic and canonical WNT signalling pathways. Scatter plot visualising the correlation between the predicted score from (**A**) Model 1, using the 11 features of the mechanistic model (*y*-axis), and the PROGENy scores (*x*-axis). (**B**) Model 2, using the nine features of the canonical WNT model and significant genes (*y*-axis), and the PROGENy scores (*x*-axis). (**C**) Model 3, using the seven features of the canonical WNT model and significant proteins (*y*-axis), and the PROGENy scores (*x*-axis). Each blue dot represents 1 of 79 patients in the cohort. The black line indicates perfect predictions. "R" denotes the Pearson correlation coefficient, and "RMSE" is the root mean square error. The second row of figures represents the corresponding linear regression correlation coefficients for Models 1–3. Presented are the regression coefficients for the (**D**) Model 1, (**E**) Model 2, and (**F**) Model 3. The features used for each machine learning model are shown on the vertical axis, with the associated correlation coefficients shown on the horizontal axis. The blue dot represents the regression correlation coefficient, and the blue line (error bar) is the associated standard error. The dashed vertical yellow line represents the midline for the linear regression coefficients at 0.

Table 6. Overview of the significant genes and proteins and the linear regression machine learning models. (**A**) List of the top six significant features from RNA sequencing and proteomics from TCGA Legacy datasets. (**B**) Model 1: An overview of the features used for the Mechanistic WNT pathway model. Models 2 and 3: An overview of the features used for the canonical WNT pathway, from CPTAC RNA and proteomics datasets, respectively. The number of features used and associated RMSE for each model is listed in the first column. "RMSE" = root mean squared error. Significant features are highlighted in blue.

(A) Top Six Prognostic Genes and Proteins	Features
mRNA Sequencing TCGA Legacy	DVL3
	VANGL2
	WNT6
	TCF7L1
	CER1
	SOX17
Proteomics TCGA Legacy	PRKCA
	ROCK2
	CSNK2A1
	LRP1
	CSNK1A1
	GPC4

(B) Linear Regression Models	Features
Model 1: Mechanistic Model Model Input: CPTAC RNA RMSE: 1.0167 11 Features	APC
	AXIN1
	CTNNB1
	DKK2
	DVL2
	DVL3
	GSK3B
	LRP6
	SFRP1
	SFRP2
	WNT3
Model 2: Canonical WNT model with top two prognostic genes and proteins Model Input: CPTAC RNA RMSE: 0.6416 9 Features	DKK3
	FZD5
	NKD1
	NOTUM
	WNT11
	DVL3
	PRKCA
	ROCK2
	VANGL2
Model 3: Canonical WNT model with top two prognostic proteins Model Input: CPTAC proteomics RMSE: 1.0509 7 Features	CTBP1
	CTBP2
	GPC4
	PLCB4
	PRKCA
	RAC1
	ROCK2

2.4. Summary

By incorporating the results from the event-free survival analysis, the PROGENy analysis, and the development of the linear regression machine learning models together, it is apparent that each finding paves the way to a valuable tool for predicting the heterogeneity of the WNT STN activity on an individual patient level. Concretely, each finding from each analysis builds sequentially upon another. The event-free survival analysis brings forth critical and novel features for the WNT pathway that have not yet been accounted for mechanistically. Interesting, from the PROGENy analysis, we found that the WNT signalling pathway was the most significant pathway associated with event-free patient survival. This finding solidifies the idea that there is a relationship between this pathway and patient survival that should be continued to be researched mechanistically. Finally, the motivation for using linear regression machine learning was particularly because it is an appropriate tool to predict WNT pathway activity from specified features (i.e., prognostic genes and proteins) on an induvial patient level. Precisely, the machine learning models additionally identified several features that were not only significantly associated with event-free patient survival, but also important for predicting the patient-specific WNT pathway activity scores. Thus, by incorporating such features into an updated mechanistic model of the canonical WNT STN activation, one would expect to better understand the patient-specific differences in the control of pathway activity, but also to relate pathway activity to clinical outcomes on the level of individual patients.

3. Discussion

Our experimental workflow employed numerous Kaplan–Meier survival scans, a PROGENy analysis, and developed three machine learning models based on mechanistic WNT pathway models and LASSO identified genes and proteins of the extant WNT network. The models can serve as a platform for improving current mechanistic models and WNT pathway activity-based patient stratification. Conversely, all models of the WNT signalling pathway to date have focussed on a few genes and proteins of the core pathway [27–32]. Building upon this work that describes the initial steps of WNT pathway activation, our work builds upon this by identifying new features and regression models that describe the heterogeneity of the WNT pathway activity scores across patients from two CRC cohorts [27]. We have also shown that the activity scores of the WNT pathway were the most significant in terms of the association to event-free patient survival compared to four other key pathways of CRC; the MAPK, PI3K, TGF-Beta, and p53 pathways.

3.1. Event-Free Survival Analysis

This study determined novel genes and proteins significantly associated with event-free survival in CRC across the major five CRC signalling pathways, WNT, PI3K-Akt, TP53, MAPK, and TGF-Beta. In total, 389 genes and 75 proteins were found to be significant across all signalling pathways from a sum of 872 genes and 284 proteins. A total of 59 of these features were significant in the WNT signalling pathway, and interestingly, only seven of these significant features, DVL3, DKK1, SFPR2, WNT3A, WNT3, SFPR1, and GSK3B, are found within previous mechanistic WNT models [27–32]. Thus, in order to build new and improved mechanistic models of the WNT pathway with the potential to fulfil the promise of precision medicine, future modelling should focus on these significant genes and proteins.

For the WNT signalling pathway, the number of significant genes determined from RNA sequencing was 53. The top two significant genes found were DVL3 and VANGL2. The most significant gene was DVL3, with a corresponding hazard ratio of 2.8158 and a p-value of 2.58×10^{-4}. WNT ligands bind to FZD receptors and the LRP5/6 co-receptor, which leads to the recruitment of cytosolic DVL3, then relays this signal to downstream signalling events that result in the translocation of ß-catenin to the nucleus and target gene expression. Thus, DVL is recruited by the receptor Frizzled and prevents the fundamental destruction of cytosolic β-catenin [43]. Previous studies demonstrated that a high expression of DVL3 in CRC acts as an unfavourable prognostic marker [44]. Additionally, Zhao et al. suggest that DVL3 is a key regulator in CRC chemoresistance and targeting it may be a potential strategy for CRC therapy [45]. Our study solidifies this further, as shown in the Kaplan–Meier curve in Figure 3, in which a high expression of DVL3 was associated with a lower event-free patient survival. VANGL2, the second most significant gene in the WNT signalling pathway, has a hazard ratio and *p*-value of 2.7692 and 1.08×10^{-3}, respectively. Studies show that VANGL2 can be activated by WNT through Frizzled receptors [46]. Furthermore, within the WNT pathway, six proteins were significantly associated with event-free patient survival. The most significant protein was PRKCA. High PRKCA protein expression correlated with longer event-free survival (Figure 5). Interestingly, the PRKCA protein was also important for explaining the WNT pathway activity scores in Model 3, with a significant negative regression coefficient (Figure 8), which is in line with the known function of PRKCA, which is to inhibit WNT signalling via several mechanisms, including the phosphorylation of β-catenin, and enhance CRC cell death, concretely fitting to the survival data presented in Figure 5 [47]. Together, these results suggest a model in which PRKCA exerts its positive effects on patient survival via repressing the WNT activity.

Finally, the results for the WNT pathway analysis identified several significant genes and proteins that were associated with event-free survival but have not yet been accounted for in current WNT mechanistic models. Most, if not all, models to date are based on a few intracellular and extracellular components of the core WNT pathway [27–32]. Focusing on describing the general mechanistic details of WNT pathway activation, these models were not intended to model patient specific differences and describe the WNT signalling heterogeneity across patients. In contrast, our analysis and machine learning models aimed at identifying genes and proteins with significant event-free survival associations and describing patient-specific differences of WNT pathway activation. In total, the remaining 52 significant features found in this study are not found in these mechanistic models. Thus, we have identified many genes that were associated with event-free survival and that constitute prime candidates to improve current WNT mechanistic models. These improved models could then be used for patient stratification and predicting response to therapy [9,33].

3.2. A PROGENy Analysis

The second part of this study consisted of a PROGENy pathway activity score analysis. There were several reasons why PROGENy and not classical GSEA was performed. One major reason PROGENy does not require predefined groups for comparisons, unlike GSEA, which determines whether a defined set of genes shows statistically significant, concurring differences between two groups, for example, normal and tumour tissue [14,48]. PROGENy, on the other hand, predicts specific pathway activity scores for each individual patient; thus, allowing us to build models that can explain these scores [35,49].

An advantage of the PROGENy analysis is that it reveals patient-specific differences, showing that different pathways are active in different patients (Figure 6). Interestingly, the PROGENy analysis did not reveal a striking hyperactivation of the WNT pathway (Figure 6); most values were found to be between 0 and 0.5 despite it being the main driving event in tumorigeneses within most CRC patients. This finding is explained by the fact that the pathway analysis is a differential analysis. PROGENy results are *z*-scores quantifying standardised (mean normalised and scaled) relative differences, yet all patients might have high WNT activity in absolute terms.

The results correlating the PROGENy pathway activity scores to patient event-free survival show that the WNT, PI3K, TGF-Beta, and MAPK pathways, when activated, were significantly associated with event-free patient survival. It is very interesting to note that the WNT signalling pathway was the most significant pathway associated with event-free patient survival. These results are in line with the literature, where high WNT pathway activation was associated with shorter survival [27–32]. The activation of the WNT pathway in CRC increases the levels of β-catenin within the cytosol, causing it to translocate into the nucleus and express WNT target genes that drive cell-proliferation [41]. How can the patient-specific differences of the WNT pathway activity be explained?

3.3. Development of Linear Regression Models

The final part of this study developed three linear regression machine learning models with the aim to predict the PROGENy WNT pathway activity scores. Current mechanistic models describe the pathway components and provide deterministic insights into how the WNT pathway is activated but neglect patient-specific differences [27–32]. Thus, combining machine learning that focuses on patient-specific differences with mechanistic models has great potential for building explainable machine learning models.

Overall, Model 2 was found as the most predictive model. In comparison, the RMSE of Model 2 was 0.3751 times smaller than Model 1. Five features were identified that were both prognostic of event-free patient survival and predictive of WNT pathway activity, including DVL3, PRKCA, VANGL2, GPC4, and ROCK2. Of these, DVL3 was the only feature found in the mechanistic model, Model 1. The finding also highlights a lack of features in current mechanistic models that can explain patient-specific differences. Concretely, solidifying the need to propose such a strategy to incorporate them. In particular, our work identified the four overlapping features above (PRKCA, VANGL2, GPC4, and ROCK2) that should be the focus of mechanistic modelling in the future. Machine learning, Model 2, accounting for such features, outperformed the current mechanistic model and can accurately predict the WNT pathway activity scores.

Interestingly, the protein PRKCA was found to be both significantly associated with event-free survival and a significant predictive feature of the WNT pathway activity score (Table 6). The results in Figure 8 show that most features used in the machine learning models have nonzero regression coefficients, in particular: DVL3, FZD5, RAC1, ROCK2, GSK3B, CTB2, CBT1, and PRKCA. These features provide valuable WNT network nodes to be included in future mechanistic models of the WNT network. The presented machine learning models incorporate patient specific significant features that describe pathway activity and can thus be termed "explainable". Despite this, these prognostic features have not yet been accounted for in current WNT models. The finding that a LASSO-based machine learning model (Model 2) identified new features outperforms the current mechanistic model, and can accurately predict the WNT pathway activity scores, highlights the need to improve current mechanistic models by incorporating these features.

4. Conclusions

In conclusion, we developed machine learning models of the activity of the canonical WNT signalling pathway and incorporated significant genes and proteins that are associated with event-free patient survival. This model is a valuable tool for predicting the heterogeneity of WNT pathway activity on an individual patient level. In this study, we identified several features (DVL3, FZD5, RAC1, ROCK2, GSK3B, CTB2, CBT1, and PRKCA) that were not only significantly associated with event-free patient survival but also important for predicting the patient-specific WNT pathway activity scores. We expect that integrating these features into mechanistic models of WNT pathway activation in the future will not only help to better understand the patient-specific differences in the control of pathway activity but also to relate pathway activity to clinical outcomes on the level of individual patients.

5. Materials and Methods

5.1. Data Acquisition from the Literature

All data used throughout this study was open-source data. The first datasets were from the 2016 TCGA Colorectal Adenocarcinoma, GDAC Firehose Legacy study, previously known as TCGA provisional. The datasets used for this study included Genomic from RNA sequencing (20,532 genes), proteomic from mass spectrometry (5562 proteins), and the associated clinical dataset [39,40]. The TCGA legacy datasets, due to the minimum overlap between patients, were only used to determine prognostic genes and proteins within such patients. Pre-processing across the datasets was performed before the survival analyses were complete. Across all datasets, patients who did not have the associated censoring status or disease-free survival months available were removed. The second set of data was from the 2019 prospective CPTAC-COAD colon adenocarcinoma studies, genomic from RNA sequencing (13,482 genes), proteomic from mass spectrometry (6422 proteins), and clinical datasets were all downloaded from LinkedOmics [39,40,50,51]. Minor pre-processing across the CPTAC datasets was performed before the following analyses were complete. The pre-processing included analysing the datasets, RNA, proteomics, and clinical to ensure that the same patients were overlapping across all and, if not, removing these patients. In total, 79 patients remained across all datasets. Additionally, for the CPTAC proteomics dataset, several NAN values were present. These values were imputed for the machine learning LASSO regression analysis of the WNT STN. This was achieved in the application Perseus using a low-shifted distribution of width 0.3 and a downshift of 1.8 across the total matrix [52]. Furthermore, proteins with data points of 10% or less were removed. In total, 27/39 proteins in the WNT STN remained after imputation. All datasets analysed and pre-processed for this study can be found in the supplementary data; consequently, Table 7 and Figure 9 represent key metrics and pre-processing steps for all datasets used.

Table 7. Important key metrics of datasets and corresponding analyses performed for each dataset. The two studies used, TCGA Legacy and CPTAC, are located in the first column [39,40,50,51]. The "datasets" column represents the type of omic dataset used from the corresponding study in column one, with key metrics identified. The "analyses" column represents the type of analysis performed on each specified dataset. "RSEM" = RNA-sequencing by expectation-maximisation. "UQ" = upper quantile normalisation. "TMT" = tandem mass tag.

Study	Datasets	Analyses
2016 TCGA Colorectal Adenocarcinoma, GDAC Firehose Legacy	1. **Genomic from RNA Sequencing** 20,532 genes Cohort of 329 patients 2. **Proteomic from Mass Spectrometry** 5562 proteins Cohort of 74 patients 3. **Corresponding Clinical Dataset** Event Free Survival	1. Kaplan Meier Survival analysis for genes and proteins in pathways to find prognostic features 2. PROGENy analysis on the TCGA RNA sequencing legacy dataset to find a common core of CRC pathway activities
2019 Prospective CPTAC-COAD Colon Adenocarcinoma	1. **Genomic from RNA Sequencing** RNA Expression (RSEM-UQ, Log2(Val+1)) 13,482 genes Cohort of 106 patients 2. **Proteomic from Mass Spectrometry** Protein Expression (TMT, Log2ratio) 6422 proteins Cohort of 96 patients 3. **Corresponding Clinical Dataset** Cohort of 110 patients 79 patients analysed Event Free Survival	1. Determine the associations between PROGENy pathway activity scores and event free survival 2. Linear Regression Machine Learning models to predict WNT pathway activity

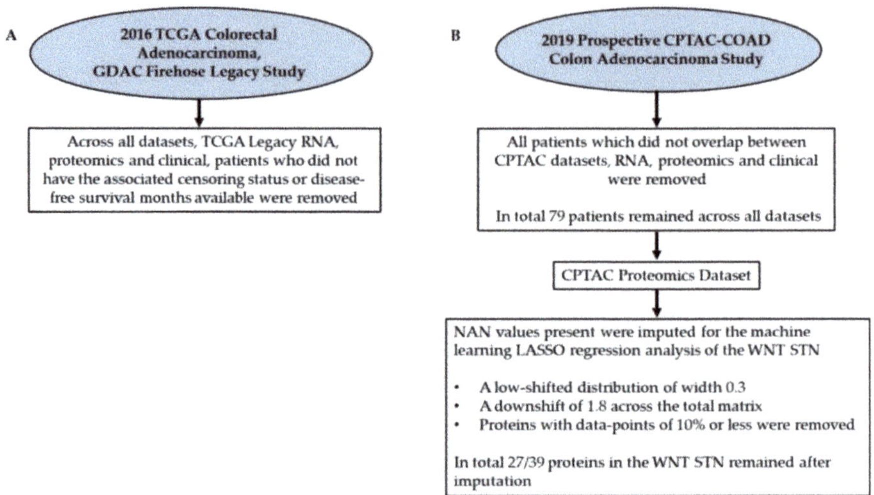

Figure 9. Pre-processing flow chart detailing all pre-processing steps completed prior to each analysis. Pre-processing flowchart for, (**A**) the 2016 TCGA Colorectal Adenocarcinoma, GDAC Firehose Legacy Study and (**B**) the 2019 Prospective CPTAC-COAD Colon Adenocarcinoma Study.

The five critical CRC STNs used in this study stemmed from the literature [3,11,23]. The network databases WikiPathways, KEGG, and GSEA, were used to create a list of gene-sets for each pathway including WNT, PI3K-Akt, TP53, MAPK, and TGF-Beta [21, 22,37,38,53–62]. Each gene-set consisted of every gene listed for each homo-sapiens STN pathway on WikiPathways. These gene-sets were then used as input scripts to determine which genes and proteins are prognostic and correlate to patient event-free survival. All gene sets are found in the supplementary data.

5.2. Kaplan–Meier Survival Analysis

The Kaplan–Meier survival analysis was simulated three times, the first to find the prognostic genes from RNA sequencing, the second to find the prognostic proteins from proteomics, both using the TCGA legacy datasets, and finally to determine the associations between PROGENy pathway activity scores and event-free survival using the RNA CPTAC dataset. The optimum cut-off for stratifying the patient populations into low and high groups was identified by scanning the group sizes from 10–90 to 90–10 percent splits, where 10–90 means that 10% of the patients were in the low group and 90% of the patients were in the high group and calculating the *p*-value for the overall event-free survival difference between the groups using a log-rank test with Yates' correction. The cut-offs were based on the TCGA legacy datasets: RNA sequencing (329 patients) and proteomics (73 patients), and the CPTAC RNA dataset (79 patients). The inputs for the first two simulations were based on the gene or protein sets for each CRC STN identified above. The output was a Kaplan–Meier curve for the gene or protein expressed, indicating the number of patients in the high expression or low expression group with the corresponding statistical values apparent. The inputs for the final simulation were the PROGENy CRC pathway activity scores, and the output was 5 Kaplan–Meier curves for each CRC STN. The number of patients in the high activity or low activity group was identified with the corresponding statistical values. All statistical computations and Kaplan–Meier analyses were performed in MATLAB (version R2020b Update 5 (9.9.0. 1592791), The MathWorks, Inc., Natick, MA, USA) using the statistics toolbox and the log-rank (www.mathworks.com/matlabcentral/fileexchange/22317-logrank (accessed on 7 April 2021)) and kmplot (www.mathworks.com/matlabcentral/fileexchange/22293-kmplot (accessed on 7 April 2021)) functions from the MATLAB (version R2020b Update 5 (9.9.0. 1592791), The MathWorks, Inc., Natick, MA, USA) file exchange [63].

5.3. PROGENy Analysis

PROGENy is a machine learning based tool installed from Bioconductor as an RStudio package (RStudio Team (2020). RStudio: Integrated Development for R. RStudio, PBC, Boston, MA URL http://www.rstudio.com/) [35,49]. A PROGENy analysis was performed on the TCGA RNA sequencing legacy dataset to find a common core of CRC pathway activities. Despite PROGENy's composition of 14-cancer relevant pathways, specifically for this study, only the CRC pathways were analysed in depth [35,49]. The available CRC pathways using PROGENy were PI3K, MAPK, TGF-Beta, WNT, and p53. The associated PROGENy CRC pathway activity scores for the entire TCGA legacy patient cohort is found in the supplementary data. PROGENy (version 1.12.0) was used in the study [35,49].

5.4. Developing Linear Regression Machine Learning Models

The three linear regression machine learning models developed in the study were performed in MATLAB using the regression learner app [64]. The three trained models were a variety of linear, stepwise linear, and robust linear regression models. Each model consisted of 79 observations, i.e., the entire CPTAC patient cohort. Cross-validation of 10-fold was used for each model. All models were developed solely from the CPTAC datasets from RNA and proteomics. The selected input features for the models were obtained through LASSO regression of both datasets. For both LASSO regression analyses, the lambda value with minimal mean squared error plus one standard deviation was applied. A threshold (p-value) of 0.05 was used to determine if a feature was significant. For the proteomics dataset, this resulted in 5/27 features (proteins) found to be significant, including CTBP1, CTBP2, PLCB4, PRKCA, and RAC1. Similarly, for the RNA sequencing dataset, 5/89 features (genes) were significant, DKK3, FZD5, NKD1, NOTUM, and WNT11. All associated datasets are found within the supplementary data.

Supplementary Materials: The following are available online at https://www.mdpi.com/article/10.3390/ijms22189970/s1.

Author Contributions: Conceptualisation, A.N. and D.F.; Data curation, A.N.; Formal analysis, A.N.; Funding acquisition, D.F.; Investigation, D.F.; Methodology, A.N.; Project administration, A.N.; Software, A.N.; Supervision, D.F.; Validation, D.F.; Visualisation, A.N.; Writing—original draft, A.N.; Writing—review and editing, A.N. and D.F. All authors have read and agreed to the published version of the manuscript.

Funding: This project has received funding from the UCD Ad Astra Fellowship No 64998 and the European Union's Horizon 2020 research and innovation programme under grant agreement No 754923. The material presented and views expressed here are the responsibility of the author(s) only. The EU Commission takes no responsibility for any use made of the information set out.

Institutional Review Board Statement: Not applicable.

Informed Consent Statement: Not applicable.

Data Availability Statement: The first dataset supporting this study were from the 2016 TCGA Colorectal Adenocarcinoma, GDAC Firehose Legacy study, previously known as TCGA provisional (cBioPortal for Cancer Genomics. Available Online: https://www.cbioportal.org/study/summary?id=coadread_tcga (accessed on 1 March 2021)) [39,40]. The second and last dataset was from the 2019 prospective CPTAC-COAD colon adenocarcinoma studies. All data were downloaded from LinkedOmics. (Datasets for Colon adenocarcinoma: Prospective CPTAC-COAD. Available Online: http://linkedomics.org/cptac-colon/ (accessed on 8 April 2021)).

Conflicts of Interest: The authors declare no conflict of interest.

References

1. Dekker, E.; Tanis, P.J.; Vleugels, J.L.A.; Kasi, P.M.; Wallace, M.B. Colorectal cancer. *Lancet* **2019**, *394*, 1467–1480. [CrossRef]
2. Xie, Y.H.; Chen, Y.X.; Fang, J.Y. Comprehensive review of targeted therapy for colorectal cancer. In *Signal Transduction and Targeted Therapy*; Springer Nature: Basingstoke, UK, 2020; Volume 5, pp. 1–30. [CrossRef]
3. Tyagi, A.; Sharma, A.K.; Damodaran, C. A Review on Notch Signaling and Colorectal Cancer. *Cells* **2020**, *9*, 1549. [CrossRef]
4. Dienstmann, R.; Vermeulen, L.; Guinney, J.; Kopetz, S.; Tejpar, S.; Tabernero, J. Consensus molecular subtypes and the evolution of precision medicine in colorectal cancer. *Nat. Rev. Cancer* **2017**, *17*, 79–92. [CrossRef] [PubMed]
5. Stintzing, S.; Wirapati, P.; Lenz, H.-J.; Neureiter, D.; von Weikersthal, L.F.; Decker, T.; Kiani, A.; Kaiser, F.; Al-Batran, S.; Heintges, T.; et al. Consensus molecular subgroups (CMS) of colorectal cancer (CRC) and first-line efficacy of FOLFIRI plus cetuximab or bevacizumab in the FIRE3 (AIO KRK-0306) trial. *Ann. Oncol.* **2019**, *30*, 1796–1803. [CrossRef] [PubMed]
6. Okita, A.; Takahashi, S.; Ouchi, K.; Inoue, M.; Watanabe, M.; Endo, M.; Honda, H.; Yamada, Y.; Ishioka, C. Consensus molecular subtypes classification of colorectal cancer as a predictive factor for chemotherapeutic efficacy against metastatic colorectal cancer. *Oncotarget* **2018**, *9*, 18698–18711. [CrossRef] [PubMed]
7. Mooi, J.; Wirapati, P.; Asher, R.; Lee, C.; Savas, P.S.; Price, T.; Townsend, A.; Hardingham, J.; Buchanan, D.; Williams, D.; et al. The prognostic impact of consensus molecular subtypes (CMS) and its predictive effects for bevacizumab benefit in metastatic colorectal cancer: Molecular analysis of the AGITG MAX clinical trial. *Ann. Oncol.* **2018**, *29*, 2240–2246. [CrossRef]
8. Roelands, J.; Kuppen, P.J.K.; Vermeulen, L.; Maccalli, C.; Decock, J.; Wang, E.; Marincola, F.M.; Bedognetti, D.; Hendrickx, W. Immunogenomic classification of colorectal cancer and therapeutic implications. *Int. J. Mol. Sci.* **2017**, *18*, 2229. [CrossRef]
9. Kolch, W.; Fey, D. Personalized computational models as biomarkers. *J. Pers. Med.* **2017**, *7*, 9. [CrossRef]
10. Kolch, W.; Halasz, M.; Granovskaya, M.; Kholodenko, B.N. The dynamic control of signal transduction networks in cancer cells. *Nat. Rev. Cancer* **2015**, *15*, 515–527. [CrossRef]
11. The Cancer Genome Atlas Network. Comprehensive molecular characterization of human colon and rectal cancer. *Nature* **2012**, *487*, 330–337. [CrossRef]
12. McDonald, G.T.; Sullivan, R.; Paré, G.C.; Graham, C.H. Inhibition of phosphatidylinositol 3-kinase promotes tumor cell resistance to chemotherapeutic agents via a mechanism involving delay in cell cycle progression. *Exp. Cell Res.* **2010**, *316*, 3197–3206. [CrossRef]
13. Temiz, T.K.; Altun, A.; Turgut, N.H.; Balcı, E.; Turgut, H.; Balcı, E. Investigation of the effects of drugs effective on PI3K-AKT signaling pathway in colorectal cancer alone and in combination Kolorektal kanserde PI3K-AKT sinyal yolağı üzerinden etki gösteren ilaçların tek başlarına ve kombinasyonlarının etkilerinin araştırılması. *Cumhur. Med. J.* **2014**, *36*, 167–177. [CrossRef]
14. Subramanian, A.; Tamayo, P.; Mootha, V.K.; Mukherjee, S.; Ebert, B.L.; Gillette, M.A.; Paulovich, A.; Pomeroy, S.L.; Golub, T.R.; Lander, E.S.; et al. Gene set enrichment analysis: A knowledge-based approach for interpreting genome-wide expression profiles. *Proc. Natl. Acad. Sci. USA* **2005**, *102*, 15545–15550. [CrossRef] [PubMed]
15. Xu, G.; Shi, Y. Apoptosis signaling pathways and lymphocyte homeostasis. *Cell Res.* **2007**, *17*, 759–771. [CrossRef] [PubMed]
16. Cell Cycle Pathway | Aviva Systems Biology—Bio-Connect. Available online: https://www.bio-connect.nl/cell-cycle-pathway/cnt/page/4814 (accessed on 9 April 2021).
17. Wang, X.W.; Zhang, Y.J. Targeting mTOR network in colorectal cancer therapy. *World J. Gastroenterol.* **2014**, *20*, 4178–4188. [CrossRef]
18. Koveitypour, Z.; Panahi, F.; Vakilian, M.; Peymani, M.; Forootan, F.S.; Esfahani, M.H.N.; Ghaedi, K. Signaling pathways involved in colorectal cancer progression. In *Cell and Bioscience*; BioMed Central Ltd.: London, UK, 2019; Volume 9. [CrossRef]
19. Previs, R.A.; Coleman, R.L.; Harris, A.L.; Sood, A.K. Molecular Pathways: Translational and Therapeutic Implications of the Notch Signaling Pathway in Cancer. *Physiol. Behav.* **2017**, *176*, 139–148. [CrossRef] [PubMed]
20. Liao, W.; Li, G.; You, Y.; Wan, H.; Wu, Q.; Wang, C.; Lv, N. Antitumor activity of Notch-1 inhibition in human colorectal carcinoma cells. *Oncol. Rep.* **2018**, *39*, 1063–1071. [CrossRef] [PubMed]
21. Goto, M.K.S. KEGG: Kyoto Encyclopedia of Genes and Genomes. In *Nucleic Acids Research*; Oxford University Press: Oxford, UK, 2000; Volume 28, pp. 27–30. [CrossRef]
22. Kanehisa, M.; Goto, S.; Kawashima, S.; Nakaya, A. Thed KEGG databases at GenomeNet. *Nucleic Acids Res.* **2002**, *30*, 42–46. [CrossRef]
23. Iurii, M. Genetics of colorectal cancer. *J. Med. Life* **2014**, *7*, 507–511.
24. Polakis, P. Wnt signaling in cancer. *Cold Spring Harb. Perspect. Biol.* **2012**, *4*, a008052. [CrossRef]
25. Zhan, T.; Rindtorff, N.; Boutros, M. Wnt signaling in cancer. *Oncogene* **2017**, *36*, 1461–1473. [CrossRef]
26. Schatoff, E.M.; Leach, B.I.; Dow, L.E. Wnt Signaling and Colorectal Cancer. *Curr. Color. Cancer Rep.* **2017**, *13*, 101–110. [CrossRef] [PubMed]
27. Kogan, Y.; Halevi-Tobias, K.E.; Hochman, G.; Baczmanska, A.K.; Leyns, L.; Agur, Z. A new validated mathematical model of the Wnt signalling pathway predicts effective combinational therapy by sFRP and Dkk. *Biochem. J.* **2012**, *444*, 115–125. [CrossRef]
28. Wolf, B.K.J. Mathematical modelling of Wnt/β-catenin signalling. *Biochem. Soc. Trans.* **2010**, *38*, 1281–1285. [CrossRef]
29. Lee, E.; Salic, A.; Krüger, R.; Heinrich, R.; Kirschner, M.W. The roles of APC and axin derived from experimental and theoretical analysis of the Wnt pathway. *PLoS Biol.* **2003**, *1*, e10. [CrossRef]
30. Lee, M.; Chen, G.T.; Puttock, E.; Wang, K.; Edwards, R.A.; Waterman, M.L.; Lowengrub, J. Mathematical modeling links Wnt signaling to emergent patterns of metabolism in colon cancer. *Mol. Syst. Biol.* **2017**, *13*, 912. [CrossRef] [PubMed]

31. Shukla, A.; Singh, T.R. Network-based approach to understand dynamic behaviour of Wnt signaling pathway regulatory elements in colorectal cancer. *Netw. Modeling Anal. Health Inform. Bioinform.* **2018**, *7*, ra130. [CrossRef]
32. Heinrich, R. Mathematical modelling of the Wnt-pathway. *Syst. Biol.* **2005**, *13*, 259–275. [CrossRef]
33. Fey, D.; Halasz, M.; Dreidax, D.; Kennedy, S.P.; Hastings, J.F.; Rauch, N.; Munoz, A.G.; Pilkington, R.; Fischer, M.; Westermann, F.; et al. Signaling pathway models as biomarkers: Patient-specific simulations of JNK activity predict the survival of neuroblastoma patients. *Sci. Signal.* **2015**, *8*, ra130. [CrossRef]
34. Golumbeanu, M.; Yang, G.; Camponovo, F.; Stuckey, E.M.; Hamon, N.; Mondy, M.; Rees, S.; Chitnis, N.; Cameron, E.; Penny, M.A. Combining machine learning and mathematical models of disease dynamics to guide development of novel disease interventions. *medRxiv* **2021**. [CrossRef]
35. Schubert, M.; Klinger, B.; Klünemann, M.; Sieber, A.; Uhlitz, F.; Sauer, S.; Garnett, M.J.; Blüthgen, N.; Saez-Rodriguez, J. Perturbation-response genes reveal signaling footprints in cancer gene expression. *Nat. Commun.* **2018**, *9*, 20. [CrossRef]
36. Holland, C.H.; Tanevski, J.; Perales-Patón, J.; Gleixner, J.; Kumar, M.P.; Mereu, E.; Joughin, B.A.; Stegle, O.; Lauffenburger, D.A.; Heyn, H.; et al. Robustness and applicability of transcription factor and pathway analysis tools on single-cell RNA-seq data. *Genome Biol.* **2020**, *21*, 1–19. [CrossRef]
37. Kanehisa, M.; Furumichi, M.; Sato, Y.; Ishiguro-Watanabe, M.; Tanabe, M. KEGG: Integrating viruses and cellular organisms. *Nucleic Acids Res.* **2021**, *49*, D545–D551. [CrossRef]
38. Kanehisa, M. Toward understanding the origin and evolution of cellular organisms. In *Protein Science*; Blackwell Publishing Ltd.: Hoboken, NJ, USA, 2019; Volume 28, pp. 1947–1951. [CrossRef]
39. Cerami, E.; Gao, J.; Dogrusoz, U.; Gross, B.E.; Sumer, S.O.; Aksoy, B.A.; Jacobsen, A.; Byrne, C.J.; Heuer, M.L.; Larsson, E.; et al. The cBio Cancer Genomics Portal: An open platform for exploring multidimensional cancer genomics data. *Cancer Discov.* **2012**, *2*, 401–404. [CrossRef]
40. Gao, J.; Aksoy, B.A.; Dogrusoz, U.; Dresdner, G.; Gross, B.; Sumer, S.O.; Sun, Y.; Jacobsen, A.; Sinha, R.; Larsson, E.; et al. Integrative analysis of complex cancer genomics and clinical profiles using the cBioPortal. *Sci. Signal.* **2013**, *6*, pl1. [CrossRef] [PubMed]
41. Gao, C.; Xiao, G.; Hu, J. Regulation of Wnt/β-catenin signaling by posttranslational modifications. *Cell Biosci.* **2014**, *4*, 13. [CrossRef]
42. Darras, S.; Fritzenwanker, J.H.; Uhlinger, K.R.; Farrelly, E.; Pani, A.M.; Hurley, I.A.; Norris, R.P.; Osovitz, M.; Terasaki, M.; Wu, M.; et al. Anteroposterior axis patterning by early canonical Wnt signaling during hemichordate development. *PLoS Biol.* **2018**, *16*, e2003698. [CrossRef] [PubMed]
43. Gao, C.; Chen, Y.G. Dishevelled: The hub of Wnt signaling. *Cell. Signal.* **2010**, *22*, 717–727. [CrossRef]
44. DVL3 Protein Expression Summary—The Human Protein Atlas. Available online: https://www.proteinatlas.org/ENSG0000016 1202-DVL3 (accessed on 12 April 2021).
45. Zhao, Q.; Zhuang, K.; Han, K.; Tang, H.; Wang, Y.; Si, W.; Yang, Z. Silencing DVL3 defeats MTX resistance and attenuates stemness via Notch Signaling Pathway in colorectal cancer. *Pathol. Res. Pract.* **2020**, *216*, 152964. [CrossRef] [PubMed]
46. Shafer, B.; Onishi, K.; Lo, C.; Colakoglu, G.; Zou, Y. Vangl2 Promotes Wnt/Planar Cell Polarity-like Signaling by Antagonizing Dvl1-Mediated Feedback Inhibition in Growth Cone Guidance. *Dev. Cell* **2011**, *20*, 177–191. [CrossRef]
47. Dupasquier, S.; Blache, P.; Lasorsa, L.P.; Zhao, H.; Abraham, J.-D.; Haigh, J.J.; Ychou, M.; Prevostel, C. Modulating PKCα activity to target Wnt/β-catenin signaling in colon cancer. *Cancers* **2019**, *11*, 693. [CrossRef] [PubMed]
48. Mootha, V.K.; Lindgren, C.M.; Eriksson, K.F.; Subramanian, A.; Sihag, S.; Lehar, J.; Puigserver, P.; Carlsson, E.; Ridderstrale, M.; Laurila, E.; et al. PGC-1α-responsive genes involved in oxidative phosphorylation are coordinately downregulated in human diabetes. *Nat. Genet.* **2003**, *34*, 267–273. [CrossRef]
49. Holland, C.H.; Szalai, B.; Saez-Rodriguez, J. Transfer of regulatory knowledge from human to mouse for functional genomics analysis. *Biochim. Biophys. Acta (BBA)—Gene Regul. Mech.* **2019**, *1863*, 194431. [CrossRef] [PubMed]
50. LinkedOmics: Data Download. Datasets for Colon Adenocarcinoma (Prospective CPTAC-COAD). Available online: http://linkedomics.org/cptac-colon/ (accessed on 8 April 2021).
51. Vasaikar, S.; Huang, C.; Wang, X.; Petyuk, V.A.; Savage, S.R.; Wen, B.; Dou, Y.; Zhang, Y.; Shi, Z.; Arshad, O.A.; et al. Proteogenomic Analysis of Human Colon Cancer Reveals New Therapeutic Opportunities. *Cell* **2019**, *177*, 1035–1049. [CrossRef] [PubMed]
52. Tyanova, S.; Temu, T.; Sinitcyn, P.; Carlson, A.; Hein, M.Y.; Geiger, T.; Mann, M.; Cox, J. The Perseus computational platform for comprehensive analysis of (prote)omics data. In *Nature Methods*; Nature Publishing Group: Berlin, Germany, 2016; Volume 13, pp. 731–740. [CrossRef]
53. Nicioli da Silva, G.; Hanspers, K.; Pico, A.; Waagmeester, A. TP53 Network (Homo Sapiens)—WikiPathways. Available online: https://www.wikipathways.org/index.php/Pathway:WP1742 (accessed on 8 April 2021).
54. Hanspers, K.; Riutta, A.; Willighagen, E. PI3K-Akt Signaling Pathway (Homo Sapiens)—WikiPathways. Available online: https://www.wikipathways.org/index.php/Pathway:WP4172 (accessed on 8 April 2021).
55. Pandey, A.; Hansper, K.; Pico, A.; Salomonis, N. TGF-beta Signaling Pathway (Homo Sapiens)—WikiPathways. Available online: https://www.wikipathways.org/index.php/Pathway:WP366 (accessed on 8 April 2021).
56. Lieberman, M.; Hanspers, K.; Pico, A.; Kelder, T. Wnt Signaling (Homo Sapiens)—WikiPathways. Available online: https://www.wikipathways.org/index.php/Pathway:WP428 (accessed on 8 April 2021).

57. Martens, M.; Ammar, A.; Riutta, A.; Waagmeester, A.; Slenter, D.N.; Hanspers, K.; Miller, R.A.; Digles, D.; Lopes, E.N.; Ehrhart, F.; et al. WikiPathways: Connecting communities. *Nucleic Acids Res.* **2021**, *49*, 613–621. [CrossRef] [PubMed]
58. Kelder, T.; Pico, A.R.; Hanspers, K.; van Iersel, M.P.; Evelo, C. Mining Biological Pathways Using WikiPathways Web Services. *PLoS ONE* **2009**, *4*, 6447. [CrossRef]
59. Team Reactome; Miller, R. MTOR Signalling (Homo Sapiens)—WikiPathways. Available online: https://www.wikipathways.org/index.php/Pathway:WP3318 (accessed on 8 April 2021).
60. Zambon, A.C.; Alexander, P.; Coort, S. Apoptosis (Homo Sapiens)—WikiPathways. Available online: https://www.wikipathways.org/index.php/Pathway:WP254 (accessed on 8 April 2021).
61. Conklin, B.; Sach, I.C.; Coort, S.; Salomonis, N. Cell Cycle (Homo Sapiens)—WikiPathways. Available online: https://www.wikipathways.org/index.php/Pathway:WP179 (accessed on 8 April 2021).
62. Burel, S.; Hanspers, K.; Pico, A. MAPK Signaling Pathway (Homo Sapiens)—WikiPathways. Available online: https://www.wikipathways.org/index.php/Pathway:WP382 (accessed on 8 April 2021).
63. Giuseppe, C. Logrank—File Exchange—MATLAB Central. 2021. Available online: https://uk.mathworks.com/matlabcentral/fileexchange/22317-logrank (accessed on 7 April 2021).
64. Regression Learner App—MATLAB & Simulink—MathWorks United Kingdom. Available online: https://uk.mathworks.com/help/stats/regression-learner-app.html (accessed on 9 April 2021).

International Journal of
Molecular Sciences

Review

Treatment of Brain Metastases of Non-Small Cell Lung Carcinoma

Agnieszka Rybarczyk-Kasiuchnicz *, Rodryg Ramlau and Katarzyna Stencel 

Department of Chemotherapy, Poznan University of Medical Sciences, Clinical Hospital of Lord Transfiguration, 61-848 Poznan, Poland; rramlau@gmail.com (R.R.); k.stencel@post.pl (K.S.)
* Correspondence: agnieszka.rybarczyk3@wp.pl

Abstract: Lung cancer is one of the most common malignant neoplasms. As a result of the disease's progression, patients may develop metastases to the central nervous system. The prognosis in this location is unfavorable; untreated metastatic lesions may lead to death within one to two months. Existing therapies—neurosurgery and radiation therapy—do not improve the prognosis for every patient. The discovery of Epidermal Growth Factor Receptor (EGFR)—activating mutations and Anaplastic Lymphoma Kinase (ALK) rearrangements in patients with non-small cell lung adenocarcinoma has allowed for the introduction of small-molecule tyrosine kinase inhibitors to the treatment of advanced-stage patients. The Epidermal Growth Factor Receptor (EGFR) is a transmembrane protein with tyrosine kinase-dependent activity. EGFR is present in membranes of all epithelial cells. In physiological conditions, it plays an important role in the process of cell growth and proliferation. Binding the ligand to the EGFR causes its dimerization and the activation of the intracellular signaling cascade. Signal transduction involves the activation of MAPK, AKT, and JNK, resulting in DNA synthesis and cell proliferation. In cancer cells, binding the ligand to the EGFR also leads to its dimerization and transduction of the signal to the cell interior. It has been demonstrated that activating mutations in the gene for EGFR-exon19 (deletion), L858R point mutation in exon 21, and mutation in exon 20 results in cancer cell proliferation. Continuous stimulation of the receptor inhibits apoptosis, stimulates invasion, intensifies angiogenesis, and facilitates the formation of distant metastases. As a consequence, the cancer progresses. These activating gene mutations for the EGFR are present in 10–20% of lung adenocarcinomas. Approximately 3–7% of patients with lung adenocarcinoma have the echinoderm microtubule-associated protein-like 4 (EML4)/ALK fusion gene. The fusion of the two genes EML4 and ALK results in a fusion gene that activates the intracellular signaling pathway, stimulates the proliferation of tumor cells, and inhibits apoptosis. A new group of drugs—small-molecule tyrosine kinase inhibitors—has been developed; the first generation includes gefitinib and erlotinib and the ALK inhibitor crizotinib. These drugs reversibly block the EGFR by stopping the signal transmission to the cell. The second-generation tyrosine kinase inhibitor (TKI) afatinib or ALK inhibitor alectinib block the receptor irreversibly. Clinical trials with TKI in patients with non-small cell lung adenocarcinoma with central nervous system (CNS) metastases have shown prolonged, progression-free survival, a high percentage of objective responses, and improved quality of life. Resistance to treatment with this group of drugs emerging during TKI therapy is the basis for the detection of resistance mutations. The T790M mutation, present in exon 20 of the EGFR gene, is detected in patients treated with first- and second-generation TKI and is overcome by Osimertinib, a third-generation TKI. The I117N resistance mutation in patients with the ALK mutation treated with alectinib is overcome by ceritinib. In this way, sequential therapy ensures the continuity of treatment. In patients with CNS metastases, attempts are made to simultaneously administer radiation therapy and tyrosine kinase inhibitors. Patients with lung adenocarcinoma with CNS metastases, without activating EGFR mutation and without ALK rearrangement, benefit from immunotherapy. This therapeutic option blocks the PD-1 receptor on the surface of T or B lymphocytes or PD-L1 located on cancer cells with an applicable antibody. Based on clinical trials, pembrolizumab and all antibodies are included in the treatment of non-small cell lung carcinoma with CNS metastases.

Keywords: brain metastases; treatment; non-small cell lung carcinoma; EGFR; ALK; immunotherapy

Citation: Rybarczyk-Kasiuchnicz, A.; Ramlau, R.; Stencel, K. Treatment of Brain Metastases of Non-Small Cell Lung Carcinoma. *Int. J. Mol. Sci.* **2021**, *22*, 593. https://doi.org/10.3390/ijms22020593

Received: 3 December 2020
Accepted: 30 December 2020
Published: 8 January 2021

Publisher's Note: MDPI stays neutral with regard to jurisdictional claims in published maps and institutional affiliations.

1. Introduction

Lung cancer is one of the most common malignant neoplasms and the main cause of death from malignant neoplasms in Poland. Each year, more than 22,000 new lung cancer cases are recorded [1].

Metastases of lung cancer to the brain occur in 18–61% of patients [2–4]. Improving the effectiveness of oncological treatment leads to a higher survival rate but also increases the population of patients at risk of this complication [5,6]. Metastases to the central nervous system (CNS) are different from other metastases due to the occurrence of neurological disorders and often require discontinuing systemic treatment in order to carry out palliative care [7].

2. The Mechanism of Brain Metastasis Formation Is Similar to Other Organ Locations

As a result of mutations in cancer cells, the degree of invasiveness increases [8,9] (Figure 1). The cells detach themselves from the primary tumor and penetrate the blood vessels, reaching other organs through the bloodstream. Being a very well-vascularized organ, the brain is often subjected to metastases [10]. Metastatic cells arrest at distinct sites and extravasate through vascular walls into the brain parenchyma. Cancer cells proliferate at the metastatic niche, form colonies in this parenchyma, and the subsequent proliferation of cells leads to clinically detectable metastatic lesions [8,9].

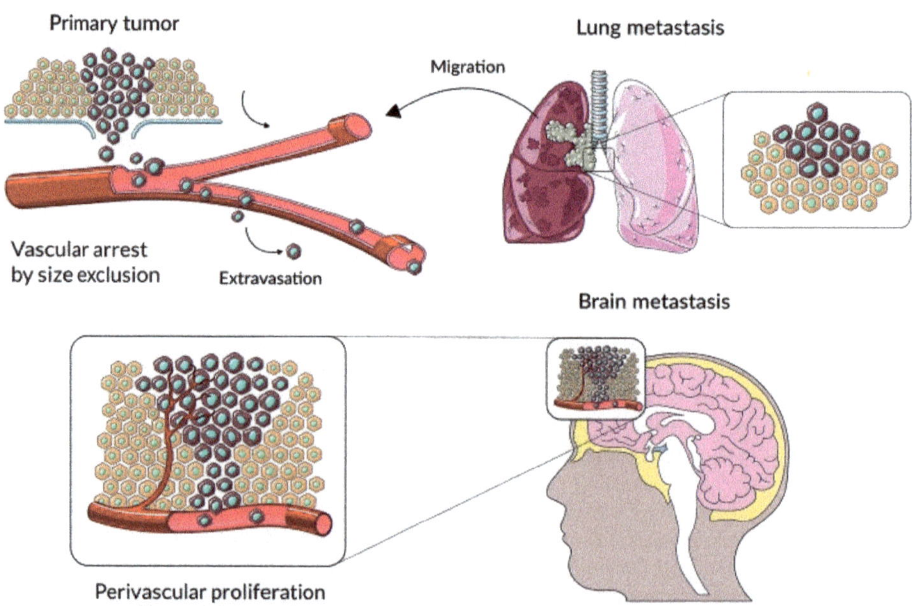

Figure 1. The main stages of cancer cell colonization of brain parenchyma (source: You H. et al., Front Immunol., 2019 [8]).

Untreated metastases to the CNS lead to a gradual deterioration of the patient's performance and to death within one to two months as a result of increased intracranial pressure. Avoiding or delaying these complications requires expertise in the radical and adjunctive treatment of brain metastases [11].

3. Non-Small Cell Lung Carcinoma without Activating the Mutation of Epidermal Growth Factor Receptor (EGFR) or Anaplastic Lymphoma Kinase (ALK)

A treatment method is chosen for lung carcinoma patients with CNS metastases based on their prognosis. It is determined by Radiation Therapy Oncology Group recursive partitioning analysis—RPA [12] (Table 1).

Table 1. Prognostic stages in patients with central nervous system metastases according to the Radiation Therapy Oncology Group (RPA). KPS: Karnofsky Performance Status.

Prognostic Class	Characteristic	Median Survival (Months)
I	KPS $\geq$ 70, < 65 years, controlled primary tumor and no extracranial metastases	7.1
II	KPS $\geq$ 70, primary tumor not controlled	4.2
	KPS $\geq$ 70, controlled primary tumor $\geq$ 65 years	
	KPS $\geq$ 70, controlled primary tumor	
	<65 years and extracranial metastases	
III	KPS < 70	2.3

The eligibility criteria are the patient's age, general performance, and the presence of metastases outside the CNS. The first group (class I) includes patients in good general condition, with a KI (Karnofsky Index) of 70% or more, less than 65 years old, without extracerebral metastases, and with good primary tumor control. The third class are patients in poor general condition, with a KI below 70%. The second class includes the remaining patients. In class III, radiotherapy of metastases to the CNS is not recommended due to a very bad prognosis. It is optimal to implement the best adjunctive treatment. The average period of survival is about two months [12].

In the case of a single metastasis to the brain up to four metastatic lesions to the CNS, either surgical removal or stereotactic radiotherapy (SRS) is recommended by the RPA in class I or II patients [7,13].

There is no evidence that the addition of whole-brain radiation therapy (WBRT) to stereotactic radiotherapy or surgery affects the overall survival of patients [14].

Data from a prospective study by Japanese researchers (JLGK0901) indicates that stereotaxis may be relevant in patients with more than three CNS metastases [15]. The observational trial lasted three years and included 1194 patients [15] with 1–10 newly diagnosed CNS metastases. The largest tumor volume was <10 mL and <3 cm in the longest dimension. The total cumulative volume did not exceed 15 mL, and the Karnofsky performance status was 70% or higher. All patients qualified in this way received stereotactic CNS radiotherapy. The overall survival (OS) of patients after stereotaxis was 13.9 months in 455 patients with a single metastasis, 10.8 months in 531 patients with two–four metastatic lesions, and 10.8 months in 208 patients with 5–10 metastatic lesions. An equal overall survival (OS) in patients with two–four metastatic lesions and 5–10 lesions indicates that stereotactic radiosurgery (SRS) is an important alternative to whole-brain radiotherapy in selected patients in good general condition.

Stereotaxis with or without whole-brain radiotherapy was analyzed in a third phase trial in which patients with one–four brain metastases of lung carcinoma were randomized [16]. Three hundred and sixty-four patients meeting the criteria for inclusion in the trial were analyzed. Fifty-one percent of patients received only stereotactic radiotherapy, while 49% received SRS followed by whole-brain radiation therapy (WBRT). It was shown that the age of patients significantly affects their survival. Stereotactic radiotherapy as a stand-alone treatment improves survival in patients aged 50 years or younger, with no difference in the age group over 50 years. Patients with a single metastasis experienced significantly longer survival than patients with two–four metastases. In the assessment of cognitive disorders during treatment, patients under 50 years of age tolerated the therapy better in both arms of the trial. Patients with a single CNS metastasis, compared to patients with two–four lesions, had less severe cognitive impairment. Local disease control was better in the arm with SRS plus WBRT in both age groups.

In the third-phase QUARTZ trial [17], the role of whole-brain radiation therapy (WBRT) was assessed in patients with non-small cell lung carcinoma with inoperable CNS metastases. The patients were randomized into two groups. In one group, they received radiation therapy—WBRT 20 Gy in four fractions—and steroid therapy, and in the other group, the best adjunctive treatment without radiotherapy. The mean survival duration of patients in the radiotherapy arm was 49 days and 51 days in the optimal adjunctive treatment arm. In both groups, there were no differences in the quality of life and the use of steroids. The entire brain can be subjected to radiation therapy in a 20-Gy regimen in five fractions or 30 Gy in 10 fractions [18]. Alternative fractionation: 40 Gy in 20 fractions twice a day does not affect patients' survival times. Attempts have been made to use chemotherapy as a radiosensitizer without improving patients' survival times [19].

In patients with asymptomatic CNS metastases who have not yet received systemic treatment, the therapy sequence should be considered. In a study published in 2014 [20], patients with non-small cell lung carcinoma with asymptomatic CNS metastases (one to four lesions) received either stereotactic radiosurgery (SRS) followed by a two-drug cisplatin-based chemotherapy or chemotherapy alone. The average age of the patients was 58 years, with a mean total survival time of 14.6 months in the arm with stereotactic radiotherapy and chemotherapy; in the arm with chemotherapy alone—15.3 months. The average time to progression in the CNS was 9.4 months in the arm with SRS; in the arm with chemotherapy alone—6.6 months. The symptomatic progression of CNS lesions was more frequently observed in patients without stereotactic radiotherapy [19].

In the phase 3 trial [21], patients with CNS metastases of non-small cell lung carcinoma (NSCLC) received chemotherapy—cisplatin with vinorelbine (days 1, 8, 15, and 22)—courses every 28 days, with a maximum of six courses. Whole-brain radiation therapy (WBRT)—30 Gy in 10 fractions—took place early in some patients—on days 1–12 of the first chemotherapy course and, in the second arm, after two chemotherapy courses (56 days). The objective response rate was 20% in the early radiotherapy arm and 21% in the delayed radiotherapy arm. The average survival duration in patients with delayed radiotherapy was 24 weeks and, in patients with early radiotherapy, 21 weeks. The results indicate that, during chemotherapy treatment, the implementation time of CNS palliative radiotherapy in patients with asymptomatic NSCLC brain metastases does not affect patients' survival duration [21].

In patients with symptomatic metastases of lung cancer to the CNS, the recommended dose of corticosteroids used long term in the prevention of cerebral edema is 4 mg of dexamethasone daily. Increasing the dose of the steroid to 16 mg daily does not improve the disease control but generates treatment toxicity [22].

4. Non-Small Cell Lung Carcinoma with Present Epidermal Growth Factor Receptor (EGFR) Mutation and ALK Rearrangement

The discovery of the EGFR mutation [23,24] and ALK rearrangement [24] and then the introduction of first-generation tyrosine kinase inhibitors (gefitinib—Figure 2A and erlotinib—Figure 2B) to the treatment of non-small cell lung carcinoma allowed, compared to platinum-based dual-drug chemotherapy, for longer progression-free survival (PFS), higher objective response rates (ORR), and better disease control rates (DCR) in comparison with two-drug platinum-derivative-based chemotherapy [23–25].

EGFR is one of the four members [25] of the HER family receptors, which comprise [26] EGFR/HER1/erbB1, HER2/erbB2, HER3/erbB3, and HER4/erbB4 [27]. EGFR signaling [25] is triggered by the binding of growth factors, such as Epidermal Growth Factor (EGF) [25,28], resulting in the dimerization of EGFR molecules [27]. Autophosphorylation and transphosphorylation of the receptors through their tyrosine kinase domains [25] leads to the recruitment of downstream effectors and the activation of proliferative and cell survival signals [25]. In recent years, intensive research has been dedicated to the Epidermal Growth Factor Receptor (EGFR) [27] due to its significant role in the pathogenesis [27,29] of malignant tumors. In many types of cancers, intracellular pathways modulated by EGFR have been identified [25,28] as crucial factors influencing tumor survival and development [30]. On the other hand, EGFR has also been shown to be a promising molecular target [25–27] for potential therapeutic agents. Attempts to modify the signal transduction exerted by EGF have been made either by blocking [25] the activity of certain elements of the EGFR pathway or by direct inhibition of the EGF receptor itself [27]. Gefitinib and erlotinib target the ATP cleft [31] within the tyrosine kinase Epidermal Growth Factor Receptor (EGFR). Specific activating mutations within the tyrosine kinase domain of the EGFR molecularly correlate to the responses [23–25] to gefitinib or erlotinib (Figure 3A,B).

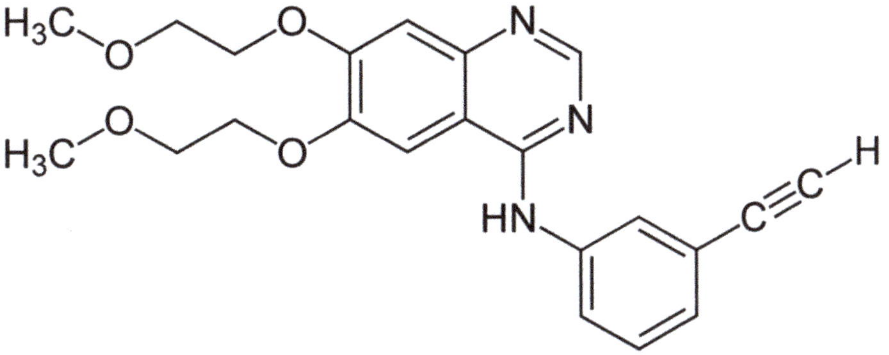

(A)

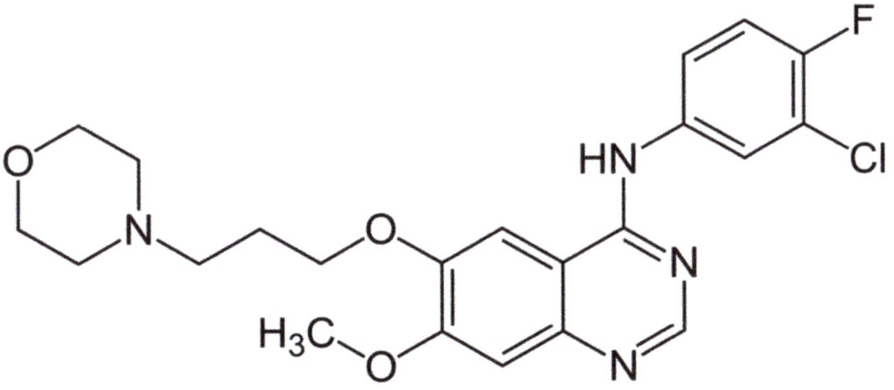

(B)

Figure 2. (**A**) Gefitinib—first-generation tyrosine kinase inhibitor; (**B**) Erlotinib—first-generation tyrosine kinase inhibitor.

In Figure 3A, the inhibitor (dark blue), representing gefitinib, occupies the ATP cleft. The locations of the two missense mutations are shown within the activating loop of the tyrosine kinase (light blue); the three in-frame deletions are all present within another loop (shown in red), which flanks the ATP cleft. Figure 3B shows a close-up view of the EGFR tyrosine kinase domain, with the critical amino acids implicated in binding the inhibitor. Specifically, 4-anilinoquinazoline compounds such as gefitinib inhibit catalysis by occupying the ATP-binding site, where they form hydrogen bonds with methionine 769 (M769) and cysteine 751 (C751) residues, whereas their anilino ring is close to the methionine 742 (M742), lysine 721 (K721), and leucine 764 (L764) residues (all shown in green). In-frame deletions within the loop that is targeted by mutations (shown in red) are predicted to alter the positions of these amino acids relative to that of the inhibitor. Mutated residues (red) are shown within the activation loop of the tyrosine kinase (light blue).

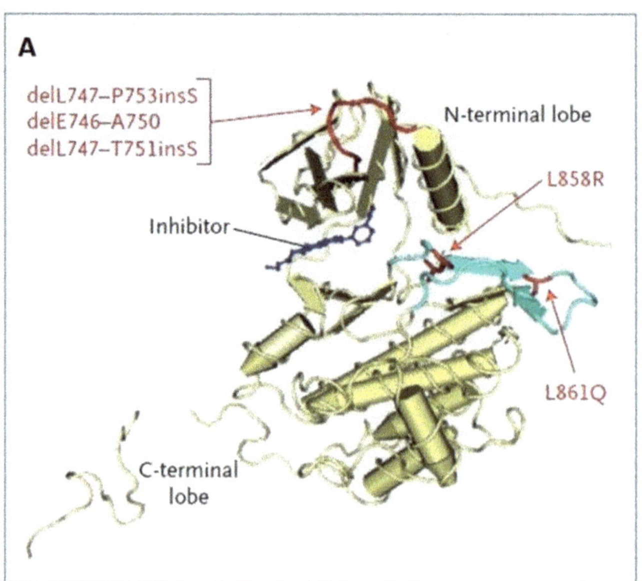

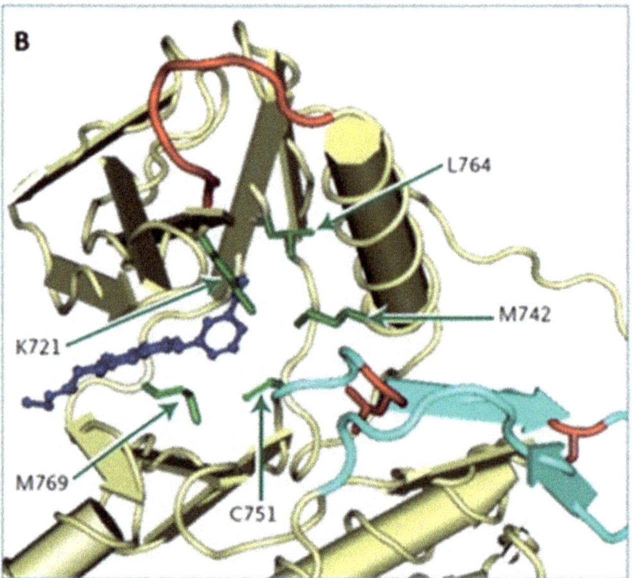

Figure 3. (**A,B**) Clustering of mutations in the Epidermal Growth Factor Receptor (EGFR) gene (adapted from Lynch T.J. et al., The New England Journal of Medicine, 2004) [23].

Gain-of-function mutations [32] in the tyrosine kinase domain of the EGFR gene markedly increase the sensitivity to EGFR tyrosine kinase inhibitors (TKIs) [33]. It has been shown that 10–30% of all lung adenocarcinomas [34,35] contain an EGFR-activating mutation. EGFR mutations occur mostly in adenocarcinoma, younger women and girls [26], and never-smokers [23–25]. The increased prevalence [32] of EGFR mutations in the metastatic disease (early stage—14, 2% and metastatic—30, 3%) in the dataset may partially reflect referral bias [26] (Figure 4).

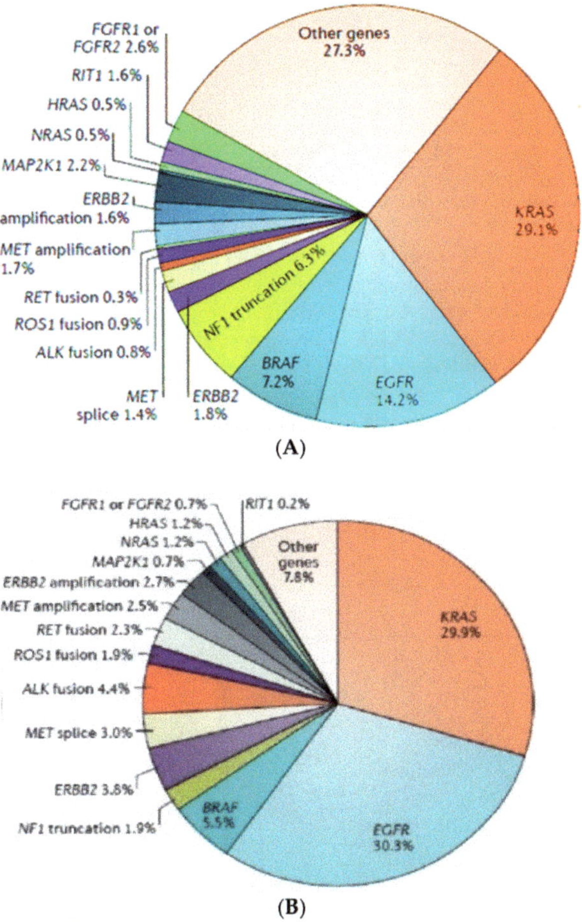

Figure 4. Distribution of the oncogenic driver mutations in non-small cell lung carcinoma (NSCLC) (adapted from Skoulidis F et al., 2019). (**A**) Early stage and (**B**) metastatic disease.

The most common oncogenic mutations are deletion in exon 19 (45–50% of all somatic EGFR mutations) and a point mutation (L858R) in exon 21 (35–45% of mutations) [25,36,37]. Ex20Ins mutations are the third-most common EGFR-activating mutations in NSCLC [38], which collectively account for approximately 4% to 10% of all EGFR mutations [35]. These mutations are predictive of the clinical activity of the EGFR TKIs [39], which yield a superior RR (response rate) [39,40] and PFS [40,41], as well as a better QoL (quality of life) [39–41] scores when compared with combination chemotherapy in the first-line setting [23]. The discovery of EGFR mutations and ALK rearrangements also contributed to the development of a new scale [37] of prognostic factors in patients with brain metastases of non-small cell lung carcinoma, taking into account the presence of EGFR mutations or ALK rearrangements. The Lung Cancer Molecular Markers Graded Prognostic Assessment

(Lung-molGPA) index facilitates making clinical decisions in this group of patients. In addition to the previous parameters [12], such as the patient's age, general performance, presence or absence of cancer outside the CNS, the number of brain metastases (one–four or >four), it also takes into account the gene status of the EGFR and ALK mutations. The higher the number of points obtained on this scale, the better the prognosis and longer survival of patients [37] (Table 2).

Table 2. Lung-molGPA (Lung Cancer Molecular Markers Graded Prognostic Assessment).

Prognostic Factor	Age (Years)	KPS	Extracranial Metastases	Number of BM	Gene Status
0	≥70	<70	Present	>4	EGFR neg/unk and ALK neg/unk
0.5	<70	70–80	-	1–4	NA
1	-	90–100	Absent	NA	EGFR-pos or ALK-pos

KPS—Karnofsky Performance Status, NA—not applicable, neg/unk—negative or unknown, pos—positive, BM—brain metastases, EGFR—Epidermal Growth Factor Receptor, and ALK—Anaplastic Lymphoma Kinase.

First-generation tyrosine kinase inhibitors block [38,39] the EGFR receptor in a reversible manner. A better control of neoplastic disease during treatment with gefitinib or erlotinib [40,41], and the longer lives of patients, drew attention to the problem of metastatic lesions in the CNS. Lung cancer patients treated with first-generation TKI achieved a mean survival time of 33.1 months. After the diagnosis of disease progression in the CNS or in the meninges, the average survival time was 5.5 and 5.1 months. The incomplete penetration of drugs into the CNS through the blood–brain barrier causes a worse response to the first-generation TKI treatment in the brain and meninges [42]. Despite the low molecular weights of gefitinib and erlotinib, their penetration rates into the cerebrospinal fluid (1.13% and 2.77%, respectively) and the CNS concentration rates are low (3.7 ng/mL and 28.7 ng/mL, respectively) [43] (Table 3).

Table 3. Concentrations of the EGFR and ALK tyrosine kinase inhibitors in the cerebrospinal fluid (CSF).

Compound	CSF Penetration Rate (%)	CSF Concentration ng/mL or nM/L
Gefitinib	1.13 ± 0.36%	3.7 ± 1.9 ng/mL
		8.2 ± 4.3 nM/L
Erlotinib	2.8–5.1%	28.7 ± 16.8 ng/mL
		66.9 ± 39.0 nM/L
Afatynib	<1%	0.464 ng/mL
Crizotinib	0.26%	0.616 ng/mL
Alectinib	0.86	2.69 nM/L
Ceritinib	0.15	not reported
Lorlatinib	20–30%	not reported

Attempts have been made to increase the doses of gefitinib or erlotinib [44] or to introduce the pulsatile administration of drugs in patients with metastatic lesions in the CNS. The achieved therapeutic effects were still unsatisfactory due to the fact that higher doses of the first-generation TKI [45] increased the drug concentration index in the CNS, but the obtained effect was short-lived. A prolonged administration of high doses of erlotinib or gefitinib causes unacceptable toxicity and is not used [44–46].

Afatinib (Figure 5) is a second-generation tyrosine kinase inhibitor.

Acquired resistance occurs [47] in patients who initially benefit from EGFR-targeted therapies (first-generation tyrosine kinase inhibitors) [25,26].

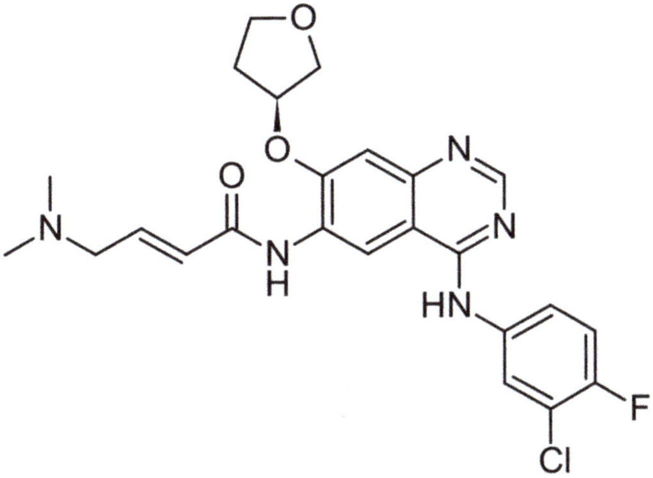

Figure 5. Afatinib—second-generation tyrosine kinase inhibitor.

A clinical definition of acquired resistance to EGFR TKIs: acquired resistance in systemic progression (by Response Evaluation Criteria in Solid Tumors (RECIST) or World Health Organization (WHO) criteria) after a complete or partial response or >six months of stable disease after treatment with targeted therapy [48]. It irreversibly binds to the EGFR receptor and also has a higher affinity for the receptor compared to first-generation drugs. The studies LUX-Lung 3 [49] (cisplatin with pemetrexed) and LUX-Lung 6 [50] (cisplatin with gemcitabine) demonstrated the superiority of TKI over platinum-based two-drug chemotherapy with new-generation drugs. In the presented studies, patients receiving TKI compared to chemotherapy benefited from longer progression-free survival (PFS). They showed higher objective response rates (ORR) and a better disease control rate (DCR). The CNS penetration rate for afatinib is below 1%, and the CNS concentration is 0.46 ng/mL [51]. The LUX-Lung 3 and LUX-Lung 6 [49,50] studies were analyzed, taking into account asymptomatic brain metastases. The progression-free time in the LUX-Lung 3 trial [52] in patients with CNS metastases was 11.1 months in the afatinib arm and 5.4 months in the chemotherapy arm. In the LUX-Lung 6 trial, patients with CNS metastases treated with afatinib [53–55] achieved a progression-free time of 8.2 months, and in the chemotherapy arm, PFS was 4.7 months. Progression-free time in the afatinib arm compared to chemotherapy was equal in patients without brain metastases and in patients with CNS metastases [50,56,57]. The LUX-Lung 7 trial compared gefitinib with afatinib and included patients with central nervous system metastases. The mean follow-up was 27.3 months; progression-free survival for the afatinib arm was 11 months and 10.9 months for the gefitinib arm. The time to treatment failure for afatinib was 13.7 months and, for gefitinib, 11.5 months. Afatinib and gefitinib in the LUX-Lung 7 trial—no difference in the overall survival (OS) [58]. Brueckl et al. (ESMO 2018 Congress, abstract 1449P) [59] presented an analysis of GIDEON, a prospective noninterventional study that was conducted in Germany to investigate the activity and tolerability of first-line afatinib in routine clinical care. Among 151 treated patients, the majority (72.8%) started treatment at an afatinib dose of $\geq$40 mg; 61.8% of them had dose reductions. In the group of patients starting at <40 mg, 46.2% had dose reductions, while dose increases were performed in 33.3%. The safety profile of afatinib was consistent with the known safety profile identified by the clinical trials. In spite of relatively high proportions of patients with brain metastases (approximately 30%) and uncommon *EGFR* mutations (approximately 13%), the results corroborated the clinical data for afatinib in the routine setting. The median PFS was 12.9 months, with a 12-month PFS rate of 54.6%. Seventy-three percent of patients responded, and 90% obtained disease control. Both the ORRs and disease control rates (DCR) were independent of the type of *EGFR* mutation, the presence of baseline brain metastases, and the starting dose (Figure 6).

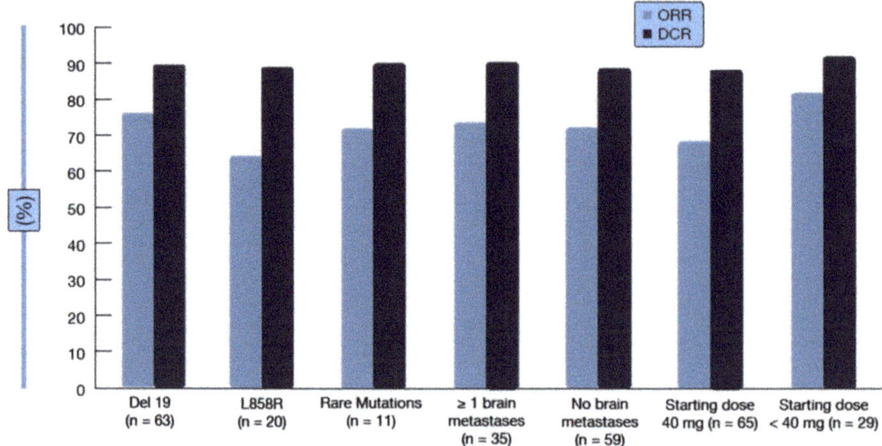

Figure 6. Overall response rates and disease control rates obtained with first-line afatinib in the noninterventional GIDEON study [59] (adapted from Brueckl et al., ESMO, 2018).

Osimertinib (Figure 7) is a third-generation tyrosine kinase inhibitor. In the AURA 3 clinical trial [60], it was compared to pemetrexed and cisplatin or carboplatin-based two-drug chemotherapy [61,62].

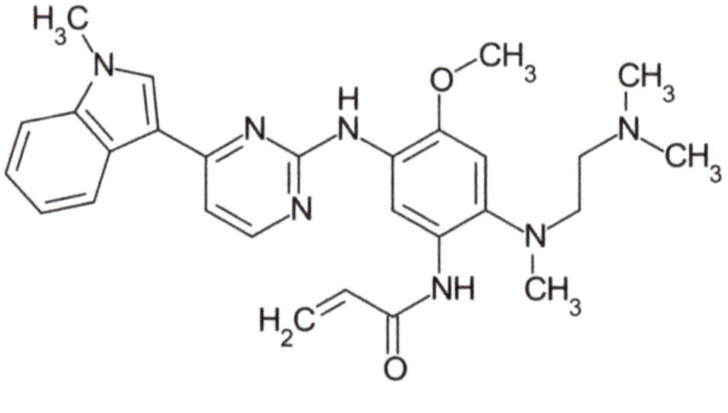

Figure 7. Osimertinib—third-generation tyrosine kinase inhibitor.

It was the second-line treatment for all patients, with the first- and second-generation EGFR TKI used in the first-line treatment. After the disease progressed, the T790M mutation determining the resistance [63] to drugs from the first- and second-generation TKI groups was determined, and patients were randomized to the Osimertinib arm or to the chemotherapy arm [64,65]. The trial also included patients with metastases to the central nervous system, without symptoms resulting from focal lesions in the CNS, who did not require treatment with steroids for at least four weeks before the start of the trial. The median treatment duration was 10.1 months for patients treated with Osimertinib (Osimertinib, *n* = 279) and 4.4 months for patients treated with chemotherapy (*n* = 140). The objective response rate (ORR) was 71% for the Osimertinib treatment and 31% for chemotherapy-treated patients [63,64]. A subgroup analysis was performed; patients with measurable CNS lesions (one or more brain lesions) were included in the first group and patients with one or more lesions measurable and nonmeasurable in the CNS in the second group. In the first group of patients [64,65], the ORR was 70% in the Osimertinib arm and 31% in the chemotherapy arm [66]. In the second group of patients [64,65,67], the ORR was 40% in the Osimertinib arm and 17% in the chemotherapy arm. In both groups of patients, the mean response time in the CNS was 8.9 months in the treatment with Osimertinib and 5.7 months in the treatment with chemotherapy [67,68]. The mean PFS in the group of patients with measurable changes in the treatment with Osimertinib was 11.7 months, while, in chemotherapy, it was 5.6 months [69].

The EGFR T790M mutation [70] is the most common mechanism of TKI first- and second-generation resistance (detected in 50–60% of patients) [25]. It is unlikely that any erlotinib combination [70–72] will overcome this specific drug resistance mechanism.

Osimertinib, a third-generation small molecule tyrosine kinase inhibitor, is recommended in patients with the T790M resistance mutation [66,67]. It is also effective in patients with metastases to the central nervous system and the meninges [67,73,74].

In the phase 3 FLAURA [71–74] clinical trial, patients receiving Osimertinib achieved a PFS of 18.9 months and 10.2 months in the control arm (gefitinib or erlotinib). Patients with CNS metastases also benefited from treatment with Osimertinib [74,75].

Osimertinib is a third-generation tyrosine kinase inhibitor and has demonstrated high tolerability [73–75]. Some patients showed resistance to this drug, and the major mutation site is C797S on the EGFR gene (discovery of genome sequencing) [75,76]. In the future, when EGFR TKI drug resistance occurs [75,76], genetic testing could be used to select the treatment method corresponding to the resistance mechanism [74–76].

Progress in the field of molecular biology in recent years has enabled the identification of potential oncogenic pathways [77,78]. In 2007, Soda and his colleagues found an echinoderm microtubule-associated protein-like 4 (EML4) ALK fusion gene from non-smell-cell lung cancers [77]. These ALK fusion proteins can induce the constitutive activation of the ALK tyrosine kinase [77,78]. The oligomerization of domains such as the coiled-coil [77] domain of the fusion partner gives stimulation [79] ALK downstream pathways as a result [79]. The P13K-AKT-Mtor, RAS-MAPK-ERK, or JAK-STAT pathways are constitutively activated [77,79].

ALK mutations are rare and can be found in approximately 3–7% of patients with the diagnosis of NSCLC [77–79]. ALK mutations are more common in young, nonsmoking men with adenocarcinoma [78,79].

Crizotinib is an ATP-competitive, orally bioavailable ALK inhibitor [80] and was first applied for the treatment of EML4 ALK-positive NSCLC [81]. Crizotinib (Figure 8) was introduced based on the phase 3 Profile 1014 study [81] as a standard of treatment in patients with ALK-positive lung cancer.

This first-generation tyrosine kinase inhibitor has a concentration rate in the cerebrospinal fluid of 0.616 ng/mL and a penetration rate to the cerebrospinal fluid of 0.26% [51,80]. In the Profile 1014 trial [81], crizotinib achieved significantly longer PFS compared to chemotherapy (nine months vs. four months), and after 12 and 24 weeks of treatment, higher intracerebral DCR of 85% and 65% was observed in the arm with crizotinib and 45% and 25% in the arm with chemotherapy. The intracerebral control of the disease was also better in patients with metastases to the CNS—23% compared to chemotherapy. However, the isolated progression of the disease in the CNS was more frequent during treatment with crizotinib; extracerebral progression was more frequent during the treatment with chemotherapy (pemetrexed cisplatin). Patients with untreated CNS metastases or progression [82,83] of the disease were not randomized for the trial, and 20% of patients participating in the trial had CNS radiotherapy.

Nearly one-third of patients treated with crizotinib had CNS metastases in the first year of therapy. In some of these patients, it was the only location of neoplastic disease progression [83,84].

Alectinib (Figure 9) is a second-generation tyrosine kinase inhibitor used in patients with ALK-positive lung tumors and is also effective in the central nervous system [85,86].

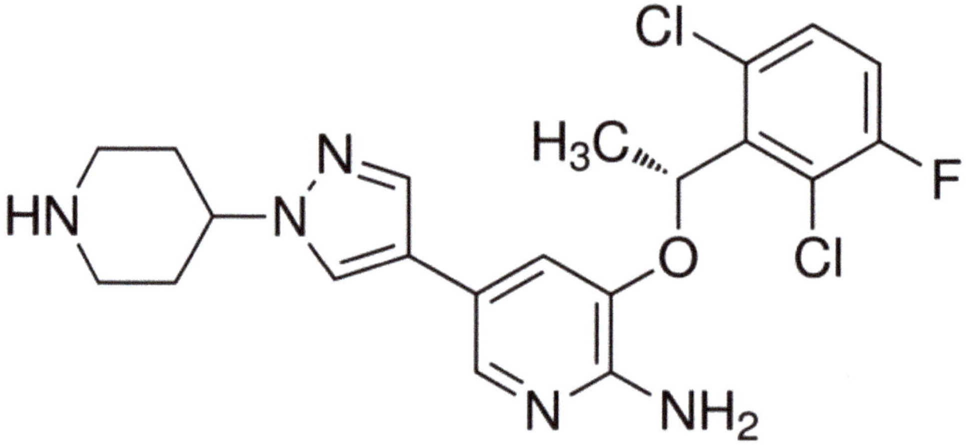

Figure 8. Crizotinib—first-generation Anaplastic Lymphoma Kinase tyrosine kinase inhibitor (ALK TKI).

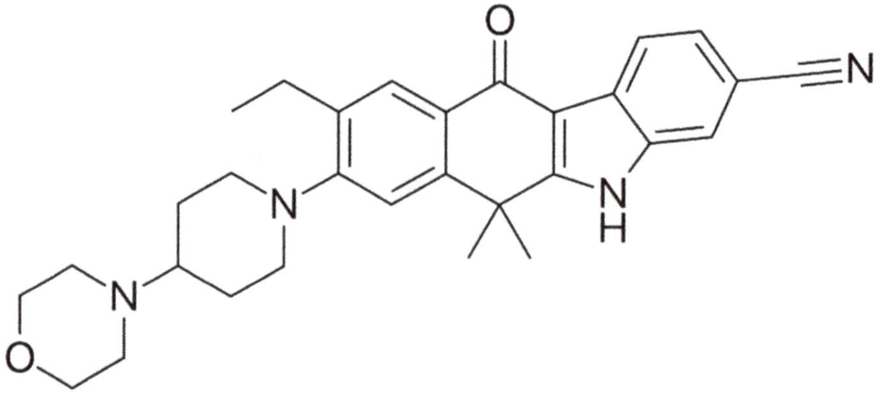

Figure 9. Alectinib—a second-generation ALK TKI.

Alectinib shows a concentration level of 2.69 nM [87,88] in the cerebrospinal fluid, and the penetration rate into the cerebrospinal fluid is 86% [51,87,88].

The ALEX phase 3 [89] clinical trial included previously untreated patients with advanced ALK-positive lung cancer. The patients received crizotinib or alectinib [89,90]. The primary endpoint was progression-free survival. The mean follow-up was 17.6 months in the crizotinib arm and 18.6 months for alectinib. The progression-free time was significantly longer in patients treated with alectinib [89,90] and was 12 months, while, in the crizotinib arm, it was 8.5 months [89,90]. In this trial, alectinib demonstrated therapeutic efficacy in patients with metastases to the central nervous system. Twelve percent of patients treated with alectinib (18 patients) and 45% (68 patients) treated with crizotinib demonstrated changes in the CNS. The one-year (12 months) CNS cumulative events (progression) level was 9.4% vs. 41.4% when comparing alectinib with crizotinib. Good intracerebral disease control coexisted with PFS—a mean average of 25.7 months for alectinib [90,91] and

10.4 months for crizotinib. Treatment toxicity was also lower in patients receiving alectinib and had a favorable safety profile [90,91].

Another ALK inhibitor was found to overcome crizotinib resistance and to better control disease in the CNS [92]. Ceritinib (Figure 10) is a second-generation ALK TKI and has a cerebrospinal fluid penetration rate of 15% [51]. Ceritinib is effective in patients with the I117N resistance mutation [92–94].

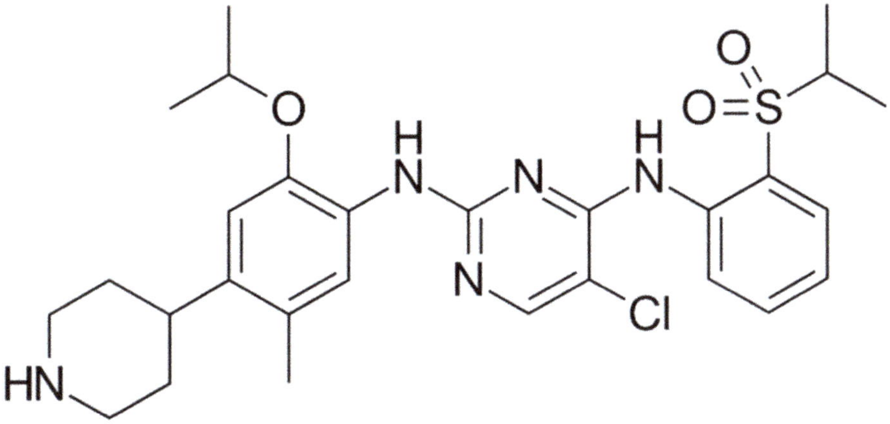

Figure 10. Ceritinib—a second-generation ALK TKI.

The phase 3 clinical trial ASCEND-4 [92] compared ceritinib with chemotherapy as the first-line treatment in patients with advanced lung cancer and ALK rearrangement. In patients with metastases to the CNS, the mean PFS was 10.7 months in the ceritinib arm and 6.7 months in the chemotherapy arm [92]. The overall intracranial response rate in patients with measurable CNS changes at the baseline was 72.7% for ceritinib and 27.3% for chemotherapy [92–94].

In the phase 1 ASCEND-1 trial [95], ceritinib achieved a total intracerebral ORR of 36% in ALK TKI previously treated patients and 63% in ALK TKI untreated patients (patients had baseline CNS measurable changes). In the ASCEND-2 trial the intracerebral ORR was almost 40%, and the intracerebral DCR was 85% [95–97].

Brigatinib (Figure 11) is a second-generation ALK TKI. Brigatinib was shown to be active against the G1202R mutation [98]. The G1202R mutation is resistant to first- and second-generation ALK inhibitors (crizotinib, alectinib, and ceritinib) [98,99].

It was noted that the G1202 mutation was discovered in about 50% of relapse patients following the use of brigatinib [98,99]. Brigatinib, another second-generation ALK inhibitor, demonstrated substantial activity in patients with crizotinib refractory *ALK*-positive NSCLC; however, its activity in the alectinib refractory setting is unknown [98].

The phase 2 ALTA trial [100,101] evaluated the efficacy of brigatinib in patients with advanced ALK-positive non-small cell lung carcinoma previously treated with crizotinib [100,101]. Patients were randomized to two arms of the trial—in one arm, the dose was 90 mg, and, in the other arm, 180 mg for seven days, then 90 mg [101]. In patients with measurable changes in the CNS, the ORR at a higher dose of the drug was 67%, and, at a lower dose, 37%. The DCR exceeded 80% in both arms. In the case of nonmeasurable CNS metastases, the ORR and DCR were higher in patients receiving the higher dose of the drug (19% vs. 6% and 87% vs. 72%). Two-thirds of patients receiving the higher dose of the drug and having measurable lesions in the CNS had an intracerebral response lasting, on average, 16.6 months [100,101]. Brigatinib was compared with crizotinib in a phase 3 trial in patients with ALK-positive [101] lung cancer who had not been previously treated with TKI. Ninety patients had baseline CNS metastases, and 39 patients had measurable CNS lesions with a diameter >10 mm. The intracerebral response to treatment in patients with measurable lesions was 78% in the brigatinib arm and 29% in the crizotinib arm. In the brigatinib group, 9% of patients had disease progression in the CNS, and, in the crizotinib group, 19% of patients [99,100]. Twelve-month PFS in the group of patients with metastatic lesions in the CNS at baseline was higher in the brigatinib arm—67% than in the crizotinib arm—21% [101].

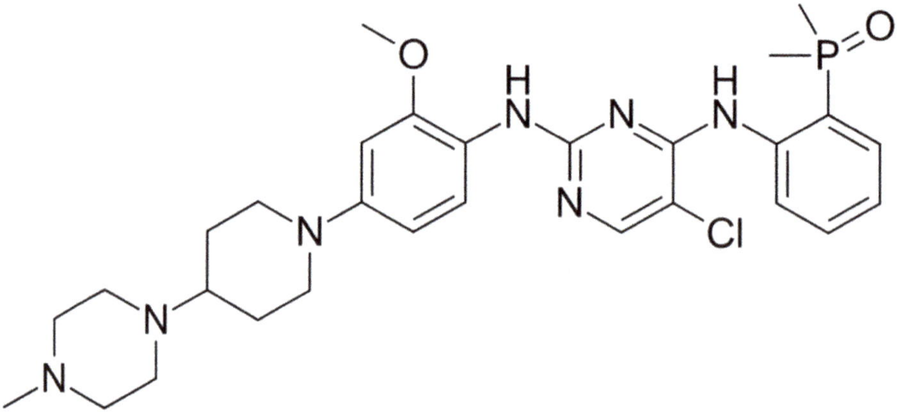

Figure 11. Brigatinib—a second-generation ALK TKI.

Lorlatinib (Figure 12) is a third-generation ALK inhibitor with a penetration rate to the CNS of 20–30% [51]. Lorlatinib is indicated for the treatment of patients with ALK-positive metastatic non-small cell lung cancer [102] whose disease progressed on crizotinib [103] and at least one other ALK inhibitor. Lorlatinib has been shown to be active against almost all of the previously identified ALK TKI resistance mutations, including G1202R [103,104]. It is supposed to overcome the resistance of cancer cells to early-generation drugs [104].

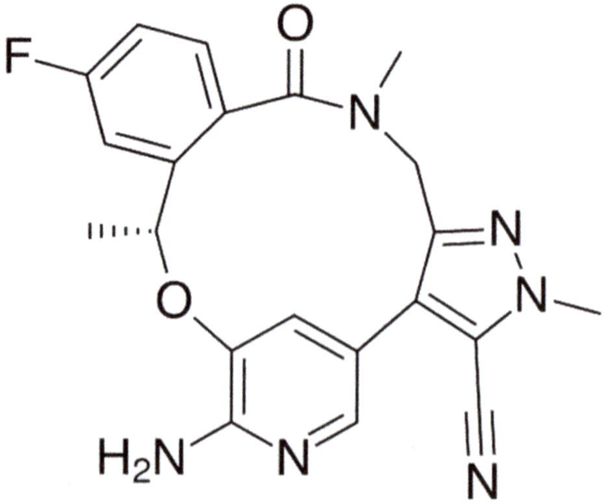

Figure 12. Lorlatinib—a third-generation ALK TKI.

In a phase 1 trial [105], an intracerebral RR of 44% was achieved in the lorlatinib arm in patients with metastatic changes in the CNS for measurable and nonmeasurable lesions and 60% for measurable lesions. Approval was based on a phase 2 study [106,107] in which lorlatinib demonstrated a substantial overall and intracranial response [106,107].

5. Simultaneous CNS Radiotherapy and TKI Therapy

It was shown that lung cancer cells with the EGFR mutation are more radiosensitive [108] than those without. At the same time, lung cancer patients with EGFR mutations have a 50–70% risk of brain metastases [109]. Before the era of targeted lung cancer treatment, patients had either neurosurgical surgery, SRS, or whole-brain radiotherapy with the occurrence of metastases to the brain.

Two hundred and thirty patients with CNS metastases and EGFR mutations were identified and divided into two groups [110]. In one group, 116 patients received TKI (gefitinib, erlotinib, or icotinib), and, in the other group (51), TKI and simultaneous radiotherapy of the whole brain. An ORR of 52% was achieved in both groups; OS in the radiotherapy and TKI arm was 26.4 months and, for the treatment with only TKI, 21.6 months. Compared with TKIs alone, EGFR TKIs plus WBRT demonstrated intracranial progression-free survival (PFS) of 6.9 vs. 7.4 months ($p = 0.232$) and systemic PFS of 7.5 vs. 7.9 months ($p = 0.546$) [110].

In a meta-analysis of seven trials [109] involving 1086 patients with brain metastases, TKI therapy alone was compared with radiotherapy used before TKI therapy. It was shown that patients with non-small cell lung carcinoma and brain metastases who received radiotherapy prior to TKI therapy had longer intracerebral PFS and longer OS [109]. The analysis in the subgroups showed that the survival time of patients was longer in the group with one–three metastatic lesions [109], and shorter OS was obtained by patients with more metastatic lesions. The analysis confirmed that radiotherapy, by damaging the blood–brain barrier, increases the effectiveness of TKI therapy. Consequently, the combined therapy reduces relapse and improves the overall survival [108,109].

In patients with lung cancer and brain metastases, attempts were made to combine up-front CNS radiation and TKI therapy. Based [109] on the current available evidence, patients of non-small cell lung cancer with brain metastases and EGFR mutations have better OS and iPFS (intracerebral progression free survival) when they receive up-front radiotherapy and TKI than TKI alone [108–110].

The subgroup analysis [109] showed that never-smokers lived longer compared to tobacco smokers, and patients diagnosed with adenocarcinoma lived longer compared to other histopathological types. Patients with a better overall performance status (ECOG) lived longer than patients in worse general condition. In the group of patients with symptomatic brain metastases who received TKI and simultaneous whole-brain radiotherapy, significantly worse intracerebral PFS was observed compared to patients treated only with TKI [109–111].

Currently, it is not recommended to discontinue TKI therapy while radiating the whole brain. For stereotaxis (SRS), it is recommended to discontinue TKI three days before SRS and restart it three days after treatment; therefore, the interval is seven days [111].

6. Immunotherapy of Lung Cancer with Brain Metastases

Pembrolizumab is the drug of choice in the first-line treatment of patients with a PD-L1 expression in >50% of tumor cells in patients with non-small cell lung carcinoma without the EGFR or ALK mutation. In the registration trial of pembrolizumab—Keynote-024 [112], 9% of patients had CNS metastases, and, in Keynote-010 [113], 15% of patients had CNS metastases. Pembrolizumab is recommended for the first-line treatment of stage IV non-small cell lung carcinoma (including patients with stable metastatic lesions in the central nervous system) [114].

In the CheckMate 057 trial [115], nivolumab was administered to patients with non-squamous lung cancer as a second-line treatment. Patients achieved an OS of 12.2 months;

in the arm with chemotherapy, the OS was 9.4 months. Patients with stable metastatic lesions in the central nervous system were randomized for the trial [115].

In the EAP (expanded access program) 1588 trial [116], nivolumab was administered to patients with the IIIB/IV tostages of non-squamous lung cancer after progression on prior systemic therapy. Four hundred and nine patients had CNS metastases. They were neurologically stable and could receive a steroid therapy of up to 10 mg of prednisone daily. In the group of patients with metastatic lesions in the brain, the mean follow-up time was 6.1 months (0.1–21.9); the DCR was 39%, and the mean OS was 8.6 months; the CNS disease stabilized in 96 patients, 64 patients achieved a partial response, and 4 complete CNS responses during the nivolumab treatment [116]. Currently, nivolumab is recommended for the second-line treatment of stage IV non-small cell lung carcinoma [114].

Lung cancer metastases to the central nervous system pose a serious problem in oncological treatment. These lesions not only cause progression of the neoplastic disease but also manifest focal symptoms from the CNS, affecting the general condition of patients and worsening contact with them. Neurosurgery, stereotactic radiotherapy, and chemotherapy help to improve the clinical conditions of patients. Introducing new molecules into clinical practice gives a chance not only to improve the general condition of patients but also to prolong their lives.

Funding: This research received no external funding.

Institutional Review Board Statement: Not applicable.

Informed Consent Statement: Not applicable.

Data Availability Statement: Not applicable.

Conflicts of Interest: The authors declare no conflict of interest.

References

1. Didkowska, J.; Wojciechowska, U. *Cancer in Poland in 2017*; The Maria Sklodowska-Curie Memorial Cancer Center: Warsaw, Poland, 2017.
2. Chason, J.L.; Walker, F.B.; Landers, J.W. Metastatic carcinoma in the central nervous system and dorsal root ganglia. A prospective autopsy study. *Cancer* **1963**, *16*, 781–787. [CrossRef]
3. Nussbaum, E.S.; Djalilian, H.R.; Cho, K.H.; Hall, W.A. Brain metastases. Histology, multiplicity, surgery, and survival. *Cancer* **1996**, *78*, 1781–1788. [CrossRef]
4. Zimm, S.; Wampler, G.L.; Stablein, D.; Hazra, T.; Young, H.F. Intracerebral metastases in solid-tumor patients: Natural history and results of treatment. *Cancer* **1981**, *48*, 384–394. [CrossRef]
5. Cagney, D.N.; Martin, A.M.; Catalano, P.J.; Redig, A.J.; Lin, N.U.; Lee, E.Q. Incidence and prognosis of patients with brain metastases at diagnosis of systemic malignancy: A population-based study. *Neuro Oncol.* **2017**, *19*, 1511–1521. [CrossRef]
6. Chen, F.; Zhang, Y.; Varambally, S.; Creighton, C.J. Molecular correlates of metastasis by systematic pan-cancer analysis across the cancer genome atlas. *Mol. Cancer Res.* **2018**, *17*, 476–487. [CrossRef]
7. Patil, C.G.; Pricola, K.; Sarmiento, J.M.; Garg, S.K.; Bryant, A.; Black, K.L. Whole brain radiation therapy (WBRT) alone versus WBRT and radiosurgery for the treatment of brain metastases. *Cochrane Database Syst. Rev.* **2017**, *9*, CD006121. [CrossRef]
8. You, H.; Baluszek, S.; Kamiska, B. Immune microenvironment of brain metastases–are microglia and other brain macrophages little helpers? *Front. Immunol.* **2019**, *10*, 1941. [CrossRef]
9. Muliaditan, T.; Caron, J.; Okesola, M.; Opzoomer, J.W.; Kosti, P.; Georgouli, M.; Gordon, P.; Lall, S.; Kuzeva, D.M.; Pedro, L.; et al. Macrophages are exploited from an innate wound healing response to facilitate cancer metastasis. *Nat. Commun.* **2018**, *9*, 2951. [CrossRef]
10. Lassman, A.B.; De Angelis, L.M. Brain metastases. *Neurol. Clin.* **2007**, *25*, 1173–1192. [CrossRef]
11. Lowery, F.J.; Yu, D. Brain metastasis: Unique challenges and open opportunities. *Biochim. Biophys. Acta Rev. Cancer* **2017**, *1867*, 49–57. [CrossRef]
12. Sperduto, P.W.; Kased, N.; Roberge, D.; Xu, Z.; Shanley, R.; Luo, X.; Sneed, P.K.; Chao, S.T.; Weil, R.J.; Suh, J.; et al. Summary report on the graded prognostic assessment: An accurate and facile diagnosis-specific tool to estimate survival for patients with brain metastases. *J. Clin. Oncol.* **2012**, *30*, 419–425. [CrossRef] [PubMed]
13. Andrews, D.W.; Scott, C.B.; Sperduto, P.W.; Flanders, A.E.; Gaspar, L.E.; Schell, M.C.; Werner-Wasik, M.; Demas, W.; Ryu, J.; Bahary, J.-P. Whole brain radiation therapy with or without stereotactic radiosurgery boost for patients with one to three brain metastases: Phase III results of the RTOG 9508 randomised trial. *Lancet* **2004**, *363*, 1665–1672. [CrossRef]

14. Soon, Y.Y.; Tham, I.W.; Lim, K.H.; Koh, W.Y.; Lu, J.J. Surgery or radiosurgery plus whole brain radiotherapy versus surgery or radiosurgery alone for brain metastases. *Cochrane Database Syst. Rev.* **2014**, *3*, CD009454. [CrossRef] [PubMed]

15. Yamamoto, M.; Serizawa, T.; Shuto, T.; Akabane, A.; Higuchi, Y.; Kawagishi, J.; Yamanaka, K.; Sato, Y.; Jokura, H.; Yomo, S.; et al. Stereotactic radiosurgery for patients with multiple brain metastases (JLGK0901): A multi-institutional prospective observational study. *Lancet Oncol.* **2014**, *15*, 387–395. [CrossRef]

16. Sahgal, A.; Aoyama, H.; Kocher, M.; Neupane, B.; Collette, S.; Tago, M.; Shaw, P.; Beyene, J.; Chang, E.L. Phase 3 trials of stereotactic radiosurgery with or without whole-brain radiation therapy for 1 to 4 brain metastases: Individual patient data meta-analysis. *Int. J. Radiat. Oncol. Biol. Phys.* **2015**, *91*, 710–717. [CrossRef]

17. Mulvena, P.M.; Nankivell, M.G.; Barton, R. Whole brain radiotherapy for brain metastases from non-small lung cancer: Quality of life (QoL) and overall survival (OS) results from the UK Medical Research Council (QUARTZ) randomised clinical trial (ISRCTN 3826061). *J. Clin. Oncol.* **2015**, *33* (Suppl. 20), 8005. [CrossRef]

18. Kotecha, R.; Gondi, V.; Takakura, K.; Saeki, N.; Kunieda, E. Recent advances in managing brain metastasis. *F1000Research* **2018**, *7*, 1772. [CrossRef]

19. Tsao, M.N.; Lloyd, N.; Wong, R.K. Whole brain radiotherapy for the treatment of newly diagnosed multiple brain metastases. *Cochrane Database Syst. Rev.* **2012**, *4*, CD003869. [CrossRef]

20. Lim, S.H.; Lee, J.Y.; Lee, M.Y.; Kim, H.S.; Lee, J.; Sun, J.M.; Ahn, J.S.; Um, S.W.; Kim, H.; Kim, B.S.; et al. A randomized phase III trial of stereotactic radiosurgery (SRS) versus observation for patients with asymptomatic cerebral oligo-metastases in non-small-cell lung cancer. *Ann. Oncol.* **2015**, *26*, 62–768. [CrossRef]

21. Robinet, G.; Thomas, P.; Breton, J.L.; Lena, H.; Gouva, S.; Dabouis, G.; Bennouna, J.; Souquet, P.J.; Balmes, P.; Thiberville, L.; et al. Results of a phase III study of early versus delayed whole brain radiotherapy with concurrent cisplatin and vinorelbine combination in inoperable brain metastasis of non-small-cell lung cancer. Groupe Francais de Pneumo- Cancerologie (GFPC) Protocol 95-1. *Ann. Oncol.* **2001**, *12*, 59–67. [CrossRef]

22. Vecht, C.J.; Hovestadt, A.; Verbiest, H.B.; van Vliet, J.J.; van Putten, W.L. Dose-effect relationship of dexamethasone on Karnofsky performance in metastatic brain tumors: A randomized study of doses of 4.8 and 16mg per day. *Neurology* **1994**, *44*, 65–680. [CrossRef] [PubMed]

23. Lynch, T.J.; Bell, D.W.; Sordella, R.; Gurubhagavatula, S.; Okimoto, R.A.; Brannigan, B.W.; Harris, P.L.; Haserlat, S.M.; Supko, J.G.; Haluska, F.G.; et al. Activating mutations in the epidermal growth factor receptor underlying responsiveness of non-small-cell lung cancer to gefitinib. *N. Engl. J. Med.* **2004**, *350*, 229–239.

24. Novello, S.; Barlesi, F.; Califano, R.; Cufer, T.; Ekman, S.; Giaj Levra, M.; Kerr, K.; Popat, S.; Reck, M.; Senan, S.; et al. ESMO Clinical Practice Guidelines for diagnosis, treatment and follow-up. *Ann. Oncol.* **2016**, *27* (Suppl. 5), V1–V27. [CrossRef] [PubMed]

25. Liu, X.; Wang, P.; Zhang, C.; Ma, Z. Epidermal growth factor receptor (EGFR): A rising star in the era of precision medicine of lung cancer. *Oncotarget* **2017**, *8*, 50209–50220. [CrossRef]

26. Hirsch, F.R.; Scagliotti, V.; Mulshine, J.L.; Kwon, R.; Curran, W.J., Jr.; Wu, Y.-L.; Paz-Ares, L. Lung cancer: Current therapies and new targeted treatments. *Lancet* **2017**, *389*, 299–311. [CrossRef]

27. Sigismund, S.; Avanzato, D.; Lanzetti, L. Emerging functions of the EGFR in cancer. *Mol. Oncol.* **2018**, *12*, 3–20. [CrossRef]

28. Tanaka, T.; Zhou, Y.; Ozawa, T.; Okimoto, R.A.; Brannigan, B.W. Ligand-activated epidermal growth factor receptor (EGFR) signaling governs endocytic trafficking of unliganded receptor monomers by non-canonical phosphorylation. *J. Biol. Chem.* **2018**, *293*, 2288–2301. [CrossRef]

29. Sooro, M.A.; Zhang, N.; Zhang, P. Targeting EGFR-mediated autophagy as a potential strategy for cancer therapy. *Int. J. Cancer* **2018**, *143*, 2116–2125. [CrossRef]

30. Keppel, T.R.; Sarpong, K.; Murray, E.M.; Monsey, J.; Zhu, J.; Bose, R. Biophysical Evidence for Intrinsic Disorder in the C-terminal tails of the epidermal growth factor receptor (EGFR) and HER3 Receptor tyrosine kinases. *J. Biol. Chem.* **2017**, *292*, 597610. [CrossRef]

31. Thomas, R.; Srivastava, S.; Katreddy, R.R.; Sobieski, J.; Zhang, W. Kinase-inactivated EGFR is required for the survival of wild-type EGFR-expressing cancer cells treated with tyrosine kinase inhibitors. *Int. J. Mol. Sci.* **2019**, *20*, 2515. [CrossRef]

32. Skoulidis, F.; Heymach, J.V. Co-occurring genomic alterations in non-small cell lung cancer biology and therapy. *Nat. Rev. Cancer* **2019**, *19*, 495–509. [CrossRef] [PubMed]

33. Castellanos, E.; Feld, E.; Horn, L. Driven by mutations: The predictive value of mutation subtype in EGFR-mutated non-small cell lung cancer. *J. Thorac. Oncol.* **2017**, *12*, 612–623. [CrossRef] [PubMed]

34. Chen, K.; Yu, X.; Wang, H.; Zhang, N.; Zhang, P. Uncommon mutation types of epidermal growth factor receptor and response to EGFR tyrosine kinase inhibitors in Chinese non-small cell lung cancer patients. *Cancer Chemother. Pharmacol.* **2017**, *80*, 1179–1187. [CrossRef] [PubMed]

35. Jang, J.; Son, J.; Park, E.; Kosaka, T.; Saxon, J.A.; De Clercq, D.J.H.; Choi, H.G.; Tanizaki, J.; Eck, M.J.; Janne, P.A.; et al. Discovery of a highly potent and broadly effective epidermal growth factor receptor and HER2 exon20 insertion mutant inhibitor. *Angew. Chem. Int. Ed. Engl.* **2018**, *7*, 11629–11633. [CrossRef] [PubMed]

36. Kosaka, T.; Tanizaki, J.; Paranal, R.M.; Endoh, H.; Lydon, C.H.; Capelletti, M.; Repellin, C.E.; Choi, J.; Ogino, A.; Calles, A.; et al. Response heterogeneity of EGFR and HER2 Exon 20 insertions to covalent EGFR and HER2 inhibitors. *Cancer Res.* **2017**, *77*, 2712–2721. [CrossRef]

37. Sperduto, P.W.; Yang, T.J. Estimating survival in patients with lung cancer and brain metastases: An update of the Graded Prognostic Assessment for Lung Cancer Using Molecular Markers (Lung-molGPA). *JAMA Oncol.* **2017**, *3*, 827–831. [CrossRef]

38. Maemondo, M.; Inoue, A.; Kobayashi, K.; Sugawara, S.; Oizumi, S.; Isobe, H.; Gemma, A.; Harada, M.; Yoshizawa, H.; Kinoshita, I.; et al. North- East Japan Study Group. Gefitinib or chemiotherapy for non-small cell lung cancer with mutated EGFR. *N. Engl. J. Med.* **2010**, *362*, 2380–2388. [CrossRef]

39. Inoue, A.; Kobayashi, K.; Maemondo, M. Updated overall survival results from a randomized phase III trial comparing gefitinib with carboplatin-paclitaxel for chemo-naive non-small cell lung cancer with sensitive EGFR gene mutations (NEJ002). *Ann. Oncol.* **2013**, *24*, 54–59. [CrossRef]

40. Han, J.Y.; Park, K.; Kim, S.W.; Lee, D.H.; Kim, H.Y.; Kim, H.T.; Ahn, M.J.; Yun, T.; Ahn, J.S.; Suh, C.H.; et al. First-SIGNAL: First-line single-agent iressa versus gemcitabine and cisplatin trial in never-smokers with adenocarcinoma of the lung. *J. Clin. Oncol.* **2012**, *30*, 1122–1128. [CrossRef]

41. Zhou, C.; Wu, Y.L.; Chen, G. Final overall survival results from a randomised phase III study of erlotinib versus chemotherapy as first-line treatment of EGFR mutation-positive advanced non-small-cell lung cancer (OPTIMAL, CTONG-0802). *Ann. Oncol.* **2015**, *26*, 1877–1883. [CrossRef]

42. Heon, S.; Yeap, B.Y.; Britt, J.; Costa, D.B.; Rabin, M.S.; Jackman, D.M.; Johnson, B.E. Development of Central Nervous System Metastases in patients with advanced non-small cell lung cancer and somatic EGFR mutations treated with gefitinib or erlotinib. *Cancer Therapy.* **2010**, *16*, 5873–5882. [CrossRef] [PubMed]

43. Haaland, B.; Tan, P.S.; De Castro, G., Jr.; Lopes, G. Meta-Analysis of First-Line Therapies in Advanced Non-Small-Cell Lung Cancer Harboring EGFR-Activating Mutations. *J. Thorac. Oncol.* **2014**, *9*, 805–811. [CrossRef] [PubMed]

44. Togashi, Y.; Masago, K.; Fukudo, M.; Tsuchido, Y.; Okuda, C.H.; Kim, Y.H.; Ikemi, Y.; Sakamori, Y.; Mio, T.; Katsuta, T.; et al. Efficacy of increased-dose erlotinib for central nervous system metastases in non-small cell lung cancer patients with epidermal growth factor receptor mutation. *Cancer Chemother. Pharmacol.* **2011**, *68*, 1089–1092. [CrossRef] [PubMed]

45. Gromes, C.; Oxnard, G.R.; Kris, M.G.; Miller, V.A.; Pao, W.; Holodny, A.I.; Clarke, J.L.; Lassman, A.B. Pulsatile high-dose weekly erlotinib for CNS metastases from EGFR mutant non-small cell lung cancer. *Neuro Oncol.* **2011**, *13*, 1364–1369. [CrossRef] [PubMed]

46. Clarke, J.L.; Pao, W.; Wu, N. High dose weekly erlotinib achieves therapeutic concentrations in CSF and is effective in leptomeningeal metastases from epidermal growth factor receptor mutant lung cancer. *J. Neurooncol.* **2010**, *99*, 283–286. [CrossRef]

47. Chong, C.R.; Janne, P.A. The quest to overcome resistance to EGFR-targeted therapies in cancer. *Nat. Med.* **2013**, *19*, 1389–1400. [CrossRef] [PubMed]

48. Jackman, D.; Pao, W.; Riely, G.J.; Engelman, J.A.; Kris, M.G.; Janne, P.A.; Lynch, T.; Johnson, B.E.; Miller, V.A. Clinical definition of acquired resistance to epidermal growth factor receptor tyrosine kinase inhibitors in non-small-cell lung cancer. *J. Clin. Oncol.* **2010**, *28*, 357–360. [CrossRef]

49. Yang, J.C.-H.; Sequist, L.V.; Geater, S.L.; Tsai, C.H.-M.; Mok, T.S.K.; Schuler, M.; Yamamoto, N.; Yu, C.-J.; Ou, S.-H.I.; Zhou, C.; et al. Clinical activity of afatinib in patients with advanced non-small-cell lung cancer harbouring uncommon EGFR mutations: A combined post-hoc analysis of LUX-Lung 2, Lux-Lung 3, and LUX-Lung 6. *Lancet Oncol.* **2015**, *16*, 830–838. [CrossRef]

50. Wu, Y.L.; Zhou, C.; Hu, C.P. LUX-Lung 6: A randomized, open-label, phase III study of afatinib (A) versus gemcitabine/cisplatin (GC) as first-line treatment for Asian patients with EGFR mutation-positive (EGFR M+) advanced adenocarcinoma for the lung. *J. Clin. Oncol.* **2013**, *31*, 8016. [CrossRef]

51. Wrona, A.; Dziadziuszko, R.; Jassem, J. Management of brain metastases in non-small cell lung cancer in the era of tyrosine kinase inhibitors. *Cancer Treat. Rev.* **2018**, *71*, 59–67. [CrossRef]

52. Yang, J.C.-H.; Sequist, L.V.; Zhhou, C.; Schuler, M.; Geater, S.L.; Mok, T.; Hu, C.-P.; Yamamoto, N.; Feng, J.; Byrne, K.O.; et al. Effect of dose adjustment on the safety and efficacy of afatinib for EGFR mutation-positive lung adenocarcinoma: Post hoc analyses of the randomized LUX-Lung 3 and 6 trials. *Ann. Oncol.* **2016**, *27*, 2103–2110. [CrossRef] [PubMed]

53. Miller, A.; Hirsh, V.; Cadranel, J. Afatinib versus placebo for patients with advanced, metastatic non-small-cell lung cancer after failure of erlotinib, gefitinib, or both, and one or two lines of chemotherapy (LUX-Lung 1): A phase 2b/3 randomised trial. *Lancet Oncol.* **2012**, *13*, 528–538. [CrossRef]

54. Yang, J.; Wu, Y.-L.; Schuler, M.; Sebastian, M.; Popat, S.; Yamamoto, N.; Zhou, C.; Hu, C.-P.; O'Byrne, K.; Feng, J.; et al. Afatinib versus cisplatin-based chemotherapy for EGFR mutation-positive lung adenocarcinoma (LUX-Lung 3 and LUX-Lung 6): Analysis of overall survival data from two randomised, phase 3 trials. *Lancet Oncol. Febr.* **2015**, *16*, 141–151. [CrossRef]

55. Wu, Y.-L.; Zhou, C.; Hu, C.-P.; Feng, J.; Lu, S.; Huang, Y.; Li, W.; Hou, M.; Shi, J.H.; Lee, K.Y.; et al. Afatinib versus cisplatin plus gemcitabine for first-line treatment of Asian patients with advanced non-small-cell lung cancer harbouring EGFR mutations (LUX_Lung 6): An open-label, randomised phase 3 trial. *Lancet Oncol.* **2014**, *15*, 213–222. [CrossRef]

56. Schuler, M.; Wu, Y.L.; Hirsh, V.; O'Byrne, K.; Yamamoto, N.; Mok, T.; Popat, S.; Sequist, L.V.; Massey, D.; Zazulina, V.; et al. First-line afatinib versus chemotherapy in patients with non-small cell lung cancer and common epidermal growth factor receptor gene mutations and brain metastases. *J. Thorac. Oncol.* **2016**, *11*, 380–390. [CrossRef] [PubMed]

57. Sequist, L.V.; Yang, J.C.; Yamamoto, N.; O'Byrne, K.; Hirsh, V.; Mok, T.; Geater, S.L.; Orlov, S.; Tsai, C.-M.; Boyer, M.; et al. Phase III study of afatinib or cisplatin plus pemetrexed in patients with metastatic lung adenocarcinoma with EGFR mutations. *J. Clin. Oncol.* **2013**, *31*, 3327–3334. [CrossRef] [PubMed]

58. Park, K.; Tan, E.H.; O'Byrne, K. Afatinib versus gefitinib as first-line treatment of patients with EGFR mutation-positive non-small-cell lung cancer (LUX-Lung 7): A phase 2B, open-label, randomised controlled trial. *Lancet Oncol.* **2016**, *17*, 577–589. [CrossRef]

59. Brueckl, W.M.; Laack, E.; Reck, M.; Griesinger, F.; Schafer, H.; Kortsik, C.; Gaska, T.; Rawluk, J.; Kruger, S. Effectiveness of afatinib in clinical practice—First results of the GIDEON trial: A prospective non-interventional study in EGFR-mutated NSCLC in Germany. *ESMO* **2018**, *19*, 23.

60. Odogwu, L.; Mathieu, L.; Goldberg, K.B.; Blumenthal, G.M.; Larkins, E.; Fiero, M.H.; Rodriguez, L.; Bijwaard, K.; Lee, E.Y.; Philip, R.; et al. FDA Benefit-Risk Assessment of Osimertinib for the Treatment of Metastatic Non-Small Cell Lung Cancer Harboring Epidermal Growth Factor Receptor T790M Mutation. *Oncologist* **2018**, *23*, 353–359. [CrossRef]

61. Cross, D.A.; Ashton, S.E.; Ghiorghiu, S.; Eberlein, C.; Nebhan, C.A.; Spitzler, P.J.; Orme, J.P.; Finlay, R.V.; Ward, R.A.; Mellor, M.J.; et al. AZD9291 an irreversible EGFR TKI, overcomes T790M-mediated resistance to EGFR inhibitors in lung cancer. *Cancer Discov.* **2014**, *4*, 1046–1061. [CrossRef]

62. Janne, P.A.; Yang, J.C.; Kim, D.W.; Planchard, D.; Ohe, Y.; Ramalingam, S.S.; Ahn, M.-J.; Kim, S.-W.; Su, W.-C.; Horn, L.; et al. AZD9291 in EGFR inhibitor- resistant non-small-cell lung cancer AURA 2. *N. Engl. J. Med.* **2015**, *372*, 1689–1699. [CrossRef] [PubMed]

63. Leonetti, A.; Sharma, S.; Minari, R.; Perego, P.; Giovannetti, E.; Marcello, T. Resistance mechanisms to osimertinib in EGFR-mutated non-small cell lung cancer. *Br. J. Cancer* **2019**, *121*, 725–737. [CrossRef] [PubMed]

64. Goss, G.; Tsai, C.M.; Shepherd, F.A.; Bazhenova, L.; Lee, J.S.; Chang, G.-C.; Crino, L.; Satouchi, M.; Chu, Q.; Hida, T.; et al. Osimertinib for pretreated EFGR Thr 790Met-positive advanced non-small-cell lung cancer (AURA2): A multicentre, open-label, single-arm, phase 2 study. *Lancet Oncol.* **2016**, *17*, 1643–1652. [CrossRef]

65. Yang, J.C.; Ahn, M.J.; Kim, D.W.; Ramalingam, S.S.; Sequist, L.V.; Su, W.-C.; Kim, S.-W.; Kim, J.-H.; Planchard, D.; Felip, E.; et al. Osimertinib in pretreated T790M-positive advanced non-small-cell lung cancer: AURA Study Phase II Extension Component. *J. Clin. Oncol.* **2017**, *35*, 1288–1296. [CrossRef] [PubMed]

66. Mok, T.S.; Wu, Y.-L.; Ahn, M.-J.; Garassino, M.C.; Kim, H.R.; Ramalingam, S.S.; Shepherd, F.A.; He, Y.; Akamatsu, H.; Theelen, S.M.E.; et al. Osimertinib or platinum-pemetrexed in EGFR T790M-positive lung cancer. *N. Engl. J. Med.* **2017**, *376*, 629–640. [CrossRef] [PubMed]

67. Hirashima, T.; Satouchi, M.; Hida, T.; Nishio, M.; Kato, T.; Sakai, H.; Imamura, F.; Kiura, K.; Okamoto, I.; Kasahara, K.; et al. Osimertinib for Japanese patients with T790M-positive advanced non-small-cell lung cancer: A pooled subgroup analysis. *Cancer Sci.* **2019**, *110*, 2884–2893. [CrossRef] [PubMed]

68. Ahn, M.J.; Han, J.Y.; Kim, D.W.; Cho, B.C.; Kang, J.-H.; Kim, S.-W.; Yang, J.C.-H.; Mitsudomi, T.; Lee, J.S. Osimertinib in Patients with T790M-Positive Advanced Non-small Cell Lung Cancer: Korean Subgroup Analysis from Phase II Studies. *Cancer Res. Treat.* **2020**, *52*, 284–291. [CrossRef]

69. Soria, J.C.; Ohe, Y.; Vansteekiste, J.; Reungwetwattana, T.; Chewaskulyong, B.; Lee, K.H.; Dechaphunkul, A.; Imamura, F.; Nami, N.; Kurata, T.; et al. Osimertinib in untreated EGFR-mutated advanced Non-small-cell lung cancer. *N. Engl. J. Med.* **2018**, *378*, 113–125. [CrossRef]

70. Fassunke, J.; Muller, F.; Keul, M.; Michels, S.; Dammert, M.A.; Schmitt, A.; Plenker, D.; Lategahn, J.; Heydt, C.; Bragelmann, J.; et al. Overcoming EGFR^{G724S}-mediated osimertinib resistance through unique binding characteristics of second-generation EGFR inhibitors. *Nat. Commun.* **2018**, *9*, 4655. [CrossRef]

71. Planchard, D.; Brown, K.H.; Kim, D.-W.; Kim, S.-W.; Ohe, Y.; Felip, E.; Leese, P.; Cantarini, M.; Vishwanathan, K.; Janne, P.A.; et al. Osimertinib Western and Asian clinical pharmacokinetics in patients and healthy volunteers: Implications for formulation, dose, and dosing frequency in pivotal clinical studies. *Cancer Chemother. Pharmacol.* **2016**, *77*, 767–776. [CrossRef]

72. Gray, J.E.; Okamoto, I.; Sriuranpong, V.; Vansteenkiste, J.; Imamura, F.; Lee, J.S.; Pang, Y.-K.; Cobo, M.; Kasahara, K.; Cheng, Y.; et al. Tissue and plasma EGFR mutation analysis in the FLAURA trial: Osimertinib versus comparator EGFR tyrosine kinase inhibitor as first-line treatment in patients with EGFR-mutated advanced non-small cell lung cancer. *Clin. Cancer Res.* **2019**, *25*, 6644–6652. [CrossRef] [PubMed]

73. Carlisle, J.W.; Ramalingam, S.S. Improving outcomes for brain metastases in EGFR mutated NSCLC. *Transl. Lung Cancer Res.* **2019**, *8* (Suppl. 4), S355–S359. [CrossRef] [PubMed]

74. Ohe, Y.; Imamura, F.; Nogami, N.; Okamoto, I.; Kurata, T.; Kato, T.; Sugawara, S.; Ramalingam, S.S.; Uchida, H.; Hodge, R.; et al. Osimertinib versus standard-of-care EGFR-TKI as first line treatment for EGFRm advanced NSCLC FLAURA Japanese subset. *Jpn. J. Clin. Oncol.* **2019**, *49*, 29–36. [CrossRef] [PubMed]

75. Sakai, H.; Hayashi, H.; Iwasa, T. Successful osimertinib treatment for leptomeningeal carcinomatosis from lung adenocarcinoma with the T790M mutation of EGFR. *ESMO Open* **2017**, *2* (Suppl. 1), e000104. [CrossRef] [PubMed]

76. Lemjabbar-Alaoui, H.; Hassan, O.U.; Yang, Y.-W.; Buchanan, P. Lung cancer: Biology and treatment options. *Biochim. Biophys. Acta* **2015**, *1856*, 189–210. [CrossRef] [PubMed]

77. Soda, M.; Choi, Y.L.; Enomoto, M.; Takada, S.; Yamashita, Y.; Ishikawa, S.; Fujiwara, S.-I.; Watanabe, H.; Kurashina, K.; Hatanaka, H.; et al. Identification of the transforming EML4-ALK Fusion gene in non-small-cell lung cancer. *Nature* **2007**, *448*, 561–566. [CrossRef]

78. Lin, J.J.; Riely, G.J.; Shaw, A.T. Targeting ALK: Precision medicine takes on drug resistance. *Cancer Discov.* **2017**, *7*, 137–155. [CrossRef]

79. Noe, J.; Lovejoy, A.; Ou, S.-H.I.; Yaung, S.J.; Bordogna, W.; Klass, D.M.; Cummings, C.A.; Shaw, A.T. ALK mutation status before and after alectinib treatment in locally advanced or metastatic ALK-positive NSCLC: Pooled analysis of two prospective trials. *J. Thorac. Oncol.* **2020**, *15*, 601–608. [CrossRef]

80. Metro, G.; Lunardi, G.; Floridi, P.; Pascali, J.P.; Marcomigni, L.; Chiari, R.; Ludovini, V.; Crino, L.; Gori, S. CSF concentration of crizotinib in two ALK-positive non-small-cell lung cancer patients with CNS metastases deriving clinical benefit from treatment. *J. Thorac. Oncol.* **2015**, *10*, e26–e27. [CrossRef]

81. Solomon, B.J.; Cappuzzo, F.; Felip, E.; Blackhall, F.H.; Costa, D.B.; Kim, D.-W.; Nakagawa, K.; Wu, Y.-L.; Mekhail, T.; Paolini, J.; et al. Intracranial efficacy of crizotinib versus chemotherapy in patients with advanced alk-positive non-small-cell lung cancer: Results from Profiel 1014. *J. Clin. Oncol.* **2016**, *34*, 2858–2865. [CrossRef]

82. Solomon, B.J.; Mok, T.; Kim, D.W.; Wu, Y.-L.; Nakagawa, K.; Mekhail, T.; Felip, E.; Cappuzzo, F.; Paolini, J.; Usari, T.; et al. First-line crizotinib versus chemotherapy in ALK-positive lung cancer. *N. Engl. J. Med.* **2014**, *371*, 2167–2177. [CrossRef] [PubMed]

83. Blackhall, F.; Cappuzzo, F. Crizotinib: From discovery to accelerated development to front-line treatment. *Ann. Oncol.* **2016**, *27* (Suppl. 3), i35–i41. [CrossRef] [PubMed]

84. Ou, S.H.; Janne, P.A.; Bartlett, C.H.; Tang, Y.; Kim, D.W.; Otterson, G.A.; Crino, L.; Selaru, P.; Cohen, D.P.; Clark, J.W.; et al. Clinical benefit of continuing ALK inhibition with crizotinib beyond initial disease progression in patients with advanced ALK- positive NSCLC. *Ann. Oncol.* **2014**, *25*, 415–422. [CrossRef] [PubMed]

85. Masuda, N.; Ohe, Y.; Gemma, A.; Kusumoto, M.; Yamada, I.; Ishii, T.; Yamamoto, N. Safety and effectiveness of alectinib in a real-world surveillance study in patients with ALK-positive non-small-cell lung cancer in Japan. *Cancer Sci.* **2019**, *110*, 1401–1407. [CrossRef] [PubMed]

86. Novello, S.; Mazieres, J.; Oh, I.J.; Castro, J.; de Migliorino, M.R.; Helland, A.; Dziadziuszko, R.; Griesinger, F.; Kotb, A.; Zeaiter, A.; et al. Alectinib versus chemotherapy in crizotinib-pretreated anaplastic lymphoma kinase (ALK)-positive non-small-cell lung cancer: Results from the phase III ALUR study. *Ann. Oncol.* **2018**, *29*, 1409–1416. [CrossRef]

87. Karachaliou, N.; Bruno MFBracht, J.W.P.; Rosell, R. Profile of alectinib for the treatment of ALK-positive non-small cell lung cancer (NSCLC): Patient selection and perspectives. *Onco Targets Ther.* **2019**, *12*, 4567–4575. [CrossRef]

88. Shaw, A.; Gandhi, L.; Gadgeel, S.; Riely, G.J.; Cetnar, J.; West, H.; Camidge, D.R.; Socinski, M.A.; Chiappori, A.; Mekhail, T.; et al. Alectinib in ALK-positive, crisotinib resistant, non-small-cell lung cancer: A single-group, multicentre, phase 2 trial. *Lancet Oncol.* **2016**, *17*, 234–242. [CrossRef]

89. Nishio, M.; Nakagawa, K.; Mitsudomi, T.; Yamamoto, N.; Tanaka, T.; Kuriki, H.; Zeaiter, A.; Tamura, T. Analysis of central nervous system efficacy in the J-ALEX study of alectinib versus crizotinib in ALK-positive non-small-cell lung cancer. *Lung Cancer* **2018**, *121*, 37–40. [CrossRef]

90. Camidge, D.R.; Dziadziuszko, R.; Peters, S.; Mok, T.; Noe, J.; Nowicka, M.; Gadgeel, S.M.; Cheema, P.; Pavlakis, N.; De Marinis, F.; et al. Efficacy and safety data and impact of the EML4-ALK Fusion variant of the efficacy of alectinib in untreated ALK-positive advanced Non-Small Cell Lung Cancer in the Global Phase III ALEX Study. *J. Thorac. Oncol.* **2019**, *14*, 233–1243. [CrossRef]

91. Tomasini, P.; Egea, J.; Souquet-Bressand, M.; Greillier, L.; Barlesi, F. Alectinib in the treatment of ALK-positive metastatic non-small cell lung cancer: Clinical trial evidence and experience with a focus on brain metastases. *Ther. Adv. Respir. Dis.* **2019**, *13*, 1753466619831906. [CrossRef]

92. Soria, J.C.; Tan, D.S.W.; Chiari, R.; Wu, Y.-L.; Paz-Ares, L.; Wolf, J.; Geater, S.L.; Orlov, S.; Cortinovis, D.; Yu, C.-J.; et al. First-line ceritinib versus platinum- based chemotherapy in advanced ALK-rearranged non-small-cell lung cancer (ASCEND -4): A randomised, open-label, phase 3 study. *Lancet* **2017**, *389*, 917–929. [CrossRef]

93. Okada, K.; Araki, M.; Sakashita, T.; Ma, B.; Kanada, R.; Yanagitani, N.; Horiike, A.; Koike, S.; Watanabe, K.; Tamai, K.; et al. Prediction of ALK mutations mediating ALK-TKIs resistance and drug re-purposing to overcome the resistance. *EBioMedicine* **2019**, *41*, 105–119. [CrossRef] [PubMed]

94. Barrows, S.T.; Wright, K.; Copley-Merriman, C.; Kaye, J.A.; Chioda, M.; Wiltshire, R.; Torgersen, K.M.; Masters, E. Systematic review of sequencing of ALK inhibitors in ALK-positive non-small-cell lung cancer. *Lung Cancer Targets Ther.* **2019**, *10*, 11–20. [CrossRef] [PubMed]

95. Kim, D.-W.; Mehra, R.; Tan, D.S.W.; Felip, E.; Chow, L.Q.M.; Camidge, D.R.; Vansteenkiste, J.; Sharma, S.; De Pas, T.; Riely, G.J.; et al. Activity and safety of ceritinib in patients with ALK- rearranged non-small-cell lung cancer (ASCEDN-1): Updated results from the multicentre, open- label, phase 1 trial. *Lancet Oncol.* **2016**, *17*, 452–463. [CrossRef]

96. Felip, E.; Crino, L.; Kim, D.W.; Mehra, R.; Tan, D.S.W.; Felip, E. Whole body and intracranial efficacy of ceritinib in patients (pts) with crizotinib CRZ) pretreated, ALK-rearranged (ALK+) non-small cell lung cancer (NSCLC) and baseline brain metastases (BM): Results from ASCEND-1 and ASCEND-2 trials. *J. Thorac. Oncol.* **2016**, *11*, S118–S119. [CrossRef]

97. Felip, E.; Orlov, S.; Park, K.; Chow, L.Q.M.; Wolf, J. ASCEND-3: A single- arm, open- label, multicenter phase II study of ceritinib in ALK- naive adult patients (pts) with ALK-rearranged (ALK+) non-small cell lung cancer (NSCLC). *J. Clin. Oncol.* **2015**, *33*, 8060. [CrossRef]

98. Lin, J.J.; Zhu, V.W.; Schoenfeld, A.J.; Yeap, B.Y.; Saxena, A.; Ferris, L.A.; Dagogo-Jack, I.; Farago, A.F.; Taber, A.; Traynor, A.; et al. Brigatinib in Patients with Alectinib- Refractory *ALK*-Positive Non-Small Cell Lung Cancer: A Retrospective Study. *J. Thorac. Oncol.* **2018**, *13*, 1530–1538. [CrossRef]

99. Camidge, D.R.; Kim, H.R.; Ahn, M.J.; Yang, J.C.-H.; Han, J.-Y.; Lee, J.-S.; Hochmar, M.J.; Li, J.Y.; Chang, G.C.; Lee, K.H.; et al. Brigatinib versus crizotinib in ALK-positive non-small-cell lung cancer. *N. Engl. J. Med.* **2018**, *397*, 2027–2039. [CrossRef]

100. Kim, D.W.; Tiseo, M.; Ahn, M.J.; Reckamp, K.L.; Hansen, K.H.; Kim, S.-W.; Huber, R.M.; West, H.L.; Groen, H.J.M.; Hochmair, M.J.; et al. Brigatinib in patients with crizotinib-refractory anaplastic lymphoma kinase- positive non-small cell lung cancer: A randomized, multicenter phase II trial. *J. Clin. Oncol.* **2017**, *35*, 2490–2498. [CrossRef]
101. Ou, S.H.; Tiseo, M.; Camidge, D.R. *Brigatinib (BRG) in Patients (pts) with Crizotinib (CRZ)-Refractory ALK+ Non-Small Cell Lung Cancer (NSCLC) and Brain Metastases in the Pivotal Randomized Phase 2 ALTA Trial*; Annual Meeting of the American Society of Clinical Oncology: Chicago, IL, USA, 2017.
102. LORBRENA (Lorlatinib). *U.S. Prescribing Information*; Pfizer Inc.: New York, NY, USA, 2018.
103. Shaw, A.T.; Solomon, B.J.; Besse, B.; Bauer, T.M.; Lin, C.-C.; Soo, R.A.; Riely, G.J.; Ou, S.-H.I.; Clancy, J.S.; Li, S.; et al. ALK resistance mutations and efficacy of lorlatinib in advanced anaplastic lymphoma kinase-positive non-small-cell lung cancer. *J. Clin. Oncol.* **2019**, *37*, 1370–1379. [CrossRef]
104. Zou, H.Y.; Friboulet, L.; Kodack, D.P.; Engstrom, L.D.; Li, Q.; West, M.; Tang, R.W.; Wang, H.; Tsaparikos, K.; Wang, J.; et al. PF-06463922, an ALK/ROS1 inhibitor, overcomes resistance to first- and second-generation ALK inhibitors in pre-clinical models. *Cancer Cell* **2015**, *28*, 70–81. [CrossRef] [PubMed]
105. Shaw, A.T.; Felip, E.; Bauer, T.M.; Besse, B.; Navarro, A.; Postel-Vinay, S.; Gainor, J.F.; Johnson, M.; Dietrich, J.; James, L.P.; et al. Loratynib in non-small cell lung cancer with ALK or ROS1 rearrangement: An international, multicentre, open-label, single-arm first-in-man phase 1 trial. *Lancet Oncol.* **2017**, *18*, 1590–1599. [CrossRef]
106. Solomon, B.J.; Bauer, T.M.; Felip, E.; Besse, B.; James, L.P. Safety and efficacy of loratynib from the dose escalation component of a study in patients with advanced ALK+ or ROS+ non-small cell lung cancer (NSCLC). *Am. Soc. Clin. Oncol.* **2016**. [CrossRef]
107. Solomo, B.J.; Besse, B.; Bauer, T.M.; Felip, E.; Soo, R.A.; Camidge, D.R.; Chiari, R.; Bearz, A.; Lin, C.-C.; Gadgeel, S.M.; et al. Lorlatinib in patients with ALK-positive non-small-cell lung cancer: Results from a global phase 2 study. *Lancet Oncol.* **2018**, *19*, 1654–1667. [CrossRef]
108. Martinez, P.; Mak, R.H.; Oxnard, G.R. Targed therapy as an alternative to whole-brain radiotherapy in EGFR-mutant or ALK-positive non-small-cell lung cancer with brain metastases. *JAMA Oncol.* **2017**, *3*, 1274–1275. [CrossRef]
109. Wang, C.; Lu, X.; Lu, Z.; Bi, N.; Wang, L. Comparison of up-front radiotherapy and TKI with TKI alone for NSCLC with brain metastases and EGFR mutation: A meta-analysis. *Lung Cancer* **2018**, *122*, 94–99. [CrossRef]
110. Jiang, T.; Su, C.; Li, X.; Zhao, C.; Zhou, F.; Ren, S.; Zhou, C.; Zhang, J. EGFR TKI plus WBRT demonstrated no survival benefit other than that of TKI alone in patients with NSCLC and EGFR mutation and brain metastases. *J. Thotacic. Oncol.* **2016**, *11*, 1718–1728. [CrossRef]
111. Levy, A.; Faivre-Finn, C.; Hasan, B.; De Maio, E.; Berghoff, A.S.; Girard, N.; Greillier, L.; Lantuejoul, S.; Obrien, M.; Reck, M.; et al. Young Investigators EORTC Lung Cancer Group. Diversity of brain metastases screening and management in non-small cell lung cancer in Europe: Results of the European Organisation for Research and Treatment of Cancer lung Cancer group survey. *Eur. J. Cancer* **2018**, *93*, 37–46. [CrossRef]
112. Reck, M.; Rodriguez-Abreu, D.; Robinson, A.G.; Hiu, R.; Csoszi, T.; Fulop, A.; Gottfried, M.; Peled, N.; Tafreshi, A.; Cuffe, S.; et al. Pembrolizumab versus chemotherapy for PD-L-positive non-small-cell lung cancer. *N. Engl. J. Med.* **2016**, *375*, 1823–1833. [CrossRef]
113. Herbst, R.S.; Baas, P.; Kim, D.-W.; Felip, E.; Perez-Gracia, J.L.; Han, J.-Y.; Molina, J.; Kim, J.-H.; Arvis, C.D.; Ahn, M.-J.; et al. Pembrolizumab versus docetaxel for previously treated, PD-L1-positive, advanced non-small-cell lung cancer (KEYNOTE-010): A randomised controlled trial. *Lancet Lond. Engl.* **2016**, *387*, 1540–1550. [CrossRef]
114. ESMO Guidelines Committee. Metastatic Non-Small-Cell Lung Cancer: ESMO Clinical Practice Guidelines for Diagnosis, Treatment and Follow-Up. Available online: https://www.esmo.org/guidelines/lung-and-chest-tumours/clinical-practice-living-guidelines-metastatic-non-small-cell-lung-cancer (accessed on 5 January 2021).
115. Borghaei, H.; Paz-Ares, L.; Horn, L.; Spigel, D.R.; Steins, M.; Ready, N.E.; Chow, L.Q.; Vokes, E.E.; Felip, E.; Holgado, E.; et al. Nivolumab versus docetaxel in advanced non-squamous non-small cell lung cancer. *N. Engl. J. Med.* **2015**, *373*, 1627–1639. [CrossRef] [PubMed]
116. Crino, L.; Bronte, G.; Bidoli, P.; Cravero, P.; Mineza, E.; Cortesi, E.; Garassino, M.C.; Proto, C.; Cappuzzo, F.; Grossi, F.; et al. Nivolumab and brain metastases in patients with advanced non-squamous non-small cell lung cancer. *Lung Cancer* **2019**, *129*, 35–40. [CrossRef] [PubMed]

International Journal of
Molecular Sciences

Article

Evaluating the Effect of Lenvatinib on Sorafenib-Resistant Hepatocellular Carcinoma Cells

Tingting Shi [1,*], Hisakazu Iwama [2], Koji Fujita [1], Hideki Kobara [1], Noriko Nishiyama [1], Shintaro Fujihara [1], Yasuhiro Goda [1], Hirohito Yoneyama [1], Asahiro Morishita [1], Joji Tani [1], Mari Yamada [1], Mai Nakahara [1], Kei Takuma [1] and Tsutomu Masaki [1,*]

[1] Department of Gastroenterology and Neurology, Faculty of Medicine, Kagawa University, 1750-1 Ikenobe, Miki 761-0793, Japan; 92m7v9@med.kagawa-u.ac.jp (K.F.); kobara@med.kagawa-u.ac.jp (H.K.); n-nori@med.kagawa-u.ac.jp (N.N.); joshin@med.kagawa-u.ac.jp (S.F.); goda0717@med.kagawa-u.ac.jp (Y.G.); hyoneyam@med.kagawa-u.ac.jp (H.Y.); asahiro@med.kagawa-u.ac.jp (A.M.); georget@med.kagawa-u.ac.jp (J.T.); mari-yamada@med.kagawa-u.ac.jp (M.Y.); m-nakahara@med.kagawa-u.ac.jp (M.N.); k-takuma@med.kagawa-u.ac.jp (K.T.)

[2] Life Science Research Center, Kagawa University, 1750-1 Ikenobe, Miki 761-0793, Japan; iwama@med.kagawa-u.ac.jp

* Correspondence: shitingtingc@med.kagawa-u.ac.jp (T.S.); tmasaki@med.kagawa-u.ac.jp (T.M.); Tel.: +81-87-891-2156 (T.M.); Fax: +81-87-891-2158 (T.M.)

Abstract: Hepatocellular carcinoma (HCC) is one of the major causes of cancer-related deaths worldwide. Sorafenib has been used as a first-line systemic treatment for over a decade. However, resistance to sorafenib limits patient response and presents a major hurdle during HCC treatment. Lenvatinib has been approved as a first-line systemic treatment for advanced HCC and is the first agent to achieve non-inferiority against sorafenib. Therefore, in the present study, we evaluated the inhibition efficacy of lenvatinib in sorafenib-resistant HCC cells. Only a few studies have been conducted on this topic. Two human HCC cell lines, Huh-7 and Hep-3B, were used to establish sorafenib resistance, and in vitro and in vivo studies were employed. Lenvatinib suppressed sorafenib-resistant HCC cell proliferation mainly by inducing G1 cell cycle arrest through ERK signaling. Hep-3B sorafenib-resistant cells showed partial cross-resistance to lenvatinib, possibly due to the contribution of poor autophagic responsiveness. Overall, the findings suggest that the underlying mechanism of lenvatinib in overcoming sorafenib resistance in HCC involves FGFR4-ERK signaling. Lenvatinib may be a suitable second-line therapy for unresectable HCC patients who have developed sorafenib resistance and express FGFR4.

Keywords: lenvatinib; sorafenib-resistant; hepatocellular carcinoma; FGFR4; autophagy; microRNA

Citation: Shi, T.; Iwama, H.; Fujita, K.; Kobara, H.; Nishiyama, N.; Fujihara, S.; Goda, Y.; Yoneyama, H.; Morishita, A.; Tani, J.; et al. Evaluating the Effect of Lenvatinib on Sorafenib-Resistant Hepatocellular Carcinoma Cells. *Int. J. Mol. Sci.* **2021**, *22*, 13071. https://doi.org/10.3390/ijms222313071

Academic Editor: Nam Deuk Kim

Received: 21 November 2021
Accepted: 29 November 2021
Published: 2 December 2021

1. Introduction

Hepatocellular carcinoma (HCC) is the most common primary liver cancer, with an increasing incidence over the past few decades in various populations; it is one of the major causes of cancer-related deaths worldwide [1,2]. Only a small fraction of patients is diagnosed with early stages of the disease, when curative strategies—liver resection (LR), ablative techniques, and orthotopic liver transplantation (OLT)—can be employed [3]. For patients with advanced HCC, systemic therapy is available, with prolonged overall survival (OS) rates.

Sorafenib, an oral multi-kinase inhibitor approved in 2008, targets the Raf–MEK–ERK pathway and several receptor tyrosine kinases, including vascular endothelial growth factor receptors (VEGFRs) 2 and 3, platelet-derived growth factor receptor (PDGFR), FMS-related tyrosine kinase 3 (FLT3), Ret, and c-Kit [4]. The median survival time with sorafenib, used as a first-line systemic therapy for the past decade, was nearly 3 months longer than that with the placebo (10.7 months vs. 7.9 months; hazard ratio (HR) 0.69; $p < 0.001$) [4].

However, intrinsic and acquired resistance to sorafenib remains a huge challenge, with only approximately 30% of patients responsive to sorafenib [4,5]. The primary resistance of HCC cells to sorafenib is postulated to be associated with genetic heterogeneity—the overexpression of epidermal growth factor receptor (EGFR) or ligand may lead to sustained activation of EGFR downstream signaling and drug resistance to sorafenib [6,7]. Acquired resistance to sorafenib, which often develops within six months [5], may be associated with several factors, such as the phosphatidylinositol 3-kinase (PI3K)/Akt pathway, autophagy, epithelial-mesenchymal transition (EMT), tumor microenvironment, epigenetic regulation, microRNAs (miRNAs), and "vessel co-option"—the ability of tumors to hijack the existing vasculature in organs such as the lungs or liver, thereby limiting the need for angiogenesis [7,8].

Second- and later-line systemic treatments are needed for patients who fail to respond or are intolerant to sorafenib. The multitargeted multi-kinase inhibitors regorafenib [9] and cabozantinib [10] were approved in 2017 and 2018, respectively, as second-line drugs. Further, PD-1 immune checkpoint inhibitors, nivolumab and pembrolizumab, have been approved as second-line therapies for patients with advanced HCC [11].

Lenvatinib is a multi-kinase inhibitor that targets VEGFR 1–3, fibroblast growth factor (FGF) receptors 1–4, PDGFR-α, RET, and KIT [12]. A recent clinical trial showed that the median survival time of 13.6 months with lenvatinib (95% CI 12.1–14.9) was non-inferior to that with sorafenib (12.3 months, 10.4–13.9; HR 0.92, 95% CI 0.79–1.06) in untreated advanced HCC [13]. Thus, lenvatinib has been approved as a first-line systemic treatment for unresectable advanced HCC. Further, some evidence indicates that lenvatinib may be used as a second-line treatment for patients who are intolerant to sorafenib or following sorafenib failure [14–16].

To explore the possible advantage of lenvatinib and the underlying mechanisms in overcoming sorafenib-resistance in HCC we developed sorafenib-resistant cell lines and performed in vitro and in vivo experiments; further, we identify the expression of miRNAs associated with the effect of lenvatinib. Only a few studies have been conducted on this topic.

2. Results

2.1. Lenvatinib Inhibits Sorafenib-Resistant HCC Cell Proliferation

The IC_{50} of Huh-7SR cells (6.76 ± 0.48 μM) was 2.9-fold higher than that of Huh-7 cells (2.33 ± 0.22 μM), and the IC_{50} of Hep-3BSR cells (7.73 ± 0.27 μM) was 2.81-fold higher than that of Hep-3B cells (2.75 ± 0.44 μM) when exposed to sorafenib (Figure 1A). The lenvatinib IC_{50} values were not significantly different between Huh-7 (9.91 ± 0.95 μM) and Huh-7SR (10.56 ± 0.73 μM) cells; however, the lenvatinib IC_{50} of Hep-3BSR cells (27.49 ± 3.01 μM) was 9.85-fold that of Hep-3B cells (2.79 ± 0.19 μM) (Figure 1B). The Huh-7SR and Hep-3BSR cells were treated with 0.3, 1, 3, 10, and 30 μM lenvatinib for 96 h, and the anti-proliferative effect of lenvatinib was assessed using the cell viability assay. DMSO-treated cells were used as controls. Lenvatinib inhibited cell proliferation in the Huh-7SR and Hep-3BSR cells in a dose- and time-dependent manner. Further, 1 μM lenvatinib showed a significant effect in the Huh-7SR cells; however, 10 μM lenvatinib was required in the Hep-3BSR cells (Figure 1C). In the colony formation assay, during a long culture period (14 days), a small dose (1 μM) of lenvatinib showed an obvious anti-proliferative effect in the Huh-7SR cells, but not in the Hep-3BSR cells (Figure 1D). This may be due to the partial cross-resistance of Hep-3BSR cells to lenvatinib; the anti-proliferative effect of lenvatinib in Hep-3BSR cells needs a higher dose than Huh-7SR cells. To further verify the anti-proliferative effect of lenvatinib in Huh-7SR and Hep-3BSR cells, a 3D tumor spheroid assay was performed. An in vitro dosage (10 μM) of sorafenib and lenvatinib was used in the 3D spheroid culture of both wild-type and resistant cells; this showed that lenvatinib (10 μM) has an advantageous anti-proliferative effect in all cells (Huh-7, Huh-7SR, Hep-3B, and Hep-3BSR) compared with sorafenib (10 μM). Moreover, sorafenib could partially inhibit Huh-7SR cell proliferation but not in Hep-3BSR cells (Figure 1E).

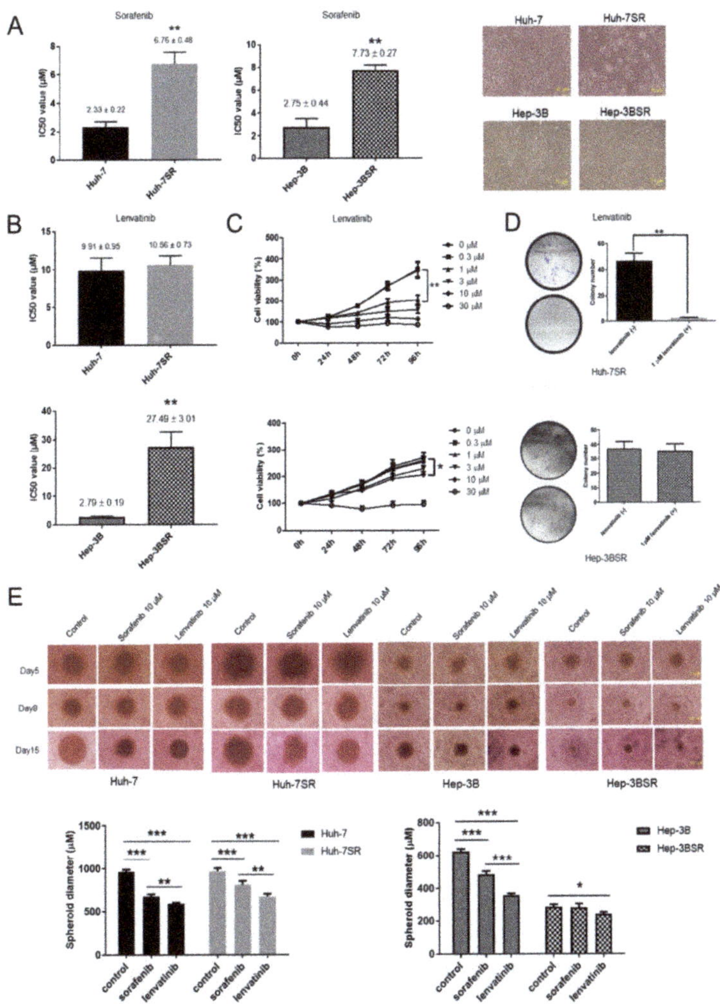

Figure 1. Lenvatinib inhibits the proliferation of sorafenib-resistant human hepatocellular carcinoma (HCC) cells. (**A**) Cytotoxic effect of sorafenib (IC_{50}): Cells were treated for 48 h with increasing concentrations of sorafenib, and dimethyl sulfoxide (DMSO) was used as a negative control. IC_{50} of Huh-7 sorafenib-resistant (Huh-7SR) cells was 2.9-fold that of Huh-7 cells, and IC_{50} of Hep-3BSR cells was 2.81-fold that of Hep-3B cells. (**B**) Cytotoxic effect of lenvatinib (IC_{50}): Cells were treated for 48 h with increasing concentrations of lenvatinib, and dimethyl sulfoxide (DMSO) was used as a negative control. Lenvatinib IC_{50} values of Huh-7 and Huh-7SR cells were not significantly different. Lenvatinib IC_{50} of Hep-3B sorafenib-resistant (Hep-3BSR) cells was 9.85-fold that of Hep-3B cells. (**C**) Anti-proliferative effect of lenvatinib measured using cell viability assay: Huh-7SR and Hep-3BSR cells were treated for 96 h with the indicated concentrations of lenvatinib or DMSO. The relative cell number was normalized with the control. (**D**) After 1 µM lenvatinib or DMSO treatment, the colonies formed were fixed and stained with crystal violet. (**E**) Three-dimensional tumor spheroid assay was performed to evaluate the effect of lenvatinib and sorafenib on the wild-type and sorafenib-resistant HCC cell proliferative ability. The representative images of spheroids are shown (scale bar: 5th day, 71 µm; 8th and 15th days, 107 µm). Spheroid diameters were measured on 15th day. Data are represented as means ± standard error of mean (SEM) or means ± standard deviation (SD) of at least three independent experiments. * $p < 0.05$, ** $p < 0.01$, *** $p < 0.001$.

2.2. Lenvatinib Induces Apoptosis and Cell Cycle G1 Phase Arrest, Decreases Invasion and Migration Ability, and Regulates the Expression of Angiogenesis-Related Proteins in Huh-7SR Cells

To determine whether lenvatinib affected the apoptosis and cell cycle in Huh-7SR cells, these cells were treated with lenvatinib, and FCM was performed. The Huh-7SR cells were treated with 10 µM lenvatinib or DMSO for 24 h, and the DMSO-treated cells were used as controls. The proportion of early apoptotic cells (lower right quadrant) among lenvatinib-treated cells was significantly higher than that among untreated cells (Figure 2A). Additionally, the expression of caspase-cleaved cytokeratin 18, cCK-18, a marker of apoptosis and necrosis, was upregulated in the lenvatinib-treated cells (Figure 2B). Furthermore, in terms of cell cycle progression, the cell population in the G0/G1 phase significantly increased, whereas cells in the S and G2/M phases decreased, suggesting that lenvatinib-treated cells were arrested in the G1 phase (Figure 2C). Invasion and wound healing assays showed that lenvatinib (10 µM) decreased invasion and migration ability after incubation for 24 h (Figure 2D,E). Further, in the Huh-7SR cells, lenvatinib affected the regulation of the expression of angiogenesis-related proteins: dipeptidyl peptidase-4 (DPPIV/CD26) [17–19] and pigment epithelium-derived factor (PEDF) [20] were upregulated, and plasminogen activator inhibitor-1 (PAI-1) [21] was downregulated, after incubation with 10 µM lenvatinib for 24 h (Figure 2F).

2.3. Lenvatinib Induces Cell Cycle G1 Phase Arrest, Decreases Migration Ability, and Regulates the Expression of Angiogenesis-Related Proteins in Hep-3BSR Cells

The Hep-3BSR cells were treated with 10 µM lenvatinib or DMSO for 24 h, and the DMSO-treated cells were used as controls. No significant differences were found between the average proportion of early apoptotic cells in the lenvatinib-treated cells and the control (Figure 3A). Moreover, no changes were observed in the expression of cCK-18 between the lenvatinib-treated and untreated cells (Figure 3B). In cell cycle progression, the cell population in the G0/G1 phase significantly increased, whereas the cells in the S and G2/M phases decreased, suggesting that the lenvatinib-treated cells were arrested in the G1 phase (Figure 3C). The invasion assay suggested no difference between the control and treatment groups (Figure 3D). The wound healing assay indicated that 10 µM lenvatinib decreased migration ability after incubation for 24 h (Figure 3E). Additionally, the expression of angiogenesis-related proteins such as endostatin [22], thrombospondin-1 (TSP-1) [23,24], interleukin-8 (IL-8) [25], along with PAI-1, were downregulated after incubation with 10 µM lenvatinib for 24 h (Figure 3F). The expression of PAI-1 was downregulated in both the Huh-7SR and Hep-3BSR cells after lenvatinib treatment.

2.4. Lenvatinib Effects against Sorafenib-Resistance in HCC Cells May Be through the FGFR4-ERK Signaling Pathway

The expression levels of EGFR, p-Akt, and p-ERK were found to be upregulated in the sorafenib-resistant cells compared to the wild-type cells (Figure 4A). These results were similar to those reported in previous studies [6–8], showing that the overexpression of EGFR activated the Akt and ERK signaling pathway to increase HCC cell survival and proliferation, which, in turn, induced sorafenib resistance. The kinase inhibition profile for lenvatinib and sorafenib indicated that lenvatinib has an advantage in FGFR4 inhibition compared with sorafenib (Table S1). Previous studies have shown that Huh-7 and Hep-3B have high FGFR4 expression [26,27]. In the Huh-7SR and Hep-3BSR cells, 10 µM sorafenib or lenvatinib decreased the expression of p-Akt, Akt, FGFR4, p-ERK, and cyclinD1, whereas the levels of p-mTOR and mTOR remained unchanged in the Hep-3BSR cells. Meanwhile, lenvatinib increased the expression of EGFR (Figure 4B). The effects of lenvatinib in the two resistant cells were mainly through FGFR4-ERK signaling; however, the difference between 10 µM sorafenib and lenvatinib treatment was not significant. After decreasing the dosage, lenvatinib showed a superior inhibition of ERK compared with the same dosage of sorafenib (Figure 4C). In addition, the low expression of FGFR4 was associated with a longer overall survival probability based on the TCGA database analysis (Figure 4D).

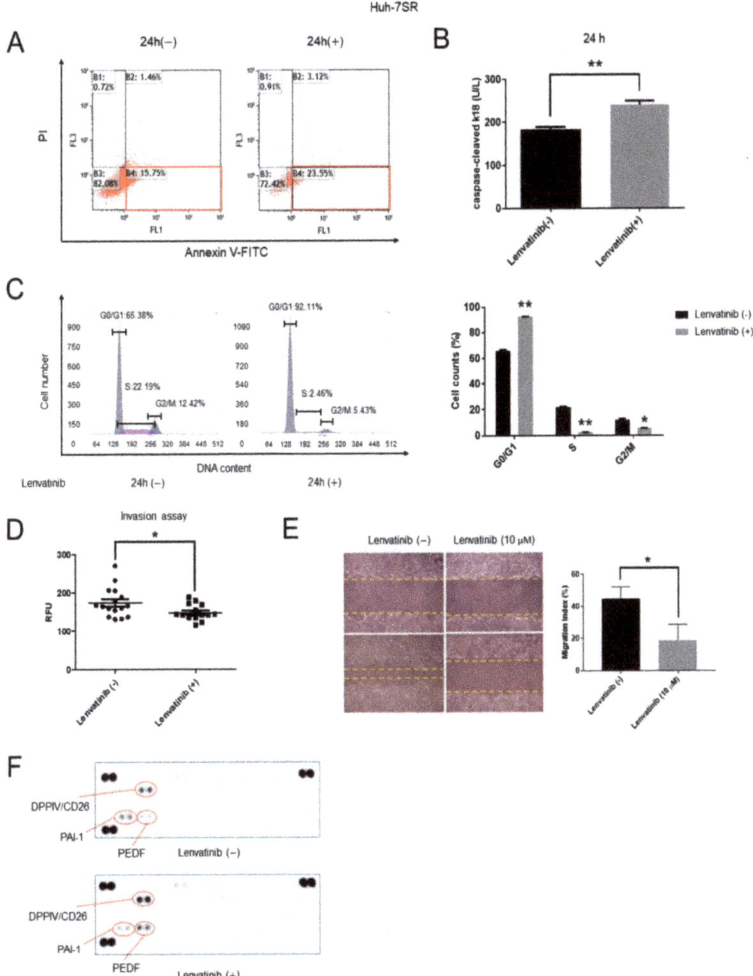

Figure 2. In Huh-7SR cells, lenvatinib induces apoptosis and cell cycle G1 phase arrest, reduces invasion and migration, and modulates the expression of proteins associated with angiogenesis. (**A**) Huh-7SR cells were treated with 10 μM lenvatinib or dimethyl sulfoxide (DMSO) for 24 h, and the level of apoptosis was measured by staining with annexin V and propidium iodide (PI) using flow cytometry. Lenvatinib treatment increased the proportion of early apoptotic cells in the Huh-7SR population. Lower right square represents early apoptosis. (**B**) The expression of caspase-cleaved cytokeratin 18 (cCK-18) was determined using Enzyme Linked Immunosorbent Assay (ELISA) after 24 h of treatment with 10 μM lenvatinib. (**C**) Huh-7SR cells treated with 10 μM lenvatinib or DMSO were analyzed using flow cytometry to determine the number of cells in each phase of the cell cycle (left panel). Representative cell cycle histograms are presented (right panel). Lenvatinib blocked the cell cycle at the G1 phase. (**D**) Invasion ability of lenvatinib-treated Huh-7SR cells was decreased. (**E**) Wound-healing assay comparing the motility of Huh-7SR cells treated with lenvatinib or DMSO. The wound-healing area was analyzed using the ImageJ software. (**F**) Representative expression of angiogenesis-related proteins: plasminogen activator inhibitor-1 (PAI-1), dipeptidyl peptidase-4 (DPPIV/CD26), and pigment epithelium-derived factor (PEDF), in Huh-7SR cells incubated with lenvatinib or DMSO for 24 h. Data are presented from three independent experiments. * $p < 0.05$, ** $p < 0.01$.

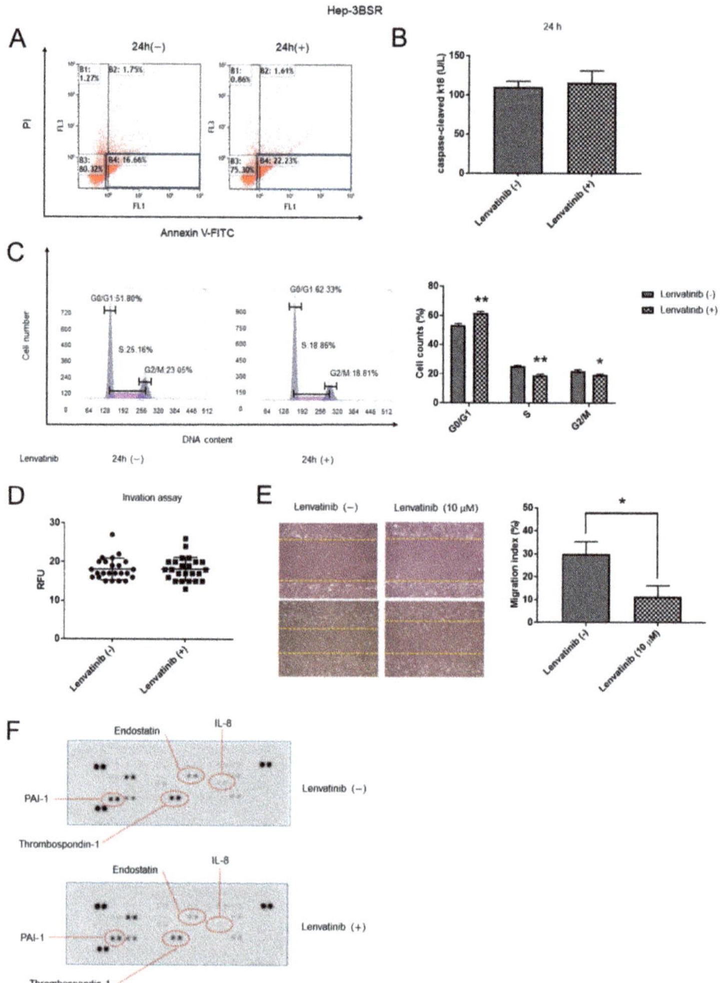

Figure 3. In Hep-3BSR cells, lenvatinib triggers cell cycle G1 phase arrest, reduces migration, and modulates the expression of proteins associated with angiogenesis. (**A**) Hep-3BSR cells were treated with 10 μM lenvatinib or dimethyl sulfoxide (DMSO) for 24 h, and the level of apoptosis was measured by staining with annexin V and propidium iodide (PI) using flow cytometry. Lenvatinib treatment did not significantly change the average proportion of early apoptotic cells in the Hep-3BSR population. Lower right square represents early apoptosis. (**B**) The expression of caspase-cleaved cytokeratin 18 (cCK-18) was determined using Enzyme Linked Immunosorbent Assay (ELISA). After 24 h of treatment with 10 μM lenvatinib, cCK-18 levels remained changed. (**C**) Hep-3BSR cells treated with 10 μM lenvatinib or DMSO were analyzed using flow cytometry to determine the number of cells in each phase of the cell cycle (left panel). Representative cell cycle histograms are presented (right panel). Lenvatinib blocked the cell cycle at the G0/G1 phase. (**D**) Invasion ability of lenvatinib-treated Hep-3BSR cells was not significantly changed. (**E**) Wound-healing assay comparing the motility of Huh-7SR cells treated with lenvatinib or DMSO. The wound-healing area was analyzed using ImageJ software. (**F**) Representative expression of angiogenesis-related proteins, endostatin, thrombospondin-1 (TSP-1), interleukin-8 (IL-8), and PAI-1, in Hep-3BSR cells incubated with lenvatinib or DMSO for 24 h. Data are presented from three independent experiments. * $p < 0.05$, ** $p < 0.01$.

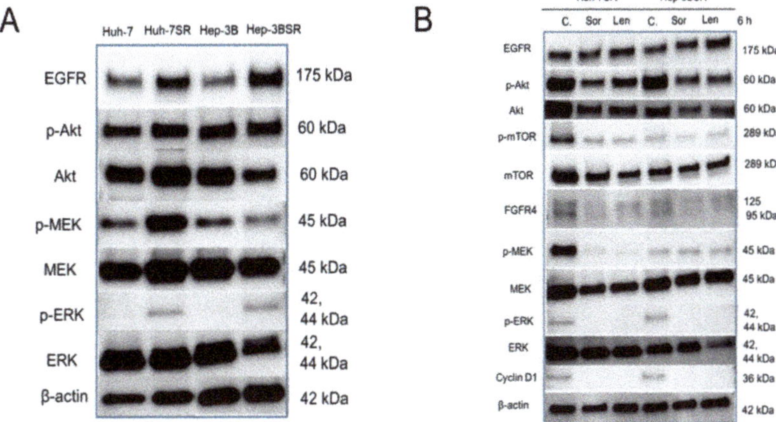

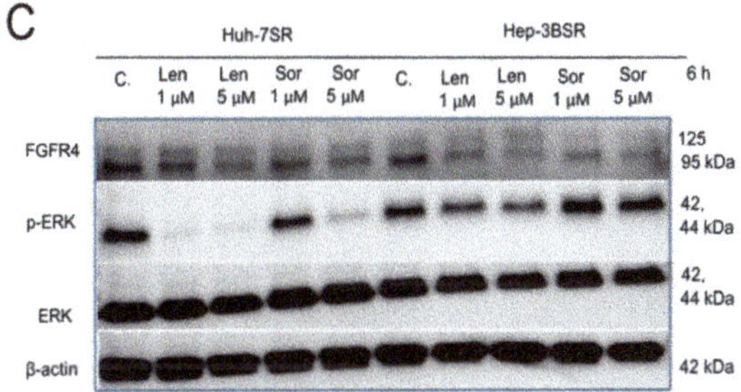

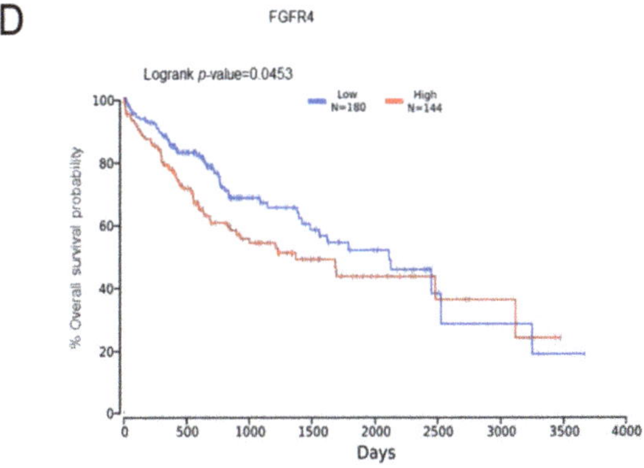

Figure 4. The underlying mechanisms of lenvatinib in sorafenib-resistant HCC cells may be through the FGFR4-ERK signaling pathway. (**A**) Proteins expressed in wild-type and sorafenib-resistant HCC cells. (**B**) Huh-7SR and Hep-3BSR cells were treated with 10 μM sorafenib or lenvatinib for 6 h. Protein expression was analyzed using Western blot. (**C**) Huh-7SR and Hep-3BSR cells were treated with 1 and 5 μM sorafenib or lenvatinib for 6 h, and the levels of FGFR4, p-ERK, and ERK were analyzed using Western blot. C.: control; Sor: sorafenib; Len: lenvatinib. (**D**) A lower FGFR4 level is associated with longer overall survival of HCC patients based on TCGA database analysis (http://www.oncolnc.org, accessed on 5 October 2021).

2.5. Different Autophagic Responsiveness between Huh-7SR and Hep-3BSR Cells

Huh-7 and Hep-3B wild-type cells are both sensitive to lenvatinib (Figure S1). The IC_{50} data indicated that the Hep-3BSR cells showed partial cross-resistance to lenvatinib compared with the Huh-7SR cells. Moreover, lenvatinib inhibited Hep-3BSR cell proliferation mainly through cell cycle arrest, whereas in Huh-7SR, cell proliferation was inhibited through both apoptosis and cell cycle arrest. In addition, although mTOR is a key molecule in the regulation of autophagy, its activation remained changed after sorafenib or lenvatinib treatment in the Hep-3BSR cells. Autophagy plays a dual role in cancer development and is also associated with multidrug resistance in cancer cells [7,28]; thus, it was important to investigate if the autophagic responsiveness to sorafenib and lenvatinib is different between the two cell lines. In the Huh-7SR cells, 10 µM sorafenib or lenvatinib significantly increased the expression of microtubule-associated protein light chain 3-II (LC3-II); however, in the Hep-3BSR cells, the expression barely changed. Furthermore, in the Huh-7SR cells, while LC3-II levels increased, the expression of p62 decreased, indicating that 10 µM sorafenib or lenvatinib altered the autophagy in the Huh-7SR cells. Meanwhile, the levels of caspase-7 and caspase-3 were decreased, and the levels of PARP and cleaved-PARP were increased in the Huh-7SR cells, but not in the Hep-3BSR cells (Figure 5A). The Huh-7SR cells showed higher autophagic responsiveness to sorafenib and lenvatinib than the Hep-3BSR cells did. Lenvatinib induced high autophagic responsiveness that may induce autophagic cell death, and this may be one of the reasons that the Hep-3BSR cells showed partial cross-resistance to lenvatinib. Moreover, a lower p62 level is associated with the longer overall survival of HCC patients based on TCGA database analysis (Figure 5B).

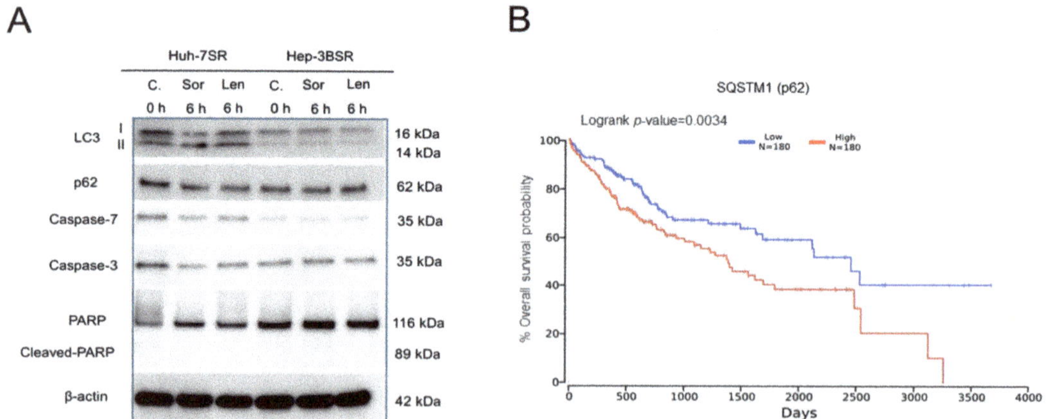

Figure 5. Autophagic response of sorafenib-resistant cells to sorafenib or lenvatinib. (**A**) Huh-7SR and Hep-3BSR cells were treated with 10 µM sorafenib or lenvatinib for 6 h. LC3, p62, caspase-7, caspase-3, and PARP expression were analyzed using Western blot. C.: control; Sor: sorafenib; Len: lenvatinib. (**B**) A lower p62 level is associated with longer overall survival of HCC patients based on TCGA database analysis (http://www.oncolnc.org, accessed on 5 October 2021).

2.6. Lenvatinib Affects microRNA Expression in Sorafenib-Resistant Cells

A customized microarray platform was used to analyze the expression of 2555 miRNAs in the lenvatinib-treated or control Huh-7SR cells. Treatment with 10 µM lenvatinib for 24 h upregulated the expression (fold change > 1.5) of 43 miRNAs, such as miR-575, miR-663a, miR-491-5p, miR-4465, miR-371b-5p, and miR-718, and suppressed the expression (fold change < 0.67) of 23 miRNAs, such as miR-4448, miR-106b-3p, miR-197-5p, and miR-130b-3p (Table 1). Unsupervised hierarchical clustering analysis was conducted by calculating Pearson's centered correlation coefficient, and the results indicated that the lenvatinib-treated Huh-7SR cells clustered together (Figure 6). Additionally, filtration

(FDR < 0.001) indicated that lenvatinib upregulated the expression of 16 miRNAs and suppressed the expression of six miRNAs (Figure S2).

Table 1. Statistical results and chromosomal locations of miRNAs that exhibited a fold change (FC) > 1.5, FC < 0.67, or $p < 0.001$ in HCC Huh-7SR cells treated with lenvatinib when compared with untreated cells.

miRNA	p-Value	Fold Change (Treated/Untreated)	FDR	Chromosomal Location
Upregulated				
has-miR-718	0.000823533	2.93	0.009177565	Xq28
hsa-miR-4787-3p	0.000378539	2.82	0.005427912	3p21.2
hsa-miR-6816-5p	1.73835×10^{-5}	2.69	0.000977428	22q11.21
hsa-miR-6790-3p	0.000369116	2.64	0.005427912	19p13.3
hsa-miR-1234-3p	2.9809×10^{-5}	2.59	0.001292429	8q24.3
hsa-miR-8063	0.000246669	2.56	0.004141378	15q14
hsa-miR-6768-5p	2.43923×10^{-6}	2.48	0.000450491	16p13.3
hsa-miR-6501-3p	1.236×10^{-5}	2.47	0.000867423	21q22.11
hsa-miR-10394-3p	0.00019664	2.40	0.003800681	19q13.43
hsa-miR-6850-5p	6.5404×10^{-5}	2.36	0.002074481	8q24.3
hsa-miR-12120	5.15411×10^{-5}	2.35	0.001850097	Yq11.221
hsa-miR-4465	1.65012×10^{-5}	2.35	0.000971998	6q24.1
hsa-miR-491-5p	1.25706×10^{-6}	2.22	0.000388745	9p21.3
hsa-miR-371b-5p	0.000874032	2.20	0.009484016	19q13.42
hsa-miR-3652	1.24497×10^{-5}	2.15	0.000867423	12q23.3
hsa-miR-921	0.000571511	2.04	0.007188459	1q24.1
hsa-miR-1237-5p	2.91345×10^{-6}	2.03	0.000450491	11q13.1
hsa-miR-1469	0.000239265	1.95	0.004141378	15q26.2
hsa-miR-1181	0.000576271	1.92	0.007188459	19p13.2
hsa-miR-6869-5p	7.72727×10^{-5}	1.88	0.00227243	20p13
hsa-miR-3665	5.65874×10^{-7}	1.85	0.000349993	13q22.3
hsa-miR-1290	0.000146648	1.85	0.003194727	1p36.13
hsa-miR-1273c	3.34339×10^{-5}	1.84	0.001292429	6q25.2
hsa-miR-4535	0.000456305	1.81	0.00613532	22q13.32
hsa-miR-10396a-5p	6.04782×10^{-5}	1.791	0.001968724	21p11.2
hsa-miR-3940-5p	2.45816×10^{-5}	1.78	0.001216297	19p13.3
hsa-miR-6808-5p	0.000904459	1.76	0.009562526	1p36.33
hsa-miR-1228-5p	0.000103027	1.76	0.002591924	12q13.3
hsa-miR-6774-5p	6.90461×10^{-6}	1.74	0.000731907	16q24.1
hsa-miR-3180-3p	0.000507427	1.74	0.006677529	16p13.11
hsa-miR-3178	2.33963×10^{-5}	1.73	0.001205886	16p13.3
hsa-miR-3126-5p	7.5417×10^{-5}	1.69	0.00227243	2p13.3
hsa-miR-4687-5p	3.28134×10^{-5}	1.67	0.001292429	11p15.4
hsa-miR-663a	8.50531×10^{-5}	1.66	0.002355633	20p11.1
hsa-miR-3158-5p	6.85365×10^{-6}	1.63	0.000731907	10q24.32
hsa-miR-762	0.000107378	1.62	0.00260445	16p11.2
hsa-miR-6715b-5p	0.00066953	1.62	0.007813291	10q25.2
hsa-miR-1247-3p	1.33234×10^{-5}	1.60	0.000867423	14q32.31
hsa-miR-4476	4.84072×10^{-5}	1.58	0.001814535	9p13.2
hsa-miR-4749-5p	8.89695×10^{-5}	1.56	0.002355633	19q13.33
hsa-miR-575	0.00018907	1.56	0.003712369	4q21.22
hsa-miR-6724-5p	8.70242×10^{-5}	1.55	0.002355633	21p11.2
hsa-miR-12114	0.000812774	1.51	0.009140012	22q13.33
Downregulated				
hsa-miR-7114-5p	0.000857372	0.65	0.009469368	9q34.3
hsa-miR-197-5p	0.000179571	0.65	0.003697546	1p13.3
hsa-miR-3907	1.10865×10^{-5}	0.65	0.000867423	7q36.1
hsa-miR-1972	0.000596743	0.63	0.00723697	16p13.11
hsa-miR-6735-5p	9.42138×10^{-5}	0.63	0.002427969	1p34.2
hsa-miR-130b-3p	7.89931E-05	0.63	0.00227243	22q11.21
hsa-miR-487b-3p	2.90019×10^{-5}	0.63	0.001292429	14q32.31
hsa-miR-6872-3p	0.000677028	0.60	0.007826946	3p21.31

Table 1. *Cont.*

miRNA	*p*-Value	Fold Change (Treated/Untreated)	FDR	Chromosomal Location
hsa-miR-10524-5p	4.7563×10^{-7}	0.60	0.000349993	6q14.1
hsa-miR-4260	0.000425744	0.60	0.00587887	1q32.2
hsa-miR-3177-3p	8.95026×10^{-5}	0.60	0.002355633	16p13.3
hsa-miR-874-5p	0.000111442	0.59	0.002651026	5q31.2
hsa-miR-4287	0.000214498	0.56	0.003919721	8p21.1
hsa-miR-4448	1.41712×10^{-5}	0.56	0.000876488	3q27.1
hsa-miR-7843-5p	5.23471×10^{-5}	0.54	0.001850097	14q24.2
hsa-miR-3189-5p	0.000538484	0.54	0.00693859	19p13.11
hsa-miR-106b-3p	3.19133×10^{-5}	0.53	0.001292429	7q22.1
hsa-miR-4521	7.10015×10^{-6}	0.47	0.000731907	17p13.1
hsa-miR-431-3p	0.00037795	0.46	0.005427912	14q32.2
hsa-miR-4451	0.000147211	0.40	0.003194727	4q21.23
hsa-miR-4632-5p	9.73972×10^{-7}	0.39	0.000388745	1p36.22
hsa-miR-6880-5p	2.78965×10^{-6}	0.31	0.000450491	12q24.31
hsa-miR-1292-5p	0.000166801	0.29	0.003557465	20p13

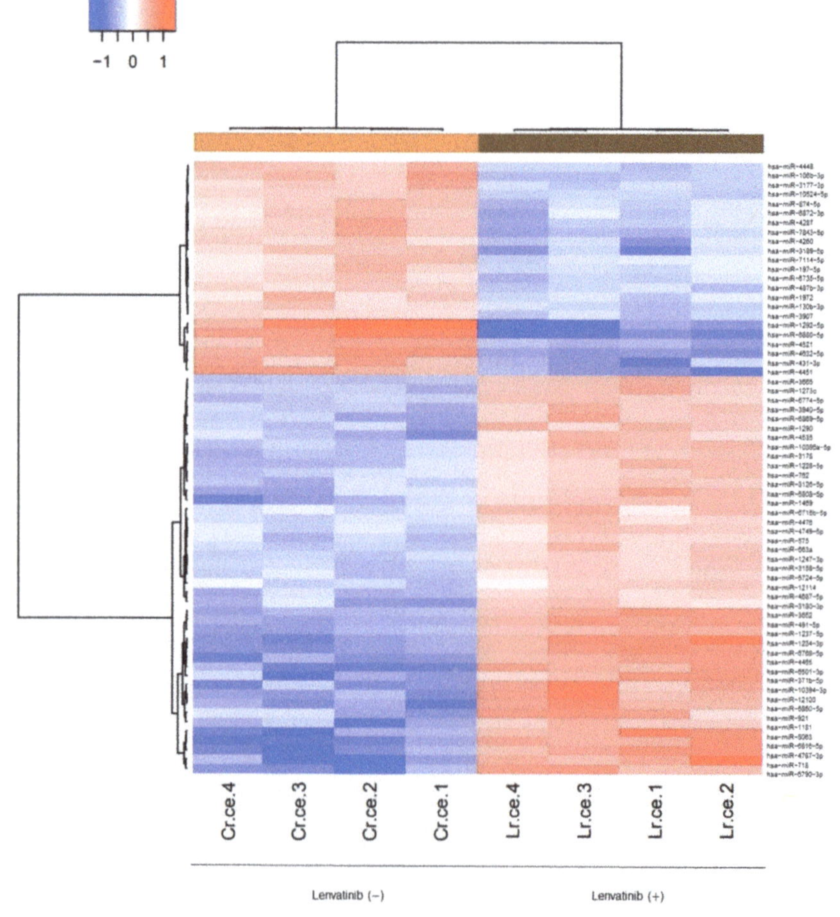

Figure 6. Lenvatinib affects miRNA expression. Hierarchical clustering of differentially expressed miRNAs from Huh-7SR cells incubated with 10 μM lenvatinib or DMSO for 24 h. Fold Change >1.5 or <0.67, *p* < 0.001.

2.7. Altered miRNA in HCC and Normal Tissues and the Relationship with Overall Survival of HCC Patients Based on TCGA Database Analysis

Based on the altered miRNA expressions, 372 HCC tissues and 50 normal tissues from The Cancer Genome Atlas Liver Hepatocellular Carcinoma (TCGA-LIHC) database were analyzed. miR-130b-3p and miR-1292-5p are highly expressed, while the levels of miR-491-5p and miR-1247-3p are decreased in HCC tissues compared with normal tissues. No differences were found in the levels of miR-1228-5p and miR-431-3p between the HCC tissues and the normal tissues (Figure 7A). After lenvatinib treatment, miR-491-5p and miR-1247-3p were upregulated, while miR-130b-3p and miR-1292-5p were downregulated in the Huh-7SR cells. In addition, lower levels of miR-130b, miR-106b, and miR-874, and higher levels of miR-487, are associated with the longer overall survival of HCC patients (Figure 7B).

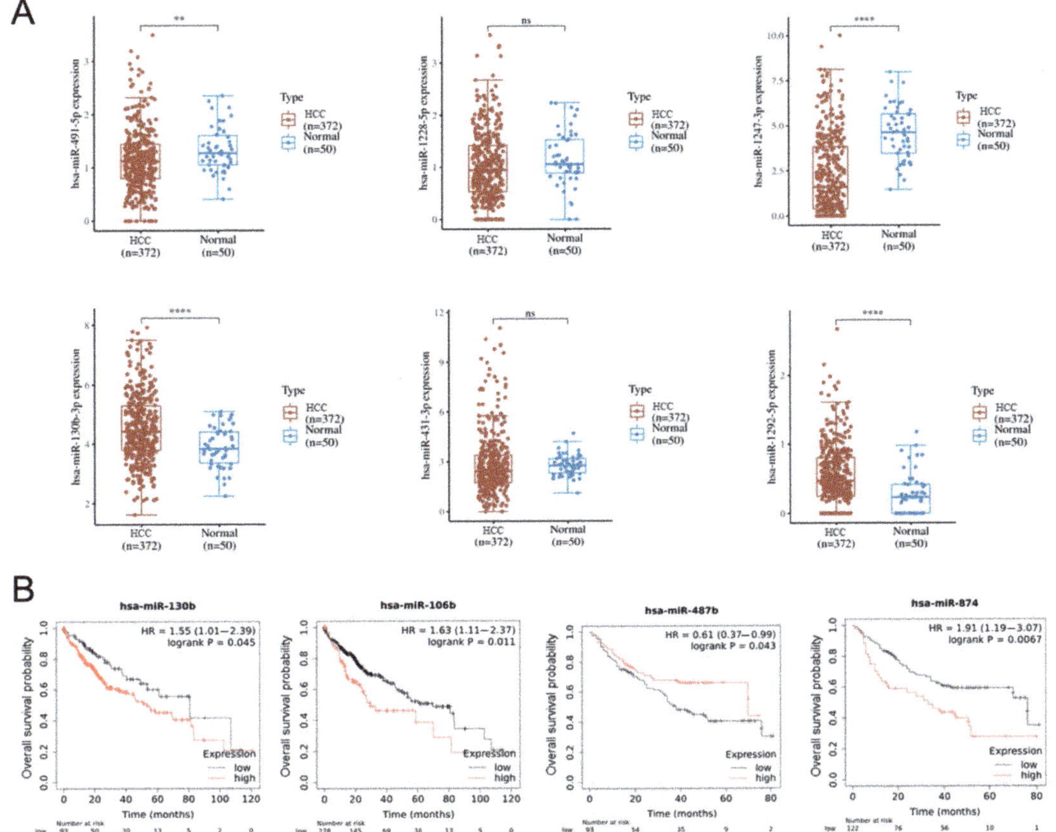

Figure 7. Altered microRNA expression and the relationship with overall survival in HCC. (**A**) Expression distribution of altered miRNAs after lenvatinib treatment in HCC and normal tissues based on TCGA database analysis, where different colors represent different groups. ** $p < 0.01$, **** $p < 0.001$, ns, no significant. (**B**) Lower levels of miR-130b, miR-106b, and miR-874, and higher levels of miR-487 are associated with longer overall survival of HCC patients (http://kmplot.com/analysis/, accessed on 5 October 2021).

2.8. Lenvatinib Inhibits Huh-7 Sorafenib-Resistant Cell Proliferation In Vivo

Next, we examined the effect of lenvatinib in a nude mice xenograft model by injecting Huh-7SR cells. Tumor growth was significantly inhibited in the group treated with lenvatinib 20 mg/kg/day (5 days/week) compared with the sorafenib 30 mg/kg/day (5 days/week) and control groups (Figure 8A). Moreover, there was no difference in body weight among these three groups. Hematoxylin and eosin (H&E) staining and immunohistochemical staining of the ki-67, cyclin D1, and CD31 proteins in the subcutaneous xenograft model suggested that lenvatinib treatment decreased the levels of ki-67, cyclin D1, and the staining area of CD31, and suppressed cell proliferation and angiogenesis (Figure 8B).

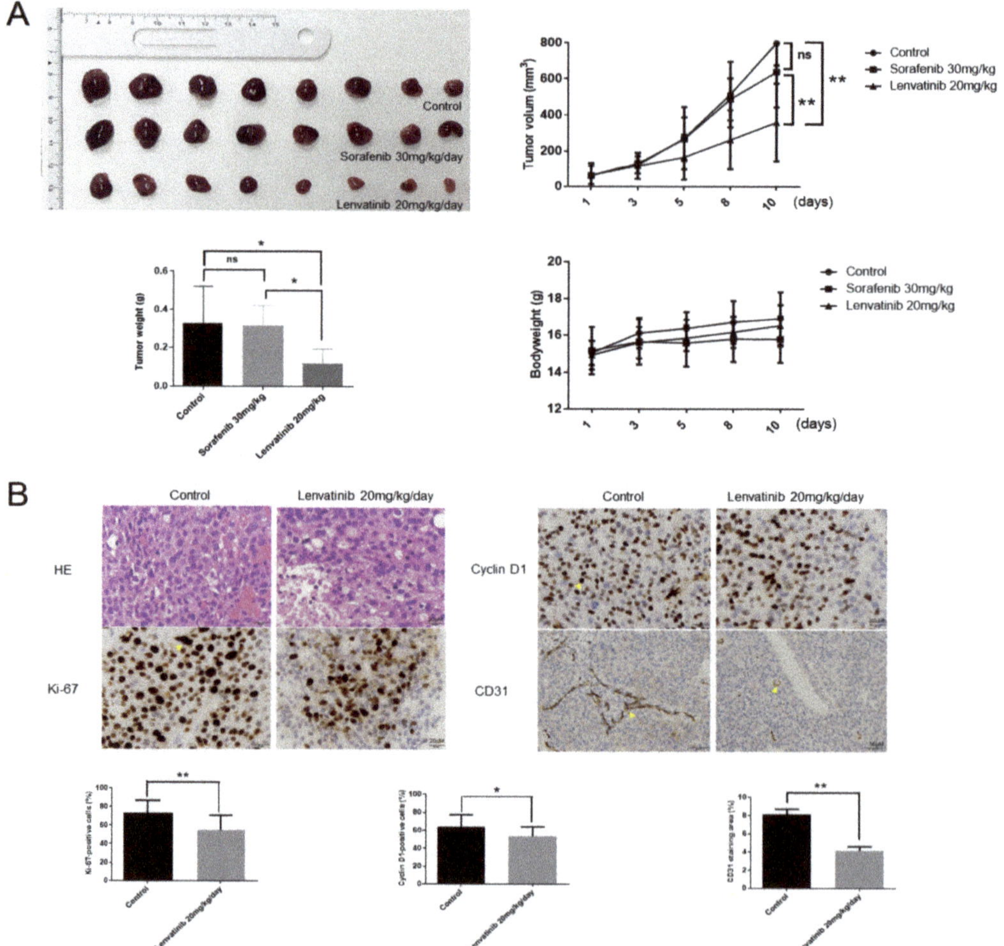

Figure 8. Lenvatinib inhibits Huh-7 sorafenib-resistant cell proliferation in vivo. (**A**) Lenvatinib 20 mg/kg/day (5 days/week) significantly inhibits tumor growth compared with control and sorafenib 30 mg/kg/day (5 days/week) (n = 8 per group). (**B**) Hematoxylin and eosin (H&E) staining and immunohistochemical staining of ki-67, cyclin D1 and CD31 proteins in the subcutaneous xenograft model. Ki-67-positive cells and cyclin D1-positive cells in the lenvatinib-treated groups were reduced in numbers compared with that in the control group. CD31 staining area in the lenvatinib-treated groups was decreased compared with that in the control group. Data are presented as mean ± standard deviation (SD). * $p < 0.05$, ** $p < 0.01$, ns, no significant.

3. Discussion

The multi-kinase inhibitor, sorafenib, has been used as first-line therapy for patients with progressive unresectable HCC for a decade. However, resistance to sorafenib limits patient response and presents a major hurdle during HCC treatment. Additionally, data from the REFLECT trial indicated that lenvatinib was the first agent to achieve non-inferiority against sorafenib [13]. Therefore, in the present study, we evaluated the inhibition efficacy of lenvatinib in sorafenib-resistant HCC cells (Huh-7SR and Hep-3BSR cells). Key points include the following: (i) lenvatinib suppressed sorafenib-resistant HCC cell proliferation, mainly by inducing G1 cell cycle arrest; (ii) the underlying advantage of lenvatinib in overcoming sorafenib resistance may occur through the FGFR4-ERK signaling pathway; (iii) along with HBV DNA, poor autophagic responsiveness may be a contributing factor toward partial cross-resistance; and (iv) miRNA alterations may contribute to the inhibition of sorafenib-resistant HCC cell growth and angiogenesis (Figure 9).

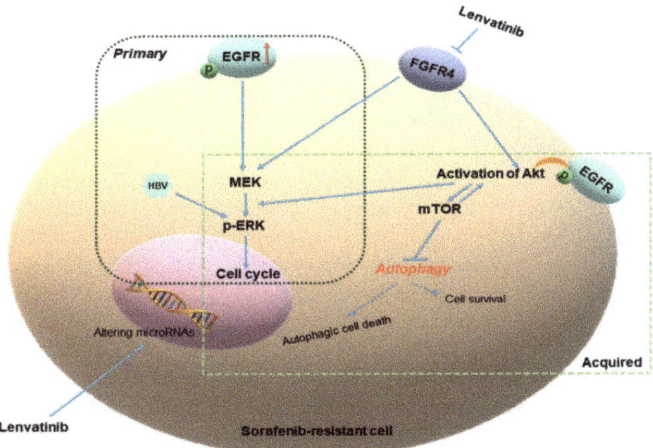

Figure 9. Proposed model for the underlying mechanism of lenvatinib in overcoming sorafenib resistance. Overexpression of EGFR leads to the activation of ERK and Akt signaling to induce sorafenib resistance and promote sorafenib-resistant cell proliferation. The underlying advantage of lenvatinib is the inhibition of FGFR4 compared with sorafenib. Huh-7SR and Hep-3BSR cells have high FGFR4 expression and the FGFR4-ERK signaling pathway is the major pathway to overcome increased EGFR-induced ERK activation. Compared with Huh-7SR cells, Hep-3BSR cells showed poor autophagic responsiveness to lenvatinib, which may contribute to their partial cross-resistance to lenvatinib.

The ERK signaling pathway plays a key role in anti-tumor effects and multi-kinase inhibitor resistance [28]. In both primary sorafenib-resistance and acquired resistance, the activation of ERK is major process that promotes cell proliferation. In addition, previous studies have revealed that HBx also activates the ERK signaling pathway in HCC [29]. The kinase inhibition profiles of lenvatinib and sorafenib indicated that lenvatinib is more effective at inhibiting FGFR4 (Table S1) [30,31]. Additionally, Huh-7SR and Hep-3BSR cells both expressed FGFR4 and EGFR. Further experiments showed that lenvatinib presented a better performance in ERK signaling inhibition compared with sorafenib. Thus, our findings suggest that lenvatinib overcomes sorafenib resistance mainly through the inhibition of the FGFR4-ERK signaling pathway. Previous studies also indicated that lenvatinib could strongly inhibit the activation of ERK, the downstream signaling molecules of FGFR4, compared with sorafenib and regorafenib [32], and high FGFR4 levels (positive immunohistochemistry >10% of tumor cells) were an independent predictor of a response to lenvatinib [33]. Moreover, lenvatinib enhanced the antitumor immune response of anti-programmed cell death-1 (PD-1) in HCC by blocking FGFR4 [34]. However, lenvatinib alone could inhibit HCC cancer stem-like cells through FGFR1-3 signaling, but not FGFR4 signaling [35]. Interestingly, lenvatinib could also increase the expression of EGFR, as previously reported [36,37]. The activation of EGFR may contribute to sorafenib or lenvatinib resistance. The combination therapy of lenvatinib and gefitinib (an EGFR inhibitor) may be an option for the approximately 50% of advanced HCC patients with high EGFR expression [37]. Combination therapy or new kinase inhibitors (such as ERK inhibitors) may represent a promising strategy for sorafenib-resistant HCC patients.

Moreover, we found that Huh-7SR cells showed higher autophagic responsiveness to sorafenib and lenvatinib than Hep-3BSR cells, and this may contribute to Hep-3BSR's partial cross-resistance to lenvatinib. Autophagy plays neutral, tumor-suppressive, or tumor-promoting roles in cancer development, and is also associated with apoptosis and multidrug-resistance [38]. Previous studies demonstrated that the autophagic responsiveness to sorafenib is distinct between Hep3B and Huh7 wild-type cells and the sensitivity to sorafenib is also different [39], where sorafenib-induced autophagy improves the death rate of HCC cells [40]. Similarly, in sorafenib-resistant cells, the different autophagic responsiveness to lenvatinib may also be associated with altered sensitivity to lenvatinib.

miRNAs are a class of endogenous, small, noncoding-RNA molecules that regulate aspects of the post-transcriptional modulation of gene expression, such as cell proliferation, differentiation, metabolism, and cell death [41,42]. Additionally, several reports have shown that miRNAs can regulate the sensitivity of HCC cells to multi-kinase inhibitor drugs, such as sorafenib, by modifying diverse molecular processes [43,44]. In this study, we reported altered cell miRNA expression in Huh-7SR cells after treatment with lenvatinib. A total of 43 miRNAs were upregulated, while 23 miRNAs were downregulated. Further data set analysis from TCGA indicated that miR-130b-3p and miR-1292-5p were highly expressed, while the levels of miR-491-5p and miR-1247-3p were decreased in HCC tissues compared with normal tissues. In addition, miRNA expression was reversed after lenvatinib treatment. In many cases, miRNAs can have oncogenic effects or act as suppressors in different cancers, and some may also have a dual function. Liao et al. indicated that miR-130b-3p was upregulated in HCC and was correlated with a poor prognosis. Overexpressed miR-130b-3p was found to enhance the angiogenesis capacity of HCC cells [45]. Decreased miR-491-5p and highly pyruvate kinase M2 (PKM2) expression were associated with unfavorable clinical features and the poor prognosis of HCC patients [46]. In addition, overexpressed miR-491-5p inhibited HCC cell proliferation and migration by targeting SEC61 translocon alpha 1 subunit1 (SEC61A) [47]. The inhibition effect of lenvatinib in sorafenib-resistant HCC cells occurred partially through miRNA regulation. However, due to the scarcity of information on some of the miRNAs, further studies are needed on their function.

Only a few clinical studies have been conducted on second-line or further-line treatment of lenvatinib. Jefremow et al. suggested that in seven patients with a later line lenvatinib treatment, partial remission (PR) was shown in four of the seven patients, stable disease (SD) in two of the seven, and mixed response with overall tolerable safety in one of the seven [48]. Chen et al. enrolled 40 patients who received lenvatinib after sorafenib. The median overall survival (OS) was 9.8 months, and the objective response rate was 27.5%; moreover, the clinical outcomes of lenvatinib treatment in later lines were similar [15]. However, a report by Tomonari et al. indicated that the objective response rate was 33.3% in the second line, and 20.0% in the third line. Additionally, sorafenib-resistant HCC cells show partial cross-resistance to lenvatinib by the decreased response to FGFR signaling pathways compared with wild-type cells [16] (Table 2). Although the number of patients was one of the limitations of the study, an effective tendency was shown.

Table 2. Lenvatinib as second-line or further-line treatment after sorafenib.

Author	Total Patients	Treatment Line		Complete Remission	Partial Remission	Stable Disease	Mixed Response	Progression
Jefremow A et al. [48]	7	Second line	2	0	1	1	0	0
		Third line	3	0	2	0	1	0
		Fourth line	2	0	1	1	0	0
Chen YY et al. [15]	40	Second line	20	1	5	7	0	7
		Third line	10	0	4	4	0	2
		Fourth line	10	0	1	5	0	4
Tomonari T et al. [16]	19	Second line	9	1	2	5	0	1
		Third line	10	0	2	8	0	0
	66				2	18	31	1	14

This study has the following several limitations: (i) We only used two sorafenib-resistant HCC cell lines; (ii) expanded clinical studies are needed, including in sorafenib-resistant HCC patients with HBV infection, sorafenib-resistant HCC patients with high or low FGFR4 expression, and an investigation of a promising strategy, such as combination therapy with lenvatinib. Hence, we plan to perform further experiments to investigate the above limitations in our future research on this topic.

In conclusion, our study shows that (i) the advantage of lenvatinib in overcoming sorafenib-resistance may be through the FGFR4-ERK signaling pathway; (ii) HBV DNA and poor autophagic responsiveness may be the reasons for partial cross-resistance; (iii) miRNA alterations may contribute to inhibiting sorafenib-resistant HCC cell growth and angiogenesis. The present study and previous clinical data (Table 2) provide evidence that lenvatinib may be a suitable second-line therapy for unresectable HCC patients who express FGFR4 and are sorafenib resistant. Combination therapy could be a promising way to expand the efficiency of first-line multi-kinase inhibitors and our study is one of the initial investigations of this topic. Drug resistance in HCC patients is a huge barrier to overcome. Attenuating first-line treatment resistance and expanding the efficiency of second-line therapy could potentially prolong the survival time of patients. Thus, further studies and clinical trials are needed.

4. Materials and Methods

4.1. Chemicals

Sorafenib (Nexavar®) was obtained from Bayer Pharmaceutical Corporation (West Haven, CT, USA) and lenvatinib from Chem Scene (Monmouth Junction, NJ, USA). Matrigel® Matrix (356230) was purchased from Corning Inc. (Lowell, MA, USA).

4.2. Cell Lines and Cell Culture

Human HCC cell lines, Huh-7 and Hep-3B, were obtained from the Japanese Research Resources Bank (Tokyo, Japan). Huh-7 cells were maintained in Dulbecco's Modified Eagle's Medium (DMEM; Gibco-Invitrogen, Carlsbad, CA, USA) supplemented with 10% fetal bovine serum (FBS; Wako, Tokyo, Japan) and penicillin/streptomycin (100 mg/L; Invitrogen, Tokyo, Japan). Hep-3B cells were cultured in Modified Eagle's Medium (MEM; Gibco-Invitrogen) supplemented with 10% FBS and penicillin/streptomycin. To establish sorafenib-resistant HCC cell lines, Huh-7 and Hep-3B cells were cultured with increasing doses of sorafenib, from 1 to 6 μM, for six months. Cells were grown in a humidified incubator at 5% CO_2 and 37 °C. Huh-7 and Hep-3B cell lines as well as Huh-7 sorafenib-resistant (Huh-7SR) and Hep-3BSR sublines were authenticated using short tandem repeat (STR) profiling (BEX Co., Ltd., Tokyo, Japan).

4.3. Cell Viability Assay

Cell viability assays were performed using the Cell Counting Kit-8 (Dojindo Laboratories, Kumamoto, Japan) according to the manufacturer's instructions. Briefly, the cells were seeded in 96-well plates at a concentration of 5000 cells/100 μL/well and incubated in a normal growth medium for 24 h. Subsequently, the cells were grown for an additional 24, 48, 72, or 96 h with sorafenib (0.3, 1, 3, 10, and 30 μM), lenvatinib (0.3, 1, 3, 10, and 30 μM), or dimethyl sulfoxide (DMSO). The medium was replaced with 100 μL of fresh medium containing 10% CCK-8 reagent, and the cells were incubated at 37 °C for 3 h. The absorbance was measured at 450 nm using a multi-grating microplate reader SH-9000Lab (CORONA Electric Co., Ltd., Ibaraki, Japan). The experiments were repeated three times.

4.4. Three-Dimensional (3D) Tumor Spheroid Assay (3D Culture)

Multicellular spheroids were generated using the liquid overlay technique (Figure S3) [49].

4.5. Flow Cytometry Analysis of Cell Cycle

Cell cycle progression was evaluated using a Cell Cycle Phase Determination Kit (Cayman Chemical Company, Ann Arbor, MI, USA). Cells (1.0×10^6 cells/100-mm dish) were treated with 10 μM lenvatinib or DMSO for 24 h. Cells were trypsinized and resuspended in phosphate-buffered saline (PBS) at a density of 10^6 cells/mL. Approximately 1.0×10^6 cells were stained in 100 μL of PBS with 10 μL of RNase A (250 μg/mL) and 10 μL of propidium iodide (PI) stain (100 μg/mL) and incubated at room temperature in the dark for 30 min. Flow cytometry (FCM) was performed to compare the proportion of lenvatinib-treated and control cells in each phase of the cell cycle. FCM was performed using a Cytomics FC 500 flow cytometer (Beckman Coulter, Brea, CA, USA) equipped with an argon laser (488 nm), and the percentage of cells was analyzed using the Kaluza software version v2.1 (Beckman Coulter). The experiments were repeated three times.

4.6. Apoptosis Analysis

Lenvatinib-mediated apoptosis was analyzed using an FCM and Annexin V-FITC Early Apoptosis Detection kit (Cell Signaling Technology, Beverly, MA, USA). Cells (1.0×10^6 cells/100-mm dish) were treated with 10 μM lenvatinib or DMSO for 24 h. Apoptotic and necrotic cells were analyzed by double staining with FITC-conjugated annexin V and PI according to the manufacturer's instructions. FCM was conducted using a Cytomics FC 500 flow cytometer equipped with an argon laser (488 nm) to compare the proportion of apoptotic cells in the lenvatinib-treated and control groups, and data were analyzed using the Kaluza software version v2.1. The experiments were repeated three times.

4.7. Apoptosis Analysis by Enzyme Linked Immunosorbent Assay (ELISA)

ELISA was performed to analyze the levels of caspase-cleaved cytokeratin 18 (cCK18) using the M30-Apoptosense ELISA kit (Peviva Ab, Bromma, Sweden). Cells (5000 cells/well) were seeded in 96-well plates and treated with 10 μM lenvatinib or DMSO for 24 h. Subsequently, the cells were lysed in polyoxyethylene octyl phenyl ether (Wako) and analyzed according to the manufacturer's instructions. The experiments were repeated three times.

4.8. Antibody Arrays to Analyze Angiogenesis-Related Proteins

Cells (1.0×10^6 cells/100-mm dish) were treated with 10 μM lenvatinib or DMSO for 24 h at 37 °C and lysed with a protease inhibitor cocktail, PRO-PREP complete protease inhibitor mixture (iNtRON Biotechnology, Seongnam, Korea). The Human Angiogenesis Array Kit (R&D Systems) was used to analyze the angiogenesis-related proteins in the lenvatinib-treated and control cells according to the manufacturer's protocol. Each array was repeated three times to validate the results.

4.9. Invasion Assay

For invasion assays, 1×10^4 cells were cultured in 100 μL of serum-free DMEM or with 10 μM lenvatinib, placed into the upper membrane chamber (Cell Biolabs, San Diego, CA, USA.) and precoated with extracellular matrix (ECM) proteins. The lower compartment was filled with 150 μL of DMEM containing 10% FBS. After incubation for 24 h in an atmosphere of 5% CO_2 at 37 °C, the cells in the upper compartment were removed, and the invested cells in the lower compartment were lysed. Fluorescence was read using an SH-9000Lab plate reader (CORONA) at 480 nm/520 nm.

4.10. Wound Healing Assay

Cells were seeded in 6-well plates and incubated in DMEM containing 10% FBS until they reached subconfluence. Scratches were introduced in the cell monolayer using a plastic pipette tip. After washing with PBS, 1.5% FBS media containing 10 µM lenvatinib or DMSO was added. The scratched area was photographed using a light microscope after 24 h.

4.11. miRNA Microarray

Total RNA of cells was extracted using the miRNeasy Mini Kit (QIAGEN) according to the manufacturer's instructions. After confirming the purity and quantity of each sample using an Agilent 2100 Bioanalyzer (Agilent Technologies, Santa Clara, CA, USA) and an RNA 6000 Nano kit (Agilent Technologies), respectively, the samples were labeled using a miRCURY Hy3 Power Labeling kit (Exiqon A/S, Vedbaek, Denmark) and hybridized to a human miRNA Oligo Chip (v.21; Toray Industries, Inc., Tokyo, Japan). Chips were scanned using a 3D-Gene Scanner 3000 (Toray Industries). The 3D-Gene extraction software version 1.2 (Toray Industries) was used to calculate the raw signal intensity of the images. The raw data were analyzed using GeneSpring GX 10.0 software (Agilent Technologies) to assess the differences in miRNA expression between the samples. Global normalization was performed on raw data obtained above the background level. Differentially expressed miRNAs were determined using Welch's *t*-test.

4.12. Colony Formation Assay

Cells were trypsinized for 3 min and resuspended at a density of 1×10^3/mL. Five hundred microliters were seeded into 6-well plates, and 1.5 mL of DMEM containing 10% FBS and 1 µM lenvatinib or DMSO was added to each well. The plates were incubated at 37 °C in 5% CO_2, and the medium was changed every three days until conspicuous colonies were observed. Colonies were fixed and stained with 0.1% crystal violet at room temperature for 5 min, and positive colony formation (>50 cells/colony) was evaluated by counting the number of colonies.

4.13. Data Set Analysis

All patients of LIHC were retrieved from the TCGA data portal (https://portal.gdc.cancer.gov/, accessed on 3 September 2021). The full clinical dataset was downloaded (up to 3 September 2021) and this study meets the publication guidelines provided by TCGA (http://cancergenome.nih.gov/publications/publicationguidelines, accessed on 3 September 2021) [50,51]. OncoLnc (http://www.oncolnc.org, accessed on 5 October 2021) [52] and KM plotter (http://kmplot.com/analysis, accessed on 5 October 2021) [53] were used for survival analysis.

4.14. Western Blot

The cells were lysed with a PRO-PREP complete protease inhibitor mixture (iNtRON Biotechnology, Korea) and collected supernatants. Protein concentration was measured using a NanoDrop 2000 spectrofluorometer (Thermo Fisher Scientific, Inc., Waltham, MA, USA). Briefly, protein aliquots (10 µg) were separated on precast protein gels (4–20% Mini-PROTEAN TGX Gels; Bio-Rad, Hercules, CA, USA) and transferred onto nitrocellulose membranes. The membranes were blocked with 2% skimmed milk (GE Healthcare) in TBST with 0.1% Tween 20 (cat. No. T9142, Takara Bio Inc., Kusatsu, Shiga, Japan) for 30 min and were then incubated overnight at 4 °C with the following primary antibodies: anti-ß-Actin (#66009-1-lg, Proteintech, dilution 1:5000); anti-LC3A/B (#12741, CST, dilution 1:1000); SQSTM1/p62 (#5114, CST, dilution 1:1000); Caspase-3 (#14220, CST, dilution 1:1000); PARP (#9542, CST, dilution 1:1000); Caspase-7 (#12827, CST, dilution 1:1000); EGFR (#PA1-1110, Thermo Fisher Scientific, dilution 1:1000); p-Akt (ser473) (#4060, CST, dilution 1:2000); Akt (#4685, CST, dilution 1:1000); p-mTOR (SER2448) (#5536, CST, dilution 1:1000); mTOR (#2983, CST, dilution 1:1000); cyclinD1 (#2978, CST, dilution 1:1000); p-ERK1/2 (#4370, CST, dilution 1:1000); ERK1/2 (#4695, CST, dilution 1:1000); p-MEK1/2 (#9154, CST, dilution 1:1000); MEK1/2 (#8727, CST, dilution 1:1000); anti-ß-Actin (#4967, CST, dilution 1:1000); FGFR4 (#8562, CST, dilution 1:1000) in 5% serum (cat. No. 9048-46-8, FUJIFILM Wako, Osaka, Japan). After washing with TBST, the membranes were incubated for 1 h with corresponding horseradish peroxidase (HRP)-conjugated secondary antibodies: anti-mouse (#7076, CST, dilution 1:2000) or anti-rabbit (#7074, CST, dilution 1:2000). The signal was visualized using a chemiluminescent (ECL) kit (cat. No. 45-000-999; Cytiva) and imaged using ImageQuant LAS 4010 (GE Healthcare).

4.15. Xenograft Model Analysis

The animal study was approved (approval No. 20627, 20640 and 21681) by, and was conducted in accordance with the guidelines set by, the Committee on Experimental Animals of the Kagawa University. Female athymic mice (BALB/c-nu/nu; 5 weeks old; 15–17 g) were purchased from Japan SLC (Shizuoka, Japan). The mice were maintained under specific pathogen-free conditions using a laminar airflow rack and had continuous free access to sterilized (γ-irradiated) food (CL-2; CLEA Japan, Inc., Tokyo, Japan) and autoclaved water. Mice were subcutaneously inoculated with 5×10^6 Huh-7SR cells in the right flank. When the xenografts were palpable with an approximate diameter of 5 mm, we randomly assigned the animals to three groups of eight mice each. These groups were orally administered 20 mg/kg/day lenvatinib, 30 mg/kg/day sorafenib, or vehicle (DMSO and saline) for five days per week. Tumor volume (mm^3) was calculated as tumor length (mm) $\times$ tumor width (mm)^2/2. All animals were sacrificed on day 10 of treatment.

4.16. Immunohistochemistry

Specimens were fixed overnight in 10% formalin, embedded in paraffin, and cut into sections 5 µm thick. Briefly, slides were deparaffinized in xylene, rehydrated through a graded alcohol series, and rinsed in PBS. Depending on the protein target to be revealed, antigen retrieval was achieved by boiling the sections in either 10 mM sodium citrate buffer (pH 6.0) or 1 mM ethylenediaminetetraacetic acid (EDTA; pH 8.5) buffer for 10 min, followed by a 20-min cool down at room temperature. After a blocking step using 5% goat serum and an Avidin-Biotin Blocking Kit (Vector Laboratories, Burlingame, CA, USA), the slides were incubated with specific primary antibodies (cyclin D1, ki-67, CD31) overnight at 4 °C. To quench the endogenous peroxidase activity, slides were incubated for 10 min with 3% hydrogen peroxide, and subsequently, the biotin-conjugated secondary antibody was applied at a 1:500 dilution for 30 min at room temperature. Immunoreactivity was visualized using the Vectastain Elite ABC kit (Vector Laboratories) and 3,3-diaminobenzidine or Vector NovaRed (Vector Laboratories) as the chromogen. Finally, the slides were counterstained with hematoxylin.

4.17. Statistical Analysis

GraphPad Prism software version 6.0 (GraphPad Software, San Diego, CA, USA) was used for all analyses. Unpaired Student's *t*-test and Wilcox test were used to determine statistical significance between different groups. Two-way analysis of variance (ANOVA) or mixed ANOVA was performed to test the comparisons and corrected using Tukey's post hoc test. One-way ANOVA was performed before the Tukey's post hoc test to test the comparisons. Values were considered statistically significant at a *p*-value < 0.05.

Supplementary Materials: The following are available online at https://www.mdpi.com/article/10.3390/ijms222313071/s1.

Author Contributions: Data curation, T.S. and H.I.; investigation, T.S., K.F., H.K., Y.G. and N.N.; methodology, T.S., M.Y., M.N. and K.T.; project administration, T.S., A.M. and T.M.; supervision, S.F., H.Y. and J.T.; writing—original draft preparation, T.S.; writing—review and editing, T.S. and T.M. All authors have read and agreed to the published version of the manuscript.

Funding: This research received no external funding.

Institutional Review Board Statement: The study was conducted according to the guidelines of the Declaration of Helsinki and approved by the Institutional Review Board of the Department of Laboratory Animal Science of Kagawa University, Kida, Japan (approval No. 20627, 20640 and 21681).

Informed Consent Statement: Not applicable.

Data Availability Statement: All data supporting the conclusions of the present study have been documented in this article.

Acknowledgments: We thank Kayo Hirose, Keiko Fujikawa, Megumi Okamura, and Fuyuko Kokado for their assistance.

Conflicts of Interest: The authors declare no conflict of interest.

References

1. Bray, F.; Ferlay, J.; Soerjomataram, I.; Siegel, R.L.; Torre, L.A.; Jemal, A. Global cancer statistics 2018: GLOBOCAN estimates of incidence and mortality worldwide for 36 cancers in 185 countries. *CA Cancer J. Clin.* **2018**, *68*, 394–424. [CrossRef] [PubMed]
2. Mattiuzzi, C.; Lippi, G. Current Cancer Epidemiology. *J. Epidemiol. Glob. Health* **2019**, *9*, 217–222. [CrossRef] [PubMed]
3. Lurje, I.; Czigany, Z.; Bednarsch, J.; Roderburg, C.; Isfort, P.; Neumann, U.P.; Lurje, G. Treatment Strategies for Hepatocellular Carcinoma-a Multidisciplinary Approach. *Int. J. Mol. Sci.* **2019**, *20*, 1465. [CrossRef] [PubMed]
4. Llovet, J.M.; Ricci, S.; Mazzaferro, V.; Hilgard, P.; Gane, E.; Blanc, J.F.; de Oliveira, A.C.; Santoro, A.; Raoul, J.L.; Forner, A.; et al. Sorafenib in advanced hepatocellular carcinoma. *N. Engl. J. Med.* **2008**, *359*, 378–390. [CrossRef]
5. Chen, J.; Jin, R.; Zhao, J.; Liu, J.; Ying, H.; Yan, H.; Zhou, S.; Liang, Y.; Huang, D.; Liang, X.; et al. Potential molecular, cellular and microenvironmental mechanism of sorafenib resistance in hepatocellular carcinoma. *Cancer Lett.* **2015**, *367*, 1–11. [CrossRef]
6. Ezzoukhry, Z.; Louandre, C.; Trécherel, E.; Godin, C.; Chauffert, B.; Dupont, S.; Diouf, M.; Barbare, J.C.; Mazière, J.C.; Galmiche, A. EGFR activation is a potential determinant of primary resistance of hepatocellular carcinoma cells to sorafenib. *Int. J. Cancer* **2012**, *131*, 2961–2969. [CrossRef]
7. Zhu, Y.J.; Zheng, B.; Wang, H.Y.; Chen, L. New knowledge of the mechanisms of sorafenib resistance in liver cancer. *Acta Pharmacol. Sin.* **2017**, *38*, 614–622. [CrossRef]
8. Méndez-Blanco, C.; Fondevila, F.; García-Palomo, A.; González-Gallego, J.; Mauriz, J.L. Sorafenib resistance in hepatocarcinoma: Role of hypoxia-inducible factors. *Exp. Mol. Med.* **2018**, *50*, 1–9. [CrossRef]
9. Finn, R.S.; Merle, P.; Granito, A.; Huang, Y.H.; Bodoky, G.; Pracht, M.; Yokosuka, O.; Rosmorduc, O.; Gerolami, R.; Caparello, C.; et al. Outcomes of sequential treatment with sorafenib followed by regorafenib for HCC: Additional analyses from the phase III RESORCE trial. *J. Hepatol.* **2018**, *69*, 353–358. [CrossRef]
10. Abou-Alfa, G.K.; Meyer, T.; Cheng, A.L.; El-Khoueiry, A.B.; Rimassa, L.; Ryoo, B.Y.; Cicin, I.; Merle, P.; Chen, Y.; Park, J.W.; et al. Cabozantinib in Patients with Advanced and Progressing Hepatocellular Carcinoma. *N. Engl. J. Med.* **2018**, *379*, 54–63. [CrossRef]
11. Jindal, A.; Thadi, A.; Shailubhai, K. Hepatocellular Carcinoma: Etiology and Current and Future Drugs. *J. Clin. Exp. Hepatol.* **2019**, *9*, 221–232. [CrossRef]
12. Hao, Z.; Wang, P. Lenvatinib in Management of Solid Tumors. *Oncologist* **2020**, *25*, e302–e310. [CrossRef]
13. Kudo, M.; Finn, R.S.; Qin, S.; Han, K.H.; Ikeda, K.; Piscaglia, F.; Baron, A.; Park, J.W.; Han, G.; Jassem, J.; et al. Lenvatinib versus sorafenib in first-line treatment of patients with unresectable hepatocellular carcinoma: A randomised phase 3 non-inferiority trial. *Lancet* **2018**, *391*, 1163–1173. [CrossRef]
14. Hiraoka, A.; Kumada, T.; Kariyama, K.; Takaguchi, K.; Atsukawa, M.; Itobayashi, E.; Tsuji, K.; Tajiri, K.; Hirooka, M.; Shimada, N.; et al. Clinical features of lenvatinib for unresectable hepatocellular carcinoma in real-world conditions: Multicenter analysis. *Cancer Med.* **2019**, *8*, 137–146. [CrossRef]
15. Chen, Y.Y.; Wang, C.C.; Liu, Y.W.; Li, W.F.; Chen, Y.H. Clinical impact of lenvatinib in patients with unresectable hepatocellular carcinoma who received sorafenib. *PeerJ* **2020**, *8*, e10382. [CrossRef]
16. Tomonari, T.; Sato, Y.; Tanaka, H.; Tanaka, T.; Fujino, Y.; Mitsui, Y.; Hirao, A.; Taniguchi, T.; Okamoto, K.; Sogabe, M.; et al. Potential use of lenvatinib for patients with unresectable hepatocellular carcinoma including after treatment with sorafenib: Real-world evidence and in vitro assessment via protein phosphorylation array. *Oncotarget* **2020**, *11*, 2531–2542. [CrossRef]
17. Enz, N.; Vliegen, G.; De Meester, I.; Jungraithmayr, W. CD26/DPP4-a potential biomarker and target for cancer therapy. *Pharmacol. Ther.* **2019**, *198*, 135–159. [CrossRef]
18. Kawaguchi, T.; Nakano, D.; Koga, H.; Torimura, T. Effects of a DPP4 Inhibitor on Progression of NASH-related HCC and the p62/Keap1/Nrf2-Pentose Phosphate Pathway in a Mouse Model. *Liver Cancer* **2019**, *8*, 359–372. [CrossRef]

19. Yu, H.; Mei, X.P.; Su, P.F.; Jin, G.Z.; Zhou, H.K. A poor prognosis in human hepatocellular carcinoma is associated with low expression of DPP4. *Braz. J. Med. Biol. Res.* **2020**, *53*, e9114. [CrossRef]

20. Li, C.; Huang, Z.; Zhu, L.; Yu, X.; Gao, T.; Feng, J.; Hong, H.; Yin, H.; Zhou, T.; Qi, W.; et al. The contrary intracellular and extracellular functions of PEDF in HCC development. *Cell Death Dis.* **2019**, *10*, 742. [CrossRef]

21. Jin, Y.; Liang, Z.Y.; Zhou, W.X.; Zhou, L. Expression, clinicopathologic and prognostic significance of plasminogen activator inhibitor 1 in hepatocellular carcinoma. *Cancer Biomark.* **2020**, *27*, 285–293. [CrossRef]

22. Baghy, K.; Tátrai, P.; Regős, E.; Kovalszky, I. Proteoglycans in liver cancer. *World J. Gastroenterol.* **2016**, *22*, 379–393. [CrossRef]

23. Li, Y.; Turpin, C.P.; Wang, S. Role of thrombospondin 1 in liver diseases. *Hepatol. Res.* **2017**, *47*, 186–193. [CrossRef]

24. Poon, R.T.; Chung, K.K.; Cheung, S.T.; Lau, C.P.; Tong, S.W.; Leung, K.L.; Yu, W.C.; Tuszynski, G.P.; Fan, S.T. Clinical significance of thrombospondin 1 expression in hepatocellular carcinoma. *Clin. Cancer Res.* **2004**, *10*, 4150–4157. [CrossRef]

25. Fousek, K.; Horn, L.A.; Palena, C. Interleukin-8: A chemokine at the intersection of cancer plasticity, angiogenesis, and immune suppression. *Pharmacol. Ther.* **2021**, *219*, 107692. [CrossRef]

26. Matsuki, M.; Hoshi, T.; Yamamoto, Y.; Ikemori-Kawada, M.; Minoshima, Y.; Funahashi, Y.; Matsui, J. Lenvatinib inhibits angiogenesis and tumor fibroblast growth factor signaling pathways in human hepatocellular carcinoma models. *Cancer Med.* **2018**, *7*, 2641–2653. [CrossRef]

27. Schmidt, B.; Wei, L.; DePeralta, D.K.; Hoshida, Y.; Tan, P.S.; Sun, X.; Sventek, J.P.; Lanuti, M.; Tanabe, K.K.; Fuchs, B.C. Molecular subclasses of hepatocellular carcinoma predict sensitivity to fibroblast growth factor receptor inhibition. *Int. J. Cancer* **2016**, *138*, 1494–1505. [CrossRef]

28. Yazdani, H.O.; Huang, H.; Tsung, A. Autophagy: Dual Response in the Development of Hepatocellular Carcinoma. *Cells.* **2019**, *8*, 91. [CrossRef] [PubMed]

29. Liao, B.; Zhou, H.; Liang, H.; Li, C. Regulation of ERK and AKT pathways by hepatitis B virus X protein via the Notch1 pathway in hepatocellular carcinoma. *Int. J. Oncol.* **2017**, *51*, 1449–1459. [CrossRef] [PubMed]

30. Stjepanovic, N.; Capdevila, J. Multikinase inhibitors in the treatment of thyroid cancer: Specific role of lenvatinib. *Biologics* **2014**, *8*, 129–139. [CrossRef] [PubMed]

31. Tohyama, O.; Matsui, J.; Kodama, K.; Hata-Sugi, N.; Kimura, T.; Okamoto, K.; Minoshima, Y.; Iwata, M.; Funahashi, Y. Antitumor activity of lenvatinib (e7080): An angiogenesis inhibitor that targets multiple receptor tyrosine kinases in preclinical human thyroid cancer models. *J. Thyroid Res.* **2014**, *2014*, 638747. [CrossRef]

32. Kanzaki, H.; Chiba, T.; Ao, J.; Koroki, K.; Kanayama, K.; Maruta, S.; Maeda, T.; Kusakabe, Y.; Kobayashi, K.; Kanogawa, N.; et al. The impact of FGF19/FGFR4 signaling inhibition in antitumor activity of multi-kinase inhibitors in hepatocellular carcinoma. *Sci. Rep.* **2021**, *11*, 5303. [CrossRef]

33. Yamauchi, M.; Ono, A.; Ishikawa, A.; Kodama, K.; Uchikawa, S.; Hatooka, H.; Zhang, P.; Teraoka, Y.; Morio, K.; Fujino, H.; et al. Tumor Fibroblast Growth Factor Receptor 4 Level Predicts the Efficacy of Lenvatinib in Patients With Advanced Hepatocellular Carcinoma. *Clin. Transl. Gastroenterol.* **2020**, *11*, e00179. [CrossRef]

34. Yi, C.; Chen, L.; Lin, Z.; Liu, L.; Shao, W.; Zhang, R.; Lin, J.; Zhang, J.; Zhu, W.; Jia, H.; et al. Lenvatinib Targets FGF Receptor 4 to Enhance Antitumor Immune Response of Anti-Programmed Cell Death-1 in HCC. *Hepatology* **2021**, *74*, 2544–2560. [CrossRef]

35. Shigesawa, T.; Maehara, O.; Suda, G.; Natsuizaka, M.; Kimura, M.; Shimazaki, T.; Yamamoto, K.; Yamada, R.; Kitagataya, T.; Nakamura, A.; et al. Lenvatinib suppresses cancer stem-like cells in HCC by inhibiting FGFR1-3 signaling, but not FGFR4 signaling. *Carcinogenesis* **2021**, *42*, 58–69. [CrossRef]

36. Rodríguez-Hernández, M.A.; Chapresto-Garzón, R.; Cadenas, M.; Navarro-Villarán, E.; Negrete, M.; Gómez-Bravo, M.A.; Victor, V.M.; Padillo, F.J.; Muntané, J. Differential effectiveness of tyrosine kinase inhibitors in 2D/3D culture according to cell differentiation, p53 status and mitochondrial respiration in liver cancer cells. *Cell Death Dis.* **2020**, *11*, 339. [CrossRef]

37. Jin, H.; Shi, Y.; Lv, Y.; Yuan, S.; Ramirez, C.F.A.; Lieftink, C.; Wang, L.; Wang, S.; Wang, C.; Dias, M.H.; et al. EGFR activation limits the response of liver cancer to lenvatinib. *Nature* **2021**, *595*, 730–734. [CrossRef]

38. Sun, T.; Liu, H.; Ming, L. Multiple Roles of Autophagy in the Sorafenib Resistance of Hepatocellular Carcinoma. *Cell Physiol. Biochem.* **2017**, *44*, 716–727. [CrossRef]

39. Fischer, T.D.; Wang, J.H.; Vlada, A.; Kim, J.S.; Behrns, K.E. Role of autophagy in differential sensitivity of hepatocarcinoma cells to sorafenib. *World J. Hepatol.* **2014**, *6*, 752–758. [CrossRef]

40. Tai, W.T.; Shiau, C.W.; Chen, H.L.; Liu, C.Y.; Lin, C.S.; Cheng, A.L.; Chen, P.J.; Chen, K.F. Mcl-1-dependent activation of Beclin 1 mediates autophagic cell death induced by sorafenib and SC-59 in hepatocellular carcinoma cells. *Cell Death Dis.* **2013**, *4*, e485. [CrossRef]

41. Wang, M.; Yu, F.; Chen, X.; Li, P.; Wang, K. The Underlying Mechanisms of Noncoding RNAs in the Chemoresistance of Hepatocellular Carcinoma. *Mol. Ther. Nucleic Acids* **2020**, *21*, 13–27. [CrossRef]

42. Oura, K.; Morishita, A.; Masaki, T. Molecular and Functional Roles of MicroRNAs in the Progression of Hepatocellular Carcinoma—A Review. *Int. J. Mol. Sci.* **2020**, *21*, 8362. [CrossRef]

43. Gramantieri, L.; Pollutri, D.; Gagliardi, M.; Giovannini, C.; Quarta, S.; Ferracin, M.; Casadei-Gardini, A.; Callegari, E.; De Carolis, S.; Marinelli, S.; et al. MiR-30e-3p Influences Tumor Phenotype through MDM2/TP53 Axis and Predicts Sorafenib Resistance in Hepatocellular Carcinoma. *Cancer Res.* **2020**, *80*, 1720–1734. [CrossRef]

44. Xu, W.P.; Liu, J.P.; Feng, J.F.; Zhu, C.P.; Yang, Y.; Zhou, W.P.; Ding, J.; Huang, C.K.; Cui, Y.L.; Ding, C.H.; et al. miR-541 potentiates the response of human hepatocellular carcinoma to sorafenib treatment by inhibiting autophagy. *Gut* **2020**, *69*, 1309–1321. [CrossRef]

45. Liao, Y.; Wang, C.; Yang, Z.; Liu, W.; Yuan, Y.; Li, K.; Zhang, Y.; Wang, Y.; Shi, Y.; Qiu, Y.; et al. Dysregulated Sp1/miR-130b-3p/HOXA5 axis contributes to tumor angiogenesis and progression of hepatocellular carcinoma. *Theranostics* **2020**, *10*, 5209–5224. [CrossRef]

46. Xu, Q.; Dou, C.; Liu, X.; Yang, L.; Ni, C.; Wang, J.; Guo, Y.; Yang, W.; Tong, X.; Huang, D. Oviductus ranae protein hydrolysate (ORPH) inhibits the growth, metastasis and glycolysis of HCC by targeting miR-491-5p/PKM2 axis. *Biomed. Pharmacother.* **2018**, *107*, 1692–1704. [CrossRef]

47. Fa, X.; Song, P.; Fu, Y.; Deng, Y.; Liu, K. Long non-coding RNA VPS9D1-AS1 facilitates cell proliferation, migration and stemness in hepatocellular carcinoma. *Cancer Cell Int.* **2021**, *21*, 131. [CrossRef]

48. Jefremow, A.; Wiesmueller, M.; Rouse, R.A.; Dietrich, P.; Kremer, A.E.; Waldner, M.J.; Neurath, M.F.; Siebler, J. Beyond the border: The use of lenvatinib in advanced hepatocellular carcinoma after different treatment lines: A retrospective analysis. *J. Physiol. Pharmacol.* **2020**, *71*. [CrossRef]

49. Nath, S.; Devi, G.R. Three-dimensional culture systems in cancer research: Focus on tumor spheroid model. *Pharmacol. Ther.* **2016**, *163*, 94–108. [CrossRef] [PubMed]

50. Zhang, J.; Chong, C.C.; Chen, G.G.; Lai, P.B. A Seven-microRNA Expression Signature Predicts Survival in Hepatocellular Carcinoma. *PLoS ONE* **2015**, *10*, e0128628. [CrossRef] [PubMed]

51. Zhou, T.; Cai, Z.; Ma, N.; Xie, W.; Gao, C.; Huang, M.; Bai, Y.; Ni, Y.; Tang, Y. A Novel Ten-Gene Signature Predicting Prognosis in Hepatocellular Carcinoma. *Front. Cell Dev. Biol.* **2020**, *8*, 629. [CrossRef]

52. Anaya, J. OncoLnc: Linking TCGA survival data to mRNAs, miRNAs, and lncRNAs. *PeerJ Comput. Sci.* **2016**, *2*, e67. [CrossRef]

53. Győrffy, B. Survival analysis across the entire transcriptome identifies biomarkers with the highest prognostic power in breast cancer. *Comput. Struct. Biotechnol. J.* **2021**, *19*, 4101–4109. [CrossRef]

International Journal of
Molecular Sciences

Article

HDAC6-Selective Inhibitor Overcomes Bortezomib Resistance in Multiple Myeloma

Sang Wu Lee [†], Soo-Keun Yeon [†], Go Woon Kim, Dong Hoon Lee, Yu Hyun Jeon, Jung Yoo, So Yeon Kim and So Hee Kwon *

College of Pharmacy, Yonsei Institute of Pharmaceutical Sciences, Yonsei University, Incheon 21983, Korea;
tkddn407@naver.com (S.W.L.); ilove_oov@hanmail.net (S.-K.Y.); goun6997@daum.net (G.W.K.);
tci30@naver.com (D.H.L.); hyun953@naver.com (Y.H.J.); jungy619@yonsei.ac.kr (J.Y.); ksy_dct@naver.com (S.Y.K.)
* Correspondence: soheekwon@yonsei.ac.kr; Tel.: +82-32-749-4513
† These authors contributed equally to this work.

Abstract: Although multiple myeloma (MM) patients benefit from standard bortezomib (BTZ) chemotherapy, they develop drug resistance, resulting in relapse. We investigated whether histone deacetylase 6 (HDAC6) inhibitor A452 overcomes bortezomib resistance in MM. We show that HDAC6-selective inhibitor A452 significantly decreases the activation of BTZ-resistant markers, such as extracellular signal-regulated kinases (ERK) and nuclear factor kappa B (NF-κB), in acquired BTZ-resistant MM cells. Combination treatment of A452 and BTZ or carfilzomib (CFZ) synergistically reduces BTZ-resistant markers. Additionally, A452 synergizes with BTZ or CFZ to inhibit the activation of NF-κB and signal transducer and activator of transcription 3 (STAT3), resulting in decreased expressions of low-molecular-mass polypeptide 2 (LMP2) and LMP7. Furthermore, combining A452 with BTZ or CFZ leads to synergistic cancer cell growth inhibition, viability decreases, and apoptosis induction in the BTZ-resistant MM cells. Overall, the synergistic effect of A452 with CFZ is more potent than that of A452 with BTZ in BTZ-resistant U266 cells. Thus, our findings reveal the HDAC6-selective inhibitor as a promising therapy for BTZ-chemoresistant MM.

Keywords: HDAC6; bortezomib-resistance; HDAC6-selective inhibitor; bortezomib; carfilzomib; multiple myeloma; LMP2; combination therapy

Citation: Lee, S.W.; Yeon, S.-K.; Kim, G.W.; Lee, D.H.; Jeon, Y.H.; Yoo, J.; Kim, S.Y.; Kwon, S.H. HDAC6-Selective Inhibitor Overcomes Bortezomib Resistance in Multiple Myeloma. *Int. J. Mol. Sci.* **2021**, 22, 1341. https://doi.org/10.3390/ijms22031341

Academic Editors: Daniela Grimm and Valentina De Falco
Received: 17 November 2020
Accepted: 26 January 2021
Published: 29 January 2021

Publisher's Note: MDPI stays neutral with regard to jurisdictional claims in published maps and institutional affiliations.

1. Introduction

Multiple myeloma (MM) is characterized by abnormally proliferating plasma cells derived from B cells [1] and ranks second as the cause of death from hematological malignancy [2]. Typical symptoms of MM include anemia, bone destruction, hypercalcemia, infection, and renal failure due to impaired immune function [3]. Recently, the survival rate of MM patients has improved due to the development of autologous stem cell transplantation and novel therapeutic agents that include proteasome inhibitors (PIs) and immunomodulatory drugs [4]. Despite recent advances in MM treatment, most patients fall into cyclic relapse and eventually develop refractory disease due to residual MM [5]. Many researchers have reported the recurrence and drug resistance in MM patients who have previously received chemotherapy, which remains a major obstacle for curing MM. Thus, because there are currently no effective therapies to treat chemotherapy-resistant MM, novel effective treatments for MM need to be identified.

Bortezomib (BTZ, Velcade) is the first therapeutic PI that was approved by the US Food and Drug (FDA) Administration in 2003 for the treatment of MM [6]. BTZ binds to the catalytic subunit of the proteasome with high affinity and inhibits the ubiquitin-proteasome system (UPS) [7,8]. Consequently, BTZ indirectly regulates the apoptotic pathway and cell cycle progression, affecting the downstream signaling of UPS [9]. The nuclear factor kappa B (NF-κB) pathway is a typical cell signaling regulated by the UPS. The NF-κB inhibitor (I-κB) masks the nuclear localization signal of NF-κB in the cytoplasm.

When I-κB is degraded by the UPS, NF-κB translocates into the nucleus and transcribes its target genes, such as anti-apoptotic genes. However, BTZ inhibits activation of the NF-κB pathway by preventing the degradation of I-κB [10,11]. BTZ also activates both intrinsic and extrinsic apoptotic pathways [12]. In addition, BTZ causes an accumulation of the ubiquitin-conjugated proteins that enhances endoplasmic reticulum stress with cytotoxicity, resulting in apoptosis induction [13].

BTZ has been approved to treat newly diagnosed and relapsed and/or refractory (R/R) MM [14,15]. Although significant improvements in the management of MM with BTZ have been reported, relapses are common, and treatment efficacy is reduced in MM patients. One of the various reasons for BTZ resistance is the abnormal expression of proteasome subunits that BTZ targets [16]. BTZ preferentially binds to the β5 subunit of the proteasome. The catalytic subunits of constitutive proteasome, β5 subunit (known as *PSMB5*), are significantly upregulated, whereas β1 (encoded by *PSMB1*) and β2 (encoded by *PSMB2*) are modestly upregulated in cancers. However, no significant changes were seen in the mRNA levels of β5, so the posttranscriptional mechanism seems to upregulate the protein level of β5 [16]. Recently, *PSMB5* mutations were also found by deep sequencing in BTZ-treated MM patients [17]. In the immunoproteasome, low-molecular-mass polypeptide 7 (LMP7, iβ5, *PSMB8*), LMP2 (iβ1, *PSMB9*), and multicatalytic endopeptidase complex-like-1 (MECL1, iβ2, *PSMB10*) subunits are catalytically active subunits that BTZ targets [18,19]. Myeloma co-expresses the constitutive proteasome and immunoproteasome [20]. A recent study has shown that the expression of LMP7 and LMP2 is modulated via suppression of EGFR/JAK1/STAT3 signaling by tight junction protein 1 in BTZ-resistant MM [21]. Moreover, SCF^{Skp2} is upregulated in BTZ-resistant MM and promotes the degradation of p27^{Kip1}, which inhibits cyclin-dependent kinase, indicating that SCF^{Skp2} is another biomarker of BTZ resistance [22]. Therefore, the discovery of biomarkers of BTZ resistance is crucial to overcome BTZ resistance.

Combined treatment of BTZ with other anti-myeloma agents is one of the therapeutic strategies for overcoming BTZ resistance. Histone deacetylase (HDAC) inhibitor is one promising approach that shows anti-myeloma effects in preclinical and clinical studies [23] and overcomes drug resistance. Suberoylanilide hydroxamic acid (SAHA, vorinostat) is the first class of pan-HDAC inhibitor approved by the US FDA in 2009 for cutaneous T cell lymphoma [24]. Another pan-HDAC inhibitor panobinostat (LBH589) was approved by the US FDA in 2015 for R/R MM in combination with BTZ and dexamethasone (DEX) [25]. The clinical trial data showed that the combination treatment of BTZ with vorinostat or panobinostat significantly showed responses in R/R MM patients [25–28]. However, pan-HDAC inhibitors are associated with significantly increased toxicity in this combination, which restricts their clinical utility. To minimize toxicity but maintain efficacy, the HDAC6-selective inhibitor, ACY-1215 (ricolinostat), was tested and showed anticancer effects in preclinical and clinical studies. Ricolinostat demonstrated anticancer activity in hematological malignancies, including MM, diffuse large B cell lymphoma, both germinal center B cell and activated B cell, follicular lymphoma, mantle cell lymphoma, and T cell lymphoma [27,29,30]. Synergistic effects have been demonstrated with ricolinostat and BTZ or carfilzomib (CFZ) in BTZ-sensitive MM [27,29,31]. Recently, Vogl et al. showed that the combination of ricolinostat with BTZ and DEX responds in patients with R/R MM (two-thirds were BTZ refractory) [32], and an ongoing trial (phase 1 and 2) is being explored (NCT 01323751).

In this study, we investigated the antimyeloma effects of HDAC6-selective inhibitors, A452 [33] and ACY-1215, in both BTZ-sensitive and BTZ-resistant MM cells. These results show that A452 and ACY-1215 efficiently inhibit the cell growth and viability in both BTZ-sensitive and BTZ-resistant MM cells. Overall, A452 is more cytotoxic than ACY-1215 in BTZ-resistant MM cells. We also observed that NF-kB, mitogen-activated protein kinase (MAPK), and protein kinase B (PKB, known as AKT) pathways involved in BTZ resistance were inactivated by HDAC6-selective inhibitor in both BTZ-sensitive and BTZ-resistant MM cells. Furthermore, treatment with HDAC6-selective inhibitor in BTZ-resistant MM cells not only reduced the activation of STAT3 and NF-kB but also decreased the expression of other BTZ-resistant markers, LMP2 and LMP7. Combination treatment of A452 and BTZ or CFZ synergistically reduced BTZ-resistant markers. Moreover, combination treatments with A452 and BTZ or CFZ restored the sensitivity of BTZ and synergistically induced apoptosis in BTZ-resistant MM cells. In summary, the HDAC6-selective inhibitor could overcome the BTZ resistance via inhibition of cell survival signaling and the STAT3–NF-κB–LMP2 pathway in BTZ-resistant MM.

2. Results

2.1. Establishment of BTZ-Resistant U266/VelR MM Cells

BTZ-resistant U266 MM cells (U266/VelR) have hyperphosphorylated NF-κB (p65) and ERK MAPK and are less sensitive to BTZ-induced cell cytotoxicity than BTZ-sensitive parental U266 cells [34]. To confirm and strengthen BTZ resistance, we performed a soft agar assay and isolated three BTZ-resistant U266 cell clones (U266/VelR-1, U266/VelR-2, and U266/VelR-3). Both BTZ-resistant U266/VelR-1 and U266/VelR-2 cells had hyperphosphorylated NF-κB and ERK compared to the parental U266 cells (Figure 1A). Next, to confirm the BTZ resistance in both BTZ-resistant U266 cells, we examined the cell viability of both BTZ-resistant U266 cells after treatment with BTZ for 72 h. BTZ-resistant U266 MM cell lines showed a 10-fold higher resistance to BTZ compared with their sensitive counterparts (IC_{50} values 0.5 nM vs. 5 nM). While the viability of BTZ-sensitive U266 cells significantly decreased in a dose-dependent manner, both BTZ-resistant U266 cells showed BTZ resistance from 2 nM BTZ to 5 nM BTZ (Figure 1B). Additionally, we tested the activation of NF-κB and ERK by BTZ. The phosphorylated levels of p65 and ERK decreased dose dependently in BTZ-sensitive U266 cells, but not in either of the BTZ-resistant U266 cells (Figure 1C). At a higher concentration (10 nM BTZ), total levels of p65 and ERK were significantly reduced in both BTZ-sensitive and BTZ-resistant MM cells. In addition, the viability of both BTZ-sensitive and BTZ-resistant U266 cells similarly decreased at 10 nM BTZ (Figure 1B), undistinguishable at a higher concentration (10 nM BTZ). We observed an inverse relationship between the levels of BTZ-resistance markers and MM sensitivity to BTZ (Figure 1A,B). Therefore, these findings indicate that active NF-κB and MAPK pathways have key roles in BTZ resistance-related cell viability in MM cells.

2.2. A452 is More Cytotoxic than ACY-1215 in Both BTZ-Sensitive and BTZ-Resistant U266 Cells

A452 is a small-molecule inhibitor with a γ-lactam that selectively inhibits HDAC6 catalytic activity in various human cancer cells [33]. Recently, A452 demonstrated significant anticancer activity in solid tumors and blood cancers, including MM [35–38]. We, therefore, sought to determine whether A452 has antimyeloma activity in both BTZ-sensitive U266 and BTZ-resistant U266/VelR cells. First, we examined the cell growth and viability in both BTZ-sensitive and BTZ-resistant U266 cells after treatment with A452 for 72 h by cell counting kit-8 (CCK-8) assays (Figure 2). Regardless of the resistance of BTZ, A452 resulted in a time- and dose-dependent decrease of the cell growth and viability in both BTZ-sensitive and BTZ-resistant U266 cells with GI_{50} and IC_{50} values ~0.25 μM (Figure 2G). The viability of both BTZ-sensitive and BTZ-resistant U266 cells decreased to less than 40% when treated with 1 μM A452 (Figure 2B,D,F), but the viability decreased to 60% when treated with 1 μM ACY-1215, the only first-in-class clinically relevant small-molecule HDAC6 inhibitor (Figure S1). The IC_{50} value of ACY-1215 (IC_{50} values > 2.5 μM) was 10 times higher than that of A452 (~0.25 μM; Figure 2G), indicating that A452 is a more sensitive drug than ACY-1215 in both BTZ-sensitive and BTZ-resistant U266 cells. The value of GI_{50} was similar to IC_{50} in both BTZ-sensitive and BTZ-resistant U266 cells. Taken together, these findings indicate that BTZ-resistant MM cells are still sensitive to HDAC6 inhibitors and that both BTZ-sensitive and BTZ-resistant U266 cells are more sensitive to A452 than ACY-1215.

2.3. A452 in Combination with BTZ or CFZ Shows Synergistic Cytotoxicity in Both BTZ-Sensitive and BTZ-Resistant U266 Cells

The HDAC6-selective inhibitor A452 showed cytotoxicity in both BTZ-sensitive and BTZ-resistant U266 cells. Next, we investigated the combined effect of A452 and BTZ in BTZ-resistant U266/VelR and BTZ-sensitive U266 cell cytotoxicity. Cells were treated with A452, BTZ, and A452 in combination with BTZ in both BTZ-resistant and BTZ-sensitive U266 cells. The CCK-8 assay was performed to measure cell viability for 72 h. BTZ treatment alone had no distinct cytotoxic effects in BTZ-resistant U266 cells, whereas notable synergistic increases in cytotoxicity with A452 were observed (Figure 3; left panel). Next, we analyzed the synergism between A452 and BTZ by applying the Chou and Talalay method [39]. The combination of A452 and BTZ showed synergistic anti-MM activity with a combination index (CI) < 1.0 (Figure 3; right panel).

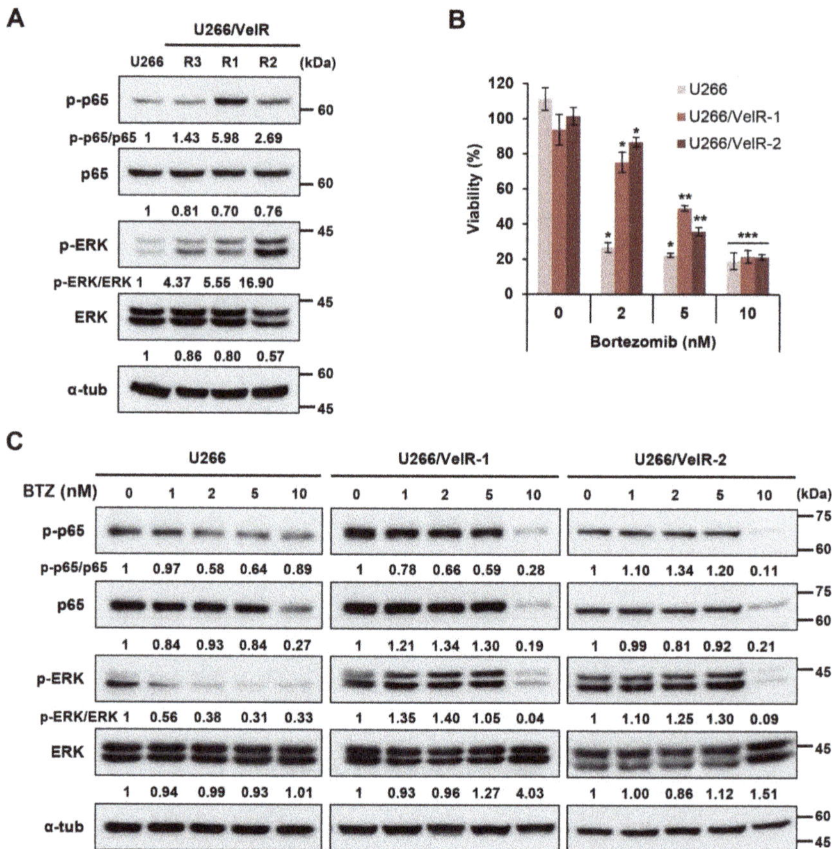

Figure 1. Establishment of bortezomib (BTZ)-resistant U266/VelR MM cells. (**A**) Immunoblotting analysis of both BTZ-sensitive U266 and BTZ-resistant U266 (U266/VelR-1, U266/VelR-2, and U266/VelR-3) MM cells. (**B**) Both BTZ-sensitive and BTZ-resistant U266 cells were treated with indicated concentrations of BTZ (2, 5, 10 nM) for 72 h, and cell counting kit-8 (CCK-8) assays were performed to analyze cell viability. Data represent mean ± SD (n = 3). Student's t-test, * $p < 0.05$, ** $p < 0.01$, and *** $p < 0.001$ vs. the DMSO control. (**C**) Immunoblotting analysis of both BTZ-sensitive U266 and BTZ-resistant U266 cells treated with 0.01% DMSO or indicated concentrations of BTZ (1, 2, 5, and 10 nM) for 24 h. Relative protein expression levels were semi-quantified by densitometric analysis of the blots. α-tubulin was used as an equal loading control. The abundance of the indicated proteins was semi-quantified relative to α-tub, and control levels were set at 1.

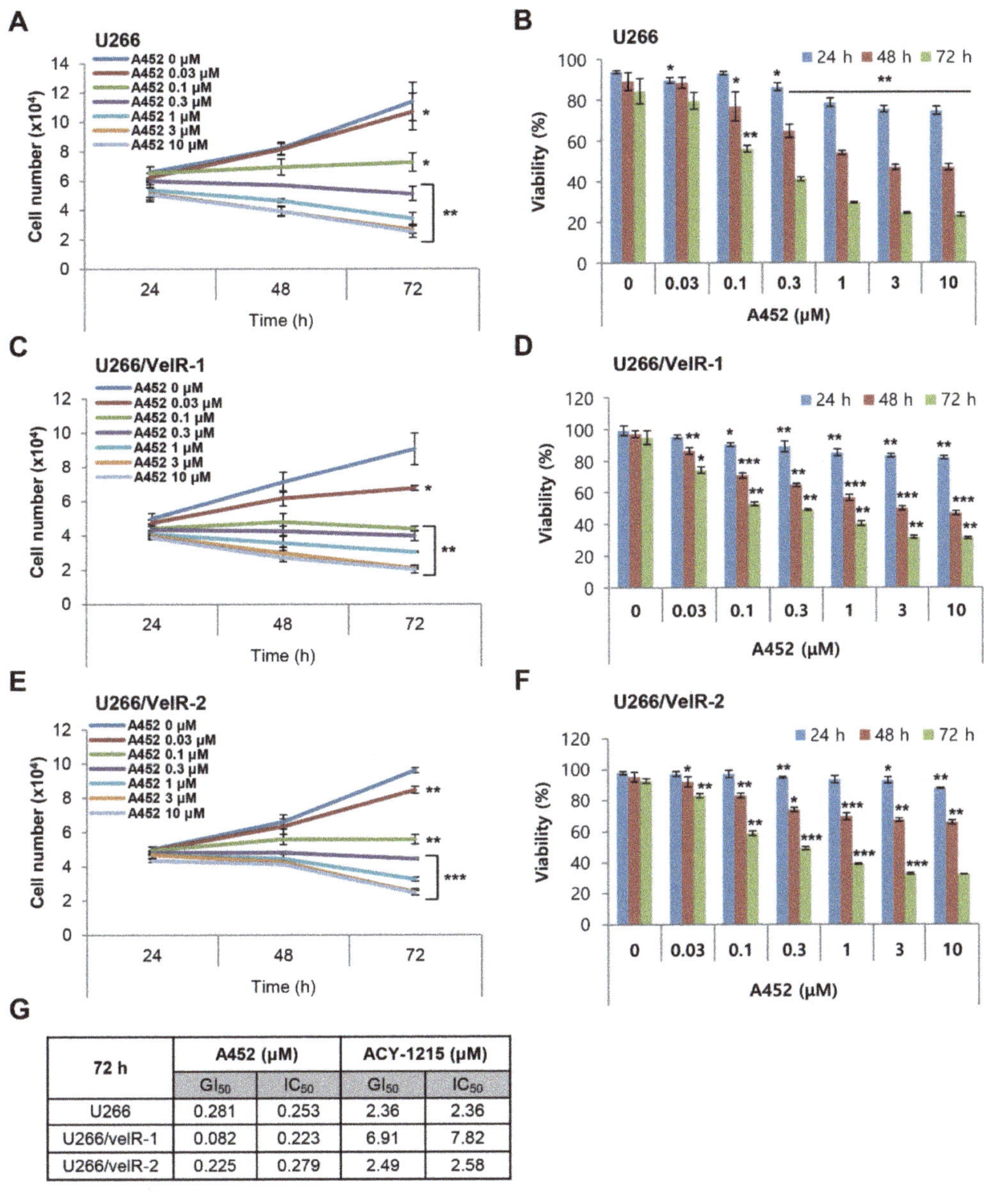

72 h	A452 (µM)		ACY-1215 (µM)	
	GI_{50}	IC_{50}	GI_{50}	IC_{50}
U266	0.281	0.253	2.36	2.36
U266/velR-1	0.082	0.223	6.91	7.82
U266/velR-2	0.225	0.279	2.49	2.58

Figure 2. Histone deacetylase 6 (HDAC6)-selective inhibitors suppress cell growth and viability in BTZ-sensitive and BTZ-resistant U266 MM cells. (**A,B**) BTZ-sensitive U266, and (**C–F**) BTZ-resistant U266 (U266/VelR-1 and U266/VelR-2) MM cells were treated with 0.1% DMSO (control) or indicated concentrations of A452 for 72 h, and CCK-8 assays were performed to analyze cell growth and viability. Cell counts were estimated indirectly from a standard curve generated using solutions of known cell counts. Absorbance was normalized to that of the negative control at each time interval. Data present as mean ± SD ($n = 3$). * $p < 0.05$, ** $p < 0.01$, and *** $p < 0.001$ vs. the DMSO control, Student's *t*-test. (**G**) The IC_{50} and GI_{50} values for A452 and ACY-1215 in BTZ-sensitive and BTZ-resistant U266 cells using GraphPad prism.

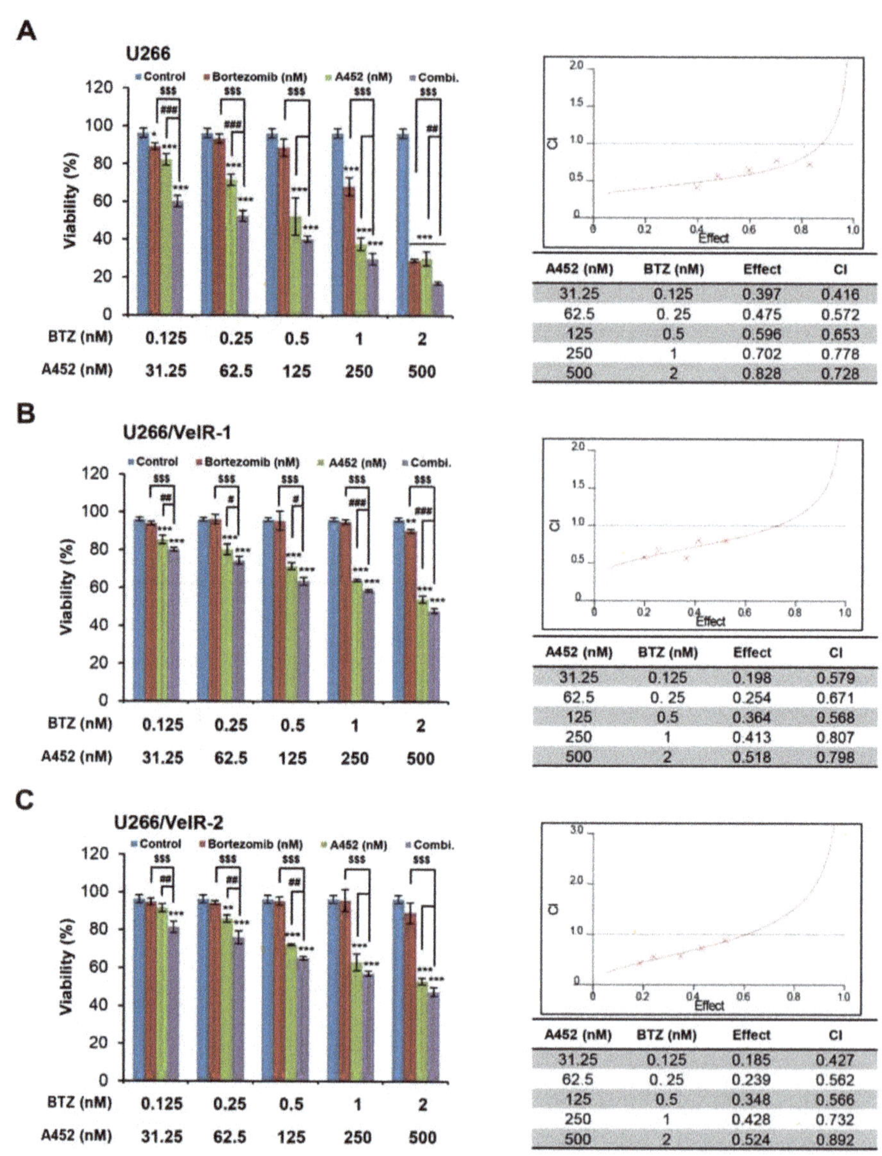

A

A452 (nM)	BTZ (nM)	Effect	CI
31.25	0.125	0.397	0.416
62.5	0.25	0.475	0.572
125	0.5	0.596	0.653
250	1	0.702	0.778
500	2	0.828	0.728

B

A452 (nM)	BTZ (nM)	Effect	CI
31.25	0.125	0.198	0.579
62.5	0.25	0.254	0.671
125	0.5	0.364	0.568
250	1	0.413	0.807
500	2	0.518	0.798

C

A452 (nM)	BTZ (nM)	Effect	CI
31.25	0.125	0.185	0.427
62.5	0.25	0.239	0.562
125	0.5	0.348	0.566
250	1	0.428	0.732
500	2	0.524	0.892

Figure 3. Cotreatment with A452 and BTZ triggers synergistic cytotoxicity in both BTZ-sensitive and BTZ-resistant U266 MM cells. (**A**) BTZ-sensitive U266 and (**B**,**C**) BTZ-resistant U266 (U266/VelR-1 and U266/VelR-2) MM cells were treated with 0.1% DMSO (control), A452, and BTZ or in combination with these compounds as indicated for 72 h. Combination treatments were performed in U266 (**A**) and BTZ-resistant U266 (**B**,**C**) cells maintaining a constant ratio between the dose of the A452 and BTZ, and cell viability were assessed at 72 h by CCK-8 assay. The combination index (CI) value and the relative fraction affected (FA) were determined at each dose combination (actual), and a simulation was run to estimate the CI value and confidence interval (—) across the entire FA range (simulation). CI < 1, CI = 1, and CI > 1 indicate synergistic, additive, and antagonistic effects, respectively. CI was calculated by the CalcuSyn software program. Data present as mean ± SD (n = 3). * $p < 0.05$, ** $p < 0.01$, and *** $p < 0.001$ vs. DMSO control, $^{\$\$\$}$ $p < 0.001$ vs. BTZ-treated group, [#] $p < 0.05$, [##] $p < 0.01$, and [###] $p < 0.001$ vs. A452-treated group; two-way analysis of variance (ANOVA) test.

Next, we examined the combined cytotoxic effect of CFZ with A452 on BTZ-resistant U266/VelR and BTZ-sensitive U266 cells. The β5/β2 co-inhibition is known as the most effective cytotoxic in PI-sensitive and PI-resistant MM [20]. Among the available PIs, only high-dose CFZ provides β5/β2 co-inhibition. Although BTZ-resistant cells were partly cross-resistant to CFZ, which is structurally and mechanistically distinct from BTZ [40], these resistant cells remained sensitive to inhibition of proliferation by CFZ (Figure S2A) [40,41] and reduced BTZ-resistant markers (Figure S2B). In addition to the combination of A452 and BTZ, the cotreatment of A452 and CFZ showed synergistic anti-MM activity in both BTZ-sensitive and BTZ-resistant U266 cells (Figure S3). Therefore, these data suggest that A452 in combination with BTZ or CFZ synergistically induces cytotoxicity in both BTZ-sensitive and BTZ-resistant U266 cells.

2.4. A452 in Combination with BTZ or CFZ Synergistically Leads to Apoptosis in BTZ-Sensitive and BTZ-Resistant U266 Cells

PI-based MM therapy induces apoptosis due to excessive proteotoxic stress [42]. The combination treatment with A452 and other anticancer agents enhanced the efficacy of anticancer activity in solid tumors [37] and hematological malignancies [30,38]. However, the effect of combination treatment with A452 and BTZ or CFZ has not been studied in hematological malignancies. To characterize the mechanism of action of synergistic cytotoxicity induced by the combination treatment, we examined the activation of the apoptotic pathway in both BTZ-sensitive and BTZ-resistant U266 cells by immunoblotting and Annexin-V/propidium iodide staining. Immunoblotting was performed to investigate the molecular mechanism of apoptosis. Combination treatment triggered synergistic cleavage of caspase-3 and poly(ADP ribose) polymerase (PARP) in both BTZ-sensitive and BTZ-resistant U266 cells (Figure 4A). In addition, combination treatment markedly increased the cleavage of caspase-8 and caspase-9. Combination treatment markedly downregulated B-cell lymphoma-extra large protein (Bcl-xL), Bcl-2, and X-linked inhibitor of apoptosis (XIAP) antiapoptotic proteins without affecting Bax and Bak proapoptotic proteins in both BTZ-sensitive and BTZ-resistant U266 cells (Figure S4A). To further confirm the apoptosis induction, we tested the Annexin-V propidium iodide staining. The population of Annexin-V-positive cells after treatment with A452 or BTZ was 18.2% and 25.8% in BTZ-sensitive and 25.2% and 15.4% in BTZ-resistant U266 cells, respectively, which increased to 54.7% in BTZ-sensitive cells and 35.0% in BTZ-resistant U266 cells, respectively, after combination treatment (Figure 4B,C and Figure S4B,C). Furthermore, more robust results were observed following combination treatment of A452 and CFZ in both BTZ-sensitive and BTZ-resistant U266 cells (Figure 5 and Figure S5A,B). Overall, these results indicate that A452 and BTZ or CFZ triggered apoptosis in BTZ-resistant MM cells by activating caspases and downregulating antiapoptotic factors.

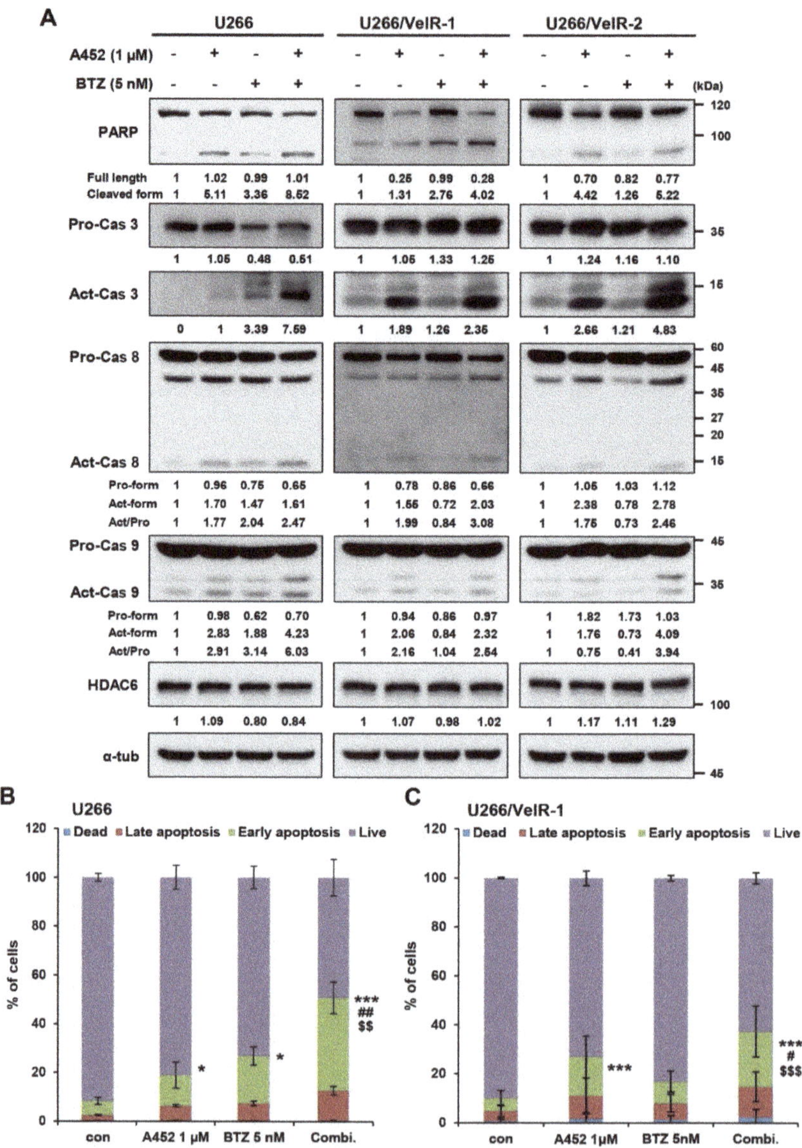

Figure 4. Cotreatment with BTZ and A452 leads to synergistic apoptosis induction in both BTZ-sensitive and BTZ-resistant U266 MM cells. (**A**) BTZ-sensitive U266 and BTZ-resistant U266 (U266/VelR-1 and U266/VelR-2) MM cells were treated with 0.1% DMSO (control), A452 (1 µM), and BTZ (5 nM) or in combination with these compounds as indicated for 24 h. Apoptotic markers were identified by immunoblotting using whole-cell lysates. α-tubulin was used as an equal loading control. The abundance of the indicated proteins was semi-quantified relative to α-tub; control levels were set at 1. (**B**) BTZ-sensitive U266 and (**C**) BTZ-resistant U266/VelR-1 cells were treated with 0.1% DMSO (control), A452 (1 µM), and BTZ (5 nM) or in combination with these compounds as indicated for 36 h. Cell death was assessed by flow cytometry and Annexin V/PI staining. Data present as mean ± SD (n = 3). * $p < 0.05$ and *** $p < 0.001$ vs. DMSO control, $^{\$\$}$ $p < 0.01$ and $^{\$\$\$}$ $p < 0.001$ vs. BTZ-treated group, $^{\#}$ $p < 0.05$ and $^{\#\#}$ $p < 0.01$ vs. A452-treated group; two-way analysis of variance (ANOVA) test.

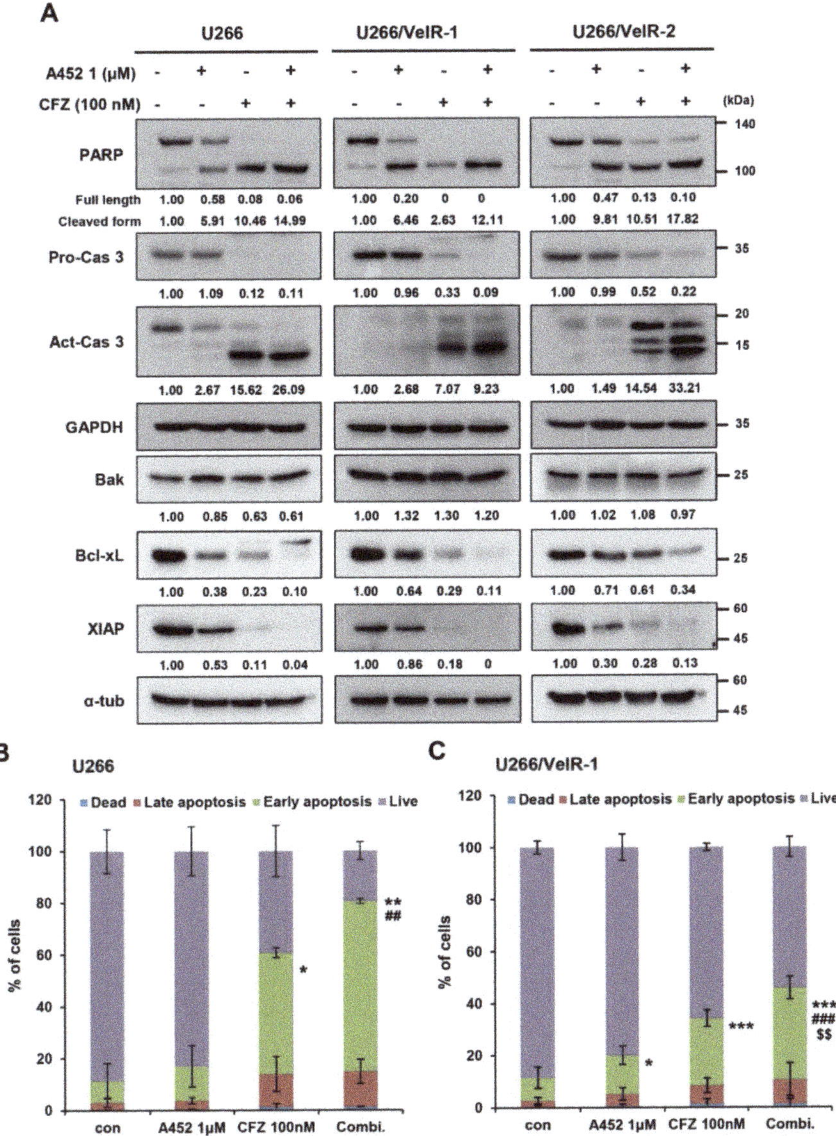

Figure 5. Cotreatment with CFZ and A452 leads to synergistic apoptosis induction in both BTZ-sensitive and BTZ-resistant U266 MM cells. (**A**) BTZ-sensitive U266 and BTZ-resistant U266 (U266/VelR-1 and U266/VelR-2) MM cells were treated with 0.1% DMSO (control), A452 (1 μM), and carfilzomib (CFZ) (100 nM) or in combination with these compounds as indicated for 24 h. Apoptotic markers were identified by immunoblotting using whole-cell lysates. Glyceraldehyde 3-phosphate dehydrogenase (GADPH) and α-tubulin were used as equal loading controls. The abundance of the indicated proteins was semi-quantified relative to GAPDH or α-tub; control levels were set at 1. (**B**) BTZ-sensitive U266 and (**C**) BTZ-resistant U266/VelR-1 MM cells were treated with 0.1% DMSO (control), A452 (1 μM), and CFZ (100 nM) or in combination with these compounds as indicated for 36 h. Cell death was assessed by flow cytometry and Annexin V/PI staining. Data present as mean $\pm$ SD (n = 3). * $p < 0.05$, ** $p < 0.01$, and *** $p < 0.001$ vs. DMSO control; [##] $p < 0.01$ and [###] $p < 0.001$ vs. A452-treated group; [$$] $p < 0.01$ vs. CFZ-treated group, two-way analysis of variance (ANOVA) test.

2.5. A452 in Combination with BTZ or CFZ Synergistically Inactivates the Cell Survival Signaling in BTZ-Sensitive and BTZ-Resistant U266 Cells

HDAC6-selective inhibitors show cytotoxicity and cause apoptosis in both BTZ-sensitive and BTZ-resistant U266 cells. From these results, we hypothesized that A452 reduces BTZ resistance. Thus, we investigated the effect of combination treatment with A452 and PI on cell survival signaling in both BTZ-sensitive and BTZ-resistant U266 cells. Regardless of the resistance of BTZ, the phosphorylated levels of ERK, AKT, and p65 were decreased by A452 treatment in both BTZ-sensitive and BTZ-resistant U266 cells (Figures 6A and 7A). Sensitivity to A452 in BTZ-resistant U266 cells was slightly lower than in BTZ-sensitive U266 cells. We also observed the changes in cell survival signaling by combined treatment of A452 and BTZ or CFZ. While the single treatment of BTZ did not affect the phosphorylated ERK, AKT, and p65, combined treatment of BTZ and A452 synergistically reduced the phosphorylated ERK, AKT, and p65 in BTZ-resistant U266 cells without changing total ERK, AKT, and p65 levels (Figures 6A and 7A). Similar to BTZ and A452, CFZ and A452 synergistically reduced the phosphorylated ERK without altering total ERK levels in BTZ-resistant U266 cells (Figure 6B). Interestingly, CFZ treatment with A452 synergistically reduced phosphorylated AKT and phosphorylated p65 and their total forms in BTZ-sensitive U266 cells and BTZ-resistant U266 cells (Figures 6B and 7B). The synergistic effect of A452 with CFZ was more potent than that of A452 with BTZ in BTZ-resistant U266 cells. Taken together, these results suggest that the HDAC6-selective inhibitor restores the sensitivity to BTZ by inactivating cell survival signaling in BTZ-resistant U266 cells.

2.6. A452 Enhances MM Sensitivity to BTZ or CFZ in BTZ-Resistant U266 Cells

Upregulation of the catalytically active immunoproteasome subunits LMP7 and LMP2 confers PI resistance by increasing proteasome capacity [21]. A recent study has demonstrated that LMP2 and LMP7 expression are related to STAT3 activity in BTZ-resistant MM cells [21], and NF-κB directly regulates the transcription of LMP2 [43]. We, therefore, examined whether the combination of BTZ or CFZ with A452 can overcome acquired BTZ resistance. The phosphorylated levels of STAT3 treated with A452 were more decreased in BTZ-sensitive U266 cells compared with BTZ-resistant U266 cells (Figure 7A). The BTZ-resistant U266 cells showed less sensitivity to the STAT3 and NF-κB inhibition of BTZ or CFZ compared with BTZ-sensitive U266 cells. A452 treatment with BTZ synergistically reduced phosphorylated STAT3 and phosphorylated p65 without changing the total forms of STAT3 and p65 in both BTZ-sensitive and BTZ-resistant U266 cells (Figure 7A). Inactive STAT3 and NF-κB, in turn, downregulated LMP2 and LMP7 in both BTZ-sensitive and BTZ-resistant U266 cells. Interestingly, A452 treatment with CFZ synergistically reduced both phosphorylated STAT3 and p65 and total STAT3 and p65, resulting in synergistically reduced expressions of LMP2 and LMP7 in both BTZ-sensitive and BTZ-resistant U266 cells (Figure 7B). Thus, these results indicate that A452 in combination with BTZ or CFZ could overcome BTZ-induced resistance in MM.

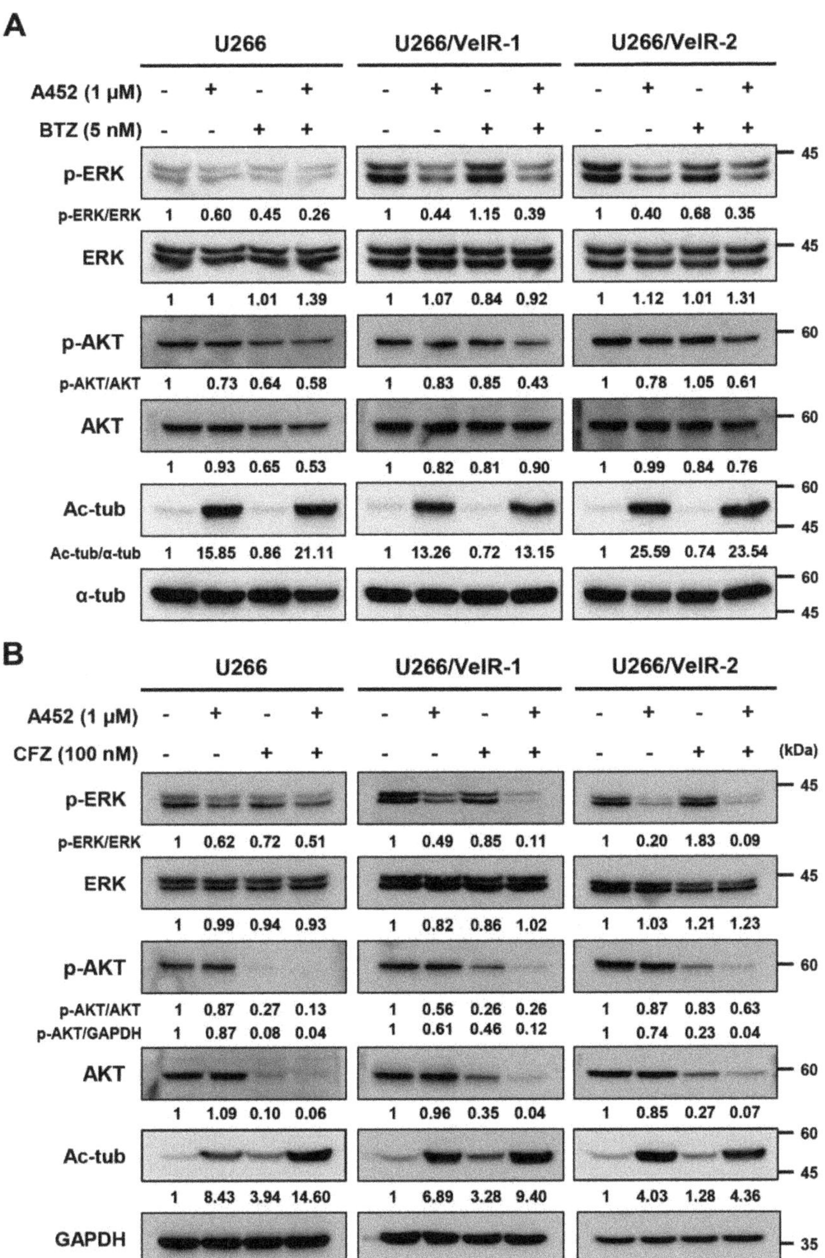

Figure 6. Cotreatment of A452 and PI synergistically inhibits activation of both mitogen-activated protein kinase (MAPK) and protein kinase B (PKB, known as AKT) pathways in BTZ-sensitive and BTZ-resistant U266 cells. (**A**,**B**) Both BTZ-sensitive U266 and BTZ-resistant U266 (U266/VelR-1 and U266/VelR-2) MM cells were treated with 0.1% DMSO (control), A452, BTZ (**A**), and CFZ (**B**) or in combination with A452 and PI as indicated for 24 h. Whole-cell lysates were subjected to immunoblotting with indicated antibodies. α-tubulin and GAPDH were used as equal loading controls. The abundance of the indicated proteins was semi-quantified relative to α-tub or GAPDH; control levels were set at 1.

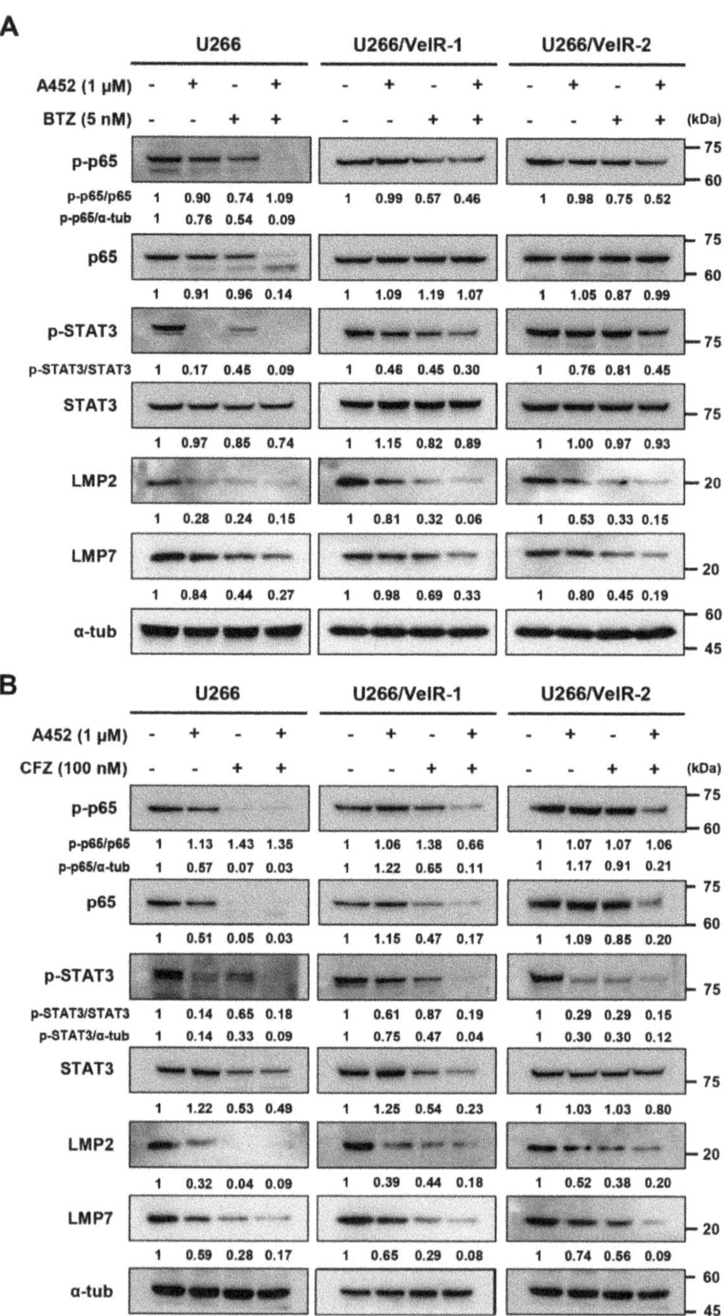

Figure 7. Cotreatment of A452 and PI synergistically decreases the level of low-molecular-mass polypeptide 2 (LMP2) and LMP7 via modulating signal transducer and activator of transcription 3

(STAT3) and nuclear factor kappa B (NF-κB) pathways in both BTZ-sensitive and BTZ-resistant U266 MM cells. (**A,B**) Both BTZ-sensitive U266 and BTZ-resistant U266 (U266/VelR-1 and U266/VelR-2) MM cells were treated with 0.1% DMSO (control), A452, BTZ (**A**), and CFZ (**B**) or in combination with A452 and PI as indicated for 24 h. Whole-cell lysates were subjected to immunoblotting with indicated antibodies. α-Tubulin was used as an equal loading control. The abundance of the indicated proteins was semi-quantified relative to α-tub; control levels were set at 1.

3. Discussion

Acquired anticancer drug resistance is a major problem to effective therapy for patients with MM [44]. Although anticancer drugs, such as BTZ, lenalidomide, and Dex, have exhibited clinical success, some MM patients fail to respond to these drugs, and their outcome is still poor due to primary refractories and acquisition of resistance [45]. Consequently, BTZ resistance causes many changes in the molecular character in cancer cells. In this study, we confirmed that several cell survival signaling pathways, including NF-κB, MAPK, and STAT3, were highly increased in BTZ-resistant MM cells. We showed that A452, an HDAC6-selective inhibitor, rescues the phosphorylated levels of NF-kB, ERK, and STAT3 upregulated by BTZ resistance and induces cytotoxicity in BTZ-resistant MM cells. Moreover, A452, combined with BTZ or CFZ, was shown to synergistically decrease pERK, pp65, and pSTAT3 levels and increase sensitivity to BTZ- and CFZ-induced apoptosis (Figure 8). Recent studies reported that single treatment of ACY-1215 or combination treatment of ACY-1215 and BTZ or CFZ markedly induce cell death in MM cells [27,29,31,46]. However, it is incompletely understood whether combination treatment of ACY-1215 and PI would lead to cell death in BTZ-resistant MM cells. In this study, we showed that HDAC6 inhibition remains effective in BTZ-resistant MM cells, which has not yet been explored. In addition, we demonstrated that A452, as well as ACY-1215, dose-dependently decreased the cell viability and cell growth rate of both BTZ-sensitive and BTZ-resistant MM cells. Additionally, combination treatment with A452 and BTZ or CFZ synergistically decreased cell viability and consequently enhanced apoptosis in both BTZ-sensitive and BTZ-resistant MM cells. Altogether, our findings suggest that A452 may be beneficial in the treatment of the PI-resistant refractory and relapsed MM patients.

Previous studies revealed that MAPK and AKT signaling correlates with MM cell survival [47]. AKT activates I-κB kinase and p38 MAPK, stimulating NF-κB transactivation. Activated AKT is known to synergize with co-activator CBP in the activation of the p65 transactivation domain. Phosphorylation of p65 at Ser 536 by IKK and acetylation of p65 at Lys 310 by CBP enhance its transcriptional activity [48]. A recent study showed that the inhibition of HDAC6 accumulated Lys 163 and Lys 377 on the kinase domain of AKT, consequently decreasing AKT kinase activity [49]. Consistent with previous findings, combination treatment with A452 and BTZ or CFZ synergistically inactivated AKT in both BTZ-sensitive and BTZ-resistant MM cells. Our results showed that the HDAC6-selective inhibitor inactivates the NF-κB pathway by modulating AKT in both BTZ-sensitive and BTZ-resistant MM cells.

LMP7 and LMP2 subunits of immunoproteasome directly bind to and inhibit BTZ. Immunoproteasome subunits can participate in the NF-κB pathway via the degradation of I-κB. Furthermore, the expression levels of proteasome subunits negatively correlate with sensitivity to PI [21,50]. As overexpressed immunoproteasome subunits can cause BTZ resistance [51], decreasing proteasome subunits is an important key to overcoming BTZ resistance. Previous studies focused on the relationship between BTZ resistance and LMP7 (overexpression and point mutation). Here, we found that LMP2 was overexpressed in BTZ-resistant cells. Then, we demonstrated that the combined treatment of A452 and BTZ or CFZ synergistically reduced the protein levels of LMP2 and, to a lesser extent, LMP7 in BTZ-resistant MM cells. The decreased level of LMP2 and LMP7 could lead to the accumulation of inactivated NF-κB in the cytoplasm by preventing the degradation of I-κB [10,11]. Overall, the combination of CFZ with A452 shows more potent synergistic effects in decreasing LMP2 and LMP7 than that of BTZ with A452 in BTZ-resistant MM cells. Although BTZ-resistant cells are also partly cross-resistant to CFZ, PI resistance is not universal across the different PI classes, and PIs possess drug-specific features [52,53]. We, therefore, suggest that LMP2 and LMP7 play an important role in overcoming BTZ resistance.

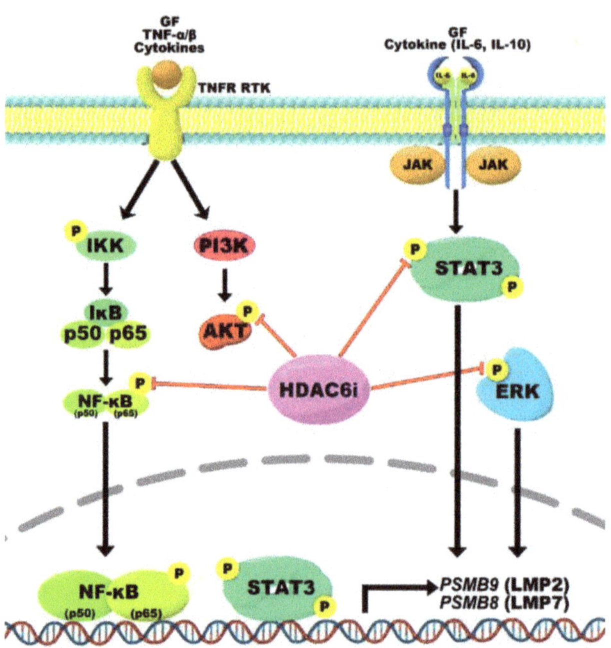

Figure 8. Working model for HDAC6 inhibitor and PI for overcoming BTZ resistance of MM. HDAC6 inhibition by A452 synergize with BTZ or CFZ to inhibit the activation of NF-κB and STAT3, resulting in decreased expressions of LMP2 and LMP7. Additionally, combination treatment of A452 and BTZ or CFZ synergistically inactivates the AKT and ERK MAPK pathways. Combining A452 with BTZ or CFZ leads to synergistic cancer cell growth inhibition and viability decreases. This combination induces apoptosis and enhances PI sensitivity in the BTZ-resistant MM cells.

Next, our study explains that STAT3 has a key role in the reduction of LMP2 and LMP7 through inhibition of HDAC6. A recent study demonstrated that LMP2 and LMP7 expression levels were related to STAT3 activity in BTZ-resistant MM cells [21]. Another study has shown that NF-κB directly regulates the transcription of LMP2 by binding on the NF-κB element (GGGACTTTCC) [43]. HDAC6 also modulates activation of STAT3 [38,54] and NF-κB [38]. Our results exhibit that inactivation of STAT3 and NF-κB by HDAC6 inhibition reduces LMP2 and LMP7 in BTZ-resistant MM cells. Furthermore, these effects are synergistically enhanced when combined with A452 and BTZ or CFZ. Thus, our findings indicate that the combination of HDAC6 inhibitor and PI can overcome BTZ resistance via regulating STAT3 and NF-κB signaling pathways.

4. Materials and Methods

4.1. Reagents

ACY-1215 (ricolinostat), Bortezomib, and Carfilzomib were purchased from Selleck Chemicals (Houston, TX, USA). Powders were solubilized in DMSO (Sigma Chemical, St. Louis, MO, USA). Antibodies against AKT (sc-8312), p-AKT (sc-7985-R), Bak (sc-832), ERK (sc-03-G), HDAC6 (sc-11420), LMP2 (sc-514345), p65 (sc-8008), STAT3 (sc-482), and α-tubulin (sc-32293) were purchased from Santa Cruz Biotechnology (Santa Cruz, CA, USA). Antibodies against GAPDH (AP0066) and p-p65 (BS4138) were from Bioworld Technology (Bloomington, MN, USA). Antibody against LMP7 (A305-229A) was from

Bethyl Laboratories (Montgomery, TX, USA). Antibodies against caspase-8 (#551244), caspase-9 (#551246), PARP (551024), and XIAP (610716) were from BD Biosciences (San Jose, CA, USA). Antibodies against Bax (#2772), Bcl-2 (#7382), Bcl-xL (#2762), caspase-3 (#9662), pERK (#2220), and pSTAT3 (49,138) were from Cell Signaling Technology (Danvers, MA, USA). Antibody against acetyl α-tubulin (T6793) was from Sigma-Aldrich (St. Louis, MO, USA).

4.2. MM Cell Lines and Culture

The BTZ-sensitive U266 and BTZ-resistant U266/VelR human MM cell lines were kindly provided by Dr. Sung-Soo Yoon (Seoul National University College of Medicine, Seoul, Korea; [34]). Cells were cultured in RPMI 1640 medium (HyClone; GE Healthcare, Logan, UT, USA) containing 10% fetal bovine serum (HyClone), 100 units/mL penicillin, and 100 µg/mL streptomycin (Gibco; Thermo Fisher Scientific, Inc., Waltham, MA, USA) in a humidified atmosphere of 5% CO_2 and 95% air at 37 °C. BTZ-resistant U266/VelR cells were maintained in RPMI 1640 medium, as described above, containing 2 nM BTZ.

4.3. MM Cell Lines and Culture

To strengthen BTZ resistance, BTZ-resistant U266/VelR MM cells provided by Dr. Sung-Soo Yoon were reselected using soft agar assay. Soft agar agarose was autoclaved in distilled water before use. In total, 5×10^5 BTZ-sensitive U266 cells in a 100-mm plate were spun down and resuspended in cooled 0.3% agar in RPMI 1640 medium containing 10% FBS with 10 nM BTZ. The cells were seeded onto a solidified base layer, including 1% agar in the culture media. The plates were kept at room temperature for 30 min before being placed into a CO_2 incubator. The agar plates were incubated for 3–4 weeks until colonies were visible on the surface. Individual colonies were picked from soft agar, and pooled colonies were grown in the culture media with 2 nM BTZ [34]. We confirmed BTZ resistance in U266/VelR MM cells relative to U266 MM cells by the cell viability assay and Western blot.

4.4. Cell Growth and Viability Assay

Cell growth and viability were assessed by measuring a water-soluble tetrazolium salt, 2-(2-methoxy-4-nitrophenyl)-3-(4-nitrophenyl)-5-(2,4-disulfophenyl)-2H tetrazolium, monosodium salt (WST-8) (Cell Counting Kit (CCK)-8 kit, Dojindo Molecular Technologies, Inc., Kumamoto, Japan) dye absorbance. Cells were pulsed with 20 µL of WST-8 to each well for the last 3 h of 72 h, and absorbance was measured at 450 nm using a multimode microplate reader (Tecan, Männedorf, Switzerland). Results are presented as the percent absorbance relative to control cultures and were generated from three independent experiments performed in triplicate.

4.5. Growth Inhibitory and Viability Inhibitory Assays

Drug concentrations that inhibited 50% of cell growth (GI_{50}) and 50% of cell viability (IC_{50}) were determined using CCK-8 assay as described elsewhere. All cell lines were treated for 72 h on day two unless otherwise stated. GI_{50} and IC_{50} were determined using Prism Version 6.0 software (GraphPad Software, Inc., La Jolla, CA, USA).

4.6. Drug Combination Analysis

Synergism between PI and HDAC6 inhibitor was evaluated using the Chou–Talalay method [39]. Fraction affected (Fa) versus the combination index (CI) plot was drawn using CalcuSyn (Biofosft). The drug combination was considered synergistic when CI was less than 1.

4.7. Apoptosis Assay

Apoptosis was assessed using Annexin V/propidium iodide double staining according to the manufacturer (BD Biosciences, Franklin Lakes, NJ, USA). After treatments, cells were trypsinized and stained with 0.5 mg/mL Annexin V in binding buffer (10 mM HEPES free acid, 0.14 M NaCl, and 2.5 mM $CaCl_2$) for 30 min. Afterward, propidium iodide (5 mg/mL final concentration) was added and incubated for another 15 min. Cells were then analyzed using a flow cytometer and BD FACSDiva software version 7 (both BD Biosciences).

4.8. Western Blot Analysis

Cells grown and treated as indicated were collected, lysed, and separated by sodium dodecyl sulfate-polyacrylamide gel electrophoresis (SDS-PAGE); Western blotting was performed as previously described [55]. The blots were semi-quantified using FusionCapt software version 16.08a (Viber Lourmat Sté, Collégien, France). The protein expression levels were semi-quantified relative to GADPH or α-tubulin, and the levels in the 0.1% DMSO-treated groups were set at 1. GADPH and α-tubulin were used as loading controls.

4.9. Statistical Analysis

All results are expressed as means ± standard deviation (SD) of three independent experiments. For the cell growth and viability test of single agents, statistical significance was determined by unpaired two-tailed Student's *t*-test. Statistical analysis for the other data was performed by GraphPad Prism software 7.0 (Graphpad Software, San Diego, CA, USA). One-way or two-way ANOVA followed by post hoc analysis with Bonferroni's multiple comparison test was used to evaluate statistical significance. $p < 0.05$ was considered statistically significant for the data.

5. Conclusions

In conclusion, our results demonstrate that PI with HDAC6-selective inhibitor induces synergistic cytotoxicity in both BTZ-sensitive and BTZ-resistant MM, associated with the inactivation of MAPK, ATK, NF-κB, and STAT3 signaling pathways. Additionally, the combination of A452 and PI synergistically reduces LMP2 and LMP7 via inhibition of the STAT3 and NF-κB in both BTZ-sensitive and BTZ-resistant MM cells. Taken together, we suggest that the HDAC6-selective inhibitor overcomes the BTZ resistance via inhibition of several different types of cell survival signaling and the STAT3–NF-κB–LMP2 pathway. Thus, our results suggest that the HDAC6-selective inhibitor may provide beneficial therapeutic opportunities for MM patients with resistance to BTZ and other PIs.

Supplementary Materials: Supplementary Materials can be found at https://www.mdpi.com/1422-0067/22/3/1341/s1.

Author Contributions: Conceptualization, S.W.L., S.-K.Y., and S.H.K.; methodology, S.W.L. and S.-K.Y.; formal analysis, S.W.L., S.-K.Y., G.W.K., J.Y., and Y.H.J.; investigation, S.W.L. and D.H.L.; data curation, S.W.L., S.-K.Y., D.H.L., G.W.K., Y.H.J., and S.Y.K.; writing—original draft, S.H.K.; writing—review & editing, G.W.K., J.Y., Y.H.J., and S.H.K. supervision, S.H.K.; project administration, S.H.K.; funding acquisition, S.H.K. All authors have read and agreed to the published version of the manuscript.

Funding: This research was supported by the Basic Science Research Program through the National Research Foundation of Korea (NRF) funded by the Ministry of Education, Science, and Technology (2018R1A6A1A03023718, 2019R1I1A1A01058601, and 2019R1I1A1A01058601).

Data Availability Statement: Data is contained within the article or supplementary material.

Conflicts of Interest: The authors declare no conflict of interest.

Abbreviations

Bcl-xL	B-cell lymphoma-extra large protein
BTZ	Bortezomib
CFZ	Carfilzomib
CI	Combination index
DEX	Dexamethasone
ERK	Extracellular signal-regulated kinase
HDAC6	Histone deacetylase 6
HDACi	Histone deacetylase inhibitor
I-κB	NF-κB inhibitor
LMP	low-molecular-mass polypeptide
MM	Multiple myeloma
NF-κB	Nuclear factor-kappa B
PARP	Poly(ADP ribose) polymerase
PI	Proteasome inhibitor
STAT3	Signal transducer and transcription
UPS	Ubiquitin-proteasome system

References

1. Bilotti, E.; Gleason, C.L.; McNeill, A. Routine Health Maintenance in Patients Living with Multiple Myeloma. *Clin. J. Oncol. Nurs.* **2011**, *15*, 25–40. [CrossRef]
2. Palumbo, A.; Anderson, K. Multiple Myeloma. *N. Engl. J. Med.* **2011**, *364*, 1046–1060. [CrossRef]
3. Özkan, M.C.; Güneş, A.E.; Sahin, F.; Saydam, G. Quality of Life and Supportive Care in Multiple Myeloma. *Turk. J. Hematol.* **2013**, *30*, 234–246. [CrossRef]
4. Kumar, S.K.; Rajkumar, S.V.; Dispenzieri, A.; Lacy, M.Q.; Hayman, S.R.; Buadi, F.K.; Zeldenrust, S.R.; Dingli, D.; Russell, S.J.; Lust, J.A.; et al. Improved survival in multiple myeloma and the impact of novel therapies. *Blood* **2008**, *111*, 2516–2520. [CrossRef]
5. Moehler, T.; Goldschmidt, H. Therapy of Relapsed and Refractory Multiple Myeloma. *Toxic. Assess.* **2011**, *183*, 239–271. [CrossRef]
6. Kane, R.C.; Bross, P.F.; Farrell, A.T.; Pazdur, R. Velcade ®: U.S. FDA Approval for the Treatment of Multiple Myeloma Progressing on Prior Therapy. *Oncologist* **2003**, *8*, 508–513. [CrossRef]
7. Chauhan, D.; Catley, L.; Li, G.; Podar, K.; Hideshima, T.; Velankar, M.; Mitsiades, C.; Mitsiades, N.; Yasui, H.; Letai, A.; et al. A novel orally active proteasome inhibitor induces apoptosis in multiple myeloma cells with mechanisms distinct from Bortezomib. *Cancer Cell* **2005**, *8*, 407–419. [CrossRef]
8. Groll, M.; Berkers, C.R.; Ploegh, H.L.; Ovaa, H. Crystal Structure of the Boronic Acid-Based Proteasome Inhibitor Bortezomib in Complex with the Yeast 20S Proteasome. *Structure* **2006**, *14*, 451–456. [CrossRef]
9. Kubiczkova, L.; Pour, L.; Sedlarikova, L.; Hajek, R.; Sevcikova, S. Proteasome inhibitors—Molecular basis and current perspectives in multiple myeloma. *J. Cell. Mol. Med.* **2014**, *18*, 947–961. [CrossRef]
10. Karin, M.; Ben-Neriah, Y. Phosphorylation Meets Ubiquitination: The Control of NF-κB Activity. *Annu. Rev. Immunol.* **2000**, *18*, 621–663. [CrossRef]
11. Hideshima, T.; Chauhan, D.; Richardson, P.G.; Mitsiades, C.S.; Mitsiades, N.; Hayashi, T.; Munshi, N.C.; Dang, L.; Castro, A.; Palombella, V.J.; et al. NF-κB as a Therapeutic Target in Multiple Myeloma. *J. Biol. Chem.* **2002**, *277*, 16639–16647. [CrossRef] [PubMed]
12. Lauricella, M.; Emanuele, S.; D'Anneo, A.; Calvaruso, G.; Vassallo, B.; Carlisi, D.; Portanova, P.; Vento, R.; Tesoriere, G. JNK and AP-1 mediate apoptosis induced by bortezomib in HepG2 cells via FasL/caspase-8 and mitochondria-dependent pathways. *Apoptosis* **2006**, *11*, 607–625. [CrossRef] [PubMed]
13. Gu, H.; Chen, X.; Gao, G.; Dong, H. Caspase-2 functions upstream of mitochondria in endoplasmic reticulum stress-induced apoptosis by bortezomib in human myeloma cells. *Mol. Cancer Ther.* **2008**, *7*, 2298–2307. [CrossRef]
14. Richardson, P.G.; Mitsiades, C.; Ghobrial, I.; Anderson, K. Beyond single-agent bortezomib: Combination regimens in relapsed multiple myeloma. *Curr. Opin. Oncol.* **2006**, *18*, 598–608. [CrossRef] [PubMed]
15. Miguel, J.S.; Schlag, R.; Khuageva, N.K.; Dimopoulos, M.A.; Shpilberg, O.; Kropff, M.; Spicka, I.; Petrucci, M.T.; Palumbo, A.; Samoilova, O.S.; et al. Bortezomib plus Melphalan and Prednisone for Initial Treatment of Multiple Myeloma. *N. Engl. J. Med.* **2008**, *359*, 906–917. [CrossRef]
16. Oerlemans, R.; Franke, N.E.; Assaraf, Y.G.; Cloos, J.; Van Zantwijk, I.; Berkers, C.R.; Scheffer, G.L.; Debipersad, K.; Vojtekova, K.; Lemos, C.; et al. Molecular basis of bortezomib resistance: Proteasome subunit β5 (PSMB5) gene mutation and overexpression of PSMB5 protein. *Blood* **2008**, *112*, 2489–2499. [CrossRef]
17. Barrio, S.; Stühmer, T.; Da-Viá, M.; Barrio-Garcia, C.; Lehners, N.; Besse, A.; Cuenca, I.; Garitano-Trojaola, A.; Fink, S.; Leich, E.; et al. Spectrum and functional validation of PSMB5 mutations in multiple myeloma. *Leukemia* **2019**, *33*, 447–456. [CrossRef]
18. Ferrington, D.A.; Gregerson, D.S. Immunoproteasomes: Structure, function, and antigen presentation. *Prog. Mol. Biol. Transl. Sci.* **2012**, *109*, 75–112. [CrossRef]

19. Miller, Z.; Lee, W.; Kim, K.B. The immunoproteasome as a therapeutic target for hematological malignancies. *Curr. Cancer Drug Targets* **2014**, *14*, 537–548. [CrossRef]
20. Besse, A.; Besse, L.; Kraus, M.; Mendez-Lopez, M.; Bader, J.; Xin, B.-T.; De Bruin, G.; Maurits, E.; Overkleeft, H.S.; Driessen, C. Proteasome Inhibition in Multiple Myeloma: Head-to-Head Comparison of Currently Available Proteasome Inhibitors. *Cell Chem. Biol.* **2019**, *26*, 340–351.e3. [CrossRef]
21. Zhang, X.-D.; Baladandayuthapani, V.; Lin, H.; Mulligan, G.; Li, B.; Esseltine, D.-L.W.; Qing, Z.; Xu, J.; Hunziker, W.; Barlogie, B.; et al. Tight Junction Protein 1 Modulates Proteasome Capacity and Proteasome Inhibitor Sensitivity in Multiple Myeloma via EGFR/JAK1/STAT3 Signaling. *Cancer Cell* **2016**, *29*, 639–652. [CrossRef] [PubMed]
22. Malek, E.; Abdel-Malek, M.A.Y.; Jagannathan, S.; Vad, N.; Karns, R.; Jegga, A.G.; Broyl, A.; Van Duin, M.; Sonneveld, P.; Cottini, F.; et al. Pharmacogenomics and chemical library screens reveal a novel SCFSKP2 inhibitor that overcomes Bortezomib resistance in multiple myeloma. *Leukemia* **2016**, *31*, 645–653. [CrossRef] [PubMed]
23. Jagannath, S.; Dimopoulos, M.A.; Lonial, S. Combined proteasome and histone deacetylase inhibition: A promising synergy for patients with relapsed/refractory multiple myeloma. *Leuk. Res.* **2010**, *34*, 1111–1118. [CrossRef] [PubMed]
24. Duvic, M.A.; Olsen, E.A.; Breneman, D.; Pacheco, T.R.; Parker, S.; Vonderheid, E.C.; Abuav, R.; Ricker, J.L.; Rizvi, S.; Chen, C.; et al. Evaluation of the Long-Term Tolerability and Clinical Benefit of Vorinostat in Patients with Advanced Cutaneous T-Cell Lymphoma. *Clin. Lymphoma Myeloma* **2009**, *9*, 412–416. [CrossRef]
25. Sivaraj, D.; Green, M.M.; Gasparetto, C. Panobinostat for the management of multiple myeloma. *Futur. Oncol.* **2017**, *13*, 477–488. [CrossRef] [PubMed]
26. Badros, A.; Burger, A.M.; Philip, S.; Niesvizky, R.; Kolla, S.S.; Goloubeva, O.; Harris, C.; Zwiebel, J.; Wright, J.J.; Espinoza-Delgado, I.; et al. Phase I Study of Vorinostat in Combination with Bortezomib for Relapsed and Refractory Multiple Myeloma. *Clin. Cancer Res.* **2009**, *15*, 5250–5257. [CrossRef]
27. Santo, L.; Hideshima, T.; Kung, A.L.; Tseng, J.-C.; Tamang, D.; Yang, M.; Jarpe, M.; Van Duzer, J.H.; Mazitschek, R.; Ogier, W.C.; et al. Preclinical activity, pharmacodynamic, and pharmacokinetic properties of a selective HDAC6 inhibitor, ACY-1215, in combination with bortezomib in multiple myeloma. *Blood* **2012**, *119*, 2579–2589. [CrossRef]
28. San-Miguel, J.F.; Richardson, P.G.; Günther, A.; Sezer, O.; Siegel, D.; Bladé, J.; Leblanc, R.; Sutherland, H.; Sopala, M.; Mishra, K.K.; et al. Phase Ib Study of Panobinostat and Bortezomib in Relapsed or Relapsed and Refractory Multiple Myeloma. *J. Clin. Oncol.* **2013**, *31*, 3696–3703. [CrossRef]
29. Amengual, J.E.; Johannet, P.M.; Lombardo, M.; Zullo, K.M.; Hoehn, D.; Bhagat, G.; Scotto, L.; Jirau-Serrano, X.; Radeski, D.; Heinen, J.; et al. Dual Targeting of Protein Degradation Pathways with the Selective HDAC6 Inhibitor ACY-1215 and Bortezomib Is Synergistic in Lymphoma. *Clin. Cancer Res.* **2015**, *21*, 4663–4675. [CrossRef]
30. Lee, N.H.; Kim, G.W.; Kwon, S.H. The HDAC6-selective inhibitor is effective against non-Hodgkin lymphoma and synergizes with ibrutinib in follicular lymphoma. *Mol. Carcinog.* **2019**, *58*, 944–956. [CrossRef]
31. Mishima, Y.; Santo, L.; Loredana, S.; Cirstea, D.; Nemani, N.; Yee, A.J.; O'Donnell, E.; Selig, M.K.; Quayle, S.N.; Arastu-Kapur, S.; et al. Ricolinostat (ACY-1215) induced inhibition of aggresome formation accelerates carfilzomib-induced multiple myeloma cell death. *Br. J. Haematol.* **2015**, *169*, 423–434. [CrossRef] [PubMed]
32. Vogl, D.T.; Raje, N.; Jagannath, S.; Richardson, P.; Hari, P.; Orlowski, R.; Supko, J.G.; Tamang, D.; Yang, M.; Jones, S.S.; et al. Ricolinostat, the First Selective Histone Deacetylase 6 Inhibitor, in Combination with Bortezomib and Dexamethasone for Relapsed or Refractory Multiple Myeloma. *Clin. Cancer Res.* **2017**, *23*, 3307–3315. [CrossRef] [PubMed]
33. Choi, E.; Lee, C.; Park, J.E.; Seo, J.J.; Cho, M.; Kang, J.S.; Kim, H.M.; Park, S.-K.; Lee, K.; Han, G. Structure and property based design, synthesis and biological evaluation of γ-lactam based HDAC inhibitors. *Bioorgan. Med. Chem. Lett.* **2011**, *21*, 1218–1221. [CrossRef] [PubMed]
34. Park, J.; Bae, E.K.; Lee, C.; Choi, J.H.; Jung, W.J.; Ahn, K.S.; Yoon, S.S. Establishment and characterization of bortezomib-resistant U266 cell line: Constitutive activation of NF-kappaB-mediated cell signals and/or alterations of ubiquitylation-related genes reduce bortezomib-induced apoptosis. *BMB Rep.* **2014**, *47*, 274–279. [CrossRef] [PubMed]
35. Ryu, H.-W.; Shin, D.-H.; Lee, D.H.; Choi, J.; Han, G.; Lee, K.Y.; Kwon, S.H. HDAC6 deacetylates p53 at lysines 381/382 and differentially coordinates p53-induced apoptosis. *Cancer Lett.* **2017**, *391*, 162–171. [CrossRef]
36. Ryu, H.-W.; Won, H.-R.; Lee, D.H.; Kwon, S.H. HDAC6 regulates sensitivity to cell death in response to stress and post-stress recovery. *Cell Stress Chaperon* **2017**, *22*, 253–261. [CrossRef]
37. Won, H.-R.; Ryu, H.-W.; Shin, D.-H.; Yeon, S.-K.; Lee, D.H.; Kwon, S.H. A452, an HDAC6-selective inhibitor, synergistically enhances the anticancer activity of chemotherapeutic agents in colorectal cancer cells. *Mol. Carcinog.* **2018**, *57*, 1383–1395. [CrossRef]
38. Won, H.; Lee, D.H.; Yeon, S.; Ryu, H.; Kim, G.W.; Kwon, S.H. HDAC6-selective inhibitor synergistically enhances the anticancer activity of immunomodulatory drugs in multiple myeloma. *Int. J. Oncol.* **2019**, *55*, 499–512. [CrossRef]
39. Chou, T.-C. Drug Combination Studies and Their Synergy Quantification Using the Chou-Talalay Method. *Cancer Res.* **2010**, *70*, 440–446. [CrossRef] [PubMed]
40. Kuhn, D.J.; Chen, Q.; Voorhees, P.M.; Strader, J.S.; Shenk, K.D.; Sun, C.M.; Demo, S.D.; Bennett, M.K.; Van Leeuwen, F.W.B.; Chanan-Khan, A.A.; et al. Potent activity of carfilzomib, a novel, irreversible inhibitor of the ubiquitin-proteasome pathway, against preclinical models of multiple myeloma. *Blood* **2007**, *110*, 3281–3290. [CrossRef] [PubMed]

41. Mitsiades, C.S.; Hayden, P.J.; Anderson, K.C.; Richardson, P.G. From the bench to the bedside: Emerging new treatments in multiple myeloma. *Best Pr. Res. Clin. Haematol.* **2007**, *20*, 797–816. [CrossRef] [PubMed]
42. Walter, P.; Ron, D. The Unfolded Protein Response: From Stress Pathway to Homeostatic Regulation. *Science* **2011**, *334*, 1081–1086. [CrossRef] [PubMed]
43. Wright, K.L.; White, L.C.; Kelly, A.; Beck, S.; Trowsdale, J.; Ting, J.P. Coordinate regulation of the human TAP1 and LMP2 genes from a shared bidirectional promoter. *J. Exp. Med.* **1995**, *181*, 1459–1471. [CrossRef] [PubMed]
44. Dalton, W.S. Drug resistance and drug development in multiple myeloma. *Semin. Oncol.* **2002**, *29*, 21–25. [CrossRef]
45. Mateos, M.-V. Management of treatment-related adverse events in patients with multiple myeloma. *Cancer Treat. Rev.* **2010**, *36*, S24–S32. [CrossRef]
46. DasMahapatra, G.; Patel, H.; Friedberg, J.; Quayle, S.N.; Jones, S.S.; Grant, S. In Vitro and In Vivo Interactions between the HDAC6 Inhibitor Ricolinostat (ACY1215) and the Irreversible Proteasome Inhibitor Carfilzomib in Non-Hodgkin Lymphoma Cells. *Mol. Cancer Ther.* **2014**, *13*, 2886–2897. [CrossRef]
47. Madrid, L.V.; Mayo, M.W.; Reuther, J.Y.; Baldwin, A.S., Jr. Akt Stimulates the Transactivation Potential of the RelA/p65 Subunit of NF-κB through Utilization of the IκB Kinase and Activation of the Mitogen-activated Protein Kinase p38. *J. Biol. Chem.* **2001**, *276*, 18934–18940. [CrossRef]
48. Chen, L.-F.; Williams, S.A.; Mu, Y.; Nakano, H.; Duerr, J.M.; Buckbinder, L.; Greene, W.C. NF-κB RelA Phosphorylation Regulates RelA Acetylation. *Mol. Cell. Biol.* **2005**, *25*, 7966–7975. [CrossRef]
49. Iaconelli, J.; LaLonde, J.; Watmuff, B.; Liu, B.; Mazitschek, R.; Haggarty, S.J.; Karmacharya, R. Lysine Deacetylation by HDAC6 Regulates the Kinase Activity of AKT in Human Neural Progenitor Cells. *ACS Chem. Biol.* **2017**, *12*, 2139–2148. [CrossRef]
50. Busse, A.; Kraus, M.; Rietz, A.; Scheibenbogen, C.; Driessen, C.; Blau, I.W.; Thiel, E.; Keilholz, U.; Na, I.-K. Sensitivity of tumor cells to proteasome inhibitors is associated with expression levels and composition of proteasome subunits. *Cancer* **2008**, *112*, 659–670. [CrossRef]
51. Sun, Y.; Peng, R.; Peng, H.; Liu, H.; Wen, L.; Wu, T.; Yi, H.; Li, A.; Zhang, Z. miR-451 suppresses the NF-kappaB-mediated proinflammatory molecules expression through inhibiting LMP7 in diabetic nephropathy. *Mol. Cell. Endocrinol.* **2016**, *433*, 75–86. [CrossRef]
52. Soriano, G.P.; Besse, L.; Li, N.; Kraus, M.; Besse, A.; Meeuwenoord, N.; Bader, J.; Everts, B.; Dulk, H.D.; Overkleeft, H.S.; et al. Proteasome inhibitor-adapted myeloma cells are largely independent from proteasome activity and show complex proteomic changes, in particular in redox and energy metabolism. *Leukemia* **2016**, *30*, 2198–2207. [CrossRef]
53. Gutman, D.A.; Morales, A.A.; Boise, L.H. Acquisition of a multidrug-resistant phenotype with a proteasome inhibitor in multiple myeloma. *Leukemia* **2009**, *23*, 2181–2183. [CrossRef]
54. Cheng, F.; Lienlaf, M.; Wang, H.-W.; Perez-Villarroel, P.; Lee, C.; Woan, K.; Rock-Klotz, J.; Sahakian, E.; Woods, D.; Pinilla-Ibarz, J.; et al. A Novel Role for Histone Deacetylase 6 in the Regulation of the Tolerogenic STAT3/IL-10 Pathway in APCs. *J. Immunol.* **2014**, *193*, 2850–2862. [CrossRef]
55. Kwon, S.; Zhang, Y.; Matthias, P. The deacetylase HDAC6 is a novel critical component of stress granules involved in the stress response. *Genes Dev.* **2007**, *21*, 3381–3394. [CrossRef]

International Journal of
Molecular Sciences

Article

Nuclear FGFR2 Interacts with the MLL-AF4 Oncogenic Chimera and Positively Regulates *HOXA9* Gene Expression in t(4;11) Leukemia Cells

Tiziana Fioretti [1,†], Armando Cevenini [1,2,†], Mariateresa Zanobio [2], Maddalena Raia [1], Daniela Sarnataro [1,2], Fabio Cattaneo [2], Rosario Ammendola [2] and Gabriella Esposito [1,2,*]

[1] CEINGE Advanced Biotechnologies s.c. a r.l., via G. Salvatore, 486, 80145 Naples, Italy;
fioretti@ceinge.unina.it (T.F.); armando.cevenini@unina.it (A.C.); raia@ceinge.unina.it (M.R.);
daniela.sarnataro@unina.it (D.S.)
[2] Department of Molecular Medicine and Medical Biotechnologies, University of Naples Federico II,
Via S. Pansini, 5, 80131 Naples, Italy; mt.zanobio@gmail.com (M.Z.); fabio.cattaneo@unina.it (F.C.);
rosario.ammendola@unina.it (R.A.)
* Correspondence: gabriella.esposito@unina.it; Tel.: +30-0817463146
† These authors contribute equally.

Abstract: The chromosomal translocation t(4;11) marks an infant acute lymphoblastic leukemia associated with dismal prognosis. This rearrangement leads to the synthesis of the MLL-AF4 chimera, which exerts its oncogenic activity by upregulating transcription of genes involved in hematopoietic differentiation. Crucial for chimera's aberrant activity is the recruitment of the AF4/ENL/P-TEFb protein complex. Interestingly, a molecular interactor of AF4 is fibroblast growth factor receptor 2 (FGFR2). We herein analyze the role of FGFR2 in the context of leukemia using t(4;11) leukemia cell lines. We revealed the interaction between MLL-AF4 and FGFR2 by immunoprecipitation, western blot, and immunofluorescence experiments; we also tested the effects of FGFR2 knockdown, FGFR2 inhibition, and FGFR2 stimulation on the expression of the main MLL-AF4 target genes, i.e., *HOXA9* and *MEIS1*. Our results show that FGFR2 and MLL-AF4 interact in the nucleus of leukemia cells and that FGFR2 knockdown, which is associated with decreased expression of *HOXA9* and *MEIS1*, impairs the binding of MLL-AF4 to the *HOXA9* promoter. We also show that stimulation of leukemia cells with FGF2 increases nuclear level of FGFR2 in its phosphorylated form, as well as *HOXA9* and *MEIS1* expression. In contrast, preincubation with the ATP-mimetic inhibitor PD173074, before FGF2 stimulation, reduced FGFR2 nuclear amount and *HOXA9* and *MEIS1* transcript level, thereby indicating that MLL-AF4 aberrant activity depends on the nuclear availability of FGFR2. Overall, our study identifies FGFR2 as a new and promising therapeutic target in t(4;11) leukemia.

Keywords: AF4; cell culture; FGFR2; *HOXA9*; MLL-AF4; nucleus; target therapy; t(4;11) leukemia

1. Introduction

The t(4;11) chromosomal translocation is a common cause of infant acute lymphoblastic leukemia (ALL) [1,2]. It fuses in-frame the *mixed-lineage leukemia* (*MLL*, aka *KMT2A*) and the *AF4/FMR2 Family Member 1* (*AFF1*) genes, located on chromosome 11 and 4 respectively, leading to fusion genes that encode chimeric oncoproteins—namely MLL-AF4 and the reciprocal AF4-MLL [1–6]. This aberration has been identified in utero and in neonatal blood, indicating it arises in the prenatal period; it is rare in adults (3–4%) and is correlated with a very poor prognosis [1,7]. Although 80% of patients exhibit both of the fusion proteins, only MLL-AF4 is essential for leukemic transformation and maintenance [8,9].

To exert its aberrant transcriptional activity, MLL-AF4 binds and deregulates the expression of key target genes involved in lymphocyte differentiation, including the homeobox A (*HOXA*) cluster genes and *MEIS1* [5,10–12]. Indeed, in many patients, survival of leukemic blasts depends on the maintenance of high expression levels of *HOXA9, HOXA8,*

HOXA7, and *MEIS1* [12–14]; consistently, *HOXA9* silencing promotes apoptotic death in t(4;11) (q21;q23) lymphoblasts [15].

The AF4 protein, which is encoded by the *AFF1* gene, takes part in the AF4 family/ENL family/P-TEFb (AEP) protein complex that is crucial for chimera's aberrant function. In addition to AF4, the AEP complex is formed by the transcriptional activators ENL, ELL, and AF5q—which are also common MLL fusion partners in human leukemia—as well as the positive elongation factor (P-TEFb) [16]. Through the direct interaction with the scaffold protein 14-3-3θ, AF4 and/or AF5q heterodimerize with MLL-AF4, thereby promoting the constitutive assembly of the AEP complex and the subsequent retrieval of RNA Pol II on target-gene promoters [17,18]. Moreover, MLL-AF4 forms a complex with DOT1L, a histone H3 lysine-79 (H3K79) methyltransferase, and other MLL fusion partners, such as AF9 and AF10 [11,18]. The recruitment of the AEP complex on target genes triggers the aberrant activity of DOT1L; in agreement, the epigenetic signature of H3K79me2, H3K27ac, and H3K4me3 marks all the MLL-AF4 target genes [11,18].

Consequently, the oncogenic potential of MLL-AF4 is mostly driven by the interaction with AF4 and its protein partners, which therefore represent promising therapeutic targets in t(4;11) leukemia. Interestingly, fibroblast growth factor receptor 2 (FGFR2) was found among the protein interactors of AF4 [19].

FGFR2 belongs to the family of receptor tyrosine kinases (RTKs), transmembrane-type receptors mainly localized on the cell surface with cytoplasmic tyrosine kinase domains [20]. It is able to recognize as ligands specific fibroblast growth factors (FGFs), with autocrine or paracrine action [21]. FGFs stimulate the intrinsic tyrosine kinase activity of the FGFRs and trigger various intracellular transduction signals that mediate multiple biological responses, including proliferation, differentiation, and cell survival [22].

The function of several RTKs is altered in different types of tumors and various drugs targeting the receptors and/or their downstream signaling pathways are already available and approved in clinical settings [23–25]. Of note, some RTK-related signaling pathways that influence cell growth and proliferation are activated in MLL-related leukemia [26–28].

Interestingly, various receptors, including RTKs and G-proteins coupled receptors (GPCRs), traffic from the cell surface to the nucleus [29–33]. In most cases, an intracellular domain fragment of the receptor translocates from the cell surface to the nucleus, whereas, for a few others, the intact receptor enters into the nucleus [28,29].

We herein analyze the function of nuclear FGFR2 in t(4;11) leukemia cells and its potential role as a molecular target for the treatment of this rare and poorly curable form of leukemia [7].

2. Results

We aimed to characterize the role of FGFR2 in t(4;11) leukemia. We analyzed the interaction between MLL-AF4 and FGFR2 and studied the effect of FGFR2 knockdown and inhibition on MLL-AF4 target gene expression.

2.1. FGFR2 Is a Nuclear Interactor of MLL-AF4

During a previous functional proteomic analysis performed in HEK293 cells, we found FGFR2 among the molecular partners of the AF4 protein [19]. Therefore, we wondered whether FGFR2 interacted also with the MLL-AF4 chimera in t(4;11) leukemia cell lines. Firstly, by flow cytofluorometry, we showed that FGFR2 was significantly represented on the cell surface of three t(4;11) leukemia cell lines that endogenously expressed MLL-AF4, namely RS4;11, SEM, and MV4-11 (Figure 1A). Therefore, we carried out coimmunoprecipitation experiments in RS4;11 and MV4-11 leukemia cells and found that endogenous FGFR2 interacted with endogenous MLL-AF4 (Figure 1B).

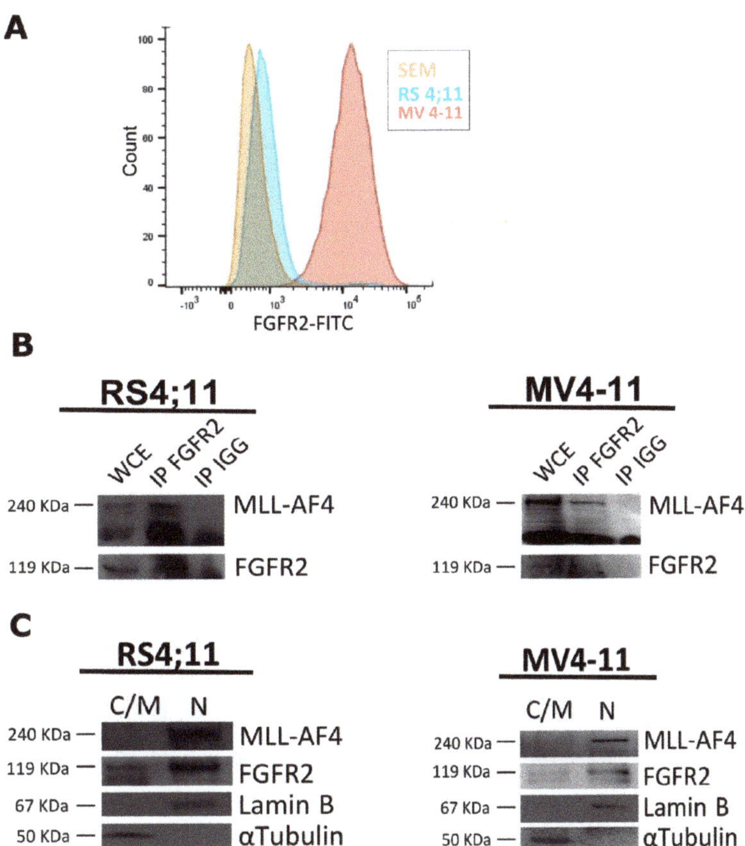

Figure 1. FGFR2 interacts with MLL-AF4 and localizes in the nucleus of t(4;11) leukemia cells. (**A**) Flow cytometry analysis carried out in different leukemia cell lines (SEM, RS4;11, and MV4-11) using anti-FGFR2 antibody and a fluorescein isothiocyanate (FITC)-conjugated secondary antibody. (**B**) Endogenous FGFR2 was immunoprecipitated from whole cellular extract (WCE) of RS4;11 and MV4-11 cells with anti-FGFR2 antibody (IP FGFR2) and with anti-IgG (IP IGG) antibodies, and isolated using A/G plus agarose beads; immunocomplexes were analyzed by western blot, with an anti-MLL antibody. (**C**) Twenty micrograms of cytosolic/membrane (C/M) and 50 μg of nuclear (N) proteins extracted from RS4;11 and MV4-11 cells were analyzed by western blot with anti-FGFR2 and anti-MLL antibodies; ⟨–tubulin and lamin B were used as cytosolic and nuclear control proteins, respectively.

In addition to being present on the cell surface, FGFR2 is known to have intracellular localization [34]. Thus, to better characterize the interaction between FGFR2 and MLL-AF4, we evaluated their cellular distribution. First, we analyzed cytosolic/membrane and nuclear protein fractions isolated from RS4;11 and MV4-11 cells by western blot. As expected, the MLL-AF4 oncoprotein was exclusively found in the nucleus, whereas FGFR2 was present in the cytosol/membrane—in agreement with its conventional function—and, significantly, also in the nuclear fraction (Figure 1C). These results indicate that FGFR2 and MLL-AF4 colocalize and therefore may interact within the nucleus of leukemia cells.

2.2. FGFR2 Silencing Affects Expression of MLL-AF4 Target Genes

Once FGFR2 was proven to be a protein partner of MLL-AF4, we wondered whether it contributed to MLL-AF4 transcriptional activity. To this aim, we first transfected RS4;11,

SEM, and MV4-11 cell lines with a pool of specific small interfering RNA (siRNA) directed against the *FGFR2* transcript to knock down expression of the receptor. The same cell lines were also transfected with a scramble siRNA as a control. Seventy-two hours after transfection, total proteins were extracted and analyzed by western blot, which demonstrated a statistically significant reduction of FGFR2 in all cell lines (Figure 2A).

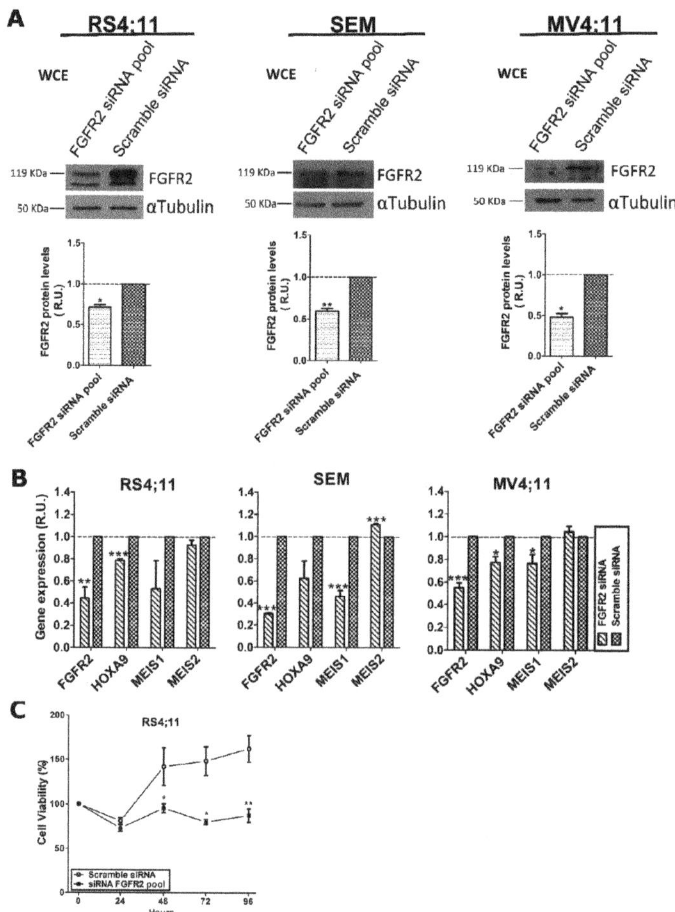

Figure 2. FGFR2 affects transcription of MLL-AF4 target genes in three t(4;11) cell lines. (**A**) Western blot analysis of total proteins (WCE) extracted 72 h after transfection with FGFR2-siRNA pool carried out with anti-FGFR2 antibody to evaluate residual FGFR2 amount; scramble siRNA represents the negative control; α-tubulin was used to normalize protein loading. (**B**) RT-qPCR analysis of *FGFR2, HOXA9, MEIS1,* and *MEIS2* transcript levels in FGFR2-silenced cells. Average values from at least three independent experiments are graphically reported as relative units (R.U.). Relative gene expression was normalized to *ACTB, POLR2A,* and *TUBA1A* genes and expression levels were determined using the $2^{-\Delta\Delta Ct}$ method. Statistical significance was calculated by one-way two-tail paired *t*-test. *p*-values are indicated as follows: * = $p < 0.05$; ** = $p < 0.01$; *** = $p < 0.005$. (**C**) Cell viability determined by MTT assay at 0, 24, 48, 72, and 96 h after transfection of cells with a FGFR2 siRNA pool and with scramble siRNA as a control. R.U., relative units.

Therefore, we extracted total RNA from silenced cells to analyze, by quantitative reverse transcription-polymerase chain reaction (RT-qPCR), transcript levels of *HOXA9* and *MEIS1*, two well-known MLL-AF4 target genes, and of *MEIS2*, which was not a chimera target gene. Results showed that *HOXA9* and *MEIS1*, but not *MEIS2*, expression was significantly reduced 72 h after FGFR2 silencing (Figure 2B). These results are in line with the assumption that MLL-AF4 transcriptional activity depends on FGFR2 availability.

To evaluate whether FGFR2 knockdown affected cell survival, we evaluated viability of RS4;11 cells, transfected with a FGFR2 siRNA pool and with scramble siRNA as a negative control, by MTT assay. Analysis was carried out 24, 48, 72, and 96 h after transfection and results were normalized to the untransfected cells (Figure 2C). Seventy-two hours after transfection, cell viability was significantly reduced in silenced cells and remained low until 96 h, whereas the viability of scrambled cells was comparable to the untransfected ones at 48, 72, and 96 h. This result strongly suggests that FGFR2 contributes to the survival of t(4;11) leukemia cells.

It is widely proven that MLL-AF4 binds to the *HOXA9* promoter and activates the transcription of this gene in MLL-related leukemia [17,35]. Since MLL-AF4 is a protein interactor of FGFR2, and FGFR2 silencing decreased expression of *HOXA9*, we evaluated the binding efficiency of MLL-AF4 to the *HOXA9* promoter (*HOXA9*pr) in FGFR2-silenced RS4;11 cells. To this aim, we performed chromatin immunoprecipitation (ChIP) assays using an anti-MLL^N and an anti-AF4 antibody; a pool of nonspecific IgG served as a negative control (Figure 3A). More specifically, the anti-MLL^N antibody was directed against an N-terminal epitope of MLL wild type and therefore recognized also the MLL-AF4 oncoprotein; the anti-AF4 antibody served to reveal AF4, which was known to colocalize with MLL-AF4 on the *HOXA9* promoter [17]. Subsequent qPCR analysis revealed that the percentage of the *HOXA9*pr sequence in chromatin precipitated with both anti-AF4 and anti-MLL^N antibodies was significantly lower in FGFR2-silenced than in scrambled siRNA cells (Figure 3A).

Interestingly, western blot analysis performed on whole protein extracts from FGFR2-silenced cells showed that AF4 and MLL-AF4 protein levels were unaffected in comparison to the control cells (Figure 3B). Overall, these results demonstrate that silencing-induced FGFR2 deficiency, which has no effect on AF4 and MLL-AF4 total protein levels, weakens the binding of these two factors to the *HOXA9* promoter, thereby confirming that *HOXA9* expression specifically depends on the availability of FGFR2.

Subsequently, we investigated whether FGFR2 participated in the MLL-AF4 transcriptional machinery assembled on the *HOXA9* promoter. To this aim, we immunoprecipitated the chromatin in RS4;11 cells with an anti-FGFR2 antibody and looked for the *HOXA9pr* sequence by real time qPCR. Chromatin was also precipitated with the anti-MLL^N antibody and with a pool of nonspecific IgG, representing positive and negative control of ChIP specificity, respectively; a sequence within the *β-actin* gene promoter (*ACTB*pr) was amplified as a negative control. Data analysis revealed that the anti-FGFR2 antibody precipitated a percentage of *HOXA9pr* not significantly different from that of *ACTBpr*. In contrast, significant amount of DNA containing the *HOXA9pr* sequence was collected with the anti-MLL^N antibody (Figure 3C). These results indicate that despite FGFR2 and MLL-AF4 being protein interactors, they do not colocalize on chimera target gene promoters.

2.3. FGF2 Stimulates Nuclear Localization of FGFR2

To exert its conventional function, FGFR2 depends on cytoplasmic domain autophosphorylation consequent to the binding of its cognate agonists, i.e., fibroblast growth factors (FGFs) [34]. Moreover, several studies demonstrated that FGFR2 is also expressed in the nucleus [36] and that its ligand-dependent autophosphorylation triggers nuclear localization [37]. Therefore, we wondered whether FGFR2 nuclear localization relied on ligand-dependent autophosphorylation also in t(4;11) leukemia cells. To this aim, we first stimulated serum-starved RS4;11 cells with increasing concentrations of FGF2 and with the vehicle as a negative control, and then we purified nuclear proteins. Western blot analysis with anti-phospho-FGFR2 and anti-FGFR2 antibodies showed that FGF2 stimulation significantly increased both total and phosphorylated (p-FGFR2) levels of nuclear FGFR2 in a concentration-dependent manner, strongly suggesting that the nuclear increase in FGFR2 and pFGFR2 depends, at least in part, on the ligand concentration (Figure 4A). Notably, values of the p-FGFR2/FGFR2 ratio were very similar among the three tested concentrations of FGF2, indicating that the agonist-mediated activation of FGFR2 enhanced total and phosphorylated levels of the nuclear receptor proportionally. In fact, the Pearson's R = 0.99 was consistent with a strong positive correlation between total and phosphorylated levels of the nuclear receptor (Figure 4A). Taken together, these results indicate that FGF2 stimulation triggers activation and nuclear translocation of FGFR2 in its phosphorylated form.

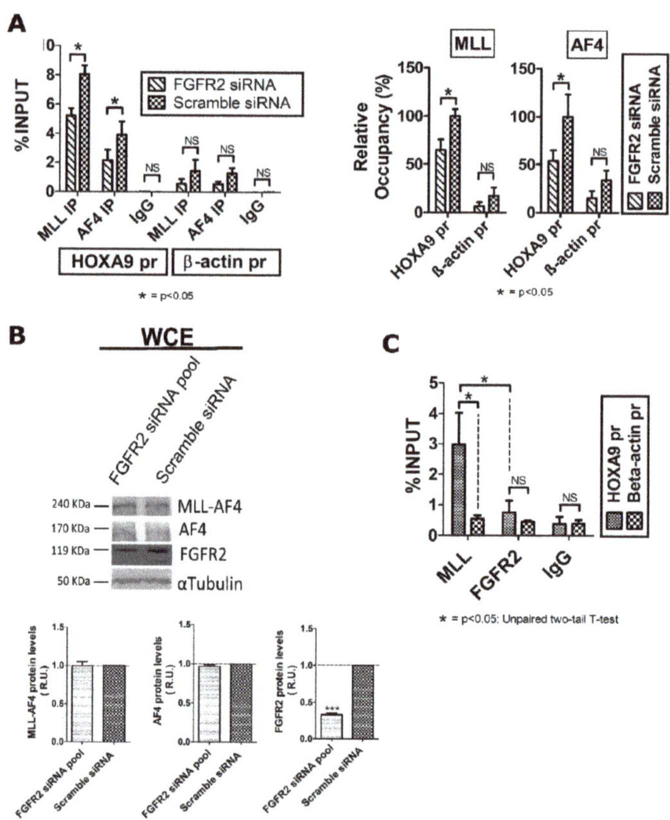

Figure 3. FGFR2 contributes to the interaction of AF4 and MLL-AF4 at the *HOXA9* promoter (HOXA9 pr), in RS4;11 leukemia cells. (**A**) ChIP assays performed to test the binding of AF4 and MLL-AF4 to *HOXA9* pr, after FGFR2 silencing. Scramble siRNA represents the negative control. (**B**) Western blot analysis of total protein (WCE) extracted 72 h after transfection of a FGFR2-siRNA pool carried out to reveal expression level of endogenous MLL-AF4, AF4, and FGFR2; scramble siRNA represents the negative control; α-tubulin was used to normalize protein loading. (**C**) Interaction of endogenous MLL and FGFR2 with the *HOXA9*pr. ChIP data are expressed as percentage of *HOXA9pr* and *β-actin* promoter (®-actin pr) in precipitated chromatin compared with the INPUT; IgG mix is the negative control. Results represent the average of three independent experiments. Error bars indicate the standard deviations. Statistical significance is calculated by one-way two tail paired *t*-test. *p*-values are indicated as follows: * = *p* < 0.05, *** *p* = < 0.005. NS, not significant. R.U., relative units.

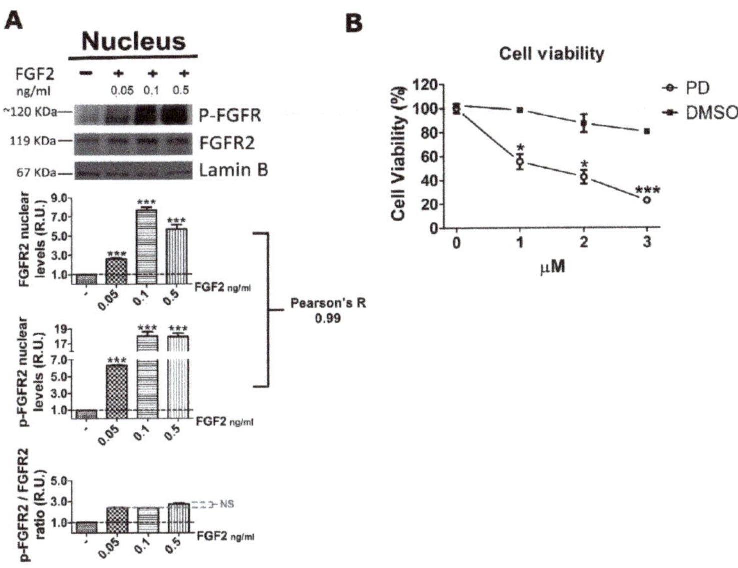

Figure 4. FGF2 promotes FGFR2 activation and nuclear translocation. (**A**) Western blot analysis of nuclear extracts obtained from serum-deprived RS4;11 cells stimulated with increasing concentrations of FGF2 performed with an anti-phospho-FGFR2 (p-FGFR2) or an anti-FGFR2 (FGFR2) antibody; lamin B was the control for protein loading; densitometric analysis of band intensity is graphically shown for FGFR2, p-FGFR2, and p-FGFR2/FGFR2 ratio. (**B**) Cell viability assay performed following treatment of RS4;11 cells with the FGFR inhibitor PD173074 (PD) and the vehicle (DMSO). Statistical significance was calculated by one-way two tail paired *t*-test. *p*-values are indicated as follows: * = $p < 0.05$, *** = $p < 0.005$. Pearson's R = 0.99 is consistent with a positive correlation between the two set of values. NS, not significant. R.U., relative units.

To further support the evidence that FGFR2 activation and nuclear translocation were ligand-dependent, we treated RS4;11 leukemia cells with PD173074, a synthetic compound belonging to the pyrido (2,3-d) pyrimidine class. PD173074 has a high selectivity for a set of RTKs, including FGFRs, and acts as an ATP mimetic; it replaces an ATP molecule in the binding pocket, thus blocking ligand-mediated autophosphorylation of FGFRs [38,39].

Preliminarily, we carried out a growth curve using increasing amount of PD173074 and, of interest, noted a dose-dependent impairment of leukemia cell viability, which therefore depended on RTK stimulation and autophosphorylation (Figure 4B).

Next, we incubated RS4;11 cells with the half inhibitory concentration of PD173074 for 1 h and then stimulated them with 0.1 ng/mL of FGF2 for 5 min. Western blot analysis performed on cytosolic/membrane and nuclear proteins showed that, differently from untreated cells (Figure 5A) and from cells preincubated with the vehicle and stimulated with FGF2 (Figure 5B), PD173074 treatment affected nuclear localization of FGFR2 (Figure 5C). These results were also confirmed by immunofluorescence assays (Figure 5A–C).

Taken together, the results in Figures 4 and 5 clearly demonstrate that FGFR2 nuclear translocation depends on its ligand-dependent autophosphorylation, so that the measure of nuclear FGFR2 is also a measure of nuclear p-FGFR2.

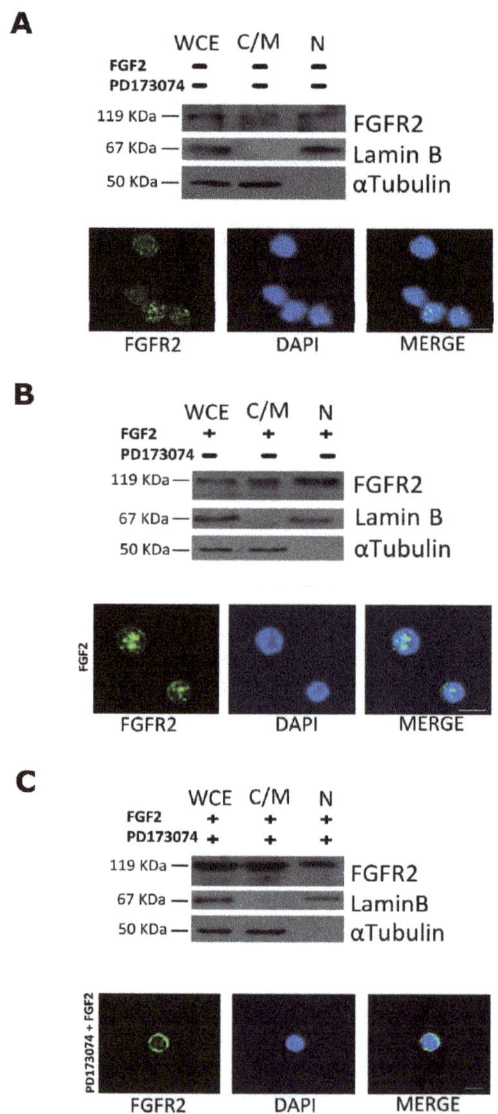

Figure 5. PD173074 inhibition affects cellular amount of FGFR2 in RS4;11 cells. RS4;11 cells were serum-starved for 12 h and preincubated for 1 h (**A**) with DMSO alone, (**B**) with DMSO and then stimulated with 0.1 ng/mL FGF2, or (**C**) with 2 µM of PD173074 and then stimulated with 0.1 ng/mL FGF2. Thirty micrograms of cytosolic/membrane proteins (C/M) and 50 µg of nuclear proteins (N) were analyzed by western blot with anti-FGFR2 antibody; αTubulin and Lamin B were used as loading control of cytosolic and nuclear proteins, respectively (upper panels); whole cellular extract (WCE). Immunofluorescence was performed with an anti-rabbit CY2 antibody to evaluate the localization of FGFR2, and DAPI, for nuclear staining (bottom panels). Scale bars: 10 µm.

Consistent with these findings and with previous evidence that the binding of MLL-AF4 to the *HOXA9* gene promoter depended on the nuclear availability of FGFR2 (Figures 2 and 3), we expected that an FGF2-mediated nuclear increase in p-FGFR2 would affect the MLL-AF4 chimera transactivity. Therefore, we carried out RT-qPCR analysis to evaluate chimera target gene expression in RS4;11 and SEM stimulated with FGF2 or pretreated with the PD173074 inhibitor. As expected, FGF2 stimulation led to increased expression of *HOXA9* and *MEIS1* (but not of *MEIS2*) and this overexpression was prevented when the cells were preincubated with PD173074 (Figure 6).

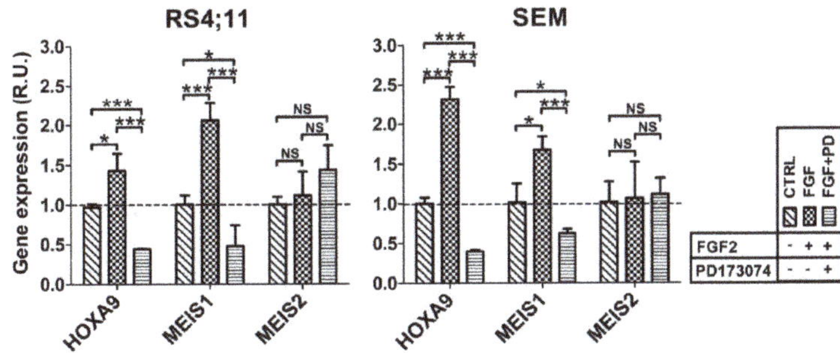

Figure 6. PD173074 inhibitor prevents FGF2-induced transcription of MLL-AF4 target genes. Graphs show transcript levels of MLL-AF4 target genes in leukemia cells, RS4;11 (left) and in SEM (right), stimulated with FGF2 (FGF) or pretreated with PD173074 (FGF + PD). Cells treated with DMSO (vehicle) are used as a control (CTRL); R.U., relative units; *p*-values are indicated as follows: * $p < 0.05$, *** $p < 0.005$.

These data definitively demonstrate that FGF2-induced nuclear translocation of p-FGFR2 promotes aberrant activity of MLL-AF4.

3. Discussion

The t(4;11) chromosomal reciprocal translocation causes a very aggressive form of ALL, which is driven by the aberrant transcriptional activity of the MLL-AF4 chimera. We previously identified a few protein partners of AF4 and demonstrated that its direct interactor 14-3-3θ enhances the aberrant activity of MLL-AF4, thereby giving proof that AF4 interactors can affect chimera function [17,19]. Among the protein partners of AF4, we identified FGFR2 as being noteworthy [19].

Significant amount of this receptor is expressed on the plasma membrane of three different t(4;11) leukemia cell lines, i.e., RS4;11, MV4-11, and SEM (Figure 1A). FGFRs are involved in multiple myeloma, in myeloproliferative disorder, and, importantly, in B-cell precursor (BCP) ALL, the latter including the MLL-rearranged leukemias, thereby supporting their crucial role in hematologic malignancies [40–42]. Interestingly, FGFR signaling can contribute to prednisolone resistance in BCP ALL cells; however, as activating mutations in this RTK family are very rare, information concerning their role in ALL is limited [41].

Herein, we showed that endogenous FGFR2 coimmunoprecipitates endogenous MLL-AF4 in t(4;11) leukemia cells, thereby demonstrating that, similarly to AF4, it participates in the MLL-AF4 protein complex [14,16,17]. Since MLL-AF4 and AF4 exert their activity in the nucleus, binding with FGFR2 is functional only if the receptor has a nuclear localization.

Similarly to other plasma membrane receptors, FGFRs have also been found in the nucleus [29,31–33,36,37]. Indeed, after its activation by extracellular ligands, FGFR1 moves to the nucleus where it regulates gene transcription in cooperation with cyclic AMP-responsive element binding protein by increasing activity of Pol II and histone acetylation [43]. Moreover, in human breast cancer tissues, FGF stimulation causes nuclear translocation of FGFR2, which interacts with the transcriptional factor STAT5 and increases expression of STAT5 target genes and proteins [33].

Besides confirming that MLL-AF4 is exclusively present in the nucleus, we revealed that an appreciable amount FGFR2 consistently localizes in the nuclear compartment of t(4;11) leukemia cells, as evidenced by western blot and immunofluorescence analyses

(Figures 1C and 5A). Therefore, the interaction between the two proteins can actually take place in the nucleus.

The interaction between FGFR2 and MLL-AF4 chimera is particularly interesting, especially based on the evidence that FGFR2 silencing reduces transcript levels of the chimera target genes, *HOXA9* and *MEIS1*, in t(4;11) leukemia cells (Figure 2). In agreement with the previous evidence that reduced expression of *HOXA9* and *MEIS1* impairs viability of leukemia cells [15,44], FGFR2 knockdown negatively affected the proliferative rate of RS4;11 leukemia cells (Figure 2C). Obviously, due to its main role in cellular growth, we cannot attribute this effect entirely to the probable reduced interaction between FGFR2 and MLL-AF4. However, our ChIP experiments demonstrated that FGFR2 deficiency reduced the binding of AF4 and of MLL-AF4 to the *HOXA9* promoter (Figure 3A). Nevertheless, further ChIP experiments performed with the anti-FGFR2 antibody did not detect FGFR2 on the *HOXA9* gene promoter (Figure 3B). Based on this evidence, we conclude that—despite FGFR2 not localizing directly on the chromatin—nuclear availability of the receptor is crucial for the binding of AF4 and MLL-AF4 to the *HOXA9* promoter, which in turn is necessary for the chimera target gene expression.

As our preliminary data suggested that reducing FGFR2 nuclear import is potentially therapeutic for t(4;11) leukemia, it is relevant to understand how FGFR2 enters into the nucleus. In human breast cancer, FGFR2 moves to the nucleus after activation by extracellular ligands; however, the commercial ATP pocket inhibitor PD173074 blocks this agonist-induced FGFR2 nuclear translocation and inhibits RTK activity and downstream pathways [30,36]. In general, PD173074 and similar drugs are able to inhibit FGF signaling in vivo [39,40].

We show that stimulation with FGF2, which is highly expressed in the hematopoietic and stromal compartments of the bone marrow [31,41], triggered phosphorylation and nuclear translocation of FGFR2 in t(4;11) leukemia cells (Figure 4B). On the other hand, when activation of FGFR2 was blocked with PD173074, a smaller amount of FGFR2 was present in the nucleus of t(4;11) leukemia cells with respect to the untreated ones, as shown by our western blot and immunofluorescence analyses (Figure 5). In agreement and of further consequence, stimulation of t(4;11) cells with FGF2 led to increased transcription of the MLL-AF4 target genes, *HOXA9* and *MEIS1*, which was prevented by pretreatment of the cells with the PD173074 inhibitor (Figure 6). Consistently, PD173074 treatment also impaired viability of the leukemia cells (Figure 4B).

Based on our overall results, we propose a model illustrating how nuclear cross-talk between FGFR2 and MLL-AF4 promotes aberrant transcription of MLL-AF4 target genes in leukemia cells (Figure 7).

In our model, the ligand FGF2 triggers phosphorylation and nuclear entry of FGFR2. In the nucleus, the activated receptor interacts with MLL-AF4 and AF4 and promotes the binding of the MLL-AF4/AF4 complex to the target gene promoter. Lastly, the MLL-AF4/AF4 complex recruits RNA Pol II on chromatin and activates gene transcription (Figure 7A). Treatment of leukemia cells with the PD173074 inhibitor, which binds the ATP pocket of FGFR2, prevents FGF2-mediated nuclear entry of pFGFR2 and, consequently, also transcriptional activation of MLL-AF4 target genes (Figure 7B). Similarly, specific silencing of FGFR2 leads to impaired chimera target gene transcription.

In conclusion, our study adds to the growing body of evidences that MLL-AF4, the oncogenic chimera typical of the t(4;11) ALL, promotes its aberrant transcriptional activity through the recruitment of nuclear molecular partners that, like FGFR2, consequently acquire an opportunistic oncogenic function. As our overall data give proof that the phosphorylation-dependent nuclear translocation of FGFR2 plays a key role in the leukemogenic mechanism, the use of FGFR-specific inhibitors or of molecules specifically able to block the nuclear import of the receptor, as well as the interaction between FGFR2 and the MLL-AF4 chimera, may be promising avenues to design new therapeutic strategies. Therefore, we realistically consider FGFR2 a novel and useful target for treatment of this very aggressive form of hematopoietic malignancy.

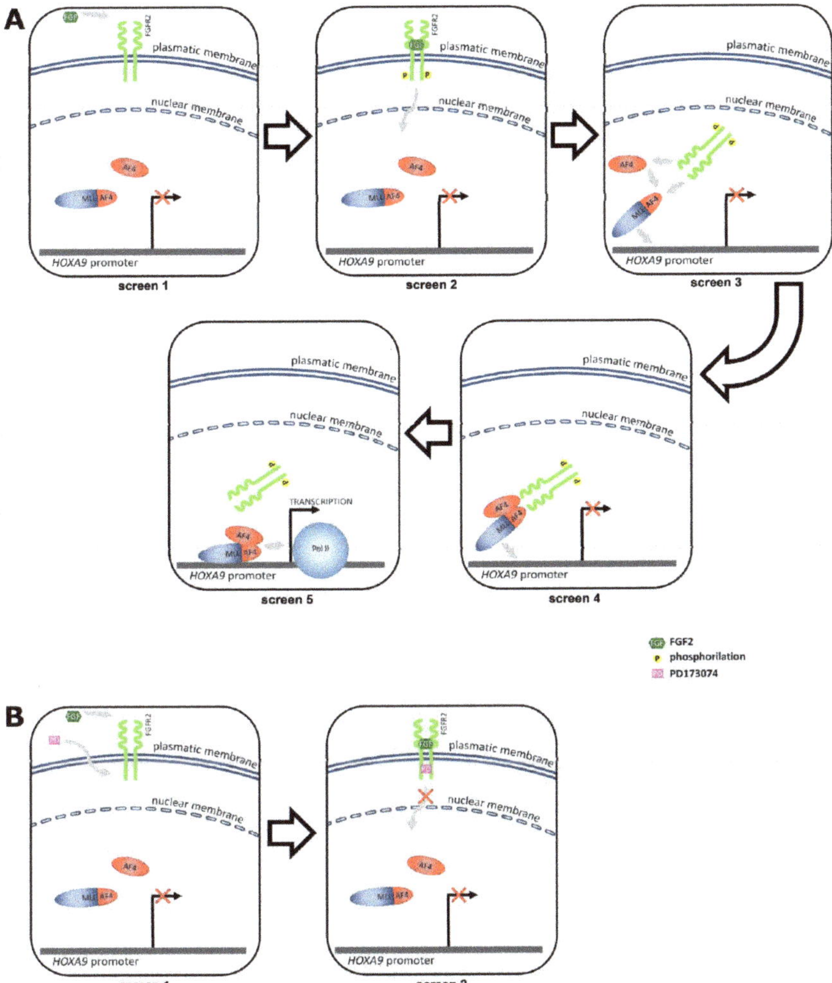

Figure 7. Nuclear cross-talk between FGFR2 and MLL-AF4 promotes aberrant transcription of *HOXA9* in leukemia cells. (**A**) screen 1: in basal condition, *HOXA9* transcription is not stimulated; screen 2: FGF2 binds FGFR2 and triggers its phosphorylation; screen 3: p-FGFR2 enters the nucleus and interacts with both the MLL-AF4 chimera and AF4; screen 4: p-FGFR2 promotes the binding of the MLL-AF4/AF4 complex to the *HOXA9* promoter; screen 5: the MLL-AF4/AF4 complex activates RNA Pol II-dependent transcription of *HOXA9*. (**B**) screen 1: the PD173074 inhibitor enters the cell, binds the ATP pocket of FGFR2, and prevents FGF2-mediated autophosphorylation of the receptor; screen 2: unphosphorylated FGFR2 does not enter the nucleus and therefore cannot promote the MLL-AF4-driven transcription of *HOXA9*.

4. Materials and Methods

4.1. Antibodies

The following antibodies were used: mouse monoclonal anti-mouse α tubulin; rabbit polyclonal anti-FGFR2; goat polyclonal anti-lamin B (Santa Cruz Biotechnology, Inc., Dallas, TX, USA); mouse monoclonal anti-MLL (Upstate Biotechnology, Inc., New York, NY, USA); human phospho-FGFR1-4 (Y653/Y654) antibody (R&D Systems, Minneapolis, MN, USA); horseradish peroxidase-conjugated anti-mouse, anti-rabbit, and anti-goat IgG secondary antibodies (GE Healthcare Italia, Milan, Italy); anti-rabbit IgG, Cy2-conjugated

(Merck KGaA, Darmstadt, Germany), and anti-rabbit IgG FITC-conjugated (Thermo Fisher Scientific, Waltham, MA, USA).

4.2. Cell Lines

Leukemia cell lines were obtained from the Cell Culture Facility of CEINGE - Advanced Biotechnologies (Naples, Italy). RS4;11, SEM, and MV4-11 harbor the t(4;11) chromosomal rearrangement and express endogenous MLL-AF4 chimera. The RS4;11 acute lymphoblastic leukemia cells were grown at 37 °C, 5% CO_2, in minimum essential medium (MEM) (Sigma-Aldrich, St. Louis, MO, USA) supplemented with 10% FBS and 10 mL/L penicillin/streptomycin (Sigma-Aldrich, St. Louis, MO, USA); SEM acute lymphoblastic leukemia cells were grown at 37 °C, 5% CO_2, in Iscove's MDM (Sigma-Aldrich) supplemented with 10% FBS and 10 mL/L penicillin/streptomycin; MV4-11 acute monocytic leukemia cells and 697 acute lymphoblastic leukemia cells were grown in RPMI (Sigma-Aldrich) supplemented with 20% FBS and 10 mL/L penicillin/streptomycin.

4.3. Flow Cytometry

Leukemia cells (1×10^6) were incubated with anti-FGFR2 antibody for 1 h. After several washes with PBS, cells were treated with fluorescein isothiocyanate (FITC)-secondary antibody for 30 min and read using a BD FACSCanto II (BD Biosciences-US, San Jose, CA, USA) flow cytometer.

4.4. Protein Extraction, Subcellular Fraction Isolation, Immunoprecipitation, and Western Blot Analysis

For immunoprecipitation (IP) experiments, RS4;11 and MV4-11 cells were lysed in IP buffer, as previously described [21], and protein extract was incubated overnight at 4 °C with anti-FGFR2 antibody (2 µg per 5 mg of total proteins). Subsequently, the protein mixture was incubated with 30 mL of protein A/G PLUS-Agarose (Santa Cruz Biotechnology, Inc., Dallas, TX, USA) for each microgram of antibody [17].

Fractioned protein extracts containing cytosolic/membrane and nuclear proteins were obtained using a Qproteome Nuclear Protein Kit (Qiagen Italia, Milan, Italy) as elsewhere described [45,46]. Briefly, cells were washed twice with ice-cold PBS, detached by scraping, and centrifuged for 5 min at 450 RCF. They were then lysed by incubation for 15 min in hypotonic nuclear lysis (NL) buffer, supplemented with protease inhibitor solution and 0.1 M DTT. Detergent solution was then added and, after brief shaking, the cell suspension was centrifuged for 5 min at 10,000 RCF. The supernatant, containing the cytosol/membrane proteins, was stored and used for further analyses. The cell nuclei, contained in the pellet, were washed by resuspension in the NL buffer and subsequent centrifugation for 5 min at 10,000 RCF. The nuclear pellet was resuspended in the buffer NX1, supplemented with protease inhibitor solution, and incubated for 30 min under shaking. The suspension was then centrifuged for 10 min at 12,000 RCF and the supernatant, containing the nuclear proteins, was stored and used for further analyses. Buffers, solutions, and reagents were provided by the Qproteome Nuclear Protein Kit (Qiagen Italia, Milan, Italy) and all the procedure steps were carried out at 4 °C. Cytosolic/membrane and nuclear proteins were analyzed with proper antibodies to verify the quality of the fractionation procedure.

Either 40 µg of WCE, or 50 µg of nuclear extract, or 20–30 µg of cytosolic/membrane extract, or 15 µL of sample from IP experiments were loaded onto SDS/PAGE. After electrophoretic separation, proteins were transferred onto nitrocellulose membrane and analyzed with appropriate primary antibodies. Protein signals were visualized with the ECL Plus detection system (GE Healthcare Italia, Milan, Italy) and protein signal intensities were quantified with the ImageJ 1.46 software. Average values from at least three independent experiments were graphically reported as relative units (R.U.).

4.5. Small Interfering RNA (SiRNA)

RS4;11, SEM, and MV4-11 leukemia cells were transfected with 100 nM FGFR2-specific siRNAs ON-TARGET plus SMARTpool (Qiagen Italia, Milan, Italy) by electroporation

(Bio-Rad Laboratories S.r.l., Milan, Italy) in MEM Eagle (Sigma-Aldrich, St. Louis, MO, USA). All Stars Negative Control siRNA scramble (Qiagen) was the nonsense control. After incubation for 15 min at room temperature, cells were cultured under standard conditions and harvested after 72 h to isolate total RNA and total proteins [17].

4.6. Total RNA Isolation, Reverse Transcription (RT), and Real Time Polymerase Chain Reaction (PCR)

Total RNA was extracted from silenced cells (RS4;11, SEM, and MV4-11) with the Nucleo Spin RNA II kit (Macherey–Nagel GmbH & Co. KG, Dueren, Germany); 200 ng of RNA were reverse transcribed using SuperScript III and random-hexamers oligo-dT (Thermo Fisher Scientific, Waltham, MA, USA). Real time PCR was carried out in an iCycler iQ Real Time PCR Thermal Cycler (Bio-Rad Laboratories, Hercules, CA, USA) using SYBR Green Master Mix (Bio-Rad Laboratories, Hercules, CA, USA) and specific primer pairs. To measure expression level of *HOXA9*, *MEIS1*, and *MEIS2*, the primer pairs used were described previously [17]; for the dosage of FGFR2 the following primers were used: FGFR2-F: 5′-GTCAGCTGGGGTCGTTTC-3′; FGFR2-R: 5′-TCATGTTTTAACACTGCCGTT-3′. Gene expression was normalized to *POLR2A*, *ACTB*, and *TUBA1A* genes and determined using the $2^{-\Delta\Delta Ct}$ method. Average values from at least three independent experiments were graphically reported as relative units (R.U.).

4.7. Cell Viability Assays

RS4;11 cells (4×10^4 per well) were transfected with FGFR2-specific siRNAs and siRNA nonsense control, and seeded in a 96-well plate. Cell viability was assessed by the 3-(4,5-dimethyl-2-thiazolyl)-2,5-diphenyl-2H-tetrazolium bromide (MTT) (Sigma Aldrich) method [17]. MTT (0.5 mg/mL of fresh media) was added to the cells (RS4;11) 24, 48, 72, and 96 h after transfection. Absorbance was read at 570 nm using a Spectramax spectrophotometer (Molecular Devices, San Jose, CA, USA).

To calculate the half-maximal inhibitory concentration (IC_{50}) of the PD173074, 2 million RS4;11 cells were plated in a 96-multiwell plate in serum-free medium and treated with various concentrations of the inhibitor. Then, 0.1 ng/mL of FGF2 was added to the medium and 72 h after treatment MTT was added (0.5 mg/mL of fresh media). Absorbance was read at 570 nm.

4.8. Chromatin Immunoprecipitation (ChIP) Assay

RS4;11 cells (30×10^6) were fixed in MEM medium containing 1% formaldehyde for 10 min at room temperature; the reaction was stopped by glycine quenching (125 mM final concentration). Nuclei were collected, digested in 50 mM Tris-HCL pH 8.1, 10 mM EDTA, 10% SDS, and then sonicated (3 cycles, consisting of 30 s with and without sonication) using a Microson XL ultrasonic cell disruptor (Misonix Inc., Farmingdale, NY, USA). Proteins tied to DNA fragments (ranging from 100–600 bp) were pulled-down overnight at 4 °C using appropriate antibodies, then mixed with protein-G magnetic beads (Santa Cruz Biotechnology Inc., Dallas, TX, USA) and incubated for 2 h. Beads were washed with ChIP buffer (10 mM Tris-HCl pH 8.1, 1 mM EDTA, 10% SDS, 0.5% EGTA, 140 mM NaCl, 10× Na-deoxycholate, 100× Triton). Immunoprecipitates were dissolved in elution buffer (0.5 M EDTA, 1 M Tris-HCl pH 8.0) and DNA was isolated by phenol/chloroform/isoamyl alcohol extraction and ethanol precipitation. Real time quantitative PCR (RT-qPCR) was performed with 1 µL of DNA using a custom-made primer set [17].

4.9. Treatment of t(4;11) Leukemia Cells with FGF2 and PD173074 Inhibitor

One million RS4;11 cells were plated in serum-free medium for 12 h and then stimulated with various concentrations of FGF2 (0.05–0.1–0.5 ng/mL). 50 µg of nuclear protein were loaded onto 10% SDS-PAGE for the western blot, performed with an anti-phospho-FGFR2 (P-FGFR2) or an anti-FGFR2 (FGFR2) antibody [47]. One million RS4;11 cells were plated in serum-free medium for 12 h, treated with 2 µM PD173074 (Sigma Aldrich) for 1 h, and then stimulated with 0.1 ng/mL of FGF2 (Sigma-Aldrich). Cells were harvested 72 h after treatment and total, cytosolic/membrane, and nuclear proteins were extracted [46,47].

4.10. Immunofluorescence Analysis

RS4;11 cells, treated with 2 µM PD173074 and stimulated with 0.1 ng/mL FGF2, as described in the previous paragraph, were fixed in 2% paraformaldehyde (Sigma Aldrich, St. Louis, MO, USA), permeabilized with 1% BSA in PBS, incubated with the anti-FGFR2 antibody, and subsequently treated with a CY2-conjugated secondary antibody. DAPI solution (Thermo Fisher Scientific) was used for nuclear staining. Cells were mounted on a slide and analyzed by confocal microscopy (LSM 510, Zeiss, München, Germany).

4.11. Statistical Analysis

All the data presented are expressed as mean $\pm$ standard error mean (SEM) and are representative of three or more independent experiments. The data of repeated experiments were analyzed using one-way Student's t-test (for independent samples).

Author Contributions: Conceptualization, A.C. and G.E.; methodology, T.F., A.C., M.Z., D.S., and G.E.; validation, G.E., T.F., and R.A.; formal analysis, T.F., A.C., M.Z., M.R., D.S., and F.C.; investigation, T.F., A.C., M.Z., M.R., F.C., and D.S.; resources, T.F., M.R., F.C., R.A., and G.E.; data curation, T.F., A.C., and M.Z.; writing—original draft preparation, T.F., A.C., and G.E.; writing—review and editing, A.C., F.C., R.A., and G.E.; supervision, A.C. and G.E.; project administration, G.E.; funding acquisition, T.F. and G.E. All authors have read and agreed to the published version of the manuscript.

Funding: This research was partially funded by the Italian Ministry of Health [RF-2011-02349269, granted to GE]. The APC was funded by CEINGE Biotecnologie Avanzate s.c. a r.l., Naples, Italy.

Data Availability Statement: Not Applicable.

Acknowledgments: We are grateful to AIL Onlus for supporting research in the field of human leukemia.

Conflicts of Interest: The authors declare no conflict of interest.

References

1. Daser, A.; Rabbitts, T.H. The versatile mixed lineage leukaemia gene *MLL* and its many associations in leukaemogenesis. *Semin. Cancer Biol.* **2005**, *15*, 175–188. [CrossRef] [PubMed]
2. Meyer, C.; Hofmann, J.; Burmeister, T.; Gröger, D.; Park, T.S.; Emerenciano, M.; Pombo de Oliveira, M.; Renneville, A.; Villarese, P.; Macintyre, E.; et al. The MLL recombinome of acute leukemias in 2013. *Leukemia* **2013**, *27*, 2165–2176. [CrossRef]
3. Bursen, A.; Schwabe, K.; Rüster, B.; Henschler, R.; Ruthardt, M.; Dingermann, T.; Marschalek, R. The AF4.MLL fusion protein is capable of inducing ALL in mice without requirement of MLL.AF4. *Blood* **2010**, *115*, 3570–3579. [CrossRef]
4. Prieto, C.; Marschalek, R.; Kühn, A.; Bursen, A.; Bueno, C.; Menéndez, P. The AF4-MLL fusion transiently augments multilineage hematopoietic engraftment but is not sufficient to initiate leukemia in cord blood CD34$^+$ cells. *Oncotarget* **2017**, *8*, 81936–81941. [CrossRef] [PubMed]
5. Wilkinson, A.C.; Ballabio, E.; Geng, H.; North, P.; Tapia, M.; Kerry, J.; Biswas, D.; Roeder, R.G.; Allis, C.D.; Melnick, A.; et al. RUNX1 Is a key target in t(4;11) leukemias that contributes to gene activation through an AF4-MLL complex interaction. *Cell Rep.* **2013**, *3*, 116–127. [CrossRef] [PubMed]
6. Bueno, C.; Calero-Nieto, F.J.; Wang, X.; Valdés-Mas, R.; Gutiérrez-Agüera, F.; Roca-Ho, H.; Ayllon, V.; Real, P.J.; Arambile, D.; Espinosa, L.; et al. Enhanced hemato-endothelial specification during human embryonic differentiation through developmental cooperation between AF4-MLL and MLL-AF4 fusions. *Haematologica* **2019**, *104*, 1189–1201. [CrossRef]
7. Britten, O.; Ragusa, D.; Tosi, S.; Kamel, Y.M. MLL-Rearranged Acute Leukemia with t(4;11)(q21;q23)—Current Treatment Options. Is There a Role for CAR-T Cell Therapy? *Cells* **2019**, *8*, 1341. [CrossRef]
8. Kowarz, E.; Burmeister, T.; Lo Nigro, L.; Jansen, M.W.; Delabesse, E.; Klingebiel, T.; Dingermann, T.; Meyer, C.; Marschalek, R. Complex MLL rearrangements in t(4;11) leukemia patients with absent AF4.MLL fusion allele. *Leukemia* **2007**, *21*, 1232–1238. [CrossRef]
9. Thomas, M.; Gessner, A.; Vornlocher, H.P.; Hadwiger, P.; Greil, J.; Heidenreich, O. Targeting MLL-AF4 with short interfering RNAs inhibits clonogenicity and engraftment of t(4;11)-positive human leukemic cells. *Blood* **2005**, *106*, 3559–3566. [CrossRef]
10. Marschalek, R. Mechanisms of leukemogenesis by MLL fusion proteins. *Br. J. Haematol.* **2011**, *152*, 141–154. [CrossRef]
11. Godfrey, L.; Kerry, J.; Thorne, R.; Repapi, E.; Davies, J.O.; Tapia, M.; Ballabio, E.; Hughes, J.R.; Geng, H.; Konopleva, M.; et al. MLL-AF4 binds directly to a BCL-2 specific enhancer and modulates H3K27 acetylation. *Exp. Hematol.* **2016**, *47*, 64–75. [CrossRef] [PubMed]

12. Collins, C.T.; Hess, J.L. Deregulation of the HOXA9/MEIS1 axis in acute leukemia. *Curr. Opin. Hematol.* **2016**, *23*, 354–361. [CrossRef] [PubMed]

13. Stam, R.W.; Schneider, P.; Hagelstein, J.A.P.; van der Linden, M.H.; Stumpel, D.J.P.M.; de Menezes, R.X.; de Lorenzo, P.; Valsecchi, M.G.; Pieters, R. Gene expression profiling-based dissection of MLL translocated and MLL germline acute lymphoblastic leukemia in infants. *Blood* **2010**, *115*, 2835–2844. [CrossRef] [PubMed]

14. Yokoyama, A. Transcriptional activation by MLL fusion proteins in leukemogenesis. *Exp. Hematol.* **2016**, *46*, 21–30. [CrossRef] [PubMed]

15. Orlovsky, K.; Kalinkovich, A.; Rozovskaia, T.; Shezen, E.; Itkin, T.; Alder, H.; Ozer, H.G.; Carramusa, L.; Avigdor, A.; Volinia, S.; et al. Down-regulation of homeobox genes MEIS1 and HOXA in MLL-rearranged acute leukemia impairs engraftment and reduces proliferation. *Proc. Natl. Acad. Sci. USA* **2011**, *108*, 7956–7961. [CrossRef]

16. Yokoyama, A.; Lin, M.; Naresh, A.; Kitabayashi, I.; Cleary, M.L. A higher-order complex containing AF4 and ENL family proteins with P-TEFb facilitates oncogenic and physiologic MLL-dependent transcription. *Cancer Cell* **2010**, *17*, 198–212. [CrossRef]

17. Fioretti, T.; Cevenini, A.; Zanobio, M.; Raia, M.; Sarnataro, D.; Salvatore, F.; Esposito, G. Crosstalk between 14-3-3θ and AF4 enhances MLL-AF4 activity and promotes leukemia cell proliferation. *Cell. Oncol.* **2019**, *42*, 829–845. [CrossRef]

18. Okuda, H.; Stanojevic, B.; Kanai, A.; Kawamura, T.; Takahashi, S.; Matsui, H.; Takaori-Kondo, A.; Yokoyama, A. Cooperative gene activation by AF4 and DOT1L drives MLL-rearranged leukemia. *J. Clin. Investig.* **2017**, *127*, 1918–1931. [CrossRef]

19. Esposito, G.; Cevenini, A.; Cuomo, A.; de Falco, F.; Sabbatino, D.; Pane, F.; Ruoppolo, M.; Salvatore, F. Protein network study of human AF4 reveals its central role in RNA Pol II-mediated transcription and in phosphorylation-dependent regulatory mechanisms. *Biochem. J.* **2011**, *438*, 121–131. [CrossRef]

20. Lemmon, M.A.; Schlessinger, J. Cell signaling by receptor tyrosine kinases. *Cell* **2010**, *141*, 1117–1134. [CrossRef]

21. Lonic, A.; Barry, E.; Quach, C.; Kobe, B.; Saunders, N.; Guthridge, M.A. Fibroblast Growth Factor Receptor 2 Phosphorylation on Serine 779 Couples to 14-3-3 and Regulates Cell Survival and Proliferation. *Mol. Cell. Biol.* **2008**, *28*, 3372–3385. [CrossRef]

22. Xie, Y.; Su, N.; Yang, J.; Tan, Q.; Huang, S.; Jin, M.; Ni, Z.; Zhang, B.; Zhang, D.; Luo, F.; et al. FGF/FGFR signaling in health and disease. *Signal Transduct. Target. Ther.* **2020**, *5*, 181. [CrossRef]

23. Montor, W.R.; Salas, A.R.O.S.E.; Melo, F.H.M. Receptor tyrosine kinases and downstream pathways as druggable targets for cancer treatment: The current arsenal of inhibitors. *Mol. Cancer* **2018**, *17*, 55. [CrossRef]

24. Yamaoka, T.; Kusumoto, S.; Ando, K.; Ohba, M.; Ohmori, T. Receptor Tyrosine Kinase-Targeted Cancer Therapy. *Int. J. Mol. Sci.* **2018**, *19*, 3491. [CrossRef] [PubMed]

25. Korc, M.; Friesel, R.E. The role of fibroblast growth factors in tumor growth. *Curr. Cancer Drug Targets* **2009**, *9*, 639–651. [CrossRef]

26. Esposito, M.T. The Impact of PI3-kinase/RAS Pathway Cooperating Mutations in the Evolution of KMT2A-rearranged Leukemia. *HemaSphere* **2019**, *3*, e195. [CrossRef]

27. Whelan, J.T.; Ludwig, D.L.; Bertrand, F.E. HoxA9 induces insulin-like growth factor-1 receptor expression in B-lineage acute lymphoblastic leukemia. *Leukemia* **2008**, *22*, 1161–1169. [CrossRef] [PubMed]

28. Nakanishi, H.; Nakamura, T.; Canaani, E.; Croce, C.M. ALL1 fusion proteins induce deregulation of EphA7 and ERK phosphorylation in human acute leukemias. *Proc. Natl. Acad. Sci. USA* **2007**, *104*, 14442–14447. [CrossRef] [PubMed]

29. Stachowiak, M.K.; Maher, P.A.; Stachowiak, E.K. Integrative nuclear signaling in cell development—A role for FGF receptor-1. *DNA Cell Biol.* **2007**, *26*, 811–826. [CrossRef] [PubMed]

30. Cerliani, J.P.; Guillardoy, T.; Giulianelli, S.; Vaque, J.P.; Gutkind, J.S.; Vanzulli, S.I.; Martins, R.; Zeitlin, E.; Lamb, C.A.; Lanari, C. Interaction between FGFR-2, STAT5, and progesterone receptors in breast cancer. *Cancer Res.* **2011**, *71*, 3720–3731. [CrossRef] [PubMed]

31. Tuzon, C.T.; Rigueur, D.; Merrill, A.E. Nuclear Fibroblast Growth Factor Receptor Signaling in Skeletal Development and Disease. *Curr. Osteoporos. Rep.* **2019**, *17*, 138–146. [CrossRef]

32. Cattaneo, F.; Parisi, M.; Fioretti, T.; Sarnataro, D.; Esposito, G.; Ammendola, R. Nuclear localization of Formyl-Peptide Receptor 2 in human cancer cells. *Arch. Biochem. Biophys.* **2016**, *603*, 10–19. [CrossRef] [PubMed]

33. Carpenter, G.; Liao, H.J. Receptor tyrosine kinases in the nucleus. *Cold Spring Harb. Perspect. Biol.* **2013**, *5*, a008979. [CrossRef]

34. Porębska, N.; Latko, M.; Kucińska, M.; Zakrzewska, M.; Otlewski, J.; Opaliński, Ł. Targeting cellular trafficking of fibroblast growth factor receptors as a strategy for selective cancer treatment. *J. Clin. Med.* **2018**, *8*, 7. [CrossRef]

35. Kerry, J.; Godfrey, L.; Repapi, E.; Tapia, M.; Blackledge, N.P.; Ma, H.; Ballabio, E.; O'Byrne, S.; Ponthan, F.; Heidenreich, O.; et al. MLL-AF4 spreading identifies binding sites that are distinct from super-enhancers and that govern sensitivity to DOT1L inhibition in leukemia. *Cell Rep.* **2017**, *18*, 482–495. [CrossRef]

36. Martin, A.J.; Grant, A.; Ashfield, A.M.; Palmer, C.N.; Baker, L.; Quinlan, P.R.; Purdie, C.A.; Thompson, A.M.; Jordan, L.B.; Berg, J.B. FGFR2 protein expression in breast cancer: Nuclear localisation and correlation with patient genotype. *BMC Res. Notes* **2011**, *4*, 72. [CrossRef]

37. Schmahl, J.; Kim, Y.; Colvin, J.S.; Ornitz, D.M.; Capel, B. Fgf9 induces proliferation and nuclear localization of FGFR2 in Sertoli precursors during male sex determination. *Development* **2004**, *131*, 3627–3636. [CrossRef]

38. Bansal, R.; Magge, S.; Winkler, S. Specific inhibitor of FGF receptor signaling: FGF-2-mediated effects on proliferation, differentiation, and MAPK activation are inhibited by PD173074 in oligodendrocyte-lineage cells. *J. Neurosci. Res.* **2003**, *74*, 486–493. [CrossRef]

39. Pardo, O.E.; Latigo, J.; Jeffery, R.E.; Nye, E.; Poulsom, R.; Spencer-Dene, B.; Lemoine, N.R.; Stamp, G.W.; Aboagye, E.O.; Seckl, M.J. The fibroblast growth factor receptor inhibitor PD173074 blocks small cell lung cancer growth in vitro and in vivo. *Cancer Res.* **2009**, *69*, 8645–8651. [CrossRef] [PubMed]

40. Chae, Y.K.; Ranganath, K.; Hammerman, P.S.; Vaklavas, C.; Mohindra, N.; Kalyan, A.; Matsangou, M.; Costa, R.; Carneiro, B.; Villaflor, V.M.; et al. Inhibition of the fibroblast growth factor receptor (FGFR) pathway: The current landscape and barriers to clinical application. *Oncotarget* **2017**, *8*, 16052–16074. [CrossRef] [PubMed]

41. Jerchel, I.S.; Hoogkamer, A.Q.; Ariës, I.M.; Boer, J.M.; Besselink, N.J.M.; Koudijs, M.J.; Pieters, R.; den Boer, M.L. Fibroblast growth factor receptor signaling in pediatric B-cell precursor acute lymphoblastic leukemia. *Sci. Rep.* **2019**, *9*, 1875. [CrossRef]

42. Kasbekar, M.; Nardi, V.; Dal Cin, P.; Brunner, A.M.; Burke, M.; Chen, Y.B.; Connolly, C.; Fathi, A.T.; Foster, J.; Macrae, M.; et al. Targeted FGFR inhibition results in a durable remission in an FGFR1-driven myeloid neoplasm with eosinophilia. *Blood Adv.* **2020**, *4*, 3136–3140. [CrossRef] [PubMed]

43. Stachowiak, M.K.; Birkaya, B.; Aletta, J.M.; Narla, S.T.; Benson, C.A.; Decker, B.; Stachowiak, E.K. Nuclear FGF receptor-1 and CREB binding protein: An integrative signaling module. *J. Cell. Physiol.* **2015**, *230*, 989–1002. [CrossRef]

44. Faber, J.; Krivtsov, A.V.; Stubbs, M.C.; Wright, R.; Davis, T.N.; van den Heuvel-Eibrink, M.; Zwaan, C.M.; Kung, A.L.; Armstrong, S.A. HOXA9 is required for survival in human MLL-rearranged acute leukemias. *Blood* **2009**, *113*, 2375–2385. [CrossRef] [PubMed]

45. Cattaneo, F.; Castaldo, M.; Parisi, M.; Faraonio, R.; Esposito, G.; Ammendola, R. Formyl Peptide Receptor 1 Modulates Endothelial Cell Functions by NADPH Oxidase-Dependent VEGFR2 Transactivation. *Oxid. Med. Cell. Longev.* **2018**, *2018*, 2609847. [CrossRef]

46. Castaldo, M.; Zollo, C.; Esposito, G.; Ammendola, R.; Cattaneo, F. NOX2-Dependent Reactive Oxygen Species Regulate Formyl-Peptide Receptor 1-Mediated TrkA Transactivation in SH-SY5Y Cells. *Oxid. Med. Cell. Longev.* **2019**, *2019*, 2051235. [CrossRef] [PubMed]

47. Cattaneo, F.; Russo, R.; Castaldo, M.; Chambery, A.; Zollo, C.; Esposito, G.; Pedone, P.V.; Ammendola, R. Phosphoproteomic analysis sheds light on intracellular signaling cascades triggered by Formyl-Peptide Receptor 2. *Sci. Rep.* **2019**, *9*, 17894. [CrossRef]

MDPI

St. Alban-Anlage 66

4052 Basel

Switzerland

Tel. +41 61 683 77 34

Fax +41 61 302 89 18

www.mdpi.com

International Journal of Molecular Sciences Editorial Office

E-mail: ijms@mdpi.com

www.mdpi.com/journal/ijms